D1218972

THE MOLECULAR BIOLOGY OF PLANT CELLS

BOTANICAL MONOGRAPHS

BOTANICAL MONOGRAPHS · VOLUME 14

THE MOLECULAR BIOLOGY OF PLANT CELLS

EDITED BY

H. SMITH

PhD, DSc, FI Biol

Professor of Plant Physiology

University of Nottingham

UNIVERSITY OF CALIFORNIA PRESS

BERKELEY AND LOS ANGELES 1977

UNIVERSITY OF
CALIFORNIA PRESS
Berkeley and Los Angeles, California

ISBN: 0 520 03465 1
Library of Congress Catalog Card Number: 77-73503

Printed in Great Britain

CONTENTS

SECTION TWO

GENE EXPRESSION AND ITS REGULATION IN PLANT CELLS

SECTION THREE

THE MANIPULATION OF PLANT CELLS

CONTRIBUTORS

W.D. BAUER Charles F. Kettering Research Laboratory, 150 East South College Street, Yellow Springs, Ohio 45387, U.S.A.

N.K. BOARDMAN Division of Plant Industry, Commonwealth Scientific and Industrial Research Organisation, Canberra 2601, Australia.

D. BOULTER Department of Botany, University of Durham, Durham DH1 3LE, U.K.

J.W. BRADBEER Department of Botany, University of London, King's College, London SE24 9JF.

R.S. CHALEFF Department of Plant Breeding and Biometry, New York State College of Agriculture and Life Sciences, Ithaca, N.Y. 14853, U.S.A.

J.H. CHERRY Horticulture Department, Purdue University, Lafayette, Indiana 47907, U.S.A.

D.T. CLARKSON ARC Letcombe Laboratory, Letcombe Regis, Wantage, Oxon OX12 9JT, U.K.

B. COLMAN Department of Biology, York University, 4700 Keele Street, Downsview, Ontario, M3J 1P2, Canada.

D.D. DAVIES School of Biological Sciences, University of East Anglia, Norwich NR4 7TJ, U.K.

R.J. ELLIS Department of Biological Sciences, University of Warwick, Coventry CV4 7AL, U.K.

D. GRIERSON Department of Physiology and Environmental Studies, University of Nottingham, School of Agriculture, Sutton Bonington, Loughborough LE12 5RD, U.K.

D. HANKE Department of Botany, University of Cambridge, Cambridge CB2 3EA, U.K.

J.W. HART Department of Botany, University of Aberdeen, Aberdeen AB9 2UD, U.K.

D.C. LEE Department of Biological Sciences, State University, College of Arts and Science, Plattsburgh, NY 12901, U.S.A.

J. POLACCO Department of Genetics, Connecticut Agricultural Experiment Station, P.O. Box 1106, New Haven, Connecticut 16504, U.S.A.

J.B. POWER Department of Botany, University of Nottingham, University Park, Nottingham NG7 2RD, U.K.

P.H. QUAIL Department of Plant Biology, Carnegie Institution of Washington, 290 Panama Street, Stanford, California 94305, U.S.A.

D.D. SABNIS Department of Botany, University of Aberdeen, Aberdeen AB9 2UD, U.K.

H. SMITH Department of Physiology and Environmental Studies, University of Nottingham, School of Agriculture, Sutton Bonington, Loughborough LE12 5RD, U.K.

H.E. STREET Department of Biological Sciences, University of Leicester, Leicester LE1 7RH, U.K.

PREFACE

The discoveries and concepts of molecular biology have, during the past 25 years, provided a radical new basis for our understanding of biological processes. The application of this theoretical framework, formulated essentially from studies on prokaryotes, to multicellular eukaryotic organisms is perhaps the major challenge in biology, offering the exciting possibility of achieving an understanding of growth, development and behaviour in mechanistic terms. Progress in the molecular biology of animals has been rapid and spectacular, with discoveries being made which lie outside the predictions derived from analogies with prokaryote studies. Up till quite recently, this has not been so far plants, largely due to the inherent difficulties of working with higher plants, but also to the "mammalian chauvinism" practised by so many biochemists. Attitudes change with fashions, however, and the generally increased awareness of the vitally important role of plants in the survival of the human race has led, in the last few years, to an enhanced availability of funds, and a heightened enthusiasm, for basic research on plant biochemistry, cell physiology and genetics. Following upon the so-called "Green Revolution", the much-publicized potential for the genetic manipulation of plant cells towards greater agricultural productivity has caught the imagination of scientists and laymen throughout the world. Unfortunately, much of the speculation has been ill-informed and sensationalist; nevertheless, the theoretical possibilities are limitless and much research investment is being injected into this area. A need exists, therefore, for the training of competent research workers with specialist knowledge of plant cells, their structure, function, biochemistry and genetics. This book aims to provide a sound factual basis for a training of this nature.

The Molecular Biology of Plant Cells is a text-book written by a number of authors, not merely a collection of unconnected articles. No one author could hope to cover the very wide subject area treated here in the necessary depth and thus each topic is dealt with by a specialist. In inviting the many distinguished scientists to contribute, however, I provided a relatively well defined "common pattern" for each chapter such that, as far as is possible in a multi-author book, uniformity of treatment and continuity of style would be achieved. I am most grateful to the authors, all of whom agreed to restrict themselves to my overall guidelines, and most of whom managed to complete their chapters within a year of my original deadline!

The book is intended as a text-book for senior undergraduate and post-graduate students in biology, biochemistry, botany, molecular biology and agricultural science. It covers the basic cellular physiology, biochemistry and genetics of plant cells, but does not deal with metabolic pathways, or with the physiology and biochemistry of the intact plant. Plant development has only been lightly touched upon in this book, since a companion volume, to be called "The Molecular Biology of Plant Development" is currently being prepared. Together, the two volumes should provide a sound textual basis for teaching and research in the molecular biology of higher plants.

In the preparation and editing of this volume I have received much support, and I would like to express my thanks for secretarial help to Miss Marjorie Bentley, Mrs. Laurel Dee, Mrs Jane Squirrell and Mrs. Elizabeth Horwood; to many scientific colleagues, particularly Professor M. B. Wilkins, Dr. D. Grierson, Dr. D. T. Clarkson and Professor D. H. Northcote for advice on content and authors; and to Mr. Robert Campbell of Blackwells for his patience and fortitude in dealing with an erratic editor.

July 1977 Harry Smith

SECTION ONE

PLANT CELL STRUCTURE AND FUNCTION

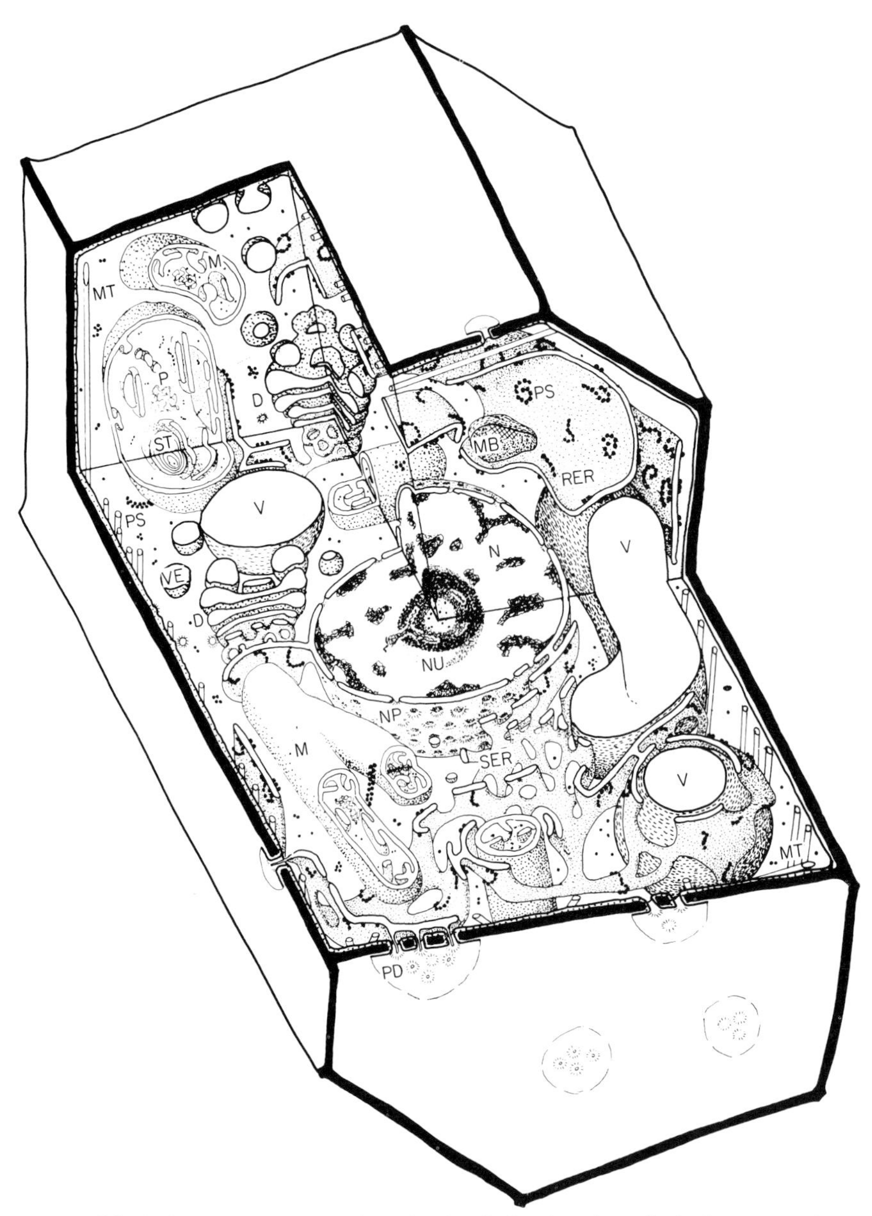

Fig. I.I. A diagram of an undifferentiated cell showing the principal components. Some of the constituents are illustrated by only a few examples (e.g. ribosomes). The components may be identified by letters which refer to those given in the text. Figure taken from *Ultrastructure and the Biology of Plant Cells* by B.E.S. Gunning and M.W. Steer, published by Edward Arnold, London. The drawing was generously provided by Dr. Steer and is reproduced with the kind permission of the authors and copyright holders.

INTRODUCTION

The term *cell*, as first used by Robert Hooke in 1665 signified an apparently empty space or lumen, surrounded by walls. We now know, of course, that the space is far from empty, and that rigid cell walls as seen by Hooke in thin slices of cork, are not ubiquitous in multicellular organisms. Indeed, the wall became to be regarded as the definitive structure of the cell, and when in the 1830s, the zoologist Schwann was able to recognise structures in cartilage resembling plant parenchymatous cell walls, the concept of the cell as the basic biological unit common to all organisms was born. Definitions have changed considerably in the subsequent century and a half, and, in particular, the cell wall is now seen in its proper perspective as being a structure, albeit of great importance, but restricted to plants and existing only outside the true cell. Nevertheless, the general concept of the cell as the basic minimum unit of life remains.

Since all organisms need to perform a number of essential functions merely in order to survive, both as individuals and as species, it should not be surprising to find a basic unity between the cells of all organisms. Each cell, at least in the early stages of its development, possesses the capacity to synthesize complex substances from simple ones, to liberate and transform the potential chemical energy of highly reduced compounds, to react to internal and external stimuli, to control the influx and efflux of materials across the limiting cell membranes and to regulate its activities in relation to the information contained in its individual store, or stores, of hereditary genetic material. Evolution has solved the problems posed by these requirements in more or less identical ways in all organisms, and thus the basic processes, activities, and structures of each individual plant cell are similar, not only to other plant cells, but also to all other eucaryotic cells. This book concentrates on the unifying features of plant cells and relates them to present knowledge and general theories of molecular biology. It should not be forgotten, however, that cells are characterised as much by their diversity as their unity. A wide range of different cell types with varying specialized functions are necessary for the life of the higher green plant; however, the origin of cell heterogeneity is a topic outside the scope of this present book.

The basic structures of an undifferentiated plant cell can be seen in Fig. 1.1. The cell proper is delimited by the *plasma membrane* (or *plasmalemma*) which is of unit membrane construction (chapters 2 and 8). Outside the plasma membrane, and thus actually *extra-cellular*, is the *cell wall* (chapter 1). The cell wall is normally closely appressed to the plasma membrane and in meristematic cells is thin and relatively weak. During differentiation various specialized wall structures develop; depending on the function of the mature cell, the walls may become relatively massive and extremely strong through the deposition of rigid, highly cross-linked polymeric substances. Adjacent protoplasts (i.e. the cells proper) are connected across the cell walls by narrow cytoplasmic channels, bounded by the plasma membrane, known as *plasmodesmata* (PD).

Within the cell a number of separate compartments, and interconnecting compartments, delimited by membranes, may be recognised (chapter 8). *Vacuoles* (V) are prominent, apparently empty spaces, spherical and numerous in the meristematic cell but irregular, very large, and coalescent in the mature expanded cell. Vacuoles serve as intracellular dust-bins—repositories for unwanted and often toxic byproducts of metabolism—and may also have functions similar to the *lysosomes* of animal cells. They are bounded by a single membrane known as the *tonoplast* (chapters 2 and 8).

The nucleus (N) (chapter 9), a major compartment in most cells, comprises a *nuclear envelope* possessing many large *nuclear pores* (NP) and *nucleoplasm*, the ground substance in which the hereditary material, *chromatin*, and the *nucleolus* (NU) lie. The nucleus is the principal site of the hereditary material of the cell, although both plastids and mitochondria also contain DNA. The material outside the nuclear envelope is commonly known as *cytoplasm*.

Ramifying throughout the cytoplasm, and occasionally connected to the outer membrane of the nuclear envelope, the cisternae of the *endoplasmic reticulum* act to integrate the biosynthetic functions of the cell (chapter 8). The endoplasmic reticulum is generally classified into two types: *rough endoplasmic reticulum* (RER), which has *ribosomes* attached to its outer face (chapter 10); and *smooth endoplasmic reticulum* (SER) which is not involved in protein synthesis. The endoplasmic reticulum may also, on occasion, be seen to be associated with stacks of vesicles (VE) known collectively as *dictyosomes* (D) or *Golgi bodies*. The endoplasmic reticulum and the dictyosomes are responsible for the formation and secretion of cellular membranes.

Three other membrane-bound compartments remain, each concerned with an aspect of energy or intermediary metabolism. *Plastids* (P), undifferentiated in meristematic cells and present only as *proplastids*, represent a general class of organelle in which the *chloroplast* is the characteristic member (chapters 3 and 4). *Mitochondria* (M) are smaller, but also bounded by a double membrane, and similarly involved in energy metabolism (chapter 5). As mentioned above, both mitochondria and plastids contain their own stores of hereditary material (chapter 11). The final compartments, in contrast, are bound by only a single membrane and do not contain hereditary material; these are known as *microbodies* (MB) and often contain dense, granular, or even crystalline contents (chapter 6). Within the cytoplasm just inside the plasma membrane lie long narrow cylinders known as microtubules (MT); microtubules function in a number of processes in which orientation of cellular components is important (chapter 7). Finally, plant cells contain many fine fibrils, known as *microfilaments*, which appear to be contractile in function and to be composed of a material similar to actin, one of the contractile components of muscle.

The 'typical' plant cell does not exist, of course, and the meristematic cell shown in Fig. 1.1 has only been chosen since it possesses all the essential characteristics of plant cells. Many of the cellular components are only present in very simple forms in meristematic cells, however, and the subsequent chapters in Section I necessarily involve a consideration of a variety of more specialized cell types.

FURTHER READING

Buvat R. (1969) *Plant Cells*. Weidenfeld and Nicolson, London.

Clowes F.A.L. & Juniper B.E. (1968) *Plant Cells*. Blackwell Scientific Publications, Oxford.

Gunning B.E.S. & Steer M.W. (1975) *Ultrastructure and the Biology of Plant Cells*. Edward Arnold, London.

Hall J.L., Flowers T.J. & Roberts R.M. (1974) *Plant Cell Structure and Metabolism*. Longman, London.

Robards A.W. (1970) *Electron Microscopy and Plant Ultrastructure*. McGraw-Hill, London.

CHAPTER I

PLANT CELL WALLS

1.1 INTRODUCTION

Plant cell walls establish a home, and indeed a city, for plant protoplasts. They serve many specialized functions in plant tissues, and form the skin, the skeleton, and the circulatory system of plants.

There are many variations in the form and substance of plant cell walls. The walls may be plastic or they may be rigid, permeable or impermeable, impregnated with plastics or coated with slime, cemented in layers to form fibres or dissolved in spots to form pores. These variations are of vital importance to the proper biological functioning of plant cells and organs, and thus the structure of the cell wall is often our best indication of the nature of the protoplast which dwells inside.

1.2 THE MOLECULAR STRUCTURE OF PLANT CELL WALLS

Polysaccharides are the principal components of all plant cell walls. The polysaccharides of the cell wall are made up of sugars which are linked to each other by glycosidic bonds to form the polymer chains. Each polysaccharide contains particular kinds of sugars which are joined to each other in characteristic patterns of linkage position and sequence. It is now known that the secondary, tertiary and quaternary structures of cell wall polysaccharides are determined by the structures of the component sugars and the linkages between them, just as the three dimensional structure of a protein is determined by the sequence and structures of its component amino acids (Rees, 1972).

The various polysaccharide chains of the plant cell wall are connected to each other in specific ways, and they form an integrated network. The properties of this network depend not only on the amounts, characteristic properties and orientations of the individual polysaccharides, but also on the nature and frequency of the interconnecting linkages between them.

The conformational structures of the nine sugars commonly found in plant cell walls are shown in Fig. 1.1. The three types of polysaccharide normally found in plant cell walls (cellulose, hemicelluloses, and pectic polysaccharides), and the structural protein of primary walls, are described briefly below.

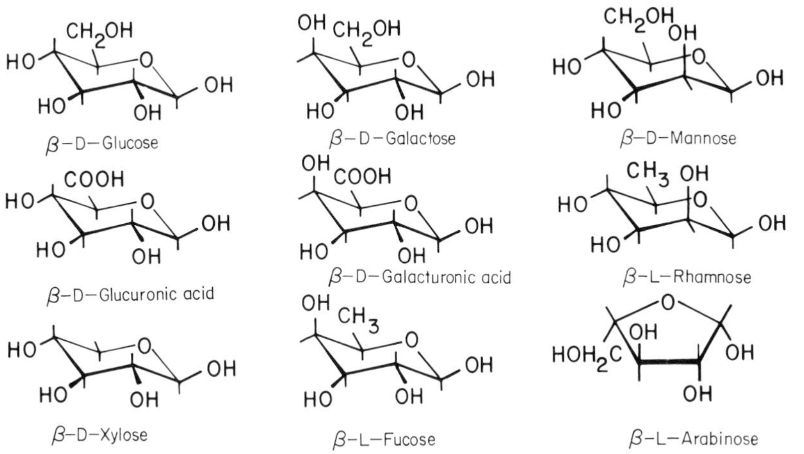

β-D-Glucose β-D-Galactose β-D-Mannose

β-D-Glucuronic acid β-D-Galacturonic acid β-L-Rhamnose

β-D-Xylose β-L-Fucose β-L-Arabinose

Fig. 1.1. Sugars of plant cell walls.

Conformational line drawings indicate approximate bond angles. β-L-arabinose is shown in its preferred planar furanose ring form. The other sugars are shown in their most stable pyranose chair form. Carbon atoms are numbered as indicated for β-D-glucose. Ring hydrogens are indicated by bonds only. Note that groups attached to a ring may be either axial (projecting above or below the ring) or equatorial (projecting to the side of the ring). Substituents at C_1 project equatorially in the β configuration, but are axial in the α configuration. All 'bulky' groups ($-OH$, $-CH_2OH$, & $-COOH$) are in equatorial positions in β-D-glucose, β-D-glucuronic acid, and β-D-xylose. Note that these sugars differ only in the group attached to C_5. Galactose, galacturonic acid and fucose are similarly related (axial $-OH$ group at C_4), as are mannose and rhamnose (axial $-OH$ group at C_2).

1.2.1 CELLULOSE

Cellulose occurs as a crystalline, fibrillar aggregate of β-1,4-linked glucan chains (Frey-Wyssling, 1969). Cellulose fibrils give plant cell walls most of their enormous strength, much as glass fibres embedded in an epoxy resin give strength to a fibreglass composite (Northcote, 1972).

The basic structure of the β-1,4-linked glucan chains of cellulose is illustrated in Fig. 1.2 by conformational line drawings and in Fig. 1.3 by molecular models. Residues of β-D-glucose (Fig. 1.1) are glycosidically linked to each other, *from* carbon 1 of one residue *to* carbon 4 of the adjacent residue. The upside-down inversion of every second residue in the chain minimizes contact between atoms of adjacent residues. Close inspection of the models in Fig. 1.3 shows that the $-OH$ groups at carbon 3 are in very close proximity to the ring oxygens (O_5) of adjacent residues. Hydrogen bonds between O_3 and O'_5 help to stabilize the flat, straight, ribbon-like structure of β-1,4-linked glucan chains.

The flat, ribbon-like structure allows the chains to fit closely together, one on top of the other, over their entire lengths. These *interchain* associations are stabilized by hydrogen bonds between O_6 of a glucose residue in one chain and

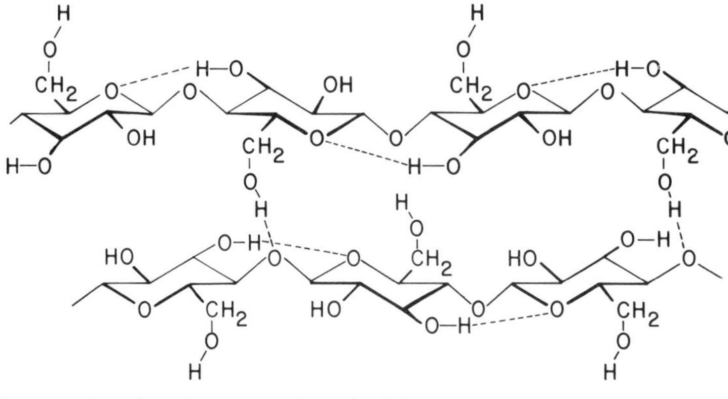

Fig. 1.2. β-1,4-linked glucan chains of cellulose.

Portions of two associated chains are illustrated by conformational line drawings. Distances between atoms are *not* accurately indicated in this illustration, but see Fig. 1.3.

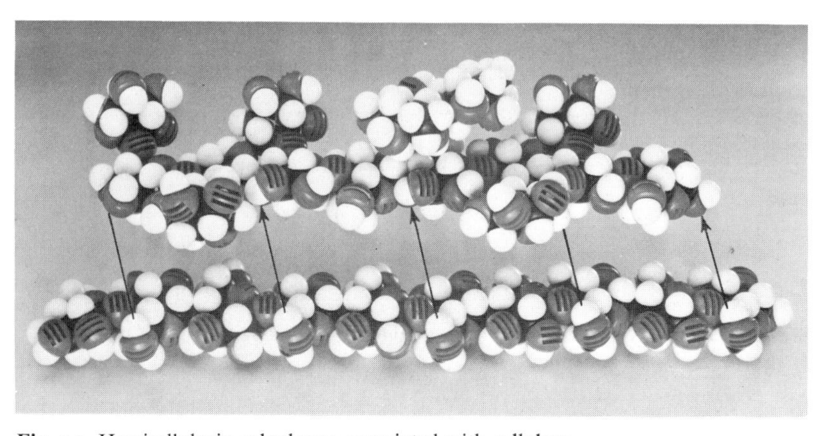

Fig. 1.3. Hemicellulosic xyloglucan associated with cellulose.

The repeating subunit of a hemicellulosic xyloglucan is shown in association with a portion of a β-1,4-linked glucan chain of cellulose (Bauer *et al.*, 1973). Molecular models have been used to accurately indicate interatomic distances and bond angles. Hydrogen bonds from the cellulosic glucan chain to the glucan backbone of the hemicellulose are indicated by arrows.

the oxygen of the glycosidic bond (O_1) between glucose residues in an adjacent chain. Since the glucan chains of cellulose are very long (8,000 to 15,000 residues), the number of hydrogen bonds between adjacent chains is very large. The resultant crystal is extremely stable and so tightly packed that there is no room for water molecules in the crystal structure.

Although there is some controversy as to whether native cellulose fibrils are 3·5 nm or 10 nm in diameter, it is clear that the glucan chains of cellulose aggregate to form stiff crystalline rods of very considerable length and mechanical strength. The molecular structure of β-1,4-linked glucose thus neatly determines the secondary and tertiary structures of cellulosic glucan chains, and establishes the quaternary, interchain associations—although not the dimensions—of the microfibrils. The stiff, crystalline rods of cellulose are clearly well suited to their biological function in the plant cell wall.

1.2.2 HEMICELLULOSES

Xylans, arabinoxylans, galactomannans, glucomannans and xyloglucans are common types of hemicelluloses (Timell, 1965). The basic repeating sequence of a hemicellulosic xyloglucan molecule is shown in Fig. 1.3, adjacent to a cellulosic glucan chain.

Although different hemicelluloses have different component sugars, all have two structural features in common which bear importantly on their biological function. (1) All hemicelluloses have straight, flat β-1,4-linked backbones. Any side chains attached to the backbone are short—usually just one sugar long—and stick out to the sides of the backbone (cf. Fig. 1.3). (2) All hemicelluloses have some structural feature which prevents the chains from extended self-aggregation of the type which exists between the β-1,4-linked glucan chains of cellulose. Xyloglucans, for example, have a β-1,4-linked glucan backbone (just as cellulose), but most of the glucose $-CH_2OH$ groups (C_6) are substituted with xylose side chains (see Fig. 1.3) and are thus not available for the formation of interchain hydrogen bonds. Similarly, since xylose is a pentose (*i.e.* a five carbon sugar), the β-1,4-linked xylose backbones of xylans and arabinoxylans have no $-CH_2OH$ groups available for interchain hydrogen bonding. The glucomannans have β-1,4-linked backbones containing both glucose and mannose. The $-CH_2OH$ groups of these sugars are unsubstituted, but the axial conformation of the $-OH$ groups at carbon 2 of the mannose residues (cf. Fig. 1.1) prevent close interchain associations wherever mannose residues occur in the backbone.

Although hemicelluloses cannot self-aggregate to form long, close-packed crystalline fibrils in the manner of cellulose, the chains of hemicellulosic polysaccharides can form important hydrogen bonded associations with each other in the cell wall, particularly between regions of the chains which have few side branches or axial hydroxyl groups (Blake & Richards, 1971; McNiel *et al.*, 1974). Even more importantly, hemicelluloses can co-crystallize with cellulosic glucan chains at the surface of the cellulose microfibrils (Bauer *et al.*, 1973, Northcote, 1972). The cocrystallization probably involves the formation of hydrogen bonds *from* the $-CH_2OH$ groups present in cellulose chains *to* the glycosidic oxygens in the adjacent hemicellulose chains (see Fig. 1.3). This association would form a tightly bound monolayer of the hemicellulose on the surface of the cellulose microfibril, and would function as part of the 'glue' which holds the microfibrils together in the cell wall.

1.2.3 THE PECTIC POLYSACCHARIDES AND STRUCTURAL PROTEIN

1.2.3.1 *Rhamnogalacturonans*

The rhamnoglacturonans are long polymers of α-1,4-linked galacturonic acid interspersed with a few residues of 1,2-linked rhamnose (Aspinall, 1973). There is some evidence that the rhamnosyl residues may occur at definite positions in the galacturonan chain, giving a subunit structure to the rhamnogalacturonan polymer (Talmadge *et al.*, 1973; see Fig. 1.4).

Fig. 1.4. Pectic rhamnogalacturonan.
CPK models illustrate the repeating subunit of a rhamnogalacturonan, with a short sequence of β-1,4-linked galactan attached to C_4 of one of the rhamnosyl residues (Talmadge *et al.*, 1973). The sequence of the subunit is $GalUA_8$ Rha GalUA Rha $GalUA_4$.

The diaxial conformation of the α-1,4 glycosidic linkages between galacturonic acid residues (see Fig. 1.1) causes the orientation of adjacent rings to be twisted. Thus, the galacturonan polymer forms a tight, stiff, rod-like helix with three residues per turn (Rees & Wight, 1971; Fig. 1.4). The insertion of 1,2-linked rhamnosyl residues in the galacturonan chain creates 'kinks' or right angle bends.

Divalent cations, particularly calcium, form complexes with the carboxyl and hydroxyl groups of galacturonic acid residues in the polymer. Complex formation of this type occurs primarily between adjacent residues in a galacturonan polymer, but could serve to create ionic ligand bridges between adjacent galacturonan chains.

1.2.3.2 *Arabinogalactans*

Two distinct types of arabinogalactans are known to occur in plant cell walls (Aspinall, 1973). The first type has a β-1,4-linked galactan backbone with highly branched arabinose side chains. The second type has a β-1,3-linked galactan backbone with many short side chains containing galactose and arabinose.

1.2.3.3 *Structural Protein*

Primary cells walls (i.e. the type of walls characteristic of actively growing cells) contain a structural protein component. While the structural protein of these walls has not yet been isolated as an intact molecule, the analysis of peptide fragments from the primary walls of dicotyledonous plants has revealed several interesting characteristics (Lamport, 1970; Lamport *et al.*, 1973). This structural protein contains over 25% hydroxyproline, which is an unusual amino acid known to break the continuity of α-helical structures. In animals, hydroxyproline occurs almost exclusively in the proteins of connective tissue (collagen and gelatin). Several tryptic peptides of the structural protein have been isolated,

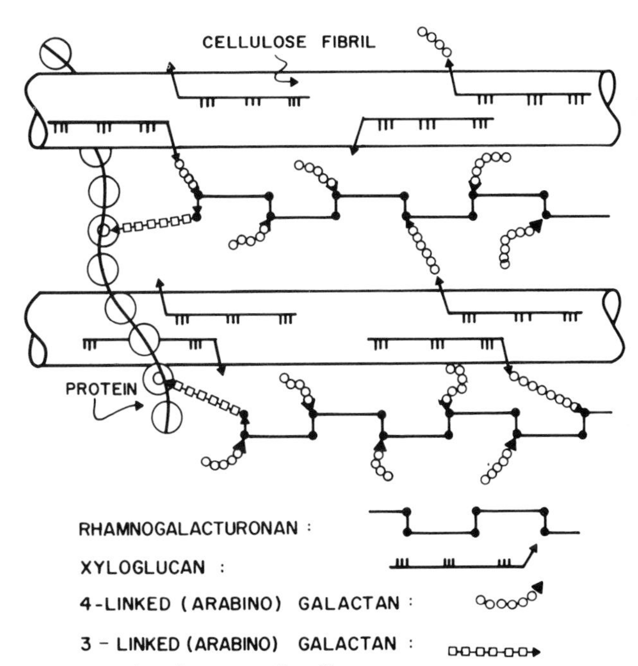

Fig. 1.5. Interconnections between cell wall components.

Schematic representation of the polymeric components of sycamore primary cell walls and their interconnections (Keegstra *et al.*, 1973). Hemicellulosic xyloglucan polymers are cocrystallized with cellulosic glucan chains on the surface of the (two) microfibrils. The reducing ends of some—but not all—of the xyloglucan chains are glycosidically attached to some—but not all—of the β-1,4-linked (arabino) galactan side chains of the rhamnogalacturonan polymers. The rhamnogalacturonan polymers and the structural protein are interconnected by 1,3-linked (arabino) galactan bridges. Arrowheads indicate the reducing ends of polysaccharide chains.

and all appear to contain the sequence Ser-Hyp$_4$. Arabinose tetrasaccharides are glycosidically linked to virtually all of the hydroxyproline residues in the protein. In addition, many of the serine residues are glycosylated with galactose. As a result of this extensive glycosylation, the structural protein is very resistant to degradation by proteases, and is likely to have an extended, rod-like shape.

1.3 INTERCONNECTIONS BETWEEN CELL WALL COMPONENTS

The polysaccharide components of the primary cell walls of dicotyledonous plants are specifically attached to one another to form an interconnected network (Keegstra *et al.*, 1973; see Fig. 1.5). The β-1,4-linked glucan chains of cellulose are aggregated to form crystalline microfibrils. The hemicellulosic xyloglucan chains cocrystallize with the cellulose chains on the surface of the microfibrils as described above (1.2.2). The reducing ends of the xyloglucan chains are covalently (glycosidically) attached to β-1,4-linked arabinogalactan. The arabinogalactan chains, in turn, are glycosidically linked to the backbone of the rhamnogalacturonan, probably to the 4 positions of the rhamnosyl residues. The 1,4 arabinogalactan is thus a side chain of the rhamnogalacturonan, and forms an interconnecting bridge between the xyloglucan and the rhamnogalacturonan. The other type of arabinogalactan (β-1,3-linked) also seems to serve as an interconnecting bridge, glycosidically linking the reducing ends of rhamnogalacturonan chains to amino acid residues in the structural protein of the cell wall.

1.4 THE UNIVERSALITY OF PLANT CELL WALL STRUCTURES

Studies of cell walls isolated from suspension-cultured plant cells have demonstrated quite clearly that dicotyledonous plants from taxonomically diverse species have very similar primary walls (Albersheim, 1974). Similar studies on the primary walls of aspen cambial tissue and potato tuber tissue have shown that these walls are quite similar to the walls of suspension-cultured cells (Timell, personal communication; Bauer, unpublished results). Thus, all dicotyledons probably have the same basic primary cell wall.

The primary cell walls of monocotyledons have a structure which is somewhat different from that of dicotyledons, although the walls from various monocotyledon species appear to be quite similar to each other. The hemicellulose of the primary monocotyledon walls is an arabinoxylan instead of a xyloglucan, and the structural protein appears to contain very little hydroxyproline (Burke *et al.*, 1974).

Most cells, after they stop active growth, synthesize 'secondary' cell wall material which is quite different from the primary wall. Secondary cell walls consist almost exclusively of large amounts of cellulose and one (or more) of the hemicellulosic polysaccharides. Lignification is very common for cells with secondary walls.

Within broad taxonomic groups, related plants appear to have very similar cell walls. However, it should be recognized that relatively few types of cell walls from relatively few plant species have been carefully analysed. The diversity of specialized cell wall functions is likely to be reflected in a diversity of cell wall structures.

1.5 CELL WALL PLASTICS

1.5.1 LIGNIN

Lignin is a biological plastic formed in plant cell walls by the enzymic dehydrogenation of coumaryl, coniferyl and synapyl alcohols (Fig. 1.6) followed by a free radical polymerization (Freundenberg, 1968; Northcote, 1972). Since the polymerization is not enzymically controlled, and the monomeric free radicals can react with each other in a variety of ways, lignin does not have a unique structure. However, the lignin from a particular plant species or tissue does usually have a characteristic monomer composition. Some covalent linkages are formed between lignin and the polysaccharides of the cell wall during lignin biosynthesis (Morrison, 1974).

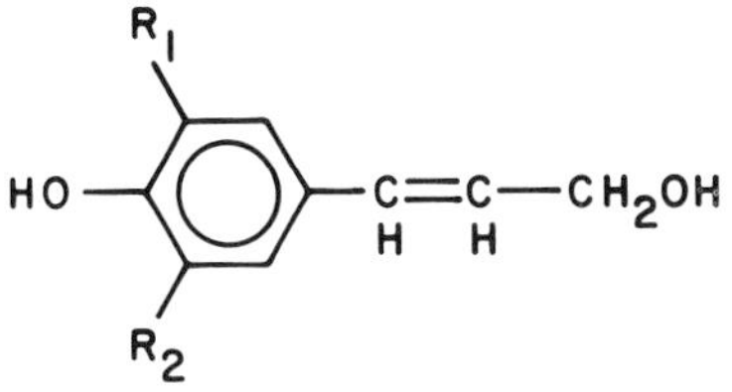

Fig. 1.6. Lignin precursors.
R_1 = H, R_2 = H, coumaryl alcohol. R_1 = H, R_2 = OCH_3, coniferyl alcohol. R_1 = OCH_3, R_2 = OCH_3, synapyl alcohol.

Lignin formation is initiated very soon after secondary wall synthesis begins, and proceeds from the region of the middle lamella (the pectin-rich layer between adjacent cell walls) inward towards the plasmalemma. Thus, both the primary and secondary walls become fully impregnated with a rigid, hydrophobic plastic which is covalently linked to the polysaccharide matrix. The resultant structure is extremely strong and resistant to degradation.

1.5.2 CUTIN

Cutin is a biological plastic which coats the cell walls on the outer surface of plant epidermal cells. Although little is known about the detailed structure of cutin, it is clear that the principal monomeric components of this material are mono-, di-, and trihydroxyfatty acids (C_{16}–C_{18}). These hydroxyfatty acids are linked to each other mainly through ester bonds, although the presence of both ether and peroxide bonds have been reported (Kolattukudy & Walton, 1972). The formation of ester linkages between monomeric hydroxyfatty acids and preformed cutin appears to involve the enzymic transacylation of the hydroxyfatty acids from a Coenzyme A intermediate to free hydroxyl groups in cutin (Croteau & Kolattukudy, 1974).

The film of cutin polymer is impregnated—and frequently coated—with a complex mixture of waxes. The cutin-wax structure (cuticle) merges gradually with the normal polysaccharide components of the epidermal cell wall. There is evidence for the existence of hydrophylic channels (ectodesmata) in the hydrophobic cuticle (Franke, 1967). These channels may occur in morphologically distinct patterns on the epidermal surface.

The cuticular waxes and the cutin polymer form a tough hydrophobic skin over the surface of the plant which is important in minimizing water loss and preventing mechanical injury or invasion by pathogens.

1.6 BIOSYNTHESIS OF PLANT CELL WALLS

It is generally believed that cell wall polysaccharides are synthesized from the appropriate sugar nucleotides (e.g. UDP-glucose) by specific enzymes or enzyme complexes. Many—though not necessarily all—cell wall polysaccharides are synthesized in the golgi bodies (dictyosomes) and transported to the wall in golgi-derived vesicles which are able to fuse with the plasmalemma (O'Brien, 1972). The cell wall polysaccharides are then presumably interconnected while in the wall.

There are many reports of isolated enzymes or particulate complexes which incorporate sugars from the sugar nucleotides into polysaccharide material. However, the amount of polysaccharide material formed by these *in vitro* enzyme systems is often a miniscule fraction of the amount formed *in vivo*. Thus the relationship between these isolated enzymes and cell wall biosynthesis is uncertain.

Isolated enzyme preparations have been reported to synthesize alkali-insoluble β-1,4 glucan ('cellulose') at rates comparable to the rate of *in vivo* cellulose synthesis (Rollit & Maclachlan, 1974). However, these enzymes have not as yet been shown to form the microfibrils characteristic of cellulose. There is some evidence that cellulose microfibrils are synthesized at the plasmalemma rather than in the golgi (Bowles & Northcote, 1972), perhaps by plasmalemma-fused vesicles which contain rows of membrane-bound, microfibril-synthesizing particles (Kiermayer & Dobberstein, 1973).

1.7 CELL WALL ULTRASTRUCTURE

1.7.1 CELLULOSE MICROFIBRILS

One of the most important and intriguing aspects of plant cell wall ultrastructure is the orientation of the cellulose microfibrils within the wall (Albersheim, 1965). Cellulose microfibrils are deposited (or possibly synthesized) at the inner surface of the wall, adjacent to the plasmalemma. They have a particular orientation when deposited. On the other side of the plasmalemma there is a thin layer of microtubules. The orientation of the microtubules in this layer exactly parallels the orientation of the cellulose microfibrils being deposited on the opposite side of the plasmalemma (Newcomb, 1969). Disruption of the microtubules (e.g. by colchicine) affects the orientation of the microfibrils.

In the primary walls of elongating cells, the most recently deposited cellulose microfibrils (those closest to the plasmalemma) lie parallel to the plasmalemma and are oriented perpendicular to the long (growth) axis of the cell, like the hoops which hold a barrel together. Cellulose microfibrils deposited at earlier times (and thus further from the plasmalemma) are still parallel to the plasmalemma, but are oriented at decreasing angles (i.e. more nearly parallel) to the growth axis of the cell. It is as though the process of cell wall elongation pulls the ends of the microfibrils towards the ends of the cell. In the end walls of a cell, or in the newly forming cell wall created by cell division (i.e. the cell plate), the cellulose microfibrils again lie parallel to the plasmalemma, but are randomly crossed and not parallel to each other.

The cellulose microfibrils in secondary walls are usually deposited in discrete, concentric layers. The microfibrils in a given layer are all parallel to each other, but are oriented at a considerable angle to both the microfibrils in adjacent layers and to the long axis of the cell. This cross-hatching of microfibrils in adjacent layers of secondary walls adds considerably to the overall strength of the wall, and is the same principle used in the manufacture of fibreglass-epoxy fishing rods, where great strength for weight is essential.

Secondary wall material may also be deposited in very localized regions to form highly specialized structures such as the rings or spirals of secondary wall found in xylem cells. In such cases the microtubules on the inside of the plasmalemma have the same localization and orientation as the cellulose microfibrils being deposited on the opposite side of the plasmalemma.

1.7.2 SPECIALIZED STRUCTURES

Many of the modifications of cell wall structure which occur are related to communication or transport between cells. In cells with primary walls, intercellular communication is facilitated by plasmadesmata. Plasmadesmata are small pores of approximately 40 nm diameter which extend through the wall between two adjacent cells (Ledbetter & Porter, 1970; Fig. 1.7). The plasmalemma

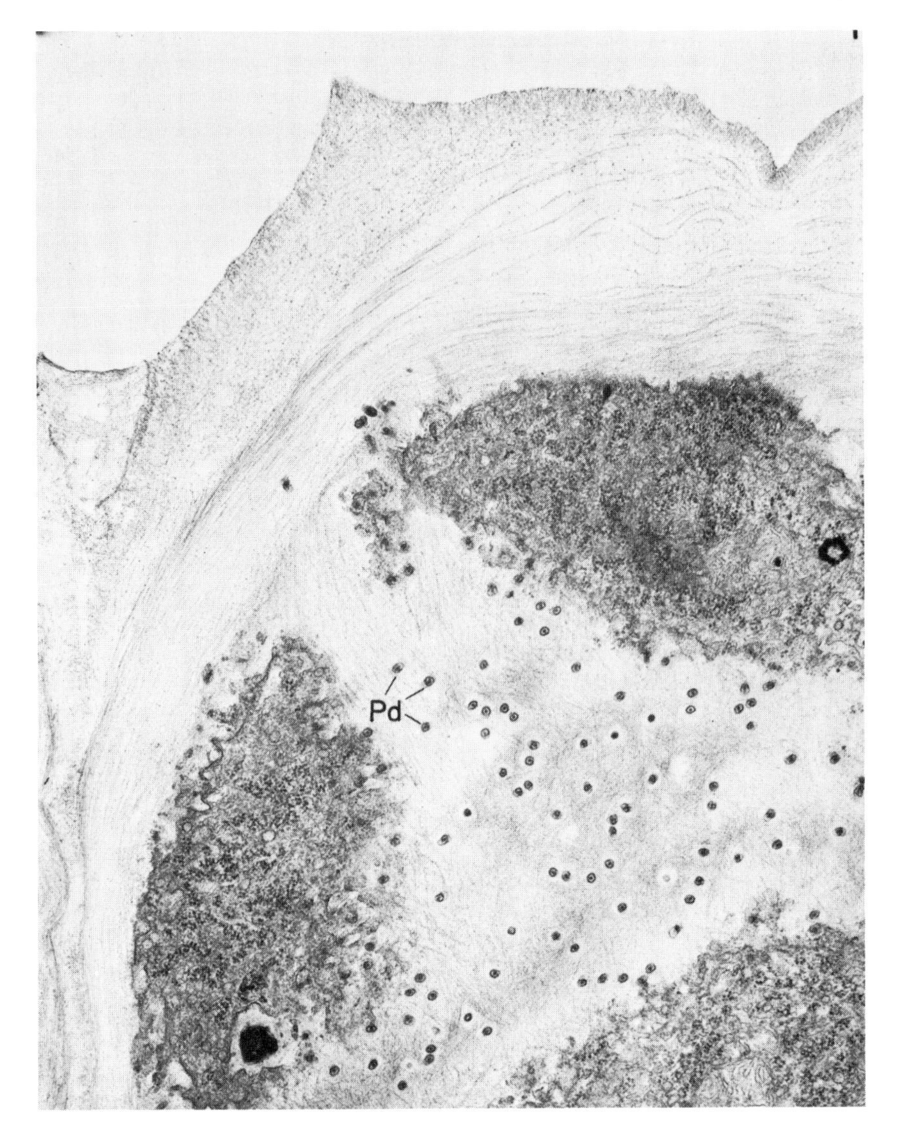

Fig. 1.7. Plasmadesmata of plant cell walls.

Plasmadesmata (Pd) in the end walls of collenchymal cells of a wheat stamen filament. The 'core' material in the plasmadesmata can be seen as an electron-dense dot in the centre of the plasmadesmatal pores. (See Ledbetter & Porter, 1970. Reproduced courtesy of the authors.)

of the cell(s) also extends through these pores, so that the cytoplasm of one cell is continuous with that of adjacent cells. There appears to be a core of electron dense material in the center of the plasmadesmatal tubes (see Fig. 1.7). This core may be a specific (though as yet speculative) structure capable

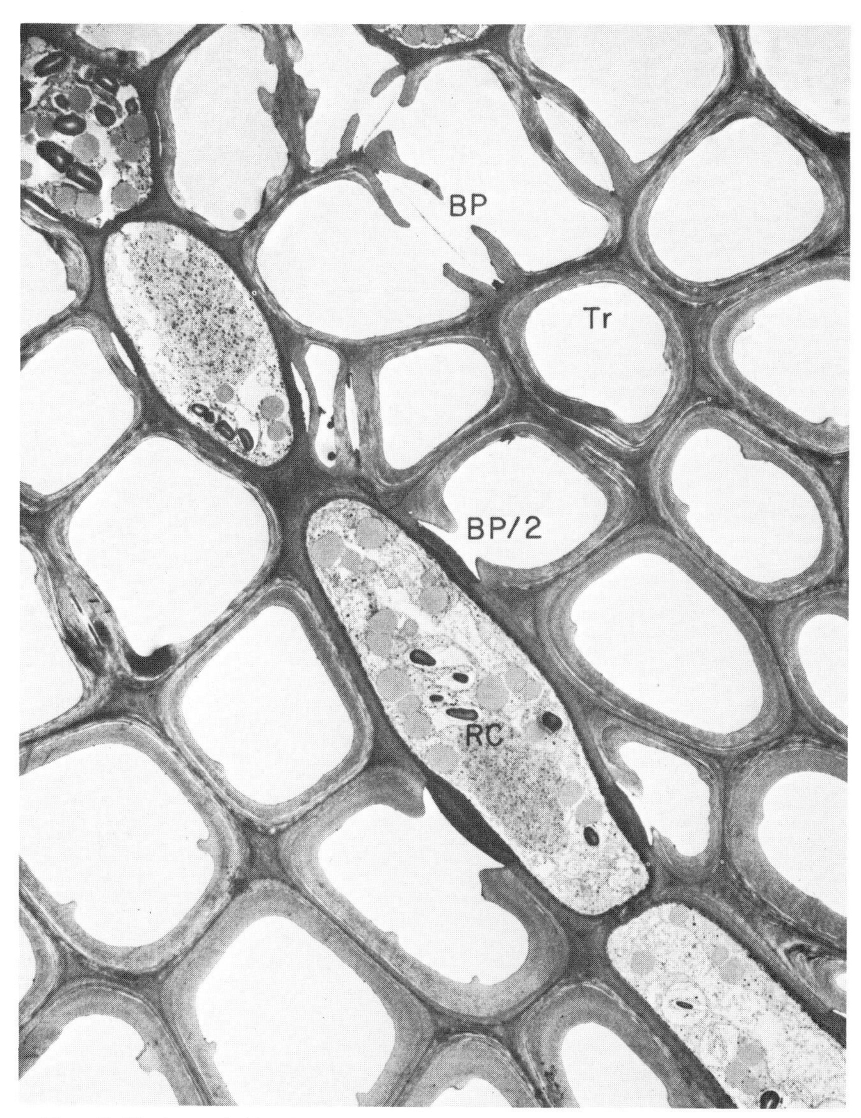

Fig. 1.8. Pits in tracheids of *Taxus canadensus*.

A full bordered pit (BP) can be seen in cross-section in the upper left portion of the figure. The faint line in the centre of the pit is the disc. The pits lie between tracheids (Tr), which are long, cylindrical, cytoplasmless tubes that carry water from the roots to the leaves. The primary cell walls and middle lamella can be seen as electron dense material running in a thin line between cells and in the corners where 3 or 4 cells meet. Layering of the thick secondary wall material is evident. Several half-bordered pits (BP/2) between tracheids and ray cells (RC) can be seen. (See Ledbetter & Porter, 1970. Reproduced courtesy of the authors.)

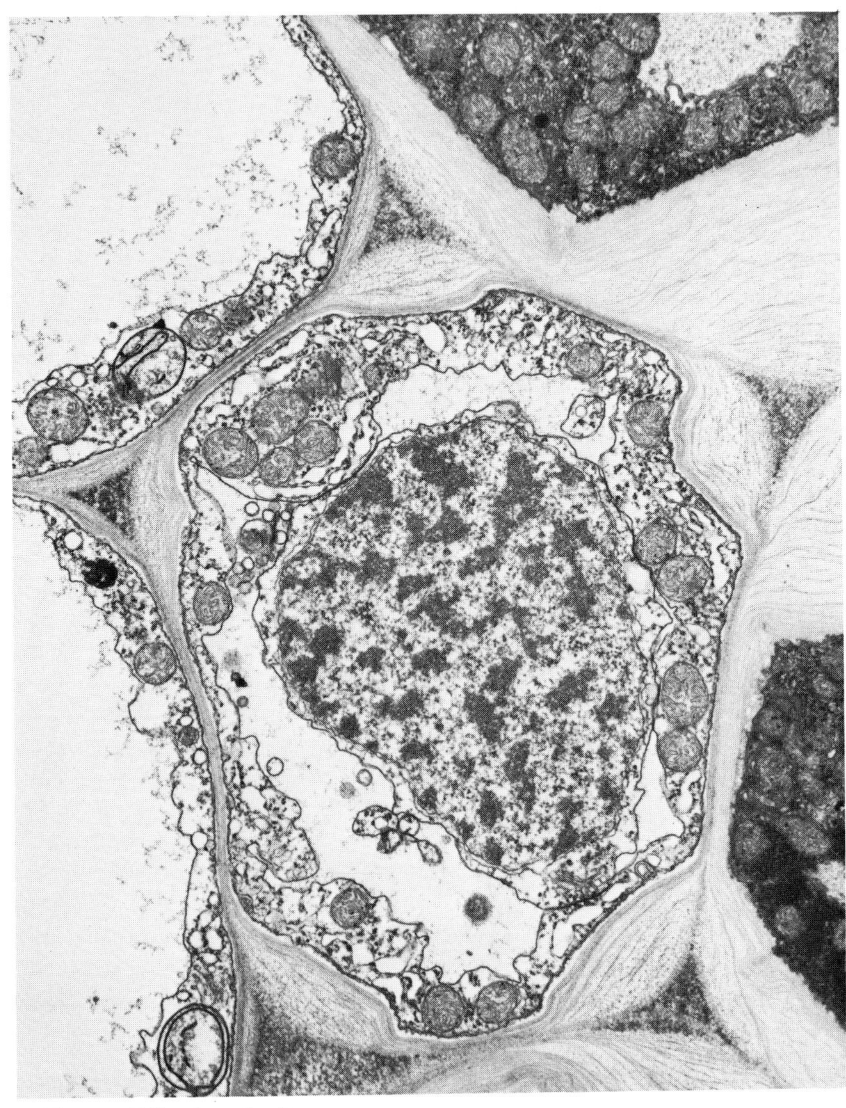

Fig. 1.9. Collenchymal cells.
The walls of these cells are unusually thick primary walls, heavily hydrated and unlignified. Cellulose microfibrils can be seen throughout the walls. These cells elongate very rapidly, and the walls are quite plastic. (See Ledbetter & Porter, 1970. Reproduced courtesy of the authors.)

of regulating what can and cannot be passed between cells. Regardless of the nature of the core, it is clear that the plasmadesmata are an aspect of plant biology which has no obvious counterpart in animal tissues. Each cell may have thousands of plasmadesmata, either randomly scattered as individual pores or grouped in distinct fields where the primary wall is often thinner than usual.

In cells with secondary walls, and particularly in tracheids, intercellular communication and transport is facilitated by pits (Ledbetter & Porter, 1970; Fig. 1.8). In the full, bordered type of pit, the primary wall and middle lamella are largely dissolved. The hole thus formed is surrounded and partly overgrown by depositions of secondary wall material which create a raised, overhanging border. The primary wall and middle lamella in the pit is replaced by a relatively impermeable disc or torus of thickened wall material. The disc is suspended by an easily permeable, radial network of cellulose microfibrils. The diameter of the disc is slightly larger than the aperture of the overhanging pit border. Thus the disc can act as a valve to close the pit when pressed against the border by a large pressure differential (Albersheim, 1965). Gas bubbles, which would break the flow of water through the tracheids, can be sealed off by the action of such valves.

Where the tracheids are adjacent to ray cells, the pits are differentiated only on the tracheid side, giving half-bordered pits (Fig. 1.8). The walls of the ray cells in the pit region appear to remain intact or only slightly modified.

A further important example of a cell wall modification which facilitates intercellular communication or transport is the sieve plate of primary phloem. The phloem cells have primary cell walls with many plasmadesmata. Some of the plasmadesmata of the crosswalls between adjacent phloem cells enlarge considerably and become lined with callose (a β-1,3 glucan) on the *wall* side of the plasmalemma. The pores of the sieve plate thus formed become filled with a fibrous protein (which may be analogous to the core material of plasmadesmata).

Many other specialized cell wall structures or modifications could be described (see also Fig. 1.9). Several examples may be mentioned just to indicate the range of possibilities: the walls of pollen cells, which are so indestructible that they are used by archeologists to characterize the flora of dwelling sites many thousands of years old; the walls of bark cells, made largely of suberin, a wall material similar to cutin that can form a protective coating over wounds so as to prevent water loss or pathogen invasion; and the walls between adjacent endodermal cells, which have very dense Casparian strips which prevent the movement of ions through the walls and into the xylem stream.

The wall structures and modifications described in this section may all be seen with the microscope—there are undoubtedly many other important differentiations of the cell wall which we cannot see.

1.8 HORMONAL CONTROL OF CELL WALL BIOSYNTHESIS AND DIFFERENTIATION

We know that cell wall biosynthesis and the modification of cell walls are important aspects of cellular differentiation, and that cellular differentiation can be controlled—or at least affected—by plant hormones. However, the mechanisms by which particular plant hormones affect the form, substance and synthesis of particular cell walls are almost wholly unknown.

Significant progress has been made recently in elucidating the role of auxins in cell wall elongation. Indoleacetic acid and several similar compounds cause a marked increase in the rate of coleoptile cell elongation (Cleland, 1971). Coleoptiles normally receive auxin from the apical region of the shoot. However, excised coleoptiles can respond to auxin exogenously supplied in solution, elongating at a rate of 10–30% per hour. The mechanism of auxin-stimulated elongation of coleoptile cells has been proposed to involve the activation of a hydrogen ion 'pump' in the plasmalemma (Cleland, 1973; Rayle, 1973). The hormonal activation of this pump results in a lower pH in the cell walls which, in turn, appears to activate enzymes in the cell wall which are capable of selectively 'loosening' the polysaccharide network so that the walls can elongate more rapidly. Auxin must also stimulate cell wall biosynthesis (by some unknown mechanism) since the walls retain a constant thickness while more than doubling their length during prolonged auxin treatment.

A further important aspect of hormonal effects on plant cell wall biosynthesis and differentiation has been revealed by studies on the changes in microtubule orientation caused by exogenously supplied hormones (Shibaoka, 1974). In bean epicotyl segments, kinetin inhibits elongation and promotes a thickening or lateral expansion of the cells. Gibberellins, on the other hand, promote elongation and inhibit lateral expansion. The microtubules adjacent to the plasmalemma in cells of epicotyl sections supplied with kinetin and auxin are found to be oriented parallel to the cell axis. However, in sections supplied with gibberellin and auxin the microtubules are oriented transverse to the cell axis. The microtubules are randomly oriented in sections supplied with auxin alone. Thus, kinetin and gibberellins (but not auxins) appear to be able to control the direction of cell growth by somehow determining the orientation of the microtubules. The orientation of the microtubules, in turn, determines the orientation of the cellulose microfibrils being deposited in the cell wall—which determines the direction in which the walls can most easily expand. (See Chapter 13 for further discussion of hormone action.)

1.9 THE ROLE OF THE PLANT CELL WALL IN INTERACTIONS WITH OTHER ORGANISMS

Plants are beset by many pests and pathogens and helped by a variety of symbionts. The walls of epidermal cells may be specialized in a number of ways to form a protective skin over the entire plant (e.g. cutin, wax, gums and mucilages, bark and thorns, etc.). Many cell walls throughout the plant can become resistant to degradation by pathogens by means of lignification.

Quite apart from such modifications, however, the structure of the cell wall

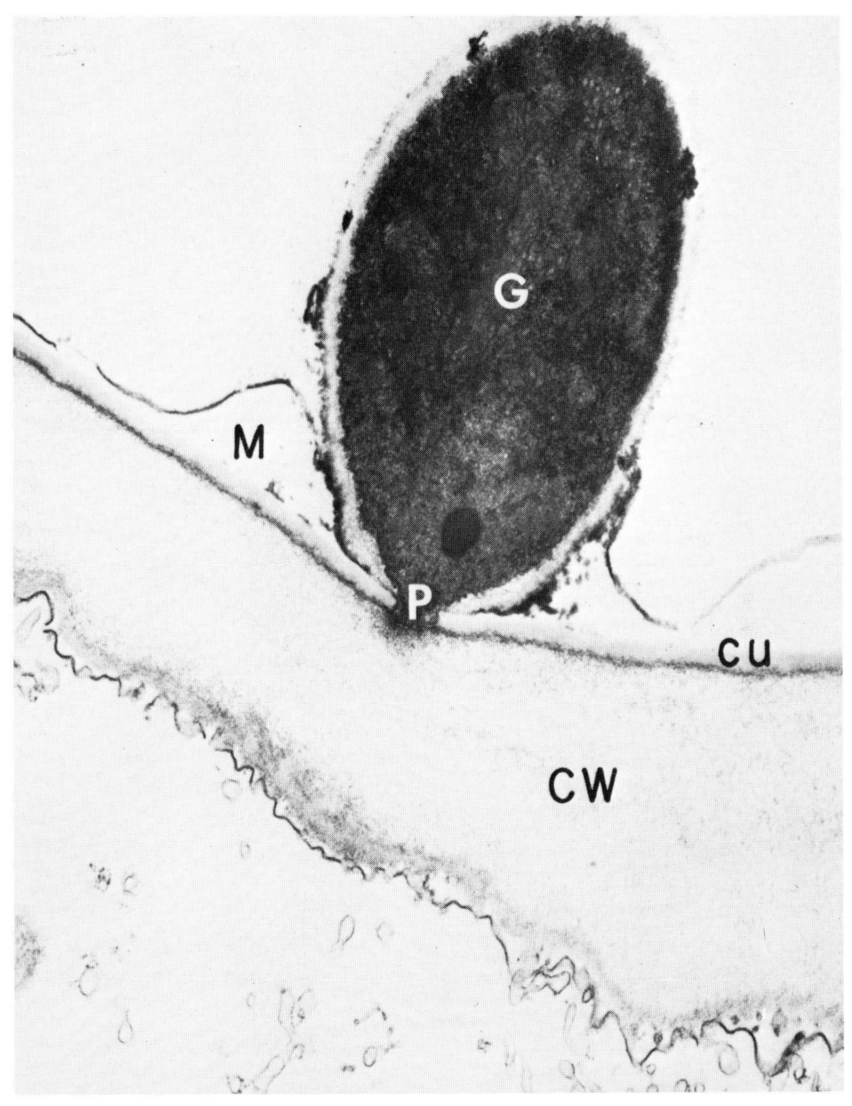

Fig. 1.10. Penetration of epidermal cell walls of broad bean by *Botrytis cinerea*.

The fungal germ tube (G) is attached to the surface of the cuticle (CU) by a layer of mucilage (M) secreted by the fungus. The fungus appears to have dissolved a hole through the cuticle and to have begun dissolving the plant cell wall (CW) beneath this hole. A membrane-bound infection peg (P) penetrates through the pore in the fungal cell wall and the cuticle. (Courtesy Prof. W. E. McKeen. See *Phytopathology* (1974), **64**, 461–67.)

can itself present a formidable barrier to pathogens. In order to penetrate the cell wall and utilize its component sugars, pathogenic fungi and bacteria have evolved sophisticated, inducible batteries of enzymes which can hydrolyze components of the plant cell wall (Bateman & Basham, 1975). Figure 1.10 illustrates the initial stages of cell wall penetration by a fungal pathogen. The pathogen appears to have dissolved a hole through the cuticle and to have started degrading the cell wall.

Plants, in turn, have evolved countermeasures to the hydrolytic enzymes of the pathogens. In primary walls, the molecular architecture is such that only the pectic polysaccharides are accessible to hydrolytic enzymes (Bauer *et al.*, 1973). It is probably for this reason that the first enzymes to be secreted by an invading pathogen are the pectin-hydrolyzing enzymes (Bateman & Basham, 1975). This is also likely to be the reason why the walls of many plants contain proteins which specifically inhibit the pectin-degrading enzymes (and only the pectin-degrading enzymes) of pathogenic microorganisms (Anderson & Albersheim, 1971). In other plants the pectic polysaccharides are heavily acetylated, and thus resistant to enzymic attack.

Plants have evolved mechanisms for counterattack as well as defence. Inducible enzymes are present in the cell walls of several plants which hydrolyze the wall polysaccharides of invading fungal pathogens (Ables *et al.*, 1970; Pegg & Vessey, 1973). The pathogens react by secreting proteins which specifically inhibit the attacking plant enzymes (Albersheim & Valent, 1974). The cell walls of a variety of other plants contain glycoside-hydrolyzing enzymes which can release hydrogen cyanide from cyanogenic glycosides. The release of hydrogen cyanide by the plant occurs in response to attack by a pathogen. The pathogen (sometimes) avoids cyanide poisoning by an inducible enzyme which converts the cyanide to harmless formamide (Fry & Munch, 1975).

From these and other examples it is clear that the cell wall is a most important battleground in the contest between plants and their pathogens. The plant cell wall is not just a strong but passive barrier to invasion. It is impregnated with a host of molecules which can recognize a pathogen, modify the defences, or mount a counterattack.

FURTHER READING

Albersheim P. (1965) Substructure and function of the cell wall. In *Plant Biochemistry* (Ed J.E. Varner) pp. 151–186. Academic Press, New York.

Aspinall G.O. (1973) Carbohydrate polymers of plant cell walls. In *Biogenesis of Plant Cell Wall Polysaccharides* (Ed. F. Loewus) pp. 95–115, Academic Press, New York.

Bateman D.F. & Basham H.G. (1975) Degradation of plant cell walls and membranes by microbial enzymes. In *Physiological Plant Pathology* Vol. I (Ed. P.H. Williams & R. Heitifus) Springer-Verlag, Berlin. In press.

Frey-Wyssling A. (1969) The ultrastructure and biogenesis of native cellulose. *Fortschr. Chem. Organ. Naturst.* **27**, 1–30.

KEEGSTRA K., TALMADGE K.W., BAUER W.D. & ALBERSHEIM P. (1973) The structure of plant cell walls III. A model of the walls of suspension-cultured sycamore cells based on the interconnections of the macromolecular components. *Plant Physiol.* **51**, 188–96.

LAMPORT D.T.A. (1970) Cell wall metabolism. *Ann. Rev. Plant Physiol.* **21**, 235–70.

LEDBETTER M.C. & PORTER K.R. (1970) *Introduction to the Fine Structure of Plant Cells.* Springer-Verlag, Berlin.

NORTHCOTE D.H. (1972) Chemistry of the plant cell wall. *Ann. Rev. Plant Physiol.* **23**, 113–32.

REES D.A. (1972) Shapely polysaccharides. *Biochem. J.* **126**, 257–73.

TIMELL T.E. (1965) Wood hemicelluloses. *Advan. Carbohyd. Chem.* **20**, 409–83.

CHAPTER 2

MEMBRANE STRUCTURE AND TRANSPORT

2.1 INTRODUCTION

The control of metabolism and the development of cells frequently depends on the right substance being present in the right amount at a specific location in the cell at the right time. This may be achieved by regulation of the passage of materials from the external environment into the cell or from one compartment of the cell to another. All compartments of the cell, and its external surface, are bounded by membranes. It is clear, therefore, that any complete understanding of control mechanisms in metabolism or development must include a precise knowledge of the structure and composition of membranes and of the mechanisms whereby materials move through them. While it would not be true to suggest that all of this knowledge is available at present, the pace at which new information and insight has been gathered in the last decade is most impressive. In this short chapter it will not be possible to trace the history of the way in which ideas about membrane structure have developed, but it is worth mentioning that a (substantially correct) view of the basic structure of biological membranes was advanced in the 1930's, long before it was possible to visualize membranes in the electron microscope or to examine their detailed structure by X-ray deffraction techniques (Danielli & Davson, 1935). The simple trilaminar appearance of biological membranes in the transmission electron microscope (Fig. 2.1) is now familar to elementary students of biology and is known as the unit membrane; its occurrence is ubiquitous and this very fact has impressed on biochemists and others that this apparently uniform structure cannot explain the diverse properties of different membranes. This chapter covers the chemical composition of membranes and how these components are arranged. From this it will become apparent that the membrane is composed of a matrix, whose design is broadly similar in all cases, and a sub-structure on which many of the specific properties of the membrane probably depend. With this picture in mind it will then be possible to explore the basic types of transport which can occur across membranes and to relate them to the structures described.

2.2 CHEMICAL COMPOSITION OF MEMBRANES

Biological membranes are composed primarily of two main classes of compounds, lipids and proteins, which interact in several ways with water to bring

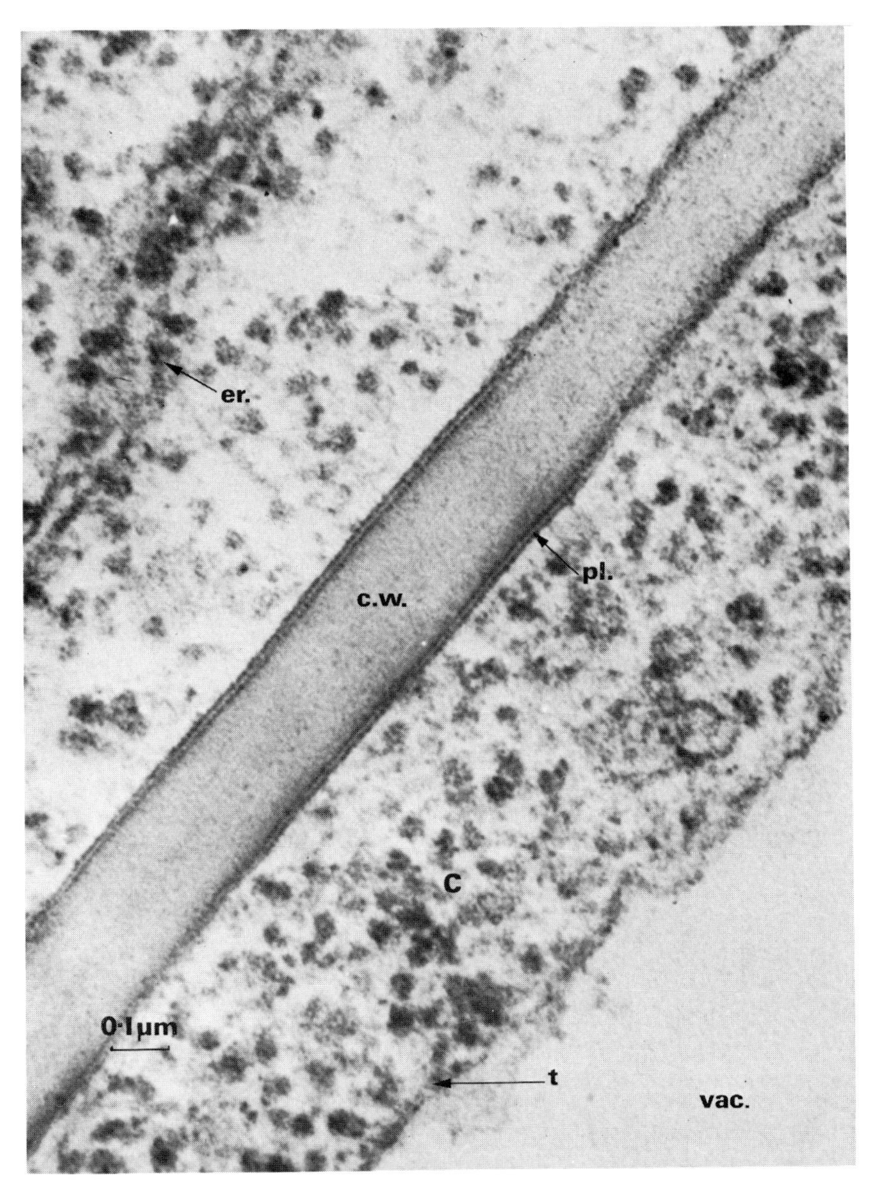

Fig. 2.1. Electronmicrograph showing the plasmalemma of two endodermal cells separated by a cell wall (c.w.) in the root of barley. The plasmalemma (pl.) is prominently stained in the central region of the picture and is clearly trilaminar. Notice that the tonoplast (t) bounding the vacuole (vac.) is much less distinct but also appears trilaminar at the point indicated by the arrow. Other symbols er= endoplasmic reticulum; c= cytoplasm. Total magnification about × 200,000. (Micrograph by courtesy of Dr. A. W. Robards).

about a characteristic trilaminar structure. As a very broad generalization it might be said that lipids make up the supporting matrix of the membrane while the proteins thus supported determine its specific properties. This is, of course, merely a convenient simplification as it is becoming plain that some of the characteristic properties of a membrane, particularly those which determine its effectiveness as a barrier to the diffusion of water and electrolytes, depend on the nature of the membrane lipids. Conversely some proteins have a structural role. For the present purposes, however, the classes of compounds will be considered separately and a synthesis attempted later on.

2.2.1 LIPIDS

Of the lipids present in plant cells the various phospholipids, glycolipids and sterols are of the greatest significance in membrane construction. The relative abundance of the components can be quite variable (Table 2.1) depending on the part of the plant analysed, the species and prevailing environmental conditions (see p. 29).

Table 2.1. Relative abundance of lipids of various classes in plant organs and organelles.

	Phospho-lipids	Glyco-lipids	Neutral lipid*	Ref.
	(% of total lipid)			
Bean leaves	22	38	40	Wilson & Crawford (1974)
Spinach chloroplasts	15	47	38	Allen *et al.* (1964)
Grape root	34	10	56	Kuiper (1968)
Oat root mitochondria	38	10	52	Keenan *et al.* (1973)

*Sterols and sterol esters usually make up more than 50% of the neutral lipid fraction.

2.2.1.1 *Phospholipids*

The phospholipid molecule can be separated into a charged or polar 'head' region and an uncharged or non-polar 'tail'. Such a molecule is described as amphipathic, and as we shall see later on, this property is of crucial importance in determining membrane structure (p. 36). Phospholipids are generally thought to be restricted to membranes but the extremely rapid rate at which membranes can be taken apart and re-assembled, as in cell plasmolysis and de-plasmolysis, makes it probable that there are stores of phospholipid within the cell.

Phospholipids are readily extracted from macerated plant tissues by a mixture of chloroform and methanol (2:1) and can be separated by thin layer chromatography using a variety of solvent systems (see Hitchcock & Nichols, 1971, for a review of techniques).

The commonest phospholipids in plant membranes are derivatives of phosphatidic acid (Fig. 2.2); thus, lecithin is the choline ester of phosphatidic acid. Other common derivatives are also shown in Fig. 2.2. Phosphatidic acid (PA) itself is generally said to occur only in minute quantities in membranes or not at all, indeed, its presence in an extract is often taken as an indication of the activity of phospholipase D (Mazliak, 1973). There is a report, however, in which phosphatidic acid is said to be one of the major constituents of the plasmalemma of oat (*Avena sativa*) root, (Keenan *et al.*, 1974). Unfortunately,

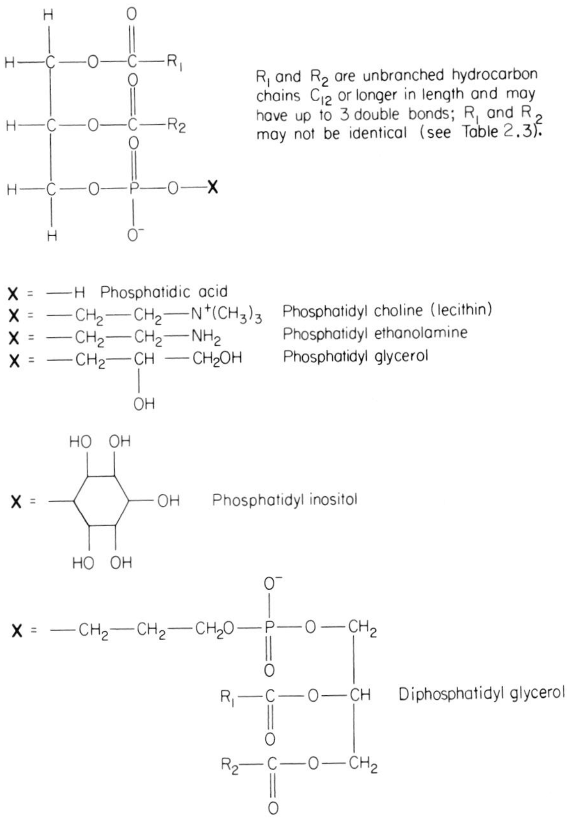

R_1 and R_2 are unbranched hydrocarbon chains C_{12} or longer in length and may have up to 3 double bonds; R_1 and R_2 may not be identical (see Table 2.3).

X = —H Phosphatidic acid

X = $—CH_2—CH_2—N^+(CH_3)_3$ Phosphatidyl choline (lecithin)

X = $—CH_2—CH_2—NH_2$ Phosphatidyl ethanolamine

X = $—CH_2—CH(OH)—CH_2OH$ Phosphatidyl glycerol

X = inositol (ring bearing five OH groups) Phosphatidyl inositol

X = $—CH_2—CH_2—CH_2O—P(O^-)(=O)—O—CH_2—CH(O—CO—R_1)—CH_2—O—CO—R_2$ Diphosphatidyl glycerol

Fig. 2.2. Structural formulae of phospholipids commonly found in plants.

detailed analyses of the plasmalemma from other plants are not available for comparison. In passing it might be noted that a great deal remains to be done, firstly in preparing pure sub-cellular fractions of the plasmalemma and of other membranes from plants and subsequently in determining their lipid composition. Table 2.2 presents some of the available information on the distribution of the different types of phospholipid. The information on the composition of the

Table 2.2. Relative abundance of phospholipids in tissues and organelles.

	PA*	PI	PC	PE	PG	DPG	Other	Ref.
			(% of total phospholipid)					
Whole leaves								
Sugar Beet	ND	11	47	23	19	ND	—	(1)
Maize	ND	8	30	16	31	15	—	(2)
Bean	5	7	42	20	26	ND	—	(3)
Chloroplasts								
Sugar Beet	10	9	35	13	33	ND	—	(1)
Tobacco	ND	4	23	16	58	ND	—	(4)
Spinach (lamellae)	trace	10	20	13	57	ND	—	(5)
Mitochondria								
Cauliflower (floret)	ND	7	44	34	3	12	—	(6)
Potato (tuber)	ND	13	44	26	ND	17	—	(6)
Microsomes								
Cauliflower (floret)	ND	6	50	35	8	1	—	(6)
Potato (tuber)	ND	19	44	18	ND	19	—	(6)
Plasmalemma								
Oat (root)	14	4	18	11	4	ND	11**	(7)

(1) Wintermans (1960) (2) Roughan & Batt (1969) (3) Wilson & Crawford (1974) (4) Ongun & Mudd (1968) (5) Allen *et al.* (1964) (6) Moreau *et al.* (1974) (7) Keenan *et al.* (1973)

Notes *possibly indicates the activity of phospholipase D during preparation (see Mazliak, 1973).

**detected lysolecithin and lysophosphatidyl-ethanolamine—indicates possible activity of phospholipase A during preparation (see Nachbaur & Vignais, 1968).

membranes of mitochondria and chloroplasts is the most detailed and reliable since these organelles can be separated with relative ease and high purity during cell fractionation. The predominant phospholipid in a given membrane may be characteristic, e.g. phosphatidylglycerol (PG) is a major component of chloroplast membranes while it is only a minor component of the inner mitochondrial membrane where diphosphatidylglycerol (DPG) is predominant. In general extracts of shoots and roots neither of these phospholipids is as abundant as lecithin (PC) or phosphatidyl ethanolamine (PE). A small quantity of phosphatidyl inositol (PI), usually less than 10% of the total phospholipid, is found in all membranes.

In Fig. 2.2 the exact chain length of the acyl groups R_1 and R_2 which make up the hydrophobic tail, is not defined precisely. In nature it can vary considerably even in one type of phospholipid from a given tissue. The chain may be made from 12 to 22 carbon atoms and may contain up to three or, rarely, six double bonds. The chain is straight in all eukaryotic organisms and has been found to be branched only in certain bacteria (Asselineau, 1966). Variation in both the length and unsaturation (i.e. the number of double bonds) of the hydrocarbon chain influences its melting point; shorter and unsaturated chains melt at much lower temperatures than longer and saturated ones. As an example of this consider the effect of double bonds on the melting of free fatty acids containing 18 carbon atoms; the saturated stearic acid ($C_{18:0}$) melts at 69°C, the mono-unsaturated oleic acid ($C_{18:1}$) at 5°C and the double unsaturated linoleic acid ($C_{18:2}$) at −12°C. Organisms which live in warm conditions and warm blooded animals are generally found to have phospholipids with an abundance of fatty acids which tend to be fully saturated (e.g. the thermophilic alga *Cyanidium caldarium*, see Kleinschmidt & McMahon, 1970). By contrast, organisms which are exposed to lower temperatures have either more unsaturated acids or ones with shorter average chain lengths or a combination of both of these (e.g. in *Acholeplasma laidlawii*, see Huang *et al.*, 1974) to give phospholipids whose tails remain fluid. The significance of the maintenance of membrane fluidity will become apparent later (p. 41). The process of hardening plants against injury from frost or chilling is accompanied by changes in the degree of unsaturation of the membrane lipids (Wilson & Crawford, 1974).

Table 2.3 shows fatty acid analyses of individual phospholipids extracted from various sources. Bearing in mind that there is a great deal of room for manoeuvre in selecting the fatty acids to suit the prevailing environmental temperature the values for the relative abundance of fatty acids should be considered only as very general guides to the types of acid found in nature. Thus, the predominant fatty acids have even numbers of carbon atoms, the saturated acids found most frequently are palmitic (16:0) and stearic (18:0), and the principal unsaturated acids are linoleic (18:2) and the triply unsaturated linolenic (18:3). The fatty acid composition of lecithin can depend very strongly on its origin. For example, the lecithin in the outer mitochondrial membrane is much richer in palmitic acid (16:0), and perhaps is a less fluid component than in the inner mitochondrial membrane where triply unsaturated linolenic (18:3) is the most abundant fatty acid.

2.2.1.2 *Glycolipids*

In several respects the glycolipids resemble phospholipids. The molecule is amphipathic, the polar group being a galactosyl derivative of a diglyceride, the non-polar part of the molecule being a pair of long, straight-chain fatty acids. Glycolipids are unusually rich in the triply unsaturated linolenic acid ($C_{18:3}$) which may make up more than 90% of the fatty acid (Table 2.3).

The two most abundant glycolipids are mono- and di-galactosyl diglyceride, the structural formulae of which are illustrated in Fig. 2.3. They are characteristic of photosynthetic tissues since they are the major lipid component of chloroplast lamellae, largely replacing the phospholipids. Ongun *et al.* (1968) showed that more than 80% of all of the glycolipid in leaf cells was present in the chloroplasts. The probable orientation of these molecules in the chloroplast lamellae is much like that described for phospholipids (see 2.2.1.1.) with the fatty acid tails inserted into the central region of the membrane with the polar

Table 2.3. Fatty acid composition of phospholipids and glycolipids extracted from various plant membranes.

	Fatty Acid* (% of total)									
	14:0	15:0	16:0	16:1	16:3	18:0	18:1	18:2	18:3	Ref.
Lecithin										
Bean leaf	—	—	27	trace	—	6	4	38	26	1
Spinach leaf	trace	—	20	trace	trace	—	11	30	40	2
Barley root	1	<1	19	1	(3)*[2]	3	7	40	25	3
Cauliflower mitochondria										
—outer membrane	trace	trace	61	trace	trace	7	17	7	8	4
—inner membrane	trace	trace	20	trace	—	1	9	13	57	4
Phosphatidyl ethanolamine										
Spinach leaf	trace	—	46	—	2	1	2	7	43	2
Barley root	9	<1	20	1	(6)*	2	6	35	21	3
Grape root	5	—	49	—	—	12	12	11	3	5
Monogalactosyl diglyceride										
Bean—whole leaf	—	—	2	trace	—	trace	trace	2	96	1
—chloroplast envelope	—	—	10	trace	—	6	11	9	62	6
—chloroplast lamellae	—	—	1	2	trace	1	1	3	94	6
Digalactosyl diglyceride										
Bean—whole leaf	—	—	5	trace	—	1	trace	1	93	1
—chloroplast envelope	—	—	9	1	—	2	6	17	65	6
—chloroplast lamellae	—	—	4	trace	—	3	2	2	89	6

(1) Sastry & Kates (1964) (2) Allen *et al.* (1964) (3) Clarkson (unpublished) (4) Moreau *et al.* (1974) (5) Kuiper (1968) (6) Mackender & Leech (1974).

* Fatty acids are given as carbon chain length and number of double bonds.

**There is uncertainty about the exact length of this component—it may possibly be 17:2.

galactosyl groups at the membrane surface protruding into the stroma (Weier & Benson, 1967). Because the fatty acid is so highly unsaturated the membranes of lamellae probably remain fluid even at sub-zero temperatures—thus any photosynthetic reaction, or molecular reorientation, which depends on membrane fluidity may have a wide temperature range in which it can occur.

A sulphur-containing glycolipid is found as a minor component of most membranes. It is known as sulpholipid (Fig. 2.3) and its structure and occurrence in chloroplasts was reported by Benson (1963); the acyl groups are mainly palmitic with a preponderance of linolenic acid, thus resembling the other

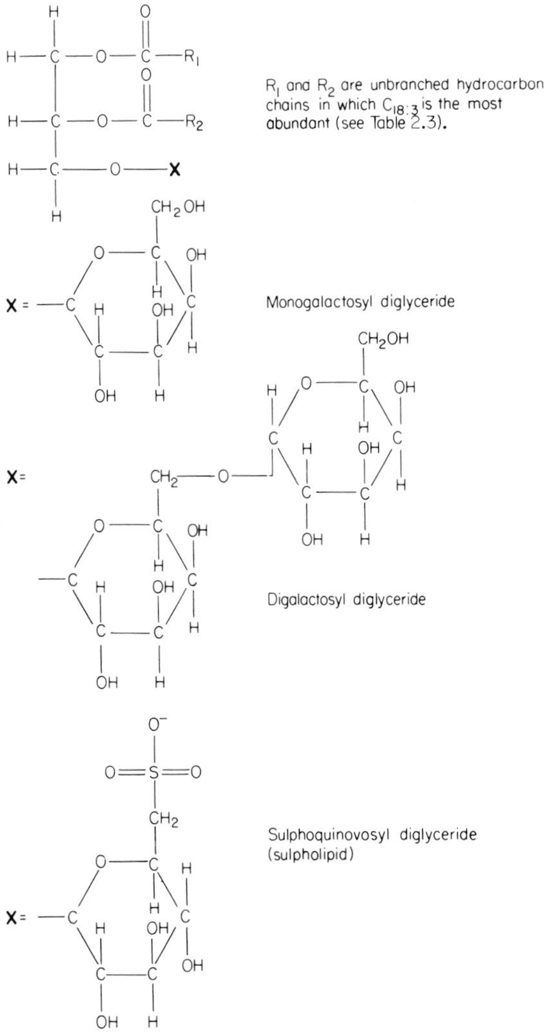

R_1 and R_2 are unbranched hydrocarbon chains in which $C_{18:3}$ is the most abundant (see Table 2.3).

Monogalactosyl diglyceride

Digalactosyl diglyceride

Sulphoquinovosyl diglyceride (sulpholipid)

Fig. 2.3. Structural formulae of glycolipids commonly found in plants.

galactosyl lipids. Sulpholipid represents only 1% of the total lipid in most tissues and organelles but in chloroplasts it may be as much as 10–15% of the lipid (Ongun & Mudd, 1968).

2.2.1.3 *Sterols*

A number of sterols can be extracted from plant tissues and fungi as well as from isolated membrane fractions. The conventional example of a sterol of common biological origin is cholesterol (Fig. 2.4); in practice plant cells contain relatively little of this sterol in comparison with animal cells. Such meagre quantitative data as is available show that sterols having 29 carbon atoms, e.g. β-sitosterol (Fig. 2.4), are the most abundant in higher plants, while in fungi the C-28 sterol, ergosterol (Fig. 2.4) is often dominant. All of these molecules have an extended concertina-like configuration (known as the 'chair' or 'boat'),

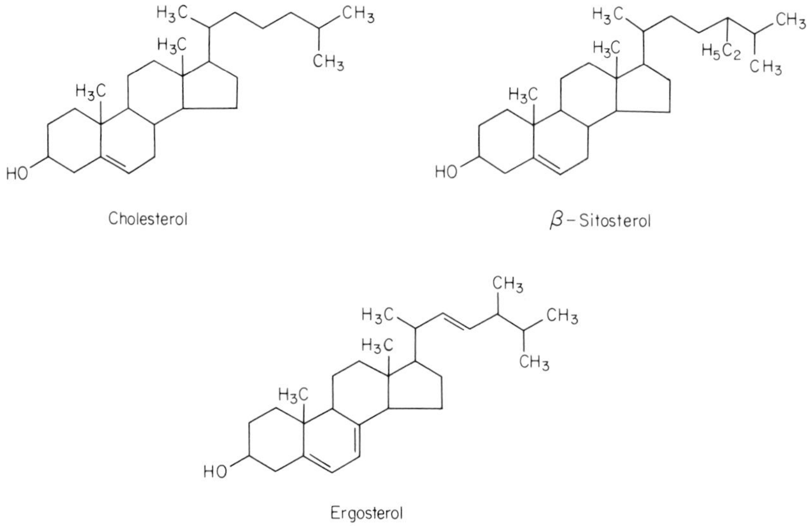

Cholesterol

β-Sitosterol

Ergosterol

Fig. 2.4. Structural formulae of sterols commonly found in biological membranes.

seem metabolically inert and are synthesized and turned over very slowly, especially in comparison with the other lipid components of membranes (Nes, 1974). Their function in membranes is not well understood but it is likely that they have an architectural role concerned with the maintenance of structure or order in the lipid domain. In this respect all of them probably function in the same was as cholesterol (see p. 38) because Butler *et al.* (1970) found that the structural order of bilayer membranes synthesized from lipids of ox brain tissue was stabilized equally well by cholesterol, β-sitosterol of plant origin and ergosterol. In the plasmalemma of the animal and plant cell there is a much higher proportion of sterols and sterol esters relative to phospholipid than in other membranes (Table 2.4). It should be noted, however, that the membranes of intracellular organelles contain much more protein than do plasmamembranes (see p. 35). To some extent this protein, much of ${}_2$which is bonded hydrophobically to the lipid, may function in a way similar to sterol in maintaining the structural order of the membrane interior.

Table 2.4. Relative proportions of sterols and phospholipids from membrane fractions of *Avena* root.

	Proportion of total lipid*(%)	
	Plasmalemma	Mitochondria
Sterol	24	18
Phospholipid	29	38
Sterol/Phospholipid	0·82	0·47

*The remaining lipid is composed largely of triglycerides and some free fatty acid and glycolipid. (Based on data from Keenan *et al.* (1973)).

2.2.2 PROTEINS

In many membranes, particularly in those of chloroplasts and mitochondria proteins make up most of the weight. The proteins found are many but as a first step in classifying them *integral* and *peripheral* proteins may be distinguished. This classification anticipates the subsequent discussion of membrane structure on page 38 but the terms clearly suggest that proteins in the two classes are associated with other components in the membrane in different ways. The recognition that certain proteins are embedded deeply in the lipid membrane represents a departure from the view, often advanced in earlier texts, that all of the membrane protein is located in the two peripheral bands which stain darkly with osmium and are visualized in the electron microscope (see Fig. 2.1). Whereas some of the protein is certainly located in this way and is probably bonded to the polar regions of the phospholipids electrostatically, it has become apparent that much protein is associated with the non-polar regions of the lipid by hydrophobic bonding. Peripherally located protein can be easily separated by washing with salt solutions or chelating agents but integral proteins are attached very strongly to the membrane and can be removed only after drastic treatment with organic solvents or detergents; even then, the isolated protein usually has some lipid attached to it.

Most of the membrane-bound enzymes, transport proteins (e.g. monovalent cation stimulated ATPase), drug and hormone receptors (in animal cells) and antigenic proteins are *integral* and are revealed when membranes are split open in freeze-fracturing (see p. 38). In many instances the enzymes arenon-functional *in vitro* in the absence of lipid. It is thought that the non-polar parts of the polypeptide chains are associated with the hydrophobic tails of the fatty acids and, this being so, several types of conformation are possible (see Singer, 1974 for a review). Proteins, like phospholipids, are amphipathic and their polar regions will arrange themselves so that their contacts with the hydrophobic regions of the membrane will be minimized; to ensure this, a protein could be arranged so that its polar, hydrophilic region lies among the phospholipid 'heads', or projects through them into the protein at the membrane periphery (Fig. 2.5).

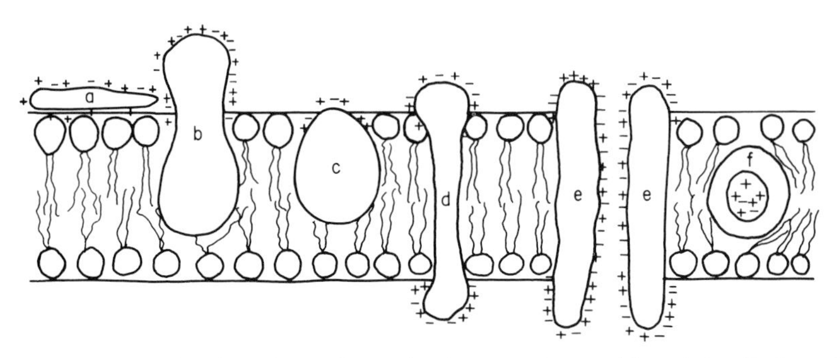

Fig. 2.5. Possible orientations of proteins in a membrane. (a) peripherally bound protein with polar groups all over its surface. (b) and (c) non-polar regions of the protein bonded hydrophobically to lipid but with different numbers of polar groups. (d) polar groups at either end of a long molecule with a non-polar central region. (e) a pair of proteins as in (d) making up a polar pore or channel in the membrane. (f) a hydrophobic globular protein wholly in the lipid domain—polar groups, if any, directed inwardly.

Long polypeptide chains with charged groups at either end may actually lie across the width of the membrane, or groups of them might lie with polar groups directed inwardly to form a hydrophilic pore across the membrane. An alternative conformation would be provided by the formation of a globular structure in which all of the polar groups would be directed towards the centre of the globule so that a hydrophobic surface would be presented to the lipid. This latter kind of conformation is probably least common.

The peripheral proteins can be attached to the polar groups of either phospholipids or integral proteins; examples which might be taken include cytochrome *c* which is located on the outer surface of the inner mitochondrial membranes (Schneider *et al.*, 1972; see also Chapter 5), the chromoprotein, phytochrome, which is thought to be attached to the plasmalemma (Marmé *et al.*, 1974; see also Chapter 12) and the sulphate and other ion-binding proteins on the outer surface of bacterial membranes (see Oxender, 1972, for a review).

From a quantitative point of view, certain generalizations about the relative abundance of peripheral and integral proteins can be made. The greater the metabolic activity which centres upon a given membrane system the greater amount of protein integrated into it. Thus, it might be anticipated that chloroplast lamellae and inner mitochondrial membranes would be relatively rich in these proteins, whereas membranes whose role is more concerned with providing a diffusion barrier, e.g. the plasmalemma and the tonoplast, would be less so; evidence from electron microscopy shows that this is so (see Table 2.5).

2.2.3 WATER

Water is an important, if neglected, constituent of membranes for several reasons. In a general way it determines their basic design since, in its presence, amphipathic lipids assume a bilayered configuration (see p. 36). There are, however, other specific associations of water molecules and membrane components which are not fully understood.

It has been estimated that water of hydration accounts for about 30% of the weight of membranes. Much of this water will almost certainly not be in a liquid state but will exist in ordered layers around the hydrophilic parts of lipids and proteins. Immobilized by hydrogen bonding these water molecules are in a liquid-crystalline condition and cannot be frozen to form ice. Water layers bound at the surface of the membrane have been estimated to have viscosity of 39 times that of pure water and to have a thickness of at least 2·2 nm (Schultz & Asunmaa, 1970). They must contribute to the mechanical stability of membranes and add significantly to their barrier properties to diffusing solutes. Expressing an extreme view, Ling (1973) has proposed that it is not lipid but these polarized multilayers of water which provide the cell with its selective surface barrier.

Hydrophobic bonding between the non-polar regions of lipids and integral proteins (see p. 34) is favoured thermodynamically by the interactions of their polar regions with water (Tait & Franks, 1971).

Much experimental evidence points to the fact that water molecules are not restricted to membrane surfaces but cross the hydrophobic regions in numerous water-filled pores. These pores are thought to conduct water and small solutes (diameter <0.4 nm) to which membranes are highly permeable (see p. 60). Some water lining these pores is fully 'organized' and should, therefore, be regarded as a structural feature but there is indirect evidence to suggest that some of it must be free water *in transit*.

2.3 MEMBRANE STRUCTURE

It is convenient to discuss membrane structure under two headings; the organization of the *membrane matrix* which is largely a matter of the relationships of the lipid components, and the *substructure* of the protein in the membrane.

2.3.1 THE MEMBRANE MATRIX

2.3.1.1 *Phospholipids*

The structure of phospholipid molecules considered earlier provides the key to understanding why it is that the membranes as seen in transverse sections have a characteristic trilaminar appearance. The hydrophilic head regions make hydrogen bonds with water and may become cross-linked to other heads and to proteins through ionic bridges, e.g. by calcium ions; thus they are organized into a lattice-like structure. By contrast the long acyl chains of the two fatty acids attached to each phospholipid are strongly hydrophobic, loosely organized and, above their melting point, are relatively fluid. If phospholipid is dispersed in water the 'tails' will take on a conformation which will minimize their contacts with water. The 'heads' will, of course, react favourably with water. If the available water surface is large relative to the amount of lipid, the molecules will arrange themselves as a film-like monolayer with the heads at the water surface and the 'tails' protruding from it at right angles (Fig. 2.6a). If more phospholipid molecules are added to this system so that there are more than can be fitted into a tightly packed monolayer over the water surface, a second type of arrangement occurs quite spontaneously. The phospholipids form two ranks with the heads facing outwards in both and the tails directed inwards to form a non-polar hydrophobic layer sandwiched between them (Fig. 2.6b). This bilayer arrangement, which is common to all biological membranes, can also be formed from mixtures of phospholipids under laboratory conditions. The synthetic membranes thus produced have helped in arriving at an understanding of many of the structure/function relationships of natural membranes (see Goldup *et al.*, 1970, for a readable review).

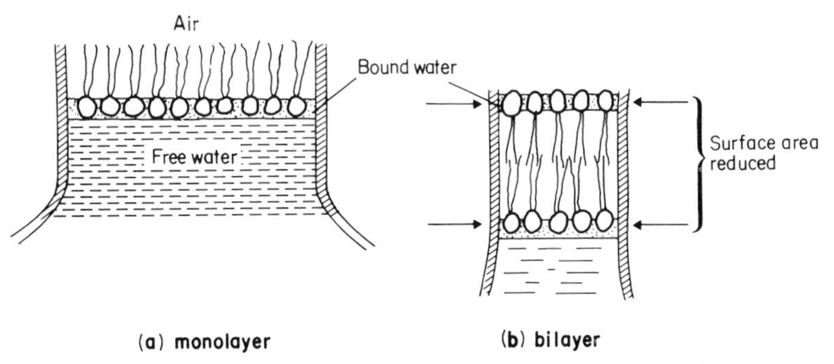

Fig. 2.6. An illustration of how a monolayer of dispersed phospholipid (a) in water, forms into a bilayer, (b) on contraction of the water surface area. The phospholipid heads have water bound to them in polarized multilayers (see p. 35).

The selected analyses in Table 2.2 show that a given membrane may contain several types of phospholipid as well as appreciable quantities of sterol. It is probable that there is a great deal more organization of phospholipids in natural membranes than can be demonstrated positively at present. Lipids of one kind may be associated into clumps so that the membrane surface may be very heterogeneous with lipids of differing physical properties arranged in a mosaic.

A mosaic of charged and uncharged areas might occur because some phospholipids carry a net electrostatic charge at normal pH values, e.g. phosphatidyl glycerol, while others, like phosphatidyl choline (lecithin) are neutral. This is of significance because it has been shown that, in synthetic bilayers, the surface charge on the phospholipid heads can partly determine both the ion-selectivity and the cation permeability of the membrane (Papahadjopoulos, 1971), and it may also be relevant in determining regions of the surface of the plasmalemma where endo-cytosis may occur (see p. 61).

Local variations in the packing of sterols may render some parts of the membrane less fluid than others and thus determine areas where diffusion may be severely restricted (Papahadjopoulos *et al.*, 1973).

More recently researchers have begun to investigate the possibility that the inner and outer halves of the bilayer may differ in their phospholipid composition. Should this prove to be the case, then it is possible that the barrier properties of the membrane to solute diffusion may be different when the membrane is approached from different sides.

In some special circumstances phospholipid molecules may become arranged into globular micelles in which the polar groups are directed towards the periphery of the sphere, the surface being hydrophilic. This state of affairs can be induced by dehydration in synthetic membrane systems, and in nature by viruses which create membrane instability, e.g. sendai virus, and by certain phospholipids (e.g. lysolecithin) with wedge-shaped head regions which tend to induce curvature of layers of closely packed phospholipids when they are introduced into a bilayer (Lucy, 1970). It has been suggested that rapid local transitions from the predominant bilayer to the micellar state are important in membrane fusion and in pinocytosis (see Lucy, 1970). If these transitions do occur then it is possible that they may cause transient gaps or pores to be created in the membrane; much physiological evidence points to the conclusion that membranes do have very fine pores in them (see p. 59).

2.3.1.2 *Sterols*

The insertion of sterol molecules into the membrane increases the structural order of the hydrophobic region. These molecules lie with their long axes parallel to the hydrocarbon chains of the fatty acids with their more rigid ring structures directed towards the outside and their open chain ends towards the centre. The mobility of the hydrocarbon chains nearest to the outside of the membrane is, therefore, restricted by these stiffening structures but they remain pliant at their ends so that the central region is fluid (Caspar & Kirschner, 1971). The rigidity conferred on the membrane by the inclusion of sterols slows down the diffusion of materials through the outer part of the lipid domain in synthetic bilayers (Papahadjopoulos *et al.*, 1973).

2.3.1.3 *A model of the membrane matrix*

Figure 2.7 provides a basic interpretation of the ideas on the membrane matrix discussed so far. A fact, which it is important to understand but which is difficult to illustrate, is that the centre of the membrane is fluid while the periphery is semi-crystalline. Although it is a stable structure, it is known that phospholipid molecules can be inserted into, and withdrawn from the matrix rapidly and that the structure illustrated in Fig 2.7 represents a dynamic steady state when it is part of a biological membrane.

2.3.2 MEMBRANE SUB-STRUCTURE

When a cell or a piece of tissue is frozen and then fractured with a suitable blade, the fracture plane will follow lines of weakness in the structure. Since the central region of the membrane matrix contains no ice it is a potential line of weakness, along which the membrane tends to fracture (Fig. 2.8). Where the fracture line passes tangentially across a cell or organelle, sheets of membrane material become apparent; since membranes tend to be cleaved down the middle it is obvious that the surface exposed is not the true membrane surface but is the membrane interior. Figure 2.9 shows a relatively smooth sheet of plasmalemma from an onion root tip on which numerous round particles and depressions can be seen. Some of the particles are arranged in files while others are randomly distributed. These particles, which are usually 6–9 nm in diameter, are embedded in the membrane and are not resting on its true surface. This was demonstrated by Pinto da Silva & Branton (1970) who etched away the ice from the fracture plane by leaving the specimens under a high vacuum for some minutes after fracturing them. Using this method, the true surface of membranes which lay obliquely to the fracture plane was eventually revealed as the ice from the surrounding cytoplasm sublimed. The true surface had a much smoother appearance, the undulations of which gave the impression of a blanket lying over the embedded particles. The particles in the membrane can be removed by treatment with proteolytic enzymes and can be re-created in synthetic membranes which have been made in the presence of a hydrophobic protein. Vail *et al.* (1974) reported that a synthetic bilayer membrane into which a hydrophobic protein had been incorporated had intercalated particles of 8.5 to 9 nm diameter occupying 12% of the internal membrane surface whereas in bilayers lacking the protein there were none.

The frequency of particles varies according to membrane type (Table 2.5); chloroplast and mitochondrial membranes have the highest frequency as one might anticipate from the many metabolic events which are centred upon them. One of the two fracture faces is usually more densely populated than the other. Bearing in mind the amphipathic nature of membrane proteins (see p. 34) it is possible that the more densely populated face may reflect the principal orientation of the polar groups of the integral proteins, i.e. if most of the polar groups were directed towards the outside of the cell or compartment, the outer fracture

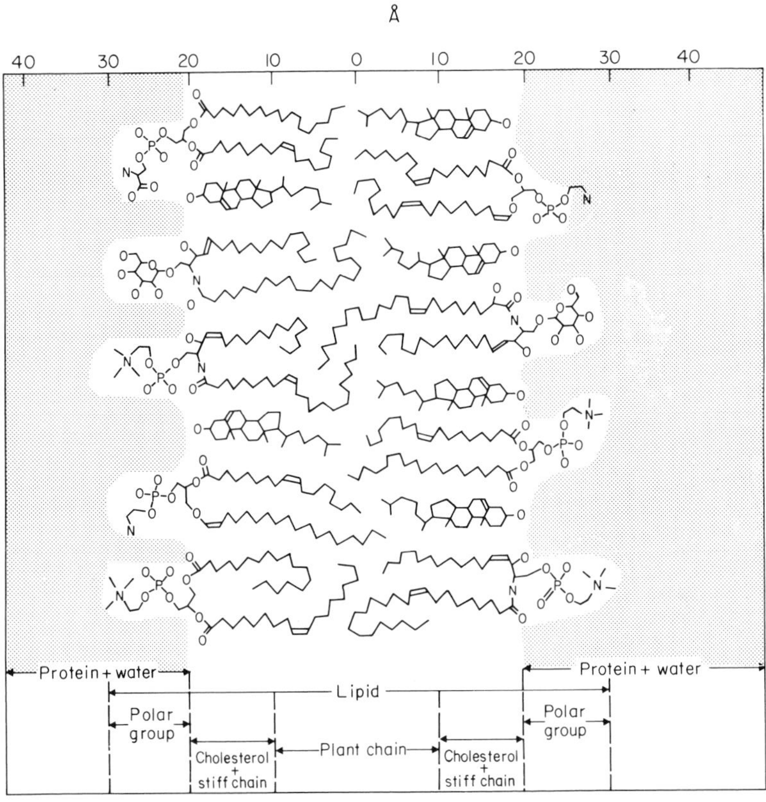

Fig. 2.7. Orientation of polar lipids, cholesterol and peripheral protein in a model of the membrane matrix. Based on X-ray diffraction analysis of nerve myelin. (From Caspar & Kirschner, 1971). In many ways myelin has been an unfortunate choice for detailed structure of cell membranes since it is not typical in having virtually no integral protein and has no intercalated membrane particles (see Table 2.5). It is, however, very appropriate as a general model of the membrane matrix.

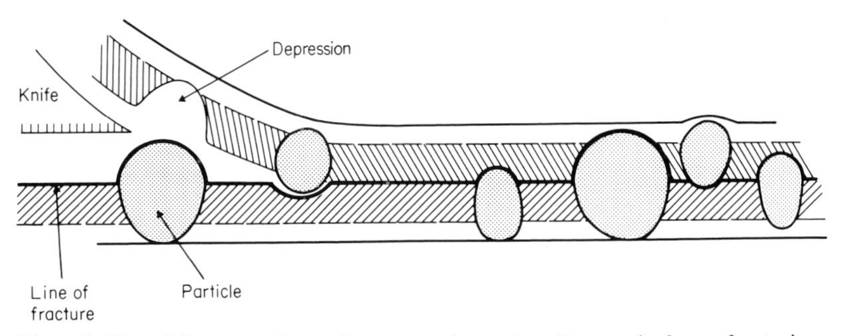

Fig. 2.8. Line of fracture when a frozen membrane is split open in freeze-fracturing. Particles embedded in the membrane become apparent in the cleavage plane either as projections or as hollows if they are removed on the upper half of the bilayer.

face would be the most densely populated. It should be noted that in synthetic lecithin membrane and in nerve myelin, which both have a very low permeability to ions, there are no particles.

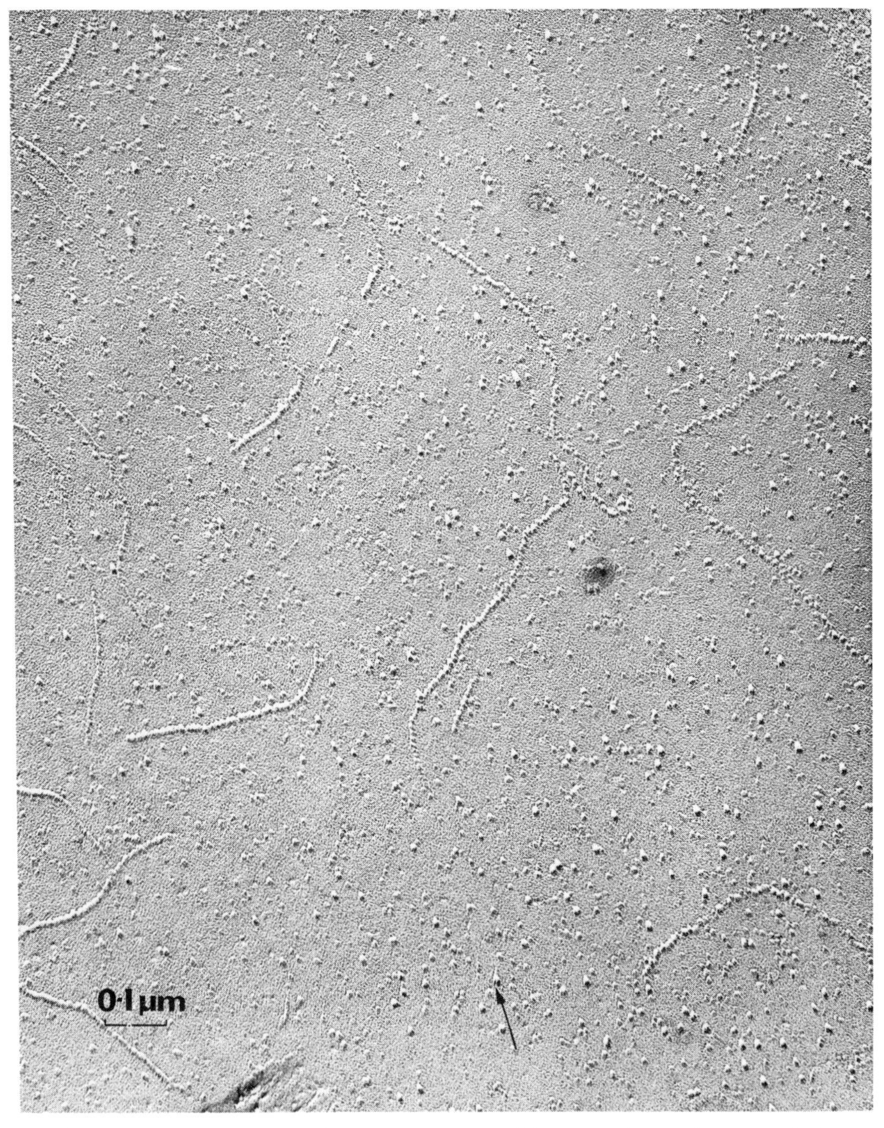

Fig. 2.9. Particles seen on the inner fracture face of the plasmalemma from a root tip cell in onion. The fracture face exposed as described in Fig. 2.8 and metal shadowed to produce a replica; the arrow represents the direction of shadowing. Note that there are distinct files of particles as well as randomly distributed individuals. The membrane matrix appears to be relatively smooth. (Micrograph by courtesy of Professor D. Branton).

Table 2.5. Density of intercalated particles in various membranes.

Type of membrane	Number of particles μm^{-2} Densely populated face	Thinly populated face	Membrane area covered with particles %	Ref.
Lecithin: synthetic	0	0	0	(1)
Nerve myelin sheath	0	0	0	(1)
Plasmalemma—cambial cells of willow	1,608	282	16	(2)
Plasmalemma—root tip of onion	2,030	550	15	(1)
Tonoplast—cambial cells of willow	695	140	7	(2)
Tonoplast—root tip of onion	3,300	2,480	32	(1)
Nuclear envelope—root tip of onion	1,790	420	12	(1)
Endoplasmic reticulum—root tip of onion	1,700	380	12	(1)
Mitochondria				
—outer membrane	2,806	770	23	(3)
—inner membrane	4,208	2,120	40	(3)
Chloroplast—lamellae*	3,860	1,800	80	(1)

(1) Data cited in Branton & Deamer (1972) (2) Parish (1974) (3) Wrigglesworth *et al.* (1970)
*The particles in the chloroplast lamellae were 16–20 nm in diameter while in most other membranes examined they were 7–9 nm.

2.3.2.1 *A model of membrane structure*

The representation of membrane structure shown in Fig. 2.10 is one which commands wide acceptance at the present time. In it we see that large masses of integral protein are inserted to varying extents in the membrane matrix while some are restricted to the surface and others protrude right across the width having a dumbell configuration. The model, which has become known as the *fluid mosaic* (Singer & Nicholson, 1972), has a disorderly appearance in contrast to the neat pictures used to illustrate the unit membrane; the membrane sub-structure has been likened to numerous protein icebergs in a sea of lipid.

2.3.2.2 *Membrane fluidity*

An essential feature of the model in Fig. 2.10 is that the various components of the membrane can move relative to one another. Rather than being fixed points, many of the proteins are best thought of as drifting around in the lipid, the viscosity of which will determine the rate at which they move. Evidence that membrane particles can move in the plane of the membrane comes from freeze

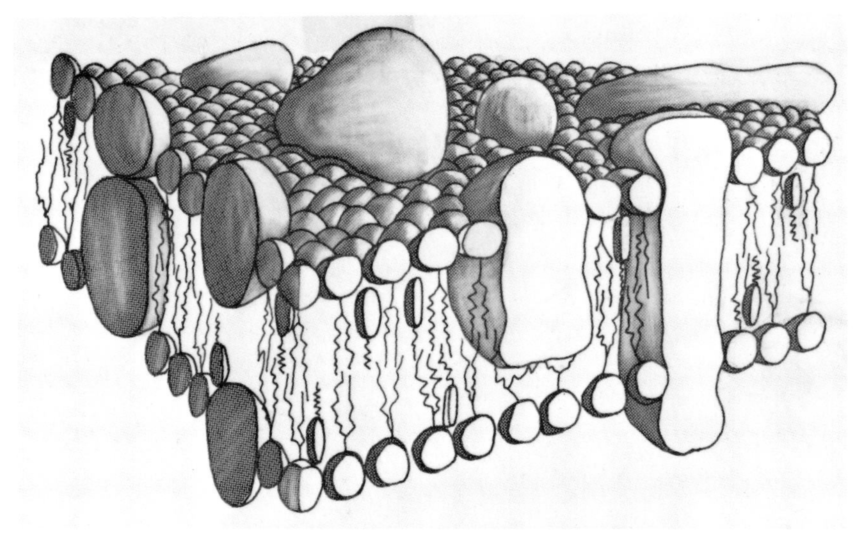

Fig. 2.10. A representation of the fluid mosaic model of membrane structure showing peripheral and integral proteins in a 'sea' of lipid molecules. Not shown are the polarized water layers which are bound to the polar regions of the lipids and proteins. The round-headed objects with twin tails represent phospholipids, the obovoid structure with the single tail represents sterol and the large irregularly shaped objects are proteins. (Based on ideas in Singer & Nicholson, 1972 and Capaldi, 1974).

fracture studies of the plasmalemma of *Mycoplasma mycoides*, in which it was seen that particles became aggregated as the membrane was cooled down but dispersed again when it was warmed up (Rottem *et al.*, 1973). A second line of the evidence comes from studies where two animal cells were induced to fuse with one another (Edidin & Farnbrough, 1973). One of the cells has a specific antigen in its plasma membrane which the other cell lacked. The distribution of the antigen could be visualized by the binding of a fluorescent antibody. The fusion of the cells to form a heterocaryon and the subsequent redistribution of the fluorescent antibody were observed using a fluorescence microscope. At first the fluorescent marker was restricted to one half of the heterocaryon, but quite quickly, and in an orderly progression the fluorescence spread right around the coat indicating that the antigenic protein, known to be integrated into the membrane matrix, had diffused in the plane of the membrane. For it to have done this it must have drifted through the lipid. If the heterocaryon was cooled below the transition temperature of its lipids, no mixing of the antigen occurred.

From what we have said above and from earlier comments (p. 29) it is clear that temperature will have a very important influence on the diffusion of materials through, and in the plane of, the membrane because of its effect on the viscosity of the lipid (Edidin, 1974). If the membrane is cooled to the extent that the lipids freeze then such diffusion will become very restricted; the biological significance of this is reflected in the fact that the rates of most physiological processes examined over a range of temperatures are found to have a sharp transition at a temperature which is close to the transition of membrane lipids from a liquid crystalline condition to the gelled condition (Simon, 1974).

Most workers agree that there must be some proteins whose position in the membrane relative to others must be maintained, e.g. components of electron-transport chains. Such proteins may require anchoring points either in parts of the membrane which are less fluid than others, or by association with some extra-membrane protein. It is possible that aggregations of sterol molecules might serve to create more viscous patches, and there is evidence, again from synthetic lipid bilayers, that cholesterol can be concentrated in association with certain phospholipids (de Kruyff *et al.*, 1974).

2.3.2.3 *Membrane synthesis and flow*

The intermediary metabolism concerned with the synthesis of the components of the membrane is beyond the scope of this chapter but it is believed that the components themselves may be centrally assembled and then distributed to the various membranes of the cell. This flow of membrane material can be detected in experiments where cells are provided with a radioactive precursor to a common membrane protein for a short while, and then returned to non-radio-active medium. This type of analysis has not been performed on plant cells but in animal cells, harvested at intervals after pulse-labelling, the radioactive label appears first in the endoplasmic reticulum and then in the Golgi cisternae. Subsequently, the radioactivity of these compartments declines followed by an increase in label associated with the plasmamembrane some hours later (see Table 2.6). Since mitochondria and chloroplasts probably do not have all of the metabolic apparatus to assemble their own membranes it is thought that they too may obtain partly finished membranes from the endoplasmic reticulum (see also chapter 8).

Table 2.6. Estimated time constants for the appearance of L-(guanido-^{14}C) arginine, administered as a pulse label to living rats, in various fractions of the liver (data of Franke *et al.* 1971).

Fraction	Labelling time (min.)	
	Half max.	Max.
Endoplasmic reticulum	5·5–6·5	10
Golgi apparatus	8·5	30
Plasmalemma	> 60	> 180

There is evidence that the cytoplasm contains numerous membrane-bounded vesicles which frequently appear to be in conjunction with the major membranes of the cell (e.g. Mahlberg *et al.*, 1974). They are particularly prominent in cells synthesizing walls and, since it is known that precursors of wall synthesis are formed inside the cell and elaborated outside, it is reasonable to conclude that the vesicles contain precursors which are discharged after fusion with the plasmalemma (Heyn, 1971). It has also been found that particulate material from the external medium can be detected in vesicles within the cytoplasm (Mayo & Cocking, 1969; Robards & Robb, 1974). These results suggest that vesicles can both fuse with, and be formed from the plasmalemma and other membranes, thus allowing materials to pass out of or into the cell without their having to cross a membrane (see p. 61); such movements are known as exo- and endo-cytosis, respectively. Membrane fusion also presents opportunities for the transfer of blocks of membrane from place to place as is implicit in the observations in Table 2.6.

Since adjacent membranes can frequently be found to be in contact over comparatively large areas and yet show no tendency to fuse with one another, it is believed that special proteins or phospholipids (e.g. lysolecithin) in the vesicle membrane may trigger fusion where they make contact with the larger membrane sheet (see Lucy, 1970).

2.4 TRANSPORT OF SUBSTANCES ACROSS MEMBRANES

In a healthy cell there is a continuous interchange of water, ions, uncharged solutes, metabolites and dissolved gases across the plasmalemma. As one might expect, not all of these substances move through the membrane in the same manner. Firstly, some substances diffuse into a cell down a gradient of potentia energy; such movement is spontaneous and is the thermodynamic equivalent o' heat passing from a warmer to a cooler body. There are, however, many substances which are accumulated by cells against a gradient of potential hence their movement into the cell is 'uphill' and is equivalent to the flow of heat from a cooler to a warmer body. 'Uphill' transport requires work to be performed and thus consumes energy. 'Downhill' transport is frequently described as passive while 'uphill' transport is described as active and directly involves the participation of cellular metabolism.

In Fig. 2.11 some further sub-divisions of transport processes have been made. Thus, passive movements may occur by at least three types of pathway, whereas active movements must be linked to some energy-consuming mechanism, referred to as a 'pump', in the membrane. The third type of movement occurs because of the undulation and vesicularization of the membrane in endo-cytosis (see p. 61). Clearly this is a process which depends at some point on metabolism but, as is discussed below, the substances which move into the cell do not necessarily cross the membrane at all. In such circumstances the observed transport is not strictly *active* in a thermodynamic sense.

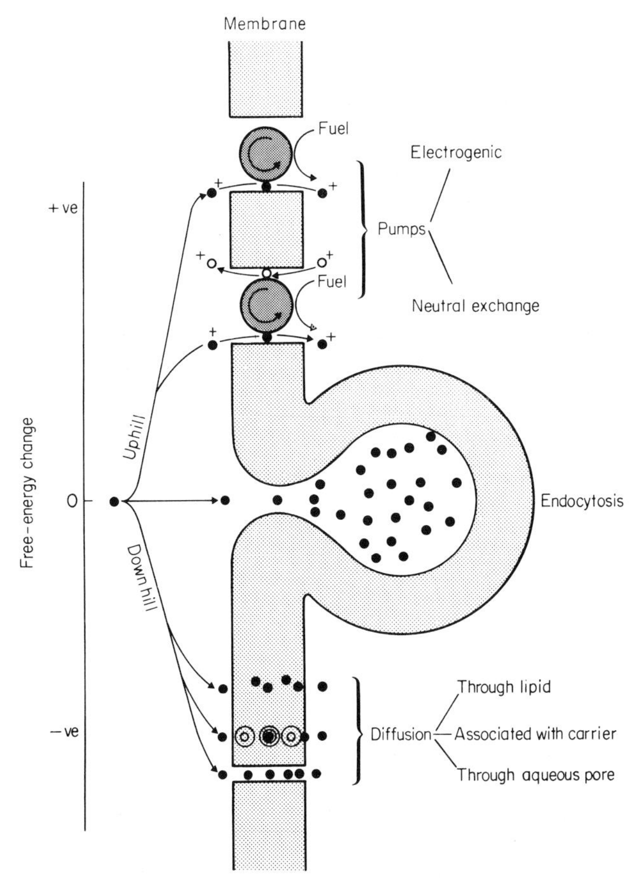

Fig. 2.11. Types of active ('uphill') and passive ('downhill') transport across a membrane. For discussion see text.

2.4.1 PASSIVE ('DOWNHILL') TRANSPORT

Solute molecules in more concentrated solutions possess more free energy than those in lower concentrations; in other words they are at a higher potential. If two solutions of different concentration are mixed, solute molecules will diffuse from areas of high to areas of low potential. In any given situation the chemical potential of an uncharged solute is dependent on its activity as shown in equation 2.1.

$$\mu_s = \mu_s^* + RT \ln a_s \qquad (2.1)$$

where μ_s = the chemical potential of the solute in joules.mole^{-1}.

μ_s^* = the chemical potential of the solute in a standard state—it is essentially a reference point and is a constant which can usually be cancelled out in calculations.

R = the ideal gas constant, usually taken as 8.314 joules °K^{-1} mole^{-1}.

T = the absolute temperature, °K.

$\ln a_s$ = the natural logarithm ($\log_e$) of the activity of the solute in moles l^{-1}.

For many practical purposes it is assumed that the activity of a solute moving freely in dilute solution is the same as its concentration so that the chemical potential is more frequently written

$$\mu_s = \mu_s^* + RT \ln C_s \qquad (2.2)$$

where C_s = the concentration in moles l^{-1}.

If the solute is charged, its movement from place to place can also be influenced by differences in electrical potential. An ion, therefore, has an electrochemical potential which is related to its concentration (strictly, its activity) and the electric potential of the medium in which it is moving. Thus,

$$\bar{\mu}_j = \bar{\mu}^*_j + RT \ln C_j + Z_j \psi F \qquad (2.3)$$

where $\bar{\mu}_j$ and $\bar{\mu}^*_j$ are the electrochemical potentials of the ion j in a given set of conditions and in a standard state respectively, C_j is the concentration of the ion j, Z_j is its algebraic valency (i.e. K^+ would be +1, and Cl^-, −1), F is the Faraday constant which defines the amount of electric charge (approx. 10^5 coulombs) carried by 1 gram equivalent of the ion and ψ is the electric potential of the medium. It is important to appreciate that ψ is a property of the medium and not of the ion j in particular.

Because C_j and ψ can influence $\bar{\mu}_j$ independently it is easy to see that a difference in electrical potential across a membrane can promote the diffusion of an ion in the absence of any difference in concentration or even against a gradient of concentration. This situation is illustrated in Fig. 2.12 where we can see that the concentration of an ion within a membrane-bounded compartment may be much greater than in the surroundings and yet be at electrochemical equilibrium providing that a sufficiently large electric potential is maintained. It is wrong, therefore, to conclude, because an ion is more concentrated within a cell than in the outside solution, that it has been actively transported.

2.4.2 CRITERIA FOR ACTIVE ('UPHILL') TRANSPORT

To decide whether or not an ion or solute is actively transported across a membrane we need to know its activity or concentration in the two solutions separated by the membrane and for ions, in addition, we must know the electrical potential difference across the membrane. It is frequently difficult to measure the concentrations, and more difficult to measure the activity, of substances within cells with any accuracy, especially in those of higher plants.

In the giant cells of several sorts of algae, which may be 5,000 to 10,000 times the volume of a parenchyma cell in a root, such measurements are made

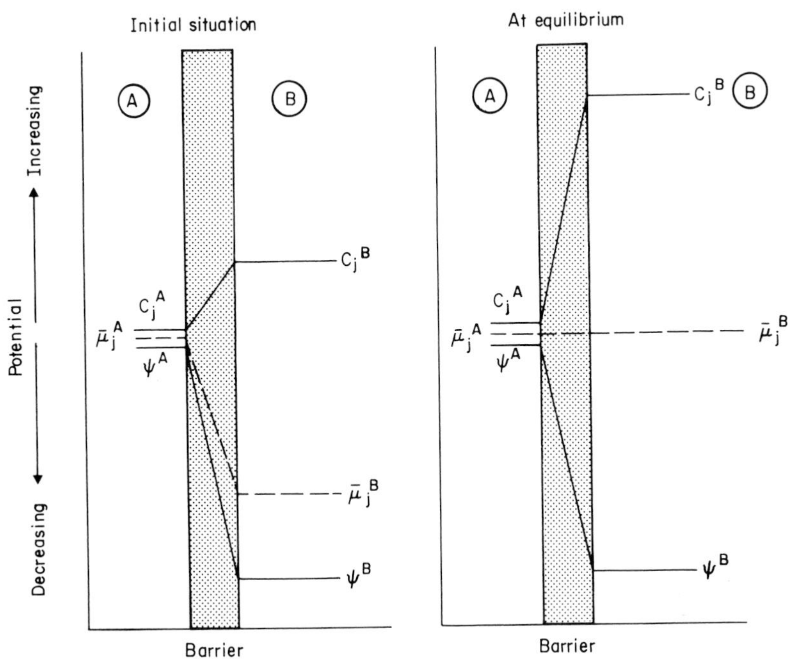

Fig. 2.12. An explanation of the way in which a decrease in electrical potential, ψ, across a membrane can result in the diffusion of an ion against a gradient of concentration. Note that, in the initial situation, in spite of concentration of j being greater in B, the electrochemical potential gradient is still directed 'downhill' towards B. As B fills with ion j, μ_j flattens out and, at equilibrium becomes zero—at this point the 'uphill' concentration gradient and the 'downhill' electrical gradient are equal and opposite.

routinely. The electrical potential difference across membranes can be measured if a small glass micro-electrode, with a tip diameter of 1 to 3μm, can be inserted into the membrane-bounded compartment (see Clarkson, 1974).

Having made the necessary measurements a simple test can be applied to see if a given ion or solute within the compartment is at a higher or lower potential than in the surrounding solution. The principal snag in this analysis is that the cell or compartment should be in a steady state and that no net movement of solute should be occurring .In nature this condition is infrequently met.

Let us suppose that the ion j is at electrochemical equilibrium between the two compartments i.e.:

$$\bar{\mu}_j^{out} = \bar{\mu}_j^{in}$$

re-writing equation 2.3 and cancelling out $\bar{\mu}^*_j$ we have

$$RT \ln C_j^{out} + Z_j \psi F^{out} = RT \ln C_j^{in} + Z_j \psi F^{in} \quad (2.4)$$

gathering the electrical terms to the left-hand side we get

$$\psi^{\text{in}} - \psi^{\text{out}} - \frac{RT}{Z_j F} \ln \frac{C_j^{\text{out}}}{C_j^{\text{in}}} \qquad (2.5)$$

$\psi^{\text{in}} - \psi^{\text{out}}$ is the electrical potential difference across the membrane where the ion j is at equilibrium and is given a special name, the Nernst Potential, and is usually symbolized Ej^N. We now compare this calculated equilibrium potential with the potential difference which is actually measured by the electrodes on either side of the membrane. If the calculated and observed values coincide we would conclude that, in spite of any differences in C_j across the membrane, the system was at equilibrium. If, however, the observed potential was lower than the equilibrium potential we would conclude that the electrical driving force was not sufficiently large to support the observed asymmetry of C_j and we would suspect that active transport was occurring. The example worked out in Table 2.8 may make this clearer. For each ion the appropriate Nernst Potential has

Table 2.8. Comparison of Nernst Potentials and equilibrium concentrations of ions with observed values from a hypothetical cell

Observed potential $E = -116$ mV Temperature 293°K (20°C)

Ion	Concentration Outside (mM)	Inside	E_j^n (mV)	E_j (mV)	Equilibrium concentration (mM)
K^+	1	100	−116	0	100
Na^+	10	50	−41	−75	1,000
Cl^-	11	150	+65	−181	0·11

The Nernst Potential for each ion was calculated from equation 2.5. The column E_j is the difference between the Nernst Potential and the observed potential. Where the observed potential E is the same as the Nernst Potential the ion is at electrochemical equilibrium. If the sign of this difference is negative it means that, for cations, the electrochemical potential is 'downhill' into the cell; for anions, however, it signifies an 'uphill' gradient. In the last column, the equilibrium concentration inside the cell, has been computed on the assumption that the observed potential was equal to the Nernst Potential for each ion. Departures between this computed value and the observed concentrations indicate how far the cell is from being in a passive equilibrium with respect to the ion.

been calculated from the observed concentrations on the outside and inside of the membrane using equation 2.5. The exact correspondence of the Nernst Potential for potassium, E_K^N, with the measured electric potential difference shows that potassium ions inside the cell are at electrochemical equilibrium with those outside. For sodium, however, E_{Na}^N is much less negative than E, hence Na^+ is at a lower electrochemical potential inside; there is, therefore, a strong tendency for sodium to diffuse into the cell, and the fact that the observed concentration is only 1/20 th of the equilibrium concentration strongly suggests that metabolic energy must be coupled to a Na^+-efflux pump. Chloride ions in the cell are a very long way indeed from being in electrochemical equilibrium with their surroundings, being more than 1,000 times greater than the equilibrium concentration, thus their movement into the cell is steeply uphill.

In theory this type of analysis can be applied to any ion, although it is difficult to apply to minor ionic constituents, e.g. trace elements, because they may be complexed with organic ligands within the cell so that their ionic activity may be very much lower than their concentration as measured by chemical analysis.

If an analysis of the kind described in Table 2.8 shows that the transport of an ion in a given direction is 'uphill', one should not conclude necessarily that the membrane is equipped with a special pumping mechanism for that ion. In some cases it may be, but in others the 'uphill' transport of the ion may be coupled with the 'downhill' transport of another *via* a common carrier; this latter possibility is described under the heading Co-transport on p.56 .

2.4.3 THE NATURE AND ORIGIN OF THE MEMBRANE POTENTIAL

It is clear that electrical potential differences across membranes are of great importance in generating driving forces on ions. It is important, therefore, to try to understand how these potentials arise and how they are maintained.

An electric potential difference arises because positive and negative charges become separated. Since the cytoplasm of most cells is electrically negative relative to the surroundings it is very slightly enriched in anions relative to cations. This can be attributed to the differential permeability of the cell membrane and to the activity of ion pumps. First let us examine how differential permeability can create an electrical potential difference.

2.4.3.1 *Diffusion potential*

Imagine a simple system of two compartments separated by a membrane which has a much higher permeability to K^+ than to Cl^- (Fig. 2.13). If the compartments are filled with potassium chloride solutions of different concentration, initially K^+ will move through the membrane out of the more concentrated compartment, and for a very brief period, the more concentrated cell will lose K^+ faster than Cl^- leaving it enriched in negative charge. The negative *diffusion potential* thus created slows down the further escape of K^+ by attracting it back into the more concentrated compartment. When the potential has developed, a large concentration difference can be maintained between the compartments. Since the membrane has a finite permeability to Cl^-, albeit a low one, over a long period of time both the electrical potential difference (the diffusion potential) and the concentration difference would run down as Cl^- leaked through the membrane. If the membrane were completely impermeable to Cl^-, the potential, once established, would be maintained indefinitely. In nature, membrane permeability to anions is a tenth to one hundredth of that for the monovalent cations; thus, the maintenance of a potential of the kind just considered depends on topping up the cell with anions at a rate comparable with their leakage into the surroundings. This is an 'uphill' transport and therefore requires the mediation of some ion pumping mechanism. Active transport is, therefore, necessary to maintain a diffusion potential.

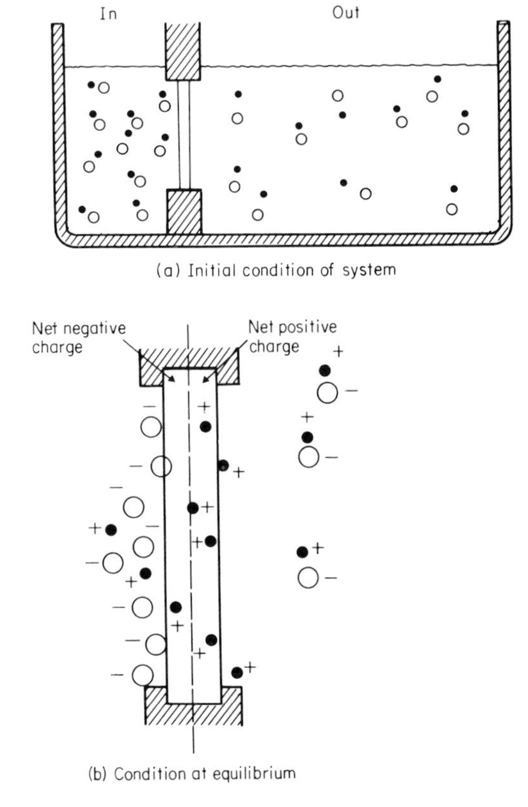

Fig. 2.13. Development of charge separation and a diffusion potential in a model system containing a membrane selectively permeable to cations. For further explanation see text. (From Clarkson, 1974.)

In nature it is frequently possible to find cells where the electrical potential difference across the plasmalemma is, indeed, a diffusion potential of the kind just described which depends very closely on the concentration of either K^+ or H^+ in the medium and in the cytoplasm. In such circumstances its value can be predicted from the Goldman equation (2.6) which relates the concentration ratios of ions across the membrane and their permeabilities (P_K, P_{Na}, P_{Cl} etc.).

$$E = \frac{RT}{F} \ln \frac{P_K[K]^o + P_{Na}[Na]^o + P_{Cl}[Cl]^i + P_x[x]^o}{P_K[K]^i + P_{Na}[Na]^i + P_{Cl}[Cl]^o + P_x[x]^i} \qquad (2.6)$$

This relationship will apply strictly only to situations where the cell and the surroundings are in a steady state and hence limits its application to mature and non-growing cells. The last pair of terms in equation 2.6 has been put in to emphasize that other diffusing ions can be added to the equation; clearly H^+ has an important effect on the membrane potential in some instances (Kitasato, 1968). Since the concentration ratio is multiplied by the permeability coefficient, the value of E will be most strongly influenced by the ionic asymmetry of the most rapidly diffusing ion. In the system considered above the value of E would have been given by

$$E = \frac{RT}{F} \ln \frac{P_K[K]^o + P_{Cl}[Cl]^i}{P_K[K]^i \quad P_{Cl}[Cl]^o} \qquad (2.7)$$

and be governed almost entirely by K^+ since P_{Cl} was very small compared with P_K. Notice that the ratio of the chloride terms is inverted relative to the cationic terms. This is explained because the chloride concentration differential will tend to reduce any negative electrical potential set up by the asymmetric distribution of the cations.

2.4.4 MEMBRANE PUMPS

Pumping mechanisms which allow cells to accumulate solutes up gradients of potential can contribute to the membrane potential in both an indirect and a direct way. The former type are referred to as neutral exchange pumps in which the cell dumps an unwanted ion of equal and like charge into the external environment in a one-to-one exchange for an ion which is more useful, e.g. there are well known exchanges of cellular Na^+ or H^+ for K^+ from the surroundings. The second type transports an ion in one direction only without coupled exchange and is known as *electrogenic* since charge is separated. They can, therefore add to, or subtract from the diffusion potential, described above, depending on which ion is carried. These two types of mechanisms are outlined in Fig. 2.11.

2.4.4.1 *Neutral ion pumps*

Neutral ion pumps are present in the plasmalemma of all cells and by their activity they create the ionic asymmetry necessary to set up the diffusion potential described above. As a much simplified illustration of this, consider a cell which, because of its synthetic and respiratory activity, is generating H^+ and HCO_3^- internally. A pair of exchange pumps could swap H^+ and HCO_3^- for K^+ and Cl^- from the surroundings quickly enriching the interior in these ions and thus setting up the conditions in which the differential rates of diffusion of K^+ and Cl^- out of the cell give rise to a membrane potential. But how does such a pump actually work? There are fewer detailed examples than one would like but the best known is of the membrane-bound ATPase which exchanges intracellular Na^+ for extracellular K^+ in many animal and plant cells (see Hall, 1971; Hodges *et al.*, 1972). In the red blood cell it is known that Na^+ is one of the cofactors which is essential for the binding of ATP to the ATP-ase enzyme. *In vivo* the active centre of the enzyme is accessible only from the cytoplasm, so that the ATP and the Na^+ must be inside the cell (Fig. 2.14). Once bound, the ATP

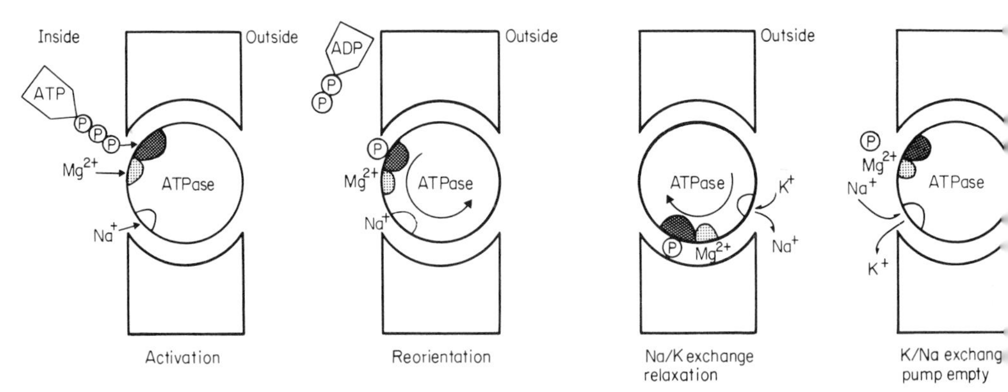

Fig. 2.14. A highly simplified illustration of the working of a sodium-potassium exchange pump based on a membrane-bound ATPase. The cross hatched area on the ATPase is its active centre. The large re-orientation of the molecule is for illustrative purposes only—quite subtle molecular re-arrangement may be all that is necessary to expose the Na^+-binding site to the outside and for the step called relaxation.

is hydrolysed, ADP is released into the cytoplasm leaving the cleaved terminal phosphorus atom attached to the active centre to form a phosphoenzyme. These reactions result in some molecular re-orientation of the phosphoenzyme and its attached Na^+ which exposes the ion-binding site to the different chemical environment of the external medium. It is proposed that this change of environment alters the ion-specificity of the binding site so that K^+ is favoured; K^+ thus replaces Na^+. This done, there is a second re-orientation (referred to as 'relaxation' in Fig. 2.14) which carries the bound K^+ to the inside. The phosphorus is released from the active centre and Na^+, which is preferentially bound on the cytoplasmic side exchanges for K^+ and the pump is ready for a second cycle. The pump has used the free energy released on hydrolysis of ATP as fuel to exchange K^+ and Na^+ against their respective electrochemical potential gradients. The two ions in the appropriate orientations are essential cofactors in the enzyme reaction; *in vitro*, ATPase of this kind will not hydrolyse ATP unless Na^+ and K^+ are both present.

In theory many pumps based on ATPase are possible with only subtle modifications of the ATPase molecule to provide binding sites of varying field strength which will select various ions, e.g. a Ca^{2+} transporting ATPase is found in mitochondria and in sarcoplasmic reticulum (Racker, 1972).

2.4.4.2 *Electrogenic pumps*

The unidirectional transport of an ion across a membrane separates charges and in so doing provides a driving force for the passive diffusion of a similarly charged ion in the opposite direction or an oppositely charged ion in the same direction. The molecular details of exactly how an electrogenic pump is put together remain uncertain although in one instance it is highly likely that an electrogenic H^+-efflux pump is based on an ATPase (Slayman *et al.*, 1973). It is possible, nevertheless, to deduce certain general consequences of their operation. If, for instance, there was an outwardly directed pump at the plasmalemma which actively pumped hydrogen ions (protons) out of the cell thus making the interior electrically negative, this could contribute to the electrical driving force on the diffusion of K^+ from the external medium. Indeed the rate at which charge is extruded and the rate at which it leaks back into the cell must be very nearly in balance unless a dangerously large potential is to accumulate. Examples of both proton extrusion pumps and anion influx pumps of the electrogenic kind are well documented from research on plant tissues (Higinbotham & Anderson, 1974; Spanswick, 1972). In the giant alga, *Acetabularia*, an electrogenic chloride influx pump contributes more than half of the potential of −170mV found across the plasmalemma when the cell is kept in the light. Almost immediately the cell is put in the dark the pump stops working (since it is closely linked with photosynthesis) and the membrane potential abruptly depolarizes to −80mV (Saddler, 1970). A similar light-dependent electrogenic pump is found in *Nitella translucens* (Spanswick, 1972, 1974). Electrogenic pumps are not, however, restricted to green tissues but have been reported in plant roots (Higinbotham *et al.*, 1970) and fungal hyphae (Slayman, 1970). In every instance, however, inhibition of the pump caused an immediate depolarization of the membrane potential, indeed this is often used to detect the activity of such a pump. The inhibition of a neutral ion pump gives rise to gradual depolarization as the ionic asymmetry runs down (see equation 2.6).

2.4.5 MEMBRANE CARRIERS

Many ions and uncharged solutes cross membranes more rapidly than could be expected if they passed through the membrane lipids without some assistance. Since pores in membranes are too narrow to accommodate many substances which are transported, it is widely believed that membranes contain carrier molecules which, in combination with the solute, facilitate diffusion. The ion pumps considered above are carriers of a special kind since they are linked to the metabolic activity of the cell; the carriers we shall now consider promote net movements of solutes only down the prevailing potential gradient and are not capable of 'uphill' transport even if in some instances (see p. 57) they appear to be doing so.

2.4.5.1 *Evidence from kinetics*

One widely used approach to gather information about carriers has been the application of kinetics first derived from enzyme reactions. The evidence for these carriers is obtained by placing a cell or tissue in a range of solute concentrations and measuring the initial rate of uptake. As illustrated in Fig. 2.15 the uptake rate shows a tendency to saturate at higher concentrations and can thus be used to calculate the maximum velocity, V_{max}, possible under the conditions used in the experiment. By making a double reciprocal plot of the data (Fig. 2.15) the concentration of solute at which half maximal velocity is achieved can

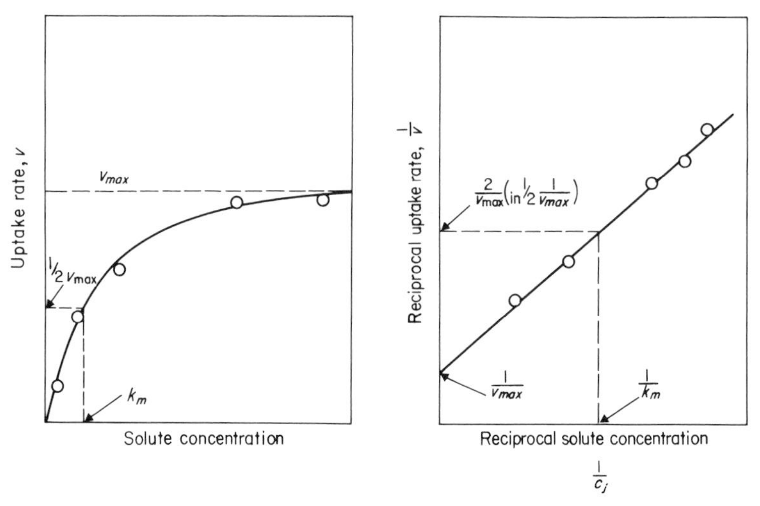

Fig. 2.15. Saturation kinetics of solute uptake versus concentration. Such results are used as evidence for the association of the solute and a carrier. The double reciprocal plot of the data gives a more accurate estimate of V_{max} and K_m when the number of points is limited.

be estimated. This is known as the Michaelis Constant, K_m. The K_m measures the affinity of the carrier for the solute it carries; if the affinity is high then the concentration, K_m will be low and *vice versa*. For most ions in plant tissue K_m is quite low, concentrations ranging from 5–100 μM, but for sugars and other metabolites K_m values are usually greater than 300 μM. Much work of this kind is summarized by Epstein (1972) who shows that at concentrations less than 0.1 mM, the uptake of a given ion is not subject to serious interference from other common ions in solution. There is, however, competitive inhibition between related ions of similar molecular dimensions, e.g. K^+ uptake is inhibited competitively by Rb^+ but not by Na^+; Ca^{2+} is inhibited by Sr^{2+} but not by Mg^{2+}. Thus the carriers which bind the major nutrient ions at low concentrations appear to be highly ion-selective. At higher concentrations (more than 1.0–10.0 mM) this selectivity begins to decline. The interpretation of this observation is contentious and beyond the scope of this chapter but can be pursued in Epstein (1972), Laties (1969) and Clarkson (1974).

The limitation of the kinetic approach is that it can tell us nothing about the nature of the carrier. One can observe similar uptake kinetics for ions whose transport into the cell must be mediated by ion pumps e.g. $H_2PO_4^-$ and Cl^- (see p. 48) as for ions which probably diffuse into the cell passively e.g. Na^+ and Ca^{2+} and for those which are completely exotic and toxic, e.g. Tl^{4+} (Barber, 1974). Indeed, it has been pointed out that saturation kinetics of this kind would also be found if salt movement was observed across a synthetic membrane containing nothing but pores (Stein & Danielli, 1956), where the system would saturate when all of the pores were filled with solute at any moment in time; V_{max} is, after all, merely a measurement of capacity to react or transport and K_m is derived from it (Fig. 2.15).

2.4.5.2 *Ionophores as lipophilic carriers*

A more illuminating approach to the nature of carriers has come from studies on the ionic conductance of synthetic membranes which have been modified in various ways. A bilayer of pure phospholipids has a very low conductance to ions, usually only 10^{-7} to 10^{-8} ohm^{-1} cm^{-2}. The addition of very small amounts of ionophores (i.e. ion-carrying antibiotics) like monactin or valinomycin to the solutions bathing the synthetic bilayer causes a huge increase in the conductance. Figure 2.16 shows that 10^{-6}M monactin changes the membrane conductance to K^+ nearly a million-fold and that even at 10^{-10}M its effect is quite strong. The conductance change for a given monactin concentration is greatest for K^+ and for Rb^+ and is much less for Na^+, Cs^+ and Li^+. Monactin is, therefore, acting as a selective carrier of K^+ and Rb^+ and valinomycin, another bacterial product, behaves similarly. Since these two compounds differ chemically it is instructive to see what they have in common. Both of them are amphipathic ring-structured molecules which have their non-polar groups on the outside of the ring and their polar groups directed towards the space at the centre of the molecule. The outside of the ring interacts favourably with lipid while the hydrophilic core, 0.7 nm in diameter, provides room for several hydrated potassium ions to be bound. Evidence from a variety of sources shows that this complex diffuses across the membrane so that the ions never leave a polar environment (Eisenman *et al.*, 1968).

Other substances are known which select for divalent cations, e.g. the unnamed compound A23187 which is a carboxylic acid antibiotic found in

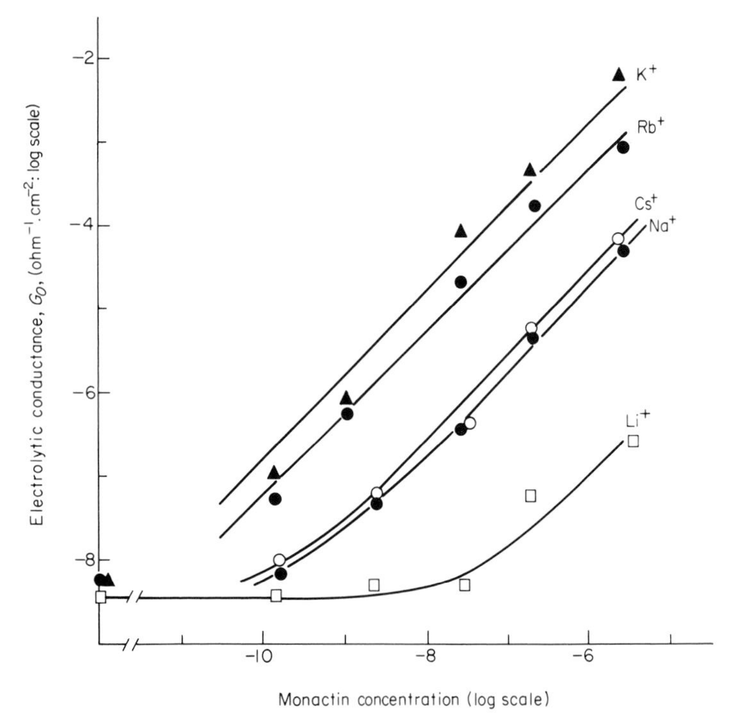

Fig.2.16. Influence of the ionophore, monactin, on the electrolytic conductance of a phospholipid bilayer in the presence of single salt solutions of alkali cations. (Redrawn from Eiseman *et al.*, 1968.)

cultures of *Streptomyces chartreusensis* (Reed & Lardy, 1972). This compound carries Ca^{2+} and Mg^{2+} across bilayers and natural membranes but has no effect on monovalent cations.

Apart from a few synthetic analogues all of the ion-carrying antibiotics are natural products of bacteria and fungi. There are many who believe that compounds of a similar kind may act as ion carriers in all membranes but the technical difficulty of isolating what are probably minute quantities of such compounds from tissues appears to be formidable and so the belief may rest on faith for some time yet.

2.4.5.3 *Co-transport*

As suggested earlier, carrier-assisted diffusion can sometimes appear to go in an 'uphill' direction, thus giving the impression of active transport. In many animal tissues and micro-organisms, sugars, amino acids, organic acids and vitamins move into the cell up a concentration gradient. This transport is, however, almost completely dependent on having Na^+ or H^+ in the external medium; other ions such as K^+, Rb^+ or Li^+ cannot be substituted. It has been found that the metabolite is carried into the cell along with an ion-carrier complex which is diffusing 'downhill' (Fig. 2.17). In both *Chlorella* and *Neurospora*, glucose is transported in this way along with protons, H^+ (Komor & Tanner, 1974; Slayman & Slayman, 1974).

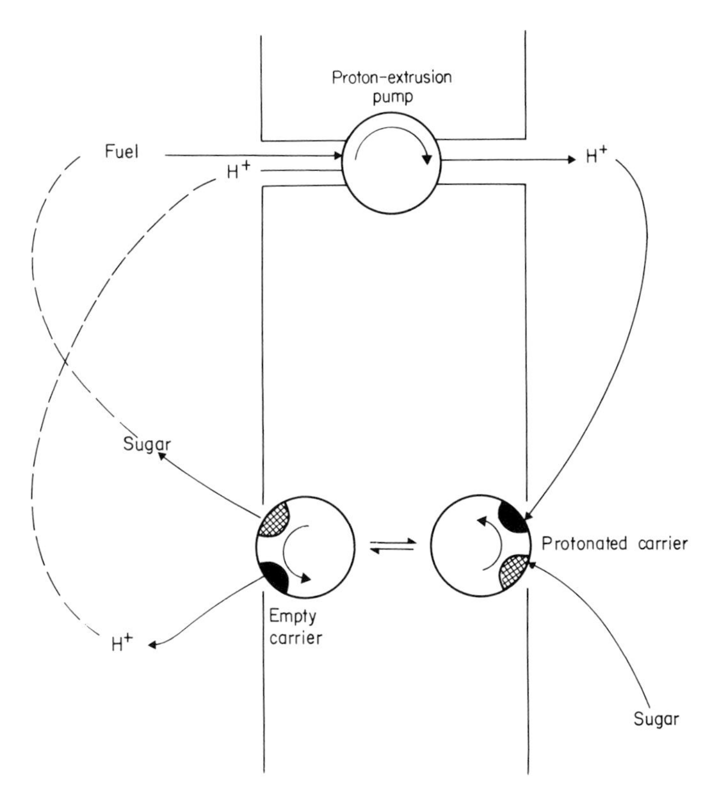

Fig. 2.17. Scheme to illustrate co-transport of protons and sugar. The proton extrusion pump is electrogenic and thus makes the inside electrically negative. Protons diffuse back into the cell passively *via* the carrier which also binds a sugar molecule. The protonated carrier plus sugar diffuses towards the inner face of the membrane where it dissociates and releases the sugar molecule.

Co-transport depends on active transport in an indirect way (as indeed does all diffusion, see p. 50) because energy-dependent extrusion pumps ensure that the cytoplasm is kept well below its equilibrium concentration in H^+ and Na^+. These ions tend to diffuse back into the cell and, in doing so, *decrease* their free energy. The energy they give up is coupled, *via* the carrier, to the co-transport of the solute whose free energy is *increased* as it moves into the cell.

Co-transport may also assist the 'uphill' movement of inorganic ions into the cell; it may be a more common process than is generally realized. Recent evidence by Lowendorf *et al.* (1974) suggests that the active transport of phosphate into the hyphae of *Neurospora* depends on (a) the activity of a proton extrusion pump at the plasmalemma which is sensitive to the pH of the external medium, and (b) the formation of a ternary complex between a proton and a phosphate ion from the external medium with a membrane carrier. The protonated phosphate carrier diffuses to the cytoplasmic side of the membrane *down* the electrochemical gradient of the proton, releasing the proton and phosphate ion into the cytoplasm. This, and other examples of inorganic ion co-transport (see Raven & Smith, 1974) suggest a reason why so little progress has been made in elucidating the molecular details of certain 'pumps' particularly of those which transport anions. Put most simply, it may be that these influx pumps do not exist and that the uphill transport is driven by a combination of active extrusion and re-entry by diffusion of protons or perhaps sodium ions.

Co-transport illustrates the ingenious way in which nature can turn necessity to its own advantage. The active excretion of H^+ and Na^+ is essential to maintain pH control and osmo-regulation in the cell, but the energy expended is partly recovered in the transport of essential metabolites and ions into the cell.

2.5 CORRELATION OF STRUCTURE AND FUNCTION

Having a picture in mind of the way membranes are assembled and of the driving forces which operate across them it is now possible to consider how their design, particularly that of the plasmalemma, manages to reconcile two conflicting sets of priorities. This conflict may be summarized as follows: the cell usually contains solutes at a far higher concentration than in the external medium; without any barrier these solutes would disperse by diffusion. Many solutes are accumulated by mechanisms which consume metabolic energy and it is sensible for nature to minimize their subsequent leakage and thereby economize in the use of 'fuel'. Insulation of this type might be most effective if the cell were surrounded by a barrier impenetrable to virtually everything except water and dissolved gases. Such a barrier would, however, isolate the cell from the outside world and render it insensitive to stimulae and incapable of growth, it might eliminate its interactions with other cells in adjacent or remote tissues and prevent the excretion of harmful excesses of substances generated during metabolism. The cell membrane must provide a balance, therefore, between minimizing losses and permitting a selective interchange of substances between the cell interior and the surroundings. The balance is achieved, in many cases, by providing channels or other more elaborate carrier mechanisms whereby materials crossing the membrane may essentially by-pass the hydrophobic region.

2.5.1 MEMBRANE PORES AND CHANNELS

Lipid or oil is a very effective electrical insulator (it is used in high voltage underground electric cables for this purpose) and it is not surprising to find that the electrical resistance of synthetic bilayers made from pure phospholipid is very high, being 10^7 to 10^9 ohms cm^2. In nature, the current conducted across membranes is carried by ions and we can see that an unmodified lipid bilayer with such a high resistance is a poor material across which to conduct this essential current of electrolyte. It is not surprising to find, therefore that the electrical resistance of natural membranes is very much less than that of bilayers, usually lying in a wide range of 10^2 to 10^5 ohms cm^2.

As described on p. 55, lipophilic carriers can greatly reduce the resistance (and hence increase the conductance) of synthetic membranes and may resemble the carrier molecules in membranes, but it is also widely believed that membranes contain water-filled pores which must contribute to their relatively high conductance by allowing water and selected solutes to by-pass the lipid domain of the membrane. Certain antibiotic molecules, such as nystatin, appear to condense cholesterol molecules in both synthetic (Holz & Finkelstein, 1970) and natural membranes (de Kruijff & Demel, 1974) to form pores with a radius of *ca.* 0.4 nm. The presence of such pores greatly increases the electrical conductance and hydraulic conductivity of the membrane. In passing we might note that many antibiotics have their effect by enormously increasing the passive permeability of cell membranes causing non-resistant cells to lose their contents or to lyse. Interest in these substances stems from the experimental evidence that cell membranes also possess pores of similar size and that the antibiotic merely induces an extreme expresssion of the normal condition.

Although the word 'pore' is often used to describe channels through which solutes and water can move, we should resist the temptation to conclude that all 'pores' are definable structural entities like the ones induced by nystatin (see above). In many instances a 'pore' may be more like a transient imperfection in membrane structure. This latter type may have a certain statistical probability but have no fixed position.

Proteins which traverse the lipid layer may give rise to hydrophilic channels or pores (see Fig. 2.5e). Indirect evidence supporting this idea comes from a study in which red blood cell membranes were exposed by deeply etching frozen cells under vacuum (Pinto da Silva, 1973). Shrinkage of the membrane surface was observed in areas overlying groups of membrane particles possibly due to sublimation of ice from within the embedded particles; for this to have occurred this water would be a free liquid in the thawed condition (cf. bound water p. 35).

The first circumstantial evidence for the existence of pores come from studies by Collander and Barlund (1933) on the permeation of the giant internodal cells of the alga, *Chara ceratophylla*, by a number of uncharged solutes and water. They found that in almost every case the rate of movement of a substance into a cell depended on its molecular weight and dimensions and on its solubility in oil relative to water; substances with high oil solubility permeated most rapidly. This strongly suggested that movement across cell membranes involved the movement of the solute out of the water, its solution in lipid and subsequent diffusion through it, and its re-entry into the aqueous phase at the inside face of the membrane. The authors found, however, that several small molecules permeated the membrane far more rapidly (more than 100 times faster in the case of water) than their relative solubility in oil suggested. The upper size limit for molecules which behaved anomalously was a radius of 0.4 nm and it was suggested that the membrane was constructed as a very fine sieve containing pores of 0·4 nm radius through which water (0.25 nm radius) and certain solutes could move. Since this early work a great deal has been done and the equivalent pore radius in many plasma membranes has been confirmed as 0.4 nm (see Solomon & Gary-Bobo, 1972). It must be said, however, that some authorities are reluctant to accept that pores can provide channels for the bulk movement of water and solutes; the arguments for and against pores have been clearly discussed in Oschman *et al.* (1974).

There has been much discussion about whether pores admit ions, and if so, which ones. Membranes are known to control very precisely the relative rates at which various ions will diffuse across them; potassium ions will diffuse ten to one-hundred times faster than sodium. Ions in solution are hydrated by binding one or more water molecules; it requires a great deal of energy to dehydrate an ion and for this reason we should think of ions in all natural circumstances as being hydrated. Sodium binds 5 water molecules, whereas potassium binds 3, the former is, therefore, the more bulky ion whose diffusion into the narrow water filled pores would be slower than for the smaller hydrated potassium ion. On the other hand the anion chloride, which has only one water molecule in its hydration shell, diffuses much more slowly than either K^+ or Na^+. This is probably due to the fact that 'pores' carry a predominantly negative charge so that cations would be attracted to them, while anions would be repelled —a small number of positively charged pores would handle the flow of anions. Polyvalent cations and anions have much more water in their hydration shells, e.g. Ca^{2+} has 10, and $SO_4{}^{2-}$ has 8 and it seems likely that these would be totally excluded from the pores.

2.5.2 ION PUMPS AND MEMBRANE SUBSTRUCTURE

It is most unlikely that the intercalated membrane particles described earlier (p. 38) are simple ion-carriers—they are much too large—but they may be ion pumps, or groups of pumps. The best evidence for this assertion comes from work on the ATPase which pumps Ca^{2+} across the membranes of sarcoplasmic reticulum. Racker (1972) was able to isolate this enzyme and put it back into a synthetic bilayer membrane made from soybean phospholipids, and thus reconstitute a membrane which could actively transport Ca^{2+} when ATP and Mg^{2+} were present in the medium. Working with a slightly different system

Packer *et al.* (1974) showed that freeze etched fractures of both the original tissue membrane and reconstituted membrane containing the purified ATPase were densely studded by numerous particles 8·5 nm in diameter. There can be little doubt that the particles were the calcium-pumping ATPase. Similar, less complete, evidence is also available for the Na^+/K^+-dependent ATPase of the red blood cell (see Branton & Deamer, 1972). There is no insuperable difficulty in repeating such observations with plant tissues but, at the time of writing there is no report that this has been done.

One should not conclude that all of the membrane substructure is concerned with solute transport, especially in mitochondria and chloroplasts where other biochemical activities are centred on membranes.

2.5.3 ENDOCYTOSIS AND VESICULAR TRANSPORT

The presence of abundant vesicles in the vicinity of the plasmalemma of plant cells has been pointed out earlier (p. 44); however, it is not yet clear what contribution endocytosis makes to the total solute transport across this, or any other membrane. There is kinetic evidence from several sources which is consistent with the notion that a measurable fraction of various ions in the cytoplasm of plant cells is sequestered into a compartment separate from the bulk of the cytoplasm. MacRobbie (1969) found that labelled Cl^- transported into the vacuole of *Nitella translucens* could be easily resolved into a fast and a slow component. The fast component was envisaged as being delivered to the vacuole formed in vesicles which had been found at the plasmalemma. The chloride ions in these vesicles would not have mixed with the unlabelled chloride in the bulk cytoplasm, whereas labelled Cl^- delivered directly to the cytoplasm *via* a pumping mechanism would have to mix with a much larger pool of unlabelled chloride ions. At the beginning of the experiment, therefore, the labelled Cl^- in the vacuole increased more rapidly than expected. There are other examples of this kind discussed in MacRobbie (1971) and Baker and Hall (1973).

If the vesicles contained nothing more than a small volume of the outside solution, their net contribution to the solute content of the cell would be very small and they could not exercise the ion-selectivity which characterizes cell membranes. Baker and Hall (1973) suggest that endocytosis becomes a much more plausible transport process if it is assumed that ions become selectively bound to the membrane surface and thus become concentrated from the dilute external medium. This assumption is entirely reasonable because both proteins and phospholipid heads can provide ion-selective binding sites. These authors also relate membrane-bound ATPase activity with the subsequent invagination of the plasmalemma to form a vesicle but this implies that vesicle formation is energy-dependent. This is against the substantial evidence from animal cells which shows that micro-vesicle formation results from Brownian movement of the membrane and does not need to be energized by ATP (Casely-Smith, 1969); ATP is required for the very large vesicles formed by re-orientation of microfibrils in phagocytosis, but the type of vesicle under discussion here is much smaller, probably 70–100 nm in diameter and is not dependent on microfibrils. There is visual evidence that these micro-vesicles can discharge their contents directly into lysosomal vacuoles (Casely-Smith & Chin, 1971).

The important advance of thinking required to deal with endocytosis will provide a severe strain on biophysical theorists accustomed to thinking of cells as homogeneous phases separated by diffusion barriers, partly because materials can reach the interior of the cell without crossing a membrane at all. Although endocytosis is probably a common phenomenon it is likely to remain an 'unknown quantity' until new experimental techniques are devised.

2.5.4 CONCLUDING REMARKS

In the light of recent discoveries we must discard the notion that membranes are neat semi-crystalline rigid lattices in favour of a more dynamic view. The diffusion of large protein molecules in the plane of the membrane emphasizes its fluid character and we see that there may be great heterogeneity both in the lipid matrix and in the proteins embedded in it. In the past it is possible that we have over-emphasized the importance of electrostatic interactions between phospholipid heads and underestimated the role of the surrounding polarized water layers in maintaining the familiar bilayered arrangement of membranes.

This more 'fluid' view of membranes has influenced the way in which we think about transport processes and diffusion, particularly those workers with interests in endo- or pino-cytosis. It seems probable that some of the processes which have hitherto been accepted as active transport may be, in reality, examples of co-transport which ultimately depend on the active transport of some other ion. This would allow a simplification of the model of the cell membrane which at present is uncomfortably cluttered by numerous hypothetical ion-pumps. Perhaps in time we shall be able to reduce this picture to one or two pumps and many smaller carrier molecules in a membrane which constantly cuts off vesicles from its undulating surface.

FURTHER READING

MEMBRANE STRUCTURE AND COMPOSITION

Branton D. & Deamer C.W. (1972) *Membrane Structure*. Springer-Verlag, New York/Wien.
Capaldi R.A. (1974) A dynamic model of cell membranes. *Scientific Amer.* **230**, 26–34.
Lucy J.A. (1974) Lipids and membranes. *FEBS Lett.* **40**, S105–S111.

SYNTHETIC LIPID BILAYERS AND ALLIED STUDIES

Eisenberg M. & McLaughlin S. (1976) Lipid bilayers as models of biological membranes. *Bioscience* **26**, 436–43.
Goldup A., Okhi S. & Danielli J.F. (1970) Black lipid films. *Recent Prog. Surface Sci.* **3**, 193–260.

MEMBRANES AND TRANSPORT PROCESSES

ANDERSON W.P. (ed.) (1973) *Ion Transport in Plants*. Academic Press, London and N.Y.
BAKER D.A. & HALL J.L. (1973) Pinocytosis, ATPase and ion uptake by plant cells. *New Phytol.* **72**, 1281–89.
HIGINBOTHAM N. (1973) The mineral absorption process in plants. *Bot. Rev.* **39**, 15–69.
RAVEN J.A. & SMITH F.A. (1974) Significance of hydrogen in transport in plant cells. *Can. J. Bot.* **52**, 1035–48.
SLEIGH M.A. & JENNINGS D.H. (eds.) (1974) *Transport at the Cellular Level*. Symposium 28, Society for Experimental Biology. Cambridge University Press.

CHAPTER 3
CHLOROPLASTS—STRUCTURE AND DEVELOPMENT

3.1 INTRODUCTION

Plastids are organelles which are bounded by double membranes and which occur, as far as is known, in all cells of eukaryotic green plants at some stage, usually becoming modified according to their function. In their undifferentiated form they may remain as proplastids, which are characteristic of epidermal and meristematic cells, for example. In the green parts of plants the proplastids normally develop into chloroplasts, which are the site of photosynthesis, while in starch-storing organs they form amyloplasts which produce the starch grains. However these two functions are not mutually exclusive as most chloroplasts will form starch under appropriate physiological conditions and the exposure of starch-storing organs to illumination results in the amyloplasts forming some thylakoids and chlorophyll. In certain plant parts, such as flowers, fruits and some leaves, the thylakoids of the chloroplasts become degraded, forming chromoplasts, which contain large amounts of carotenoids, the pigments responsible for 'autumn colouration' and the characteristic colours of certain flowers and fruits.

As the whole range of plastid structure, composition, genetics and develop ment has been extensively covered in the monograph by Kirk and Tilney Bassett (1967) this chapter will confine itself to the consideration of chloroplasts, on which recent work has centred. The work of the author's laboratory has concentrated on a study of the structure and development of chloroplasts in *Phaseolus vulgaris* (bean) and *Zea mays* (maize) and thus many of the examples are drawn from work with these plants in order to make a consistent account. This approach must induce some lack of balance, which is acknowledged, but a properly balanced account of this topic would virtually require a volume of its own. References have been selected primarily for their clarity and not necessarily for their priority.

It has been postulated that the eukaryotic cell gained its autotrophic capacity by the capture of a prokaryotic organism at an early stage in the evolutionary process (Stanier, 1970), that the trapped prokaryote became the chloroplast, and that the nucleic acid of the resultant chloroplast still retains an important genetic function (see also chapter 9). Although the genetic evidence indicates that the plastid DNA is transmitted from generation to generation independently of the nuclear genes, Bell (1970) has published electron microscopic evidence that proplastids may arise anew each generation from the nucleus. However, it is hard to reconcile this latter proposal with the evidence that both nuclear and plastid DNA contribute to the plastid genotype (see chapter 9).

3.2 CHLOROPLAST STRUCTURE

3.2.1 CHLOROPLAST DIMENSIONS AND NUMBER

Beans possess chloroplasts typical of the sun leaves of higher plants in that they are discoid or lens-shaped with a diameter of about 5 μm and maximum thickness of about 1 μm as may be seen in the electron micrograph in Fig. 3.1B, which is a vertical section through the disc. On the other hand observations of chloroplasts in living cells by phase contrast microscopy suggest that the maximum thickness of the chloroplast may be rather less (S. G. Wildman, personal communication). Under optimum conditions for growth, *Phaseolus vulgaris* averages 45 chloroplasts per palisade mesophyll cell and 32 per spongy mesophyll cell, giving about 8.3×10^8 chloroplasts per leaf and about $2{\cdot}3 \times 10^7$ chloroplasts per cm^2 of leaf. This latter number is similar to that of mature spinach leaves (Possingham & Saurer, 1969) although the number of chloroplasts per spinach mesophyll cell is more than ten times greater than for bean.

3.2.2 CHLOROPLAST FINE-STRUCTURE

The following interpretation of chloroplast fine structure is based on transmission electron microscopy of thin sections of leaf material prefixed in 3% glutaraldehyde for 24 hours at 5°C, followed by fixation and staining with osmium tetroxide and post-staining with uranyl acetate and lead citrate. This technique is the most widely used, and, as glutaraldehyde is a relatively mild reagent, is considered to produce images most closely resembling the living plastid. Heslop-Harrison (1966) has written a critical account of the interpretation of chloroplast electron micrographs while the results of freeze-etch studies are considered in chapter 4.

The chloroplast envelope consists of two membranes (Figs. 3.1 and 3.2) the outer of which is unspecifically permeable to crystalloidal solutes. In contrast the inner membrane shows very specific permeability and has so far been shown to be the site of three specific anion translocation systems; (a) the phosphate translocator, facilitating a counter exchange of inorganic phosphate, 3-phosphoglycerate and dihydroxyacetone phosphate; (b) the dicarboxylate translocator, facilitating a counter exchange of dicarboxylic acids; and (c) the ATP translocator which is less active than the previous two and may be responsible for the entry of ATP in the dark (Heldt *et al.*, 1972).

Within the inner membrane is a complex system of flattened sacs of membrane which were first termed thylakoids by Menke (1962). The thylakoids are closely associated in stacks (grana) which are shown in section for chloroplasts of bean and maize mesophyll in Fig. 3.1 (B and C). When seen from above the plane of the membrane the grana are essentially circular in outline with an average diameter of about 0·5 μm. Under optimum growing conditions bean grana

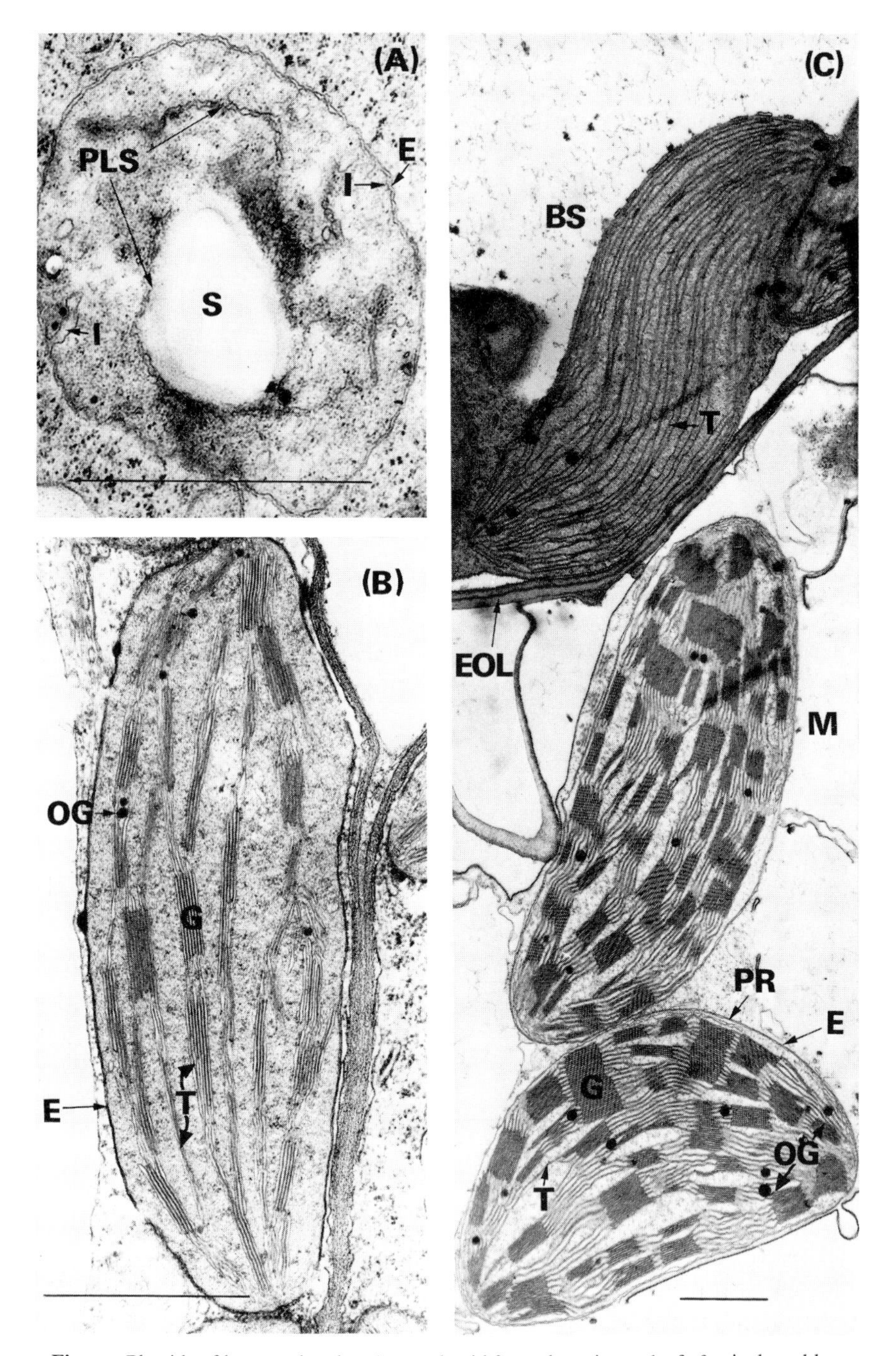

Fig. 3.1 Plastids of bean and maize. A, proplastid from the primary leaf of a six day-old dark-grown bean (× 42,000); B, chloroplast from the primary leaf of a bean grown

contain up to 8 thylakoids while those of maize have up to 40. The thylakoids in each granum are continuous with those in other grana through intergranal thylakoids. In bean the granal thylakoids average 78%, and the intergranal thylakoids 22%, of the total thylakoid membrane (Bradbeer *et al.*, 1974a). During chloroplast development the inner membrane of the envelope appears to give rise to the thylakoids by invagination (see Fig. 3.1A and p. 75) and it is possible that the space between the inner and outer membranes of the envelope is continuous with the whole of the thylakoid space of the chloroplast. Although the magnification of Figs 3.1B and C is such as to allow an overall impression of chloroplast structure, enough detail is visible to show that the intra-thylakoid space is electron-transparent, the thylakoid membrane is electron-opaque and the inter-thylakoid space seen in the grana between adjacent thylakoids is densely electron-opaque.

In the chloroplast the thylakoids are embedded, or suspended, in a matrix, the stroma, which has a somewhat granular appearance (Fig. 3.1). Within the stroma may be seen DNA fibrils and ribosomes (chapter 9), starch grains, osmiophilic globules and occasional extensive crystal-like structures. Such crystals can often be induced to appear in chloroplasts suspended in hypertonic media or in plants subjected to stress, although they have also been observed in plants grown under normal environmental conditions. It is thought likely that such crystals are composed of ribulosebisphosphate carboxylase (Larsson *et al.*, 1973).

3.2.3 THE MOBILE AND STATIONARY PHASES OF CHLOROPLASTS

From observations of chloroplasts in living cells by phase-contrast microscopy Wildman (1967) distinguished two components of the chloroplasts, a stationary phase which he equated to the thylakoid system and a mobile phase which he equated to the stroma. The mobile phase surrounds the grana and also penetrates the intergranal regions of the chloroplast. The mobile phase is always in some kind of motion though the intensity of the activity shows variability, even in the same chloroplast. In the reported investigations, observations were always made in cells which could be seen to be living by the presence of visible protoplasmic streaming. In such cells the mobile phase formed protuberances which sometimes broke away from the chloroplast and became indistinguishable from

for 14 days in the dark and then transferred to continuous illumination (3 mW. cm^{-2}) for 48 hours (× 29,000); C, bundle sheath and mesophyll chloroplasts of maize grown under normal diurnal illumination (× 12,500). Scale lines represent 1 μm. Key to lettering: BS, bundle sheath; E, plastid envelope; EOL, electron opaque layer; G, granum; I, invagination of envelope inner membrane; M, mesophyll; OG, osmiophilic globule; PLS, porous lamellar sheet; PR, peripheral reticulum; S, starch grain; T, thylakoid.

mitochondria, and also mitochondria-like bodies appeared to fuse with the mobile phase. Subsequently Wildman and coworkers (1974) have reported that mitochondria-like bodies became stationary below tobacco chloroplasts in living cells and that starch grains subsequently appeared in similar positions in the chloroplasts. In the opinion of the present writer the interpretation that mitochondria give rise to starch grains cannot be substantiated. We have found that in plants grown under low light intensities each chloroplast has one or more mitochondria embedded in deep pockets close to the chloroplast margin but that no break in the chloroplast envelope occurs and there is always at least a thin layer of cytoplasm between chloroplast and mitochondrion (Montes & Bradbeer, 1976). It is possible that some of the observations of Wildman and colleagues may be explained by the development of the latter phenomenon.

3.2.4 THE RANGE OF CHLOROPLAST STRUCTURE

Quite early in the application of electron microscopy to plant structure the algae were found to show an interesting range of structural diversity of their chloroplasts. Subsequent studies on vascular plants established that a number of structural types were to be found in both cultivated and non-cultivated plants and furthermore that structural modification might be induced by mutation or by variation of the environmental conditions.

3.2.4.1 *The Algae*

A monograph such as that of Dodge (1973) should be consulted for details of the range of chloroplast structure of this group. The characteristics of the chloroplasts have provided an important part of the basis for the taxonomic classification of the algae. In the green algae (Chlorophyceae and Prasinophyceae) the arrangement of the thylakoids tends to be basically similar to that in vascular plants and many of the differences may be associated with the non-discoid shape of the chloroplasts in most green algae. The simplest arrangement is found in the Rhodophyceae where the thylakoids occur as single sheets. In the Cryptophyceae the thylakoids tend to occur as pairs but the individual thylakoids in the pair do not appear to be fused to each other. The other algal classes have their thylakoids arranged in threes; in some cases (Dinophyceae, Euglenophyceae and Haptophyceae) the component thylakoids appear to be fused to each other while in other cases (Chrysophyceae and Phaeophyceae) the thylakoids do not appear to be fused. In two algal classes (Dinophyceae and Euglenophyceae) the chloroplast envelope consists of three membranes instead of two. In a number of classes (Bacillariophyceae, Chloromonadophyceae, Chrysophyceae, Cryptophyceae, Haptophyceae, Phaeophyceae and Xanthophyceae) the chloroplast is also surrounded by a sheath of endoplasmic reticulum which is usually continuous with the nuclear membrane. To complete this brief list of the major peculiarities of algal chloroplasts it should be noted that the chloroplasts of many algae possess pyrenoids and eye spots.

3.2.4.2 *The dimorphic chloroplasts of C_4 plants*

C_4 plants form oxalacetate, malate and aspartate as the primary products of their photosynthetic CO_2-fixation in contrast to the more 'usual' C_3 plants whose primary fixation product is 3-phosphoglycerate. Laetsch (1974) lists the following families of flowering plants as containing C_4 species: Amaranthaceae, Aizoaceae, Chenopodiaceae, Compositae, Cyperaceae, Euphorbiaceae, Gramineae, Nyctaginaceae, Portulaceae and Zygophyllaceae. All of these families also contain C_3 plants and there is the case of the genus *Atriplex* where the C_4 *A. rosea* will hybridize with the C_3 *A. patula* (Björkman *et al.*, 1970). Apart from the difference in primary photosynthetic CO_2-fixation products C_4 and C_3 plants show other substantial differences in their biochemistry, physiology, anatomy and fine structure. Maize and bean will be discussed here as typical examples of C_4 and C_3 plants respectively

In the maize leaf the chloroplasts are concentrated in two concentric sheaths of cells around each vascular bundle. The inner sheath of cells is described as the bundle sheath and consists of equal numbers of large barrel-shaped cells and smaller cells which can be divided into two sorts on the basis of their dimensions (Montes & Bradbeer, 1975). Laetsch (1974) points out that the thick walls of bundle sheath cells adjoining mesophyll cells in C_4 grasses possess an electron-opaque layer (Fig. 3.1C). The bundle sheath chloroplasts possess abundant thylakoids which do not associate into grana, (Fig. 3.1C) and they normally contain starch grains. The plant from which the material was taken for Figure 3.1C had been stored in the dark for 24 hours prior to fixation to remove the starch so as to obtain a clear electron micrograph. In most cases, the bundle sheath chloroplasts of maize have been found to be completely agranal, although by modification of the environmental conditions the formation of grana can be induced (Bradbeer & Montes, 1976). In contrast Laetsch (1974) considers it normal for bundle sheath chloroplasts to show a small amount of thylakoid appression.

The outer sheath of cells is known as the mesophyll sheath and it contains chloroplasts similar to those of C_3 plants in that they have grana. They are, however, different in that they do not normally contain starch grains. Not all C_4 plants show such structural dimorphism of their chloroplasts. However all chloroplasts of C_4 plants possess a system of tubules, the peripheral reticulum, (PR in Fig. 3.1C), which is associated with the inner membrane of the chloroplast envelope. There are reports of peripheral-reticulum-like membrane systems in chloroplasts of some cells of C_3 plants (Laetsch, 1974). Circumstantial evidence assembled by the latter author suggests that the function of the peripheral membrane in C_4 plants may be to facilitate transfer of metabolites between chloroplast and cytoplasm.

3.2.4.3 *The effect of environmental conditions on chloroplast structure*

The fact that plants grown in the complete absence of illumination form etioplasts (Fig. 3.2A and section 3.4.2) while those grown under diurnal illumination form chloroplasts establishes that light is an essential requirement for chloroplast development. When Björkman *et al.* (1972) compared plants of *Atriplex patula* which had been grown under three different irradiances: 20, 6·3 and 2 mW. cm^{-2} (in the waveband 400–700 nm), which are referred to as high, intermediate and low, they found that the high irradiance treatment yielded thicker leaves than the low, with more cells per leaf section and more chloroplasts per cell. The chloroplasts from plants grown under low irradiances were larger than those from high irradiance conditions and they contained more thylakoids and larger grana. The intermediate illumination gave intermediate results. The low irradiance-grown *Atriplex* plants were compared with plants adapted for growth on the floor of the Queensland rainforest where the daily quantum flux was about one-twentieth of that provided by the low irradiance treatment. The chloroplasts of these plants, *Alocasia macrorrhiza, Cordyline rubra* and *Lomandra longifolia* were found to be dark green, unusually large and irregular in outline and to contain very well developed grana of enormous size (Anderson *et al.*, 1973). The grana were also irregularly arranged and apparently adapted for a maximum efficiency in light-trapping.

When maize plants were grown under very low irradiances (0.3 mW. cm^{-2}), which nevertheless provided about six times the daily quantum flux received by the rainforest plants, some chlorophyll and photosynthetic CO_2-fixation developed even though the light compensation point was not exceeded and the leaves showed a net loss of CO_2 followed by senescence and premature death (Bradbeer & Montes, 1976). Grana did not develop and both bundle sheath and mesophyll chloroplasts developed closely parallel arrangements of thylakoids which did not become appressed. Transfer of plants greened under very low irradiances to higher irradiances resulted in a partial recovery towards normal structure.

Research work in Belgium has shown that exposure of etiolated seedlings to electronic flashes of 1 millisecond in duration and given at 15 minute intervals brought about greening in which the resultant plastids were agranal with parallel unfused thylakoids. These flashed leaves showed the interesting property that, on transfer to continuous illumination, they initially did not show any photosynthetic oxygen evolution, but within 2 minutes of the commencement of illumination oxygen evolution was detected and a maximum rate was found after 6 minutes (Strasser & Sironval, 1972). When etiolated leaves are transferred to continuous illumination without any pretreatment with light the onset of oxygen evolution tends to be considerably delayed. An alternative treatment which has produced almost completely agranal plastids is exposure of dark-grown seedlings to continuous far-red irradiation at wavelengths longer than 700 nm (De Greef *et al.*, 1971). For further discussion see page 80.

3.2.4.4 *Chloroplast mutants*

Many chloroplast mutants are known with defects in pigments or other components of structure or function. Some of the mutations are nuclear and show normal Mendelian inheritance whilst others show non-Mendelian inheritance and are considered to be mutations of the chloroplast DNA. Chloroplast mutations in algae can be maintained in heterotrophic culture and have been used for important research on chloroplast genetics, development and function (see e.g. Levine, 1969). In contrast, although there have been numerous investigations of individual chloroplast mutations in higher plants, the only substantial research collection of higher plant material with mutant chloroplasts is the barley mutant collection of D. von Wettstein's group in Copenhagen (von Wettstein *et al.*, 1971). This collection has so far not been made generally available for studies on chloroplast development.

3.3 ISOLATION OF CHLOROPLASTS

Although Hill in 1937, obtained chloroplast suspensions from *Stellarsia media* which were capable of O_2 evolution under illumination, it was not until 1954 that Arnon *et al.* reported the occurrence of photosynthetic phosphorylation in isolated chloroplasts. These latter chloroplast preparations were capable of rates of photophosphorylation and electron transport similar to those assumed to occur in intact leaves, but they showed very low rates of CO_2 fixation (<1 μmole CO_2 fixed/hour/mg chlorophyll) compared with the rates found in intact leaves (200 μmole CO_2 fixed/hour/mg chlorophyll). Little improvement in the rates of CO_2 fixation by chloroplast preparations was obtained until Walker (1964) devised procedures which initially gave spinach chloroplasts able to fix 24·3 μmole CO_2/hour/mg chlorophyll. Subsequent modifications gave chloroplasts with improved rates of CO_2 fixation and in the author's laboratory the following adaptation of Walker's method has been used routinely (Reeves & Hall, 1973). Washed spinach leaves are pre-illuminated for 30 minutes before use. 50 g of deribbed blades are rapidly cut up with a sharp knife to give pieces of about 2 cm^{-2} which are placed in a cooled perspex grinding vessel 6 cm × 5 cm × 25 cm. The leaves are then covered with 200 ml of fresh grinding medium which has been partially frozen to a slushy consistency. The grinding medium consists of 400 mM sorbitol, 10 mM NaCl, 5 mM $MgCl_2$, 1 mM $McCl_2$, 2 mM EDTA, 2 mM isoascorbic acid, 0·4% (w/v) bovine serum albumin and 50 mM 2-(*N*-morpholino)ethanesulphonic acid (MES) adjusted to pH 6·5 (at room temperature). The leaves are ground for 3 seconds with a Polytron PT20 with a PT35 head (Willems Kinematica GbmH, Lucerne, Switzerland) at a speed setting of 3·5. The resultant slurry is squeezed through two layers of butter muslin (cheese cloth) and the filtrate poured through eight more layers of muslin. The final filtrate is centrifuged in an MSE Super Minor bench centrifuge with a precooled head. Rapid acceleration up to 4,000 × *g*, followed by braking by hand, (only to be attempted after training and with adequate safety precautions) enables the total centrifugation time to be less than 1 minute. The supernatant is discarded and the pellet gently resuspended with the aid of cotton wool and a glass rod in about 1 ml of solution identical to the grinding solution except that the isoascorbic acid is omitted and the MES is replaced by 50 mM *N*-2-hydroxyethylpiperzaine-*N'*-2-ethanesulphonic acid (HEPES) as buffer adjusted to pH 7·5 (at room temperature). The procedures are carried out at 0–4°C and the resultant chloroplast preparation is kept on ice. The total preparation time from cutting the leaves should be about 4 minutes. Most of the chloroplasts (60–80%) in such a preparation should be complete chloroplasts with a morphologically and functionally intact envelope and a high rate of light-dependent CO_2-fixation and O_2-evolution, similar to that found *in vivo*. They are considered to possess a full complement of unimpaired photosynthetic reactions although most of the reactions cannot be measured directly as the inner membrane of the envelope is impermeable to most of the intermediates which might be added to test these reactions. S. G. Wildman has demonstrated to me, by phase contrast microscopy, that most of the chloroplasts in this sort of preparation have lost their starch grains and therefore presumably possess resealed membranes.

The literature contains a multiplicity of published methods for obtaining a range of chloroplast preparations. To bring some order to the situation, Hall (1972) devised a scheme of nomenclature for these preparations, the main features of which are presented in Table 3.1. In this scheme the complete chloroplasts of Walker are classified as type A and the remaining types represent a succession of increased degradation. To the present writer the criteria for type B chloroplasts seem to be somewhat obscure and unsatisfactory and Walker (personal communication) is also sceptical about the validity of type B. Type B seems to have been proposed by Hall on the basis of older publications which had reported the preparation of apparently unbroken chloroplasts with impaired CO_2 fixation. The other types of chloroplast preparation listed in Table 3.1 are more clearly defined and they correspond with the present state of knowledge.

For most chloroplast preparations investigators would seem to be best advised to prepare type A chloroplasts as the first step. Disadvantages of this procedure are that type A chloroplast preparations (a) contain a proportion of damaged and fragmented chloroplasts (normally 20–40% of the chloroplasts are not type A), (b) are contaminated with other organelles, (c) are contaminated with a substantial amount of cytosol, and (d) are obtained with a fairly low yield. In any further purification stages such as gradient centrifugation or washing there may be further damage to the chloroplasts and reduction of the yield. Consequently it is not yet possible to obtain a pure preparation of type A chloroplasts which is completely free from all other cellular components.

At the present time the application of similar techniques to those described above have yielded active preparations of type A chloroplasts from a very limited number of species (see e.g. Walker, 1971). For successful chloroplast

Table 3.1. Types of chloroplast preparations after the classification of Hall (1972) together with an outline of their main properties.

Chloroplast type	Description	Preparation method	Envelope	Rate of CO_2 fixation μmole CO_2/mg chlorophyll/hour	Electron transport and photophosphorylation capacity
A	Complete chloroplasts	Rapid in isotonic or hypertonic sugar, 1 centrifugation	Intact	50–250	Presumed to be unimpaired.
B	Unbroken chloroplasts	In isotonic or hypertonic sugar or salt with 2 or 3 centrifugations.	Morphologically but not functionally intact	<5	Good
C	Broken chloroplasts	Vigorous preparation in isotonic sugar or salt.	Removed during preparation	Little or none	Good, but addition of ferredoxin is necessary for NADP reduction.
D	Free-lamellar chloroplasts	Osmotic shock of type A chloroplasts immediately followed by return to isotonic medium.	Lost from chloroplasts but retained in medium.	Appreciable if carbon-pathway intermediates added.	Good
E	Chloroplast fragments	Resuspend chloroplasts in hypotonic medium.	Lost	None	Ferredoxin required for NADP reduction. Better rates of electron transport and lower rates of photophosphorylation than by types B and C.
F	Sub-chloroplast particles	By sonication or treatment with detergent or French press. Fractionation by differential centrifugation.	Lost	None	Photophosphorylation (cylic) low or absent. Limited electron transport when electron donors added.

preparation it is usually necessary to modify the standard procedures, but despite modification some species have failed to yield appreciably active chloroplasts.

There are also two very different ways of obtaining plastid preparations which are worth noting. Wellburn and Wellburn (1971) devised a method of purifying structurally intact etioplasts by re-suspending the centrifugation pellet and passing this material through a loosely packed column of coarse Sephadex G-50. *In vitro* developmental studies have been carried out with such preparations (Wellburn & Wellburn, 1973), which show structually intact envelopes, although the biochemical properties of such etioplasts and the functional intactness of the envelopes do not appear to have been established so far. Alternatively, fractions rich in chloroplast material have been obtained by the non-aqueous homogenization and fractionation of freeze-dried leaf material (Stocking, 1971). Although important data have been obtained from non-aqueous preparations, the method is both technically difficult and hazardous, the preparations are impure and many of the biochemical reactions of the chloroplast are destroyed during the preparation.

3.4 CHLOROPLAST DEVELOPMENT

Although the study of chloroplast development and the onset of photosynthesis in seedlings has attracted attention during essentially the whole of the twentieth century (Irving, 1910) it is in recent years that most interest has been shown. There have been two main methods of conducting this study, of which the first is essentially concerned with chloroplast development in plants grown under natural environmental conditions. Since natural conditions are normally neither constant nor consistent such studies have often been conducted in controlled environmental chambers set to a standard day length and standard conditions for day and night. When it became evident that illumination was the critical environmental factor controlling chloroplast development it became fashionable to study chloroplast development by allowing the seedlings to grow for an initial period of several days in complete darkness prior to their transfer to continuous illumination. Both approaches have their merits and it is clear to the present author that both are necessary to obtain an understanding of chloroplast development. Obviously seedling growth under natural environmental conditions shows the normal state of chloroplast development but it provides two main experimental difficulties. Firstly, under diurnal conditions there is a gradient of chloroplast development both within the plant and within the leaf, such that any experimental analysis is either very difficult or impossible. The second difficulty is that diurnal conditions make it difficult to distinguish the effects of illumination on chloroplast development. If seedlings are grown in continuous darkness for a sufficiently long period, their development reaches a stationary phase in which all of the developing plastids in a leaf or a section of a leaf show the same stage of growth. On transfer to illumination the subsequent development then tends to occur in a synchronous manner, thus facilitating microscopic analysis and making biochemical analysis feasible. Such treatments may enable leaf chloroplast development to show synchrony like that obtainable in certain microbial cultures. It should also be pointed out that much progress in the study of photomorphogenesis has depended on illumination treatments of such dark-grown seedlings (Mohr, 1972). The behaviour of dark-grown seedlings in response to illumination does differ in a number of respects from seedlings grown under diurnal conditions of illumination, as discussed by Schiff (1975) for example.

3.4.1 THE PROPLASTID

Figure 3.1A shows an electron micrograph of a proplastid in a section of a primary leaf of a 6-day-old dark-grown bean seedling. The invaginations of the inner membrane of the envelope are interpreted as representing the formation of porous sheets of membrane which are shown in section. The proplastid also contains a starch grain and scattered ribosomes while the cytoplasmic ribosomes are more prominent and are apparently mostly present as polysomes.

3.4.2 ETIOPLAST FORMATION

Between 4 and 14 days of dark growth of *Phaseolus vulgaris* seedlings, the primary leaf primordia, which are already present in the dry seed, show a considerable amount of growth in increasing from a fresh weight of less than 1 mg to 30–40 mg while the cell number increases by more than 10 times and plastid number increases by 18 times (Bradbeer *et al.*, 1974c). During this time the amount of membrane within the plastid increases considerably so that etioplasts like that in Fig. 3.2A are formed. The term etioplast was used first by Kirk and Tilney-Bassett (1967) and defines a structure which is typical of dark-grown seedlings, not normally found in plants grown under diurnal conditions of light and dark. In the bean etioplast about half of the membrane is organised in a regularly-arranged network of tubules called the prolamellar body with the remainder in concentrically arranged porous lamellar sheets (thylakoids). The prolamellar body shows a para-crystalline form and a knowledge of crystallography has contributed to the interpretation of its basic structure. In one plane the tubules form a mesh of hexagons, each individual hexagon being connected to the one immediately above by tubules arising from three alternate nodes, and to the one immediately below by tubules arising from the other three nodes (Weier & Brown, 1970). In planes cutting the hexagonal plane at 90° the arrangement of the tubules is approximately rectangular. Prolamellar bodies are frequently large and complex structures with evident discontinuities, but there are no published reports which account exactly for the structure of these large pro-

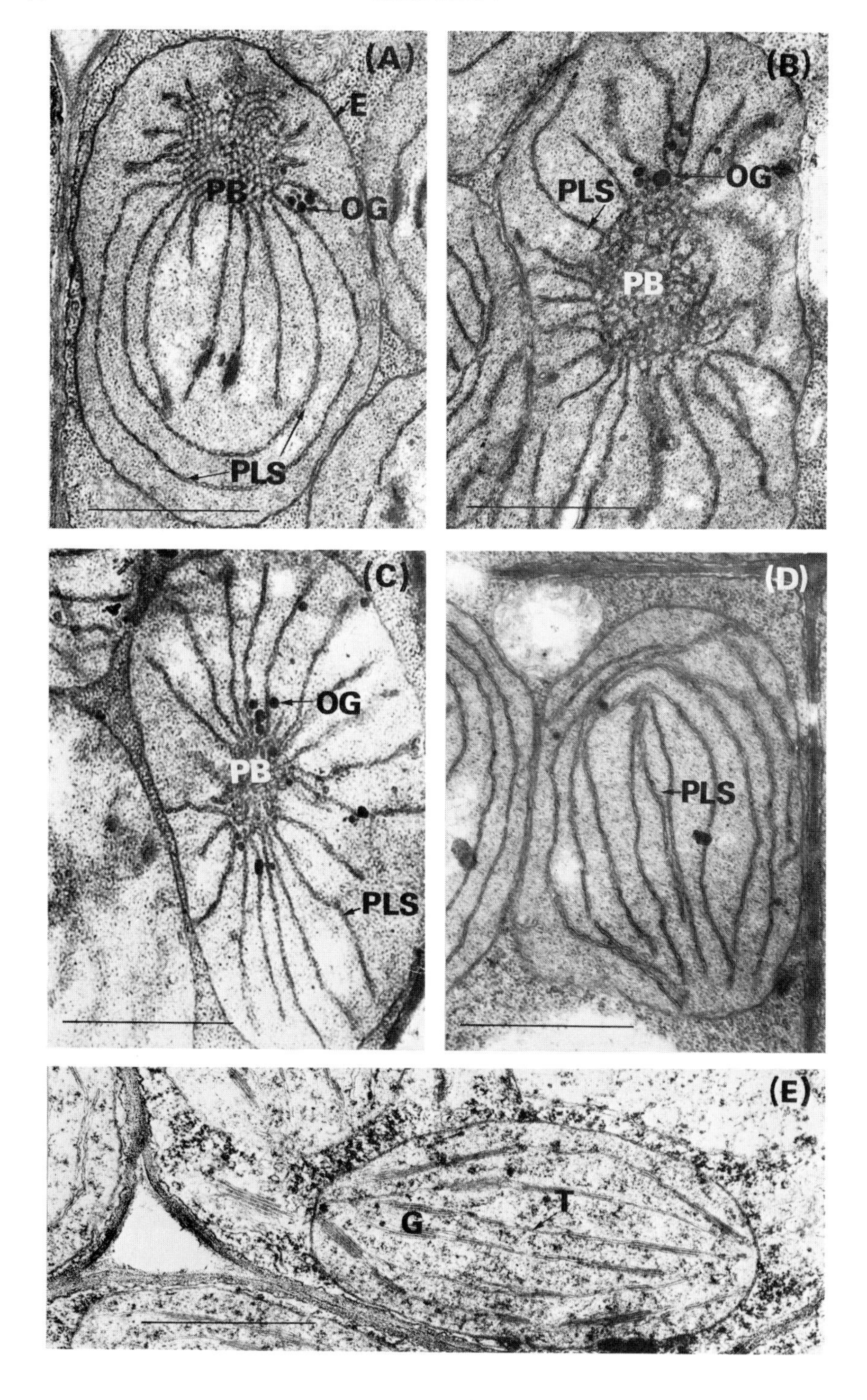
(A)
E
PB
OG
PLS
(B)
OG
PLS
PB
(C)
OG
PB
PLS
(D)
PLS
(E)
G
T

lamellar bodies. Calculations based on measurements in the author's laboratory of electron micrographs of the prolamellar body structure described above show that 1 μm^3 of the prolamellar body of 14-day-old dark-grown bean leaves should contain the equivalent of 44 μm^2 of membrane.

From measurements of electron micrographs and the dimensions of the plastids determined by light microscopy it has been possible to follow the changes in the area of the lamellar sheets and of the volume of the prolamellar bodies during etioplast development. Determination of the number of plastids in the leaf then enables quantities per plastid (Bradbeer *et al.*, 1974c) to be expressed on a per leaf basis as shown in Fig. 3.3. In Fig. 3.3 the area of membrane in the prolamellar bodies has been calculated on the basis stated above that 1 μm^3 of prolamellar body contains 44 μm^2 of membrane. As this factor is likely to vary during development according to the degree of contraction of the prolamellar body some inaccuracy is inevitable from this source for samples other than the 14-day-old one. In particular the slight apparent fall in the total membrane, shown in Fig. 3.3, after 14 days of dark growth may represent a contraction of the prolamellar body without any change in its membrane content. In the experiment shown in Fig. 3.3 membrane formation occurred between 6 and 14 days of growth. The porous lamellar sheets were formed first and the data are consistent with the presumed formation of the prolamellar bodies by some sort of condensation of these sheets (Weier & Brown, 1970). The area of the lamellar sheets reached a peak at 12 days after which the continued formation of the prolamellar bodies seems to have resulted in some consumption of the lamellar sheets.

During dark development of the etioplast most of the chemical components of the chloroplast are formed; for example all of the photosynthetic carbon cycle enzymes are present in the etioplast (Bradbeer *et al.*, 1974c). Substances not yet detected in etioplasts which have received no illumination are chlorophyll, the chloroplast pigment-protein complexes and cytochrome $b-559_{HP}$. The etioplasts of beans appear to reach the peak of their development by 14 days of dark growth while retaining an ability to form chloroplasts on illumination; however 17-day-old leaves with 75% of the etioplast membrane in the prolamellar body green only feebly and 21-day-old dark-green bean leaves fail to survive when transferred to illumination. It should also be pointed out that more rapid greening occurs in leaves younger than 14 days.

Fig. 3.2 Stages of chloroplast development during the greening of the primary leaves of 14-day-old dark-grown beans under continuous illumination of 3 mW. cm^{-2}. A, no illumination; B, 105 minutes illumination; C, 4 hours illumination; D, 5 hours illumination; E, 15 hours illumination. Magnification ×25,000. Key to lettering: PB, prolamellar body; other details as in Fig. 3.1.

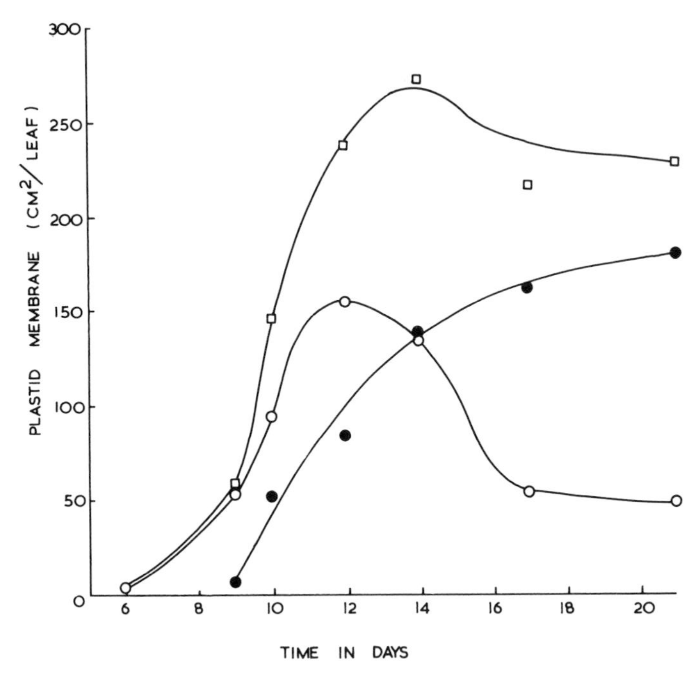

Fig. 3.3. The formation of internal membrane during etioplast development in the mesophyll of primary leaves of *Phaseolus vulgaris* during growth in continuous darkness at 23°C □, total membrane; ○, membrane in porous lamellar sheets; ●, membrane in prolamellar body.

3.4.3 THE CONVERSION OF ETIOPLASTS INTO CHLOROPLASTS

The formation of chloroplasts in dark-grown leaves requires the provision of an appropriate amount and quality of illumination. Continuous illumination of 3 mW. cm^{-2} by fluorescent tubes in a growth cabinet was used for the electron microscopic study of 14-day-old dark grown beans shown in Fig. 3.2. For this material fixed with glutaraldehyde-osmium tetroxide, the paracrystalline appearance of the prolamellar body is retained for about 30 minutes after the beginning of illumination, and then between 30 and 60 minutes there is a change from a regular to an irregular appearance like that shown in Fig. 3.2B (Bradbeer *et al.*, 1974a). Transformation of the prolamellar body appears to occur much more rapidly, usually within less than 1 minute of illumination, if the leaves are fixed with permanganate but this rapid change is commonly regarded as an artefact in structural terms. On the other hand, it may result from an early photochemical reaction in the etioplast which renders the paracrystaline nature of the prolamellar body less stable (Berry & Smith, 1971). Subsequently the volume of the reacted prolamellar body falls and the area of the thylakoid sheets increases as shown in Fig. 3.2C & D for 4 and 5 hours of illumination

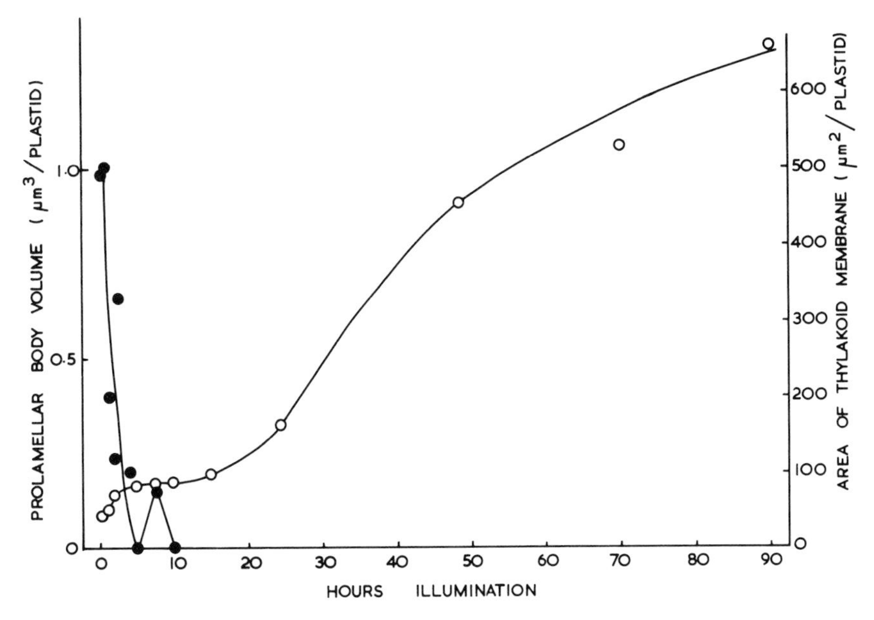

Fig. 3.4. The effects of the transfer of 14-day-old dark-grown beans to continuous illumination of 3 mW. cm^{-2} on the prolamellar body volume and the thylakoid area of the plastids of the primary leaves. ●, prolamellar body volume; ○, thylakoid area.

respectively and in Fig. 3.4. Figure 3.4 shows that the fall in the prolamellar body volume and the increase in the area of the thylakoid sheets approximately correspond with each other with the actual increase of 21·5 μm^2 thylakoid/plastid (43 μm^2 membrane) accounting for the loss of 0·99 μm^3 of prolamellar body per plastid (43·6 μm^3 membrane). After 10 hours of illumination, appression of the thylakoids becomes more obvious, the formation of grana occurs, and the further increase in thylakoid membrane may be presumed to have resulted from *de novo* membrane synthesis. Figure 3.2E shows the stage of development after 20 hours of illumination and Fig. 3.1B that after 48 hours of illumination. During the course of bean chloroplast development under these conditions, *de novo* membrane synthesis results in an 8-fold increase in thylakoid area.

The illumination of 14-day-old dark-grown beans results in a synchronous development of the etioplasts in the primary leaf mesophyll cells. There is no division of these cells, less than 10% of the plastids divide, and the rather amoeboid etioplasts double in diameter to give typical discoid chloroplasts. The small amount of chloroplast division does not seem to be typical of developing leaves in which replication of chloroplasts is usual. Greening leaves show a lag before the onset of photosynthesis. The length of this lag depends upon the nature of the plant material, the conditions and duration of dark growth, the conditions for greening, the intensity of the illumination used to assay photosynthesis and the sensitivity and nature of the method of assay. Consequently, published reports on the duration of this lag show a range of values; however, all agree that at least one hour of illumination, and usually more, is required before photosynthetic CO_2-fixation can be detected. For the bean leaves studied in the author's laboratory, the lag has been at least 5 hours, after which the photosynthesis of the developing chloroplasts becomes increasingly important as the source of the energy and the carbon requirements, of their own development (Bradbeer, 1976).

The effects of different light treatments on the form of the resultant chloroplast have been considered in an earlier section (3.2.4.3). There have been a number of studies in which light of different wavelengths has been used for the illumination of dark-grown seedlings in an attempt to determine an action spectrum of the light dependent reactions involved in chloroplast development. Henningsen (1967) obtained a sharp peak at 450 nm in the action spectrum for a stage termed vesicle dispersal in the development of bean plastids. Unfortunately this developmental stage may be another artefact of permanganate fixation as thylakoid extrusion does not involve such a stage when seen in glutaraldehyde-osmic acid fixed material and studies with this latter fixative have not so far observed any corresponding action spectrum. However Henningsen's experiments have not been exactly replicated with the latter fixative and his data may well indicate a so far undefined photoresponse. Most studies of the effects of light quality on chloroplast development have used rather wide bands of wavelengths and have concentrated on the role of phytochrome in chloroplast development. It has been shown in a number of laboratories that phytochrome has an important role in chloroplast development, and a number of aspects of chloroplast development are promoted by short treatments with red light and reversed by short exposure to far-red. In bean, the presumably-active form of phytochrome, Pfr (see chapter 12) promotes plastid expansion, plastid division, the formation of plastid membrane and the synthesis of chloroplast proteins (Bradbeer, 1971; Bradbeer *et al.*, 1974b). These reactions may be regarded as 'slow' reactions in that they require several hours to become evident. In addition to phytochrome, 'slow' reactions may possibly also be effected by a red-light-absorbing photoreceptor other than phytochrome, and by a blue-light-absorbing photoreceptor. Most of the 'rapid' changes in etioplasts, which occur within three hours of illumination, do not seem to involve phytochrome; such changes as prolamellar body transformation and loss are sensitive to a wide range of wavelengths whilst thylakoid extrusion seems to depend on a red-absorbing photoreceptor other than phytochrome. Only two 'rapid' effects of phytochrome on etioplast fine structure have so far been defined; they are the Pfr-promoted crystallization of the prolamellar body in mustard cotyledons (Kasemir *et al.*, 1975) and the Pfr-inhibited recondensation of the prolamellar body in bean primary leaves (Bradbeer & Montes, 1976). Although Wellburn and Wellburn (1973) have implicated Pfr in 'rapid' changes in isolated and *in situ* etioplasts their method of analysis does not permit them to define these changes with any precision and their conclusions seem to require reconsideration.

3.4.4 THE FORMATION OF CHLOROPLAST COMPONENTS IN GREENING LEAVES

The transfer of dark grown plants to continuous illumination results in substantial increases in the amounts of most of the etioplast constituents which are also found in the chloroplasts (see for example Bradbeer *et al.*, 1969; Gregory & Bradbeer, 1973). A few components, namely chlorophyll, the chloroplast pigment-protein complexes and cytochrome $b-559_{HP}$, which cannot be detected in dark-grown etioplasts, appear as a result of illumination.

The photosynthetic enzymes have almost exclusively been studied by determination of enzyme activity and thus any change in activity may result from either a change in the amount of enzyme protein or a change in the activity of the enzyme protein. Since, during the greening process, there is considerable protein synthesis it was considered that the simultaneous increases in enzyme activity probably resulted from protein synthesis. Subsequent investigations have established that enzyme activation is also responsible for a substantial part of the increase in the activities of certain enzymes in greening leaves.

Ribulosebisphosphate carboxylase is the most abundant protein in the chloroplast and probably consists of 8 large subunits (molecular weight about $5{\cdot}2 \times 10^4$) and 8 small subunits (molecular weight about $1{\cdot}3 \times 10^4$) which give a molecule of about $5{\cdot}2 \times 10^5$ in molecular weight. By the use of two-dimensional polyacrylamide gel electrophoresis, Arron and Bradbeer (1975) found that the commencement of the synthesis of both subunits in bean leaves coincided with the beginning of illumination but that there was a lag before enzyme activity increased. Smith *et al.* (1974) reached a similar conclusion from experiments with barley in which they measured newly-synthesized ribulosebisphosphate carboxylase by labelling it during its biosynthesis and trapping the labelled substance with a specific antibody. In both bean and barley, synthesis of the enzyme occurs early in greening with a corresponding increase in enzyme activity, while late in greening after the enzyme synthesis has ceased considerable activation occurs. The mechanism of the activation is not known although the effect of illumination may be mediated by a small (molecular weight 5×10^3) constituent (Wildner *et al.*, 1972).

In contrast, rather more is known about the activation of phosphoribulokinase and triosephosphate dehydrogenase, where a pretreatment of the extracted enzyme with 6 mM ATP prior to assay brings about activation, (Wara Aswapati *et al.*, 1977). Activation may also be brought about by NADPH and sulphydryl reagents. On illumination of etiolated leaves, the activity of each of these enzymes shows a lag of several hours before it increases; however, preincubation of these extracts with ATP shows that the increases in activity seem to commence from the beginning of illumination (Fig. 3.5). It is concluded that synthesis of

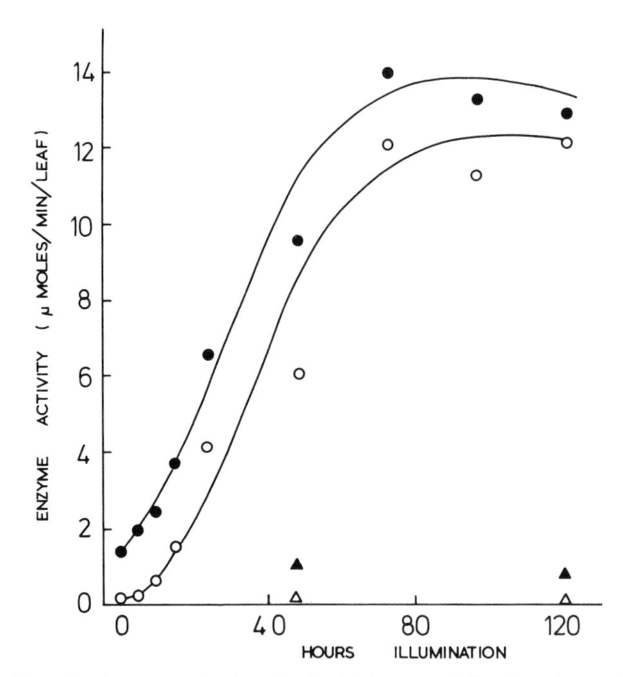

Fig. 3.5. The development of phosphoribulokinase activity in primary leaves of 14-day-old dark-grown beans on transfer to continuous illumination of 3 mW.cm^{-2}. ●, ▲, assayed after pretreatment of the extract with 6 mM ATP; ○, △, no pretreatment; ●, ○, leaves illuminated; ▲, △, leaves left in dark. After Wara-Aswapati (1973).

these two enzymes commences with the beginning of illumination but that activity does not develop until photophosphorylation commences, thus accounting for the lag in Fig. 3.5 for the extracts which were not activated prior to assay. Thus, for the phosphoribulokinase in greening bean leaves Fig. 3.5 shows that there was a 10-fold increase in enzyme protein and a 9-fold increase in the enzyme activity of the protein, thus accounting for a 90-fold increase in total enzyme activity.

There is the further example of ATPase, which is a polypeptide complex in which illumination brings about an increase in enzyme activity with comparatively little new synthesis of the protein being apparent (Gregory & Bradbeer, 1975). In this case the effect of illumination seems to be more direct and is probably necessary for the onset of photophosphorylation rather than a result of it.

3.5 SUMMARY

The chloroplasts of higher plants are normally lens-shaped organelles with a diameter of about 5 µm and a maximum thickness of about 1 µm. They are bounded by an envelope consisting of two membranes the outer of which is freely permeable to solutes while the inner shows very specific permeability. Within the inner membrane is a complex system of thylakoids (flattened sacs of membrane) which are closely associated in a number of stacks (grana), the thylakoid system itself being embedded in a matrix (stroma). Considerable variation in chloroplast structure is found in the algae. In plants which possess the C-4 pathway of photosynthetic CO_2 fixation, dimorphism of chloroplasts is usual with agranal starch-containing chloroplasts occurring in the bundle sheath, which is a single layer of cells immediately surrounding the vascular bundles, while in the next outermost layer of cells (mesophyll sheath) are found most of the remaining chloroplasts, which possess grana and do not normally accumulate starch. Chloroplast structure may also be modified by environmental conditions or mutation.

Techniques for the isolation of chloroplasts for physiological and biochemical studies have been developed and the properties of the main types of chloroplast preparation are summarized in tabular form.

Proplastids occur in meristematic cells and are small with little membrane within the envelope. The direct conversion of proplastids to chloroplasts has been studied in plants grown under diurnal conditions of illumination although in most developmental studies seedlings have been grown in complete darkness at first so that the proplastids have developed into etioplasts. Etioplasts possess a substantial amount of internal membrane, some of which occurs as porous thylakoid sheets with the remainder condensed to give one or more regular lattices of tubules, the prolamellar bodies. On transfer of dark-grown seedlings to illumination the prolamellar body loses its regular structure and becomes converted into thylakoids. Subsequently *de novo* thylakoid synthesis occurs together with the formation of grana. Most of the chemical constituents of chloroplasts are already present in etioplasts. On illumination the missing components are synthesized, some enzymes are subjected to activation, most components undergo further *de novo* synthesis and photosynthetic activity develops after a lag. Some preliminary investigations have also been made of the effects of quality and intensity of irradiation on chloroplast development.

FURTHER READING

ANDERSON J.M., GOODCHILD D.J. & BOARDMAN N.K. (1973) Composition of the photosystems and chloroplast structure in extreme shade plants. *Biochim. Biophys. Acta* **325**, 573–85.

BRADBEER J.W., GLYDENHOLM A.O., IRELAND H.M.M., SMITH J.W., REST J. & EDGE H.J.W. (1974a) Plastid development in primary leaves of *Phaseolus vulgaris*. VIII. The effects of the transfer of dark-grown plants to continuous illumination. *New Phytol.* **73**, 271–79.

BRADBEER J.W., IRELAND H.M.M., SMITH J.W., REST J. & EDGE H.J.W. (1974c) Plastid development in primary leaves of *Phaseolus vulgaris*. VII. Development during growth in continuous darkness. *New Phytol.* **73**, 263–70.

HELDT H.W., SAUER F. & RAPLEY L. (1972) Differentiation of the permeability properties of two membranes of the chloroplast envelop. In *Proceedings of the 2nd International Congress on Photosynthesis Research*, (eds G. Forti, M. Avron & A. Melandri) pp. 1345–55. Dr. W. Junk, The Hague.

HESLOP-HARRISON J. (1966) Structural features of the chloroplast. *Sci. Prog., Oxf.* **54**, 519–41.

KIRK J.T.O. & TILNEY-BASSETT R.A.E. (1967) *The Plastids*. W.H. Freeman & Co., London and San Francisco.

LAETSCH W.M. (1974) The C_4 syndrome: a structural analysis. *Ann. Rev. Plant Physiol.* **25**, 27–52.

SCHIFF J.A. (1975) The control of chloroplast differentiation in *Euglena*. In *Proceedings of the 3rd International Congress on Photosynthesis Research*, (ed. M. Avron) pp. 1691–1717. Elsevier, Amsterdam.

CHAPTER 4

CHLOROPLASTS—STRUCTURE AND PHOTOSYNTHESIS

4.1 INTRODUCTION

Chapter 3 dealt with the ultrastructure of the higher plant chloroplast and the differentiation of the chloroplast's internal membrane system into grana and stroma thylakoids. It outlined the structure of the etioplast and examined the sequence of structural and biochemical changes during the maturation of an etioplast into a chloroplast. This chapter examines the thylakoid membrane at the molecular level in relation to its prime function of converting light energy into chemical energy in the form of NADPH and ATP. The formation of a functional thylakoid from the prolamellar body membranes of the etioplast during the greening process will also be considered at the molecular level. Finally, we briefly examine the relationship of the chloroplast to cytoplasm in terms of energy metabolism.

4.2 THYLAKOID STRUCTURE

4.2.1 THIN SECTIONING AND HEAVY METAL SHADOWING

When thin sections of fixed and stained leaf material are examined in the electron microscope, the chloroplast thylakoid membrane appears as a single electron dense line or as a tripartite structure (two dense lines with an intermediate light line) characteristic of the unit membrane (chapter 2). In high resolution electron micrographs of thin sections, fixed in permanganate, the thylakoid membranes give the appearance of containing globular subunits, 9 nm in diameter. It is possible, however, that these are an artefact since it is difficult to understand how several layers of overlapping 9 nm particles can be clearly resolved in a section thicker than 40 nm (Branton, 1968).

Some of the earliest electron micrographs of thylakoid membranes were obtained simply by drying isolated chloroplast fragments of *Spirogyra* and *Mougeotia* on an electron microscope grid and shadowing the preparations with chromium (Steinmann, 1952). The membranes showed a granular or particulate structure. Similar observations were made subsequently with spinach thylakoid membranes (Park & Pon, 1961). The particles usually existed in a random array but sometimes they were seen in highly ordered arrays. Their dimensions were $15 \times 18 \times 10$ nm and they appeared to contain subunits. Park (1962) suggested that the particle, which he termed quantasome, might be the morphological expression of the so-called photosynthetic unit, defined as the minimum number of chlorophyll molecules needed to fix one molecule of carbon dioxide per intense flash of light (Emerson & Arnold, 1932). The quantasomes were seen on the inner surface of the thylakoid membrane, suggesting that there was some distortion of the thylakoid membrane during drying.

4.2.2 FREEZE-ETCHING

The freeze-etch technique is a different approach to the study of the substructure of the thylakoid membrane, because chemical fixatives are not used. Leaf material or isolated chloroplasts are frozen and fractured (Moor *et al.*, 1961). The fracture plane in the frozen sample occurs along hydrophobic regions within the thylakoid membrane exposing complementary faces. Platinum-carbon replicas of the fracture planes are made and examined in the electron microscope.

Two major sizes of particles are seen when mature plant chloroplasts are freeze-fractured; large particles of average diameter, 17.5 nm and small particles of average diameter, 11 nm (Fig. 4.1a). The small and large particles occur on different fracture faces (Fig. 4.1b). Fracture face C, which is towards the stroma

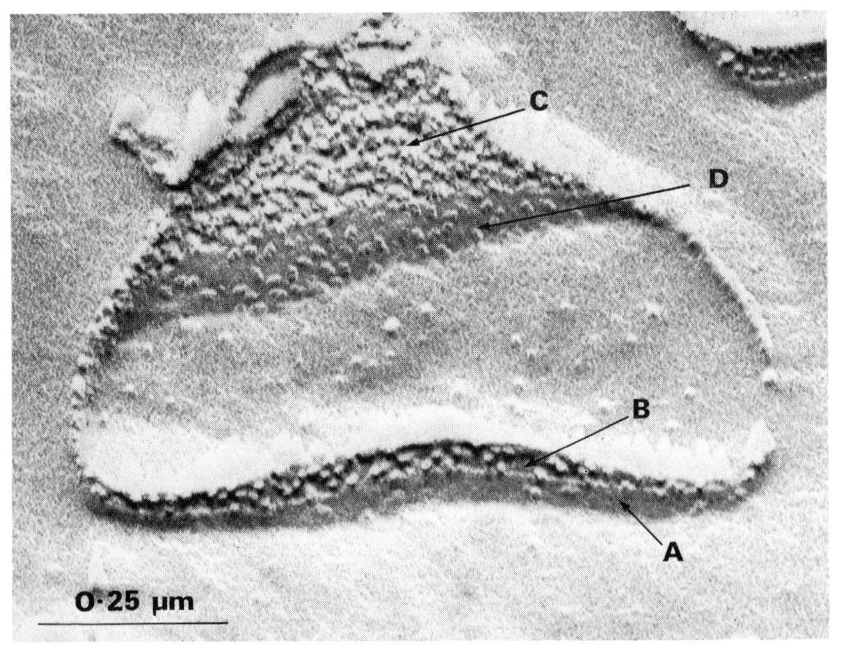

Fig. 4.1a. Freeze-fracture electron micrograph of a spinach grana thylakoid, showing the large and small particle fracture faces, B and C, and the surfaces, A′ and D. (By courtesy of Dr. D.J. Goodchild.)

of the chloroplast contains tightly packed small particles, and the complementary fracture face B contains the large particles at about half the density per μm^2. The large particles are only observed within grana stacks, where the adjacent thylakoid membranes are in close contact, while the small particles are observed in grana and stroma thylakoids. In the unstacked stroma thylakoids the complementary fracture face (designated B_u) is relatively smooth and shows only a few small particles (Park & Sane, 1971).

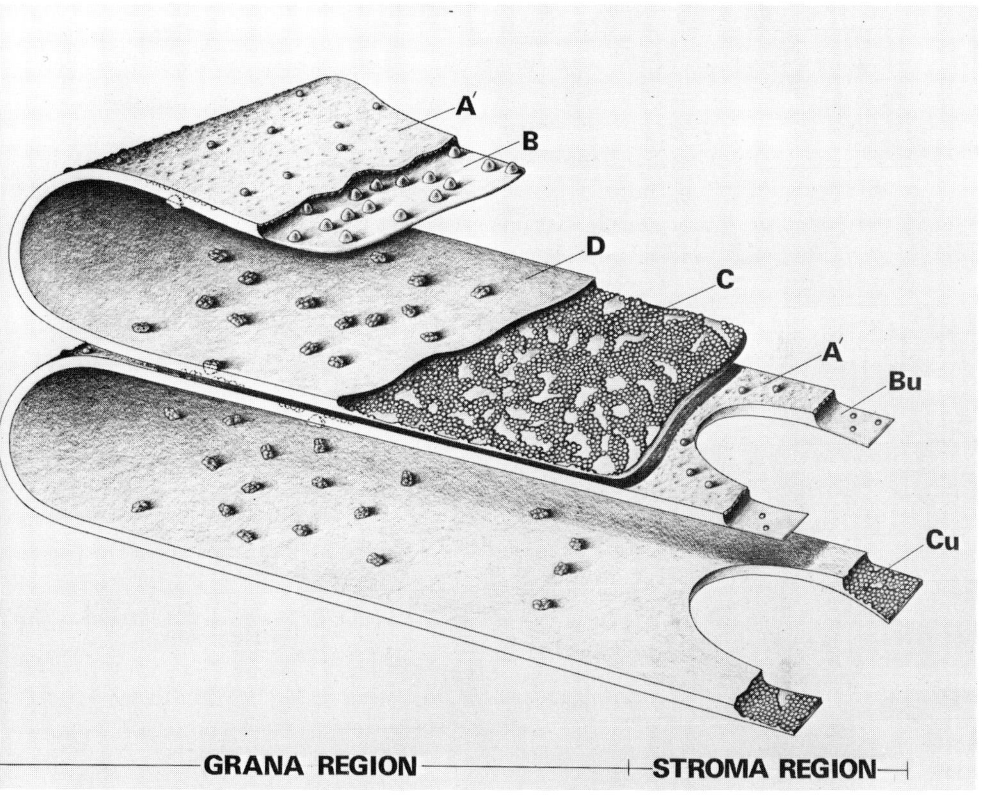

Fig. 4.1b. Model of the thylakoid membrane, showing grana and stroma regions. The particles on the A′ surface represent the coupling factor (CF_1). (By courtesy of Dr. D. J. Goodchild.)

The exterior thylakoid surface, A′, and the interior surface, D, are observable by the deep etch technique, in which the ice covering the surfaces is sublimed off before the replica is made. In well-washed thylakoids the A′ surface shows little surface relief, but is covered with proteins in unwashed lamellae. The interior surface is covered with particles about 18.5×15.5 nm in size which show slight surface relief. These particles on the D surface contain 3 or 4 subunits and they may correspond to the quantasomes as seen by metal shadowing.

The freeze-etch data suggest that the particles within the thylakoid membrane are arranged somewhat asymmetrically. The larger particles are located towards the inner surface and partly protrude into the intrathylakoid space, and the smaller particles are near the outer surface. The model of the thylakoid membrane shown in Fig. 4.2 appears to be consistent with the freeze-etching results and also with X-ray diffraction studies of thylakoid membranes (Sadler *et al.*, 1973). The membrane is viewed as a lipid bilayer in which the particles are embedded, in accordance with the fluid mosaic model for biological membranes (Singer & Nicolson, 1972; see chapter 2).

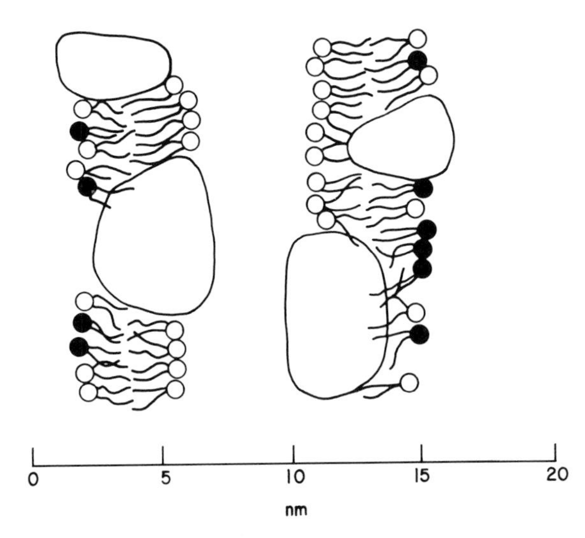

Fig. 4.2. Structural model of the thylakoid membrane showing an asymmetric distribution of large and small particles. The two membranes belong to one thylakoid. The lipid molecules are indicated as a bilayer, the circles representing the polar head groups. Lipid asymmetry is indicated by the shading of the head groups. The model appears to be consistent with freeze-etch electron microscopy and X-ray diffraction of thylakoid membranes. (From Sadler *et al.*, 1973.)

4.2.3 NEGATIVE STAINING

Two sizes of particles are observed on the surface of unwashed thylakoids by negative staining with phosphotungstic acid. The larger particles (11 nm in diameter) are removed by water washing the thylakoids and correspond to the CO_2-fixing enzyme, ribulose bisphosphate carboxylase. The smaller particles (9 nm) are removed by washing with the chelating agent, ethylenediamine tetraacetate, and are identical with the coupling factor (CF_1) which catalyses the terminal steps in the energy coupling process of ATP formation. These surface particles are not related to the membrane particles seen by freeze-etching (Park & Sane, 1971).

4.3 THYLAKOID COMPOSITION

4.3.1 LIPIDS

The thylakoid membrane consists of 50% protein and 50% lipid. The feature of its lipid composition is the high percentage of the neutral galactolipids, digalactosyl diglyceride (27%) and monogalactosyl diglyceride (13%), compared with the phospholipids (9%) (Table 4.1). Chlorophyll *a* plus chlorophyll *b* account for 21% of the lipids. The thylakoid membrane also contains an anionic sulpholipid. The other unique feature of the thylakoid lipids is their high content of the polyunsaturated fatty acid, linolenic acid (18:3) (Table 4.2).

Table 4.1. Lipid composition of spinach thylakoid membranes (adapted from Lichtenthaler & Park, 1963).

	% Weight of total lipid	Number of lipid moles per 400 chlorophyll moles
Monogalactosyl diglyceride	26.8	602
Digalactosyl diglyceride	13.4	198
Sulpholipid	4.1	42
Phospholipid*	9.1	201*
Chlorophyll *a*	15.0	290
Chlorophyll *b*	5.8	110
Carotenoids**	2.8	82**
Quinone compounds	3.3	80
Sterols	2.2	
Unidentified lipids	17.0	

*Phosphatidyl glycerol (91); phosphatidylcholine (73); phosphatidyl inositol (24); phosphatidyl ethanolamine (10); phosphatidic acid (3).
**β-carotene (24); lutein (38); violaxanthin (10); neoxanthin (10).

Table 4.2. Fatty acids of spinach thylakoid lipids (percent composition) (from Allen & Good, 1971).

Fatty acid	14:0	16:0	Δ_3-*trans* 16:1	16:2	16:3	18:0	18:1	18:2	18:3
Monogalactosyl diglyceride	—	—	—	—	25	—	—	2	72
Digalactosyl diglyceride	—	3	—	—	5	—	2	2	87
Sulpholipid	—	39	—	—	—	—	—	6	52
Phosphatidyl glycerol	1	11	32	—	2	—	2	4	47
Phosphatidyl choline	—	12	—	—	4	—	9	16	58
Phosphatidyl inositol	4	34	6	—	3	2	7	15	27

In spinach thylakoids, trans-Δ^3-hexadecenoic acid is a major constituent of phosphatidylglycerol. Lipophilic quinones, of which PQA_9 (2, 3 dimethyl-*p*-benzoquinine with a 45-carbon side chain at the 5-position) is the major constituent, account for 3% of the thylakoid lipids.

4.3.2 CHLOROPHYLL-PROTEINS

The proteins of the thylakoid membrane include the carriers of the photosynthetic electron transport chain, but these account for only a relatively small percentage. When a thylakoid preparation, from which the coupling factor, lipids and pigments have been removed, is solubilized by the anionic detergent, sodium dodecyl sulphate (SDS), and the extract examined by SDS-polyacrylamide gel electrophoresis, at least 20 polypeptides are obtained (Anderson & Levine, 1974). These range in size from 10,000 to 100,000 daltons. A major fraction of the proteins appear in the region of 25,000 daltons. If the thylakoids are extracted with SDS without prior removal of pigments, two chlorophyll-protein complexes (termed complex I and complex II) are separated by SDS polyacrylamide gel electrophoresis (Thornber, 1975). Complex I which is derived from photosystem 1 (see section 4.4) has a chlorophyll *a*/chlorophyll *b* ratio of 12:1 and accounts for about 20% of the thylakoid membrane protein. It has an apparent molecular weight of 110,000 daltons and contains 14 moles chlorophyll *a* per mole of protein. Complex II, which accounts for 40–50% of the thylakoid membrane proteins has an apparent molecular weight of 32,000 daltons and contains one mole chlorophyll *a* and one mole of chlorophyll *b* per mole of protein.

4.3.3 ELECTRON CARRIER PROTEINS

The thylakoid membrane contains four spectroscopically distinguishable cytochromes: cytochrome *f* (a *c*-type) with an α-band at 554 nm and a midpoint potential of +0.36 at pH 7, cytochrome b_6 (alternate name $b-563$) with an α-band at 563 nm and a mid-point potential below −0.1V, cytochrome $b-559_{HP}$ (high potential form) with an α-band at 559 nm and a mid-point potential (+0.37V) close to that of cytochrome *f* and cytochrome $b-559_{LP}$ (low potential form) with an α-band at 559 nm and a mid-point potential of 0.06V (Bendall *et al.*, 1971). The thylakoid cytochromes are not extracted by aqueous solvents. Cytochrome *f* is extracted into ammoniacal ethanol (Bendall *et al.*, 1971), but the *b* cytochromes are tightly associated with the thylakoid membranes and are released on extraction of the membranes with detergent (Boardman, 1975).

Plastocyanin is a copper-containing protein with a mid-point potential of +0.37V at pH 7. It is extracted by sonication of the thylakoid membranes or by detergent treatment. Its molecular weight is 10,500 with 1 gm atom of Cu per mole. The oxidized form of plastocyanin is intensely blue with a main absorption band at 597 nm.

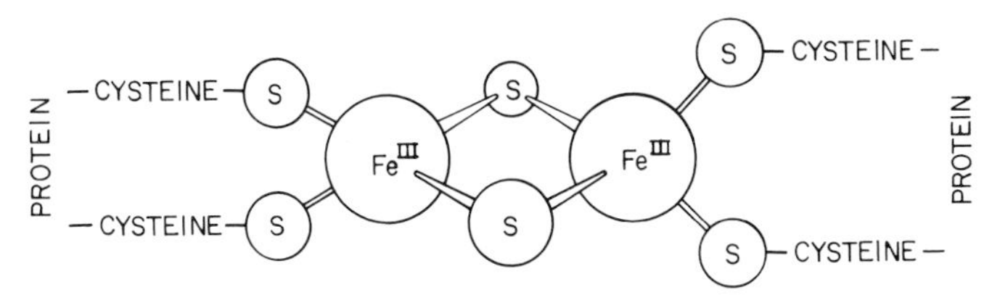

Fig. 4.3. Structure of iron-sulphur group in ferredoxin (from Hall *et al.*, 1972).

The ferredoxins are stable iron-sulphur proteins in which Fe is coordinated to the sulphur atoms of cysteine and inorganic sulphur (Fig. 4.3). Chloroplasts contain a soluble ferredoxin ($E_0' = -0.43$V at pH 7.5), which is readily released in aqueous media, and one or more membrane-bound ferredoxins. Soluble ferredoxin has a molecular weight of 12,000 and contains 2 gm atoms of Fe and 2 gm atoms of inorganic sulphur per mole. Bound ferredoxin has also been extracted and purified (Malkin *et al.*, 1974). It has a molecular weight of 8,000 and 4 gm atoms of Fe and 4 gm atoms of inorganic sulphur per mole. Its mid-point potential is 0.1V more negative than that of soluble ferredoxin.

The thylakoid membrane contains a flavoprotein, ferredoxin-NADP$^+$ reductase, which is extractable with detergents. It has a molecular weight of 40,000.

4.4 PHOTOSYNTHESIS

In the process of photosynthesis in green plants, light energy is absorbed by the pigments of the chloroplast and converted into chemical free energy in the form of ATP and NADPH, which are then used for the reduction of carbon dioxide and the synthesis of plant materials. The major organic products of photosynthesis are carbohydrate and the overall equation of photosynthesis is:

$$CO_2 + H_2O \rightarrow (CH_2O) + O_2 : \Delta G_0 = 0.47 \text{ MJ.}$$

The oxygen which is evolved is derived from water. Carbon dioxide is converted to carbohydrate via the Calvin cycle of reactions (Calvin & Bassham, 1962). In the carbon reduction cycle, one molecule of ribulose bisphosphate (RBP or RuDP) reacts with CO_2 to form two molecules of 3-phosphoglyceric acid (PGA), which are converted into phosphoglyceraldehyde (triose phosphate) in a reaction which needs two molecules of NADPH and two molecules of ATP. The regeneration of the CO_2 acceptor (RBP) involves a complex series of reactions and requires one molecule of ATP. The overall requirement of the Calvin cycle is 3 molecules of ATP and 2 molecules of NADPH for each molecule of CO_2 reduced to carbohydrate. In the C_4-dicarboxylic acid pathway of photosynthesis, in which CO_2 reacts initially with phosphopyruvate, 5 molecules of ATP and 2 molecules of NADPH are needed for each molecule of CO_2 reduced (Hatch & Slack, 1970).

The light reactions of photosynthesis and the formation of NADPH and ATP are performed by the thylakoids, and the carbon reduction cycle occurs in the stroma or soluble phase of the chloroplast.

4.4.1 THE PHOTOSYNTHETIC UNIT AND ENERGY TRANSFER

The light absorbing molecules, i.e. the chlorophylls and carotenoids, are organized into units in the thylakoid membrane. Quanta absorbed by a large number of pigment molecules are transferred by an efficient resonance mechanism to a special molecule of chlorophyll *a*, called the trap or reaction centre chlorophyll, where the primary conversion of light energy into chemical free energy takes place. This is the concept of the photosynthetic unit which was first proposed to account for the observation that the maximum yield of CO_2 fixed, or O_2 evolved, per single flash of intense light was one mole per 2,500 moles of chlorophyll. The absorption band of the reaction centre chlorophyll *a* is at a longer wavelength than the absorption bands of the light-harvesting pigments to ensure efficient trapping of the energy at the reaction centre.

The reaction centre chlorophyll is in close association with an electron acceptor molecule (A) and an electron donor molecule (D) in the thylakoid membrane. On excitation of the reaction centre chlorophyll (Chl*) by transfer of energy from the 'antenna' pigments, an electron is donated to A giving A^- and leaving the chlorophyll molecule deficient in an electron (eq. 4.1). The positively charged chlorophyll then receives an electron from the donor D, and is restored to its ground state energy level (eq. 4.2). The net result is that the energy of a photon is used to transfer an electron from D to A, and thus the primary photochemical event of photosynthesis is an oxidation-reduction process.

$$\mathrm{Chl}a^* + \mathrm{A} \rightarrow \mathrm{Chl}a^+ + \mathrm{A}^- \qquad (4.1)$$

$$\mathrm{Chl}a^+ + \mathrm{D} \rightarrow \mathrm{Chl}a + \mathrm{D}^+ \qquad (4.2)$$

4.4.2 TWO PHOTOSYSTEMS AND THE Z-SCHEME OF ELECTRON TRANSPORT

Investigations in the 1950's and early 1960's established that the thylakoid membrane contains two types of photosynthetic units, designated photosystem 1 (PS – 1) and photosystem 2 (PS – 2), which cooperate in a sequential manner to transfer electrons from water to $NADP^+$ (Boardman, 1968). Figure 4.4 depicts the electron transport pathway known as the Z-scheme first put forward by Hill and Bendall (1960). Both photosystems contain chlorophyll *a*, chlorophyll *b* and the four carotenoids, but in different proportions. Each unit of PS – 1 or PS – 2 contains 200 light-harvesting chlorophylls and one reaction centre.

Quanta of light absorbed by PS – 2 are transferred to a form of chlorophyll absorbing at 682 nm and termed chl *a* – 682. Excitation of chl *a* – 682 catalyses the transfer of an electron from Y to Q, giving a strong oxidant, Y^+, and a weak reductant, Q^-. Oxidation of water and the release of a molecule of O_2 requires the sequential absorption of four quanta and the accumulation of four oxidizing equivalents (Cheniae, 1970). The mechanism of water oxidation is unknown, although it is established that manganese, probably in the form of a manganese-protein complex, and Cl^- are required (Boardman, 1975). The primary acceptor of PS – 2, (Q), has a redox potential around zero volts, and is incapable of reducing $NADP^+$, without an imput of energy into PS – 1. Quanta absorbed by PS – 1 are transferred to P – 700, a form of chlorophyll *a* absorbing at 700 nm. On excitation, P – 700 donates an electron to an acceptor Z, resulting in the formation of P – 700^+, a weak oxidant with a mid-point potential of +0.43V, and Z^-, a strong reductant with a redox potential in the vicinity of –0.6V. P – 700^+ interacts with the weak reductant, Q^-, generated by PS – 2 via an electron transport chain, which includes plastoquinone A, cytochrome *f* and plastocyanin. The reduction of $NADP^+$ by Z^- is mediated by soluble ferredoxin and ferredoxin-NADP reductase. It seems possible that the primary acceptor of PS – 1, Z, is identical to bound ferredoxin, since the latter is photoreduced at very low temperatures (25°K) (Bearden & Malkin, 1974).

Cytochrome b_6 appears to function on a cyclic electron transport pathway around PS – 1 and it may play a role in cyclic phosphorylation. There is conflicting evidence concerning the role of cytochrome $b-559_{HP}$ (Boardman, 1975). This cytochrome is oxidized by PS – 2 at liquid nitrogen temperature and under certain conditions at room temperature, but at high light intensity it can also be reduced by PS – 2. The function of cytochrome $b-559_{LP}$ is unknown at present.

Much evidence for the scheme of photosynthetic electron transport depicted in Fig. 4.4 has come from difference spectroscopy of chloroplasts. In this method, the change in the absorbance of various chloroplast components is measured on illumination of the chloroplasts with monochromatic light of various wavelengths. By following the change in the spectrum of individual components e.g. cytochrome *f* or P – 700, the role of these components can be deduced. For example, if chloroplasts are illuminated with far-red light of wavelength 720 nm, which is only absorbed by PS – 1, there is a decrease in the absorbance of the chloroplasts in the region of 554 nm, due to the oxidation of cytochrome *f*. If the chloroplasts are then illuminated with 650 nm light, absorbed by PS – 2 (as well as by PS – 1) there is an increase in absorbance due to the reduction of cytochrome *f*. The oxidation of P – 700 is followed by a decrease in absorption at 700 nm.

Electron transport from water to $NADP^+$ in isolated chloroplasts may be intercepted by the addition of artificial electron acceptors (oxidants), which accept electrons from Z^- in PS – 1 or from PQ in PS – 2. For example, methyl viologen accepts electrons at PS – 1, while *p*-phenylenediamine intercepts the chain at PQ. Ferricyanide or dichlorophenolindophenol can interact at either PS – 1 or PS – 2, depending on the experimental conditions (Trebst, 1974).

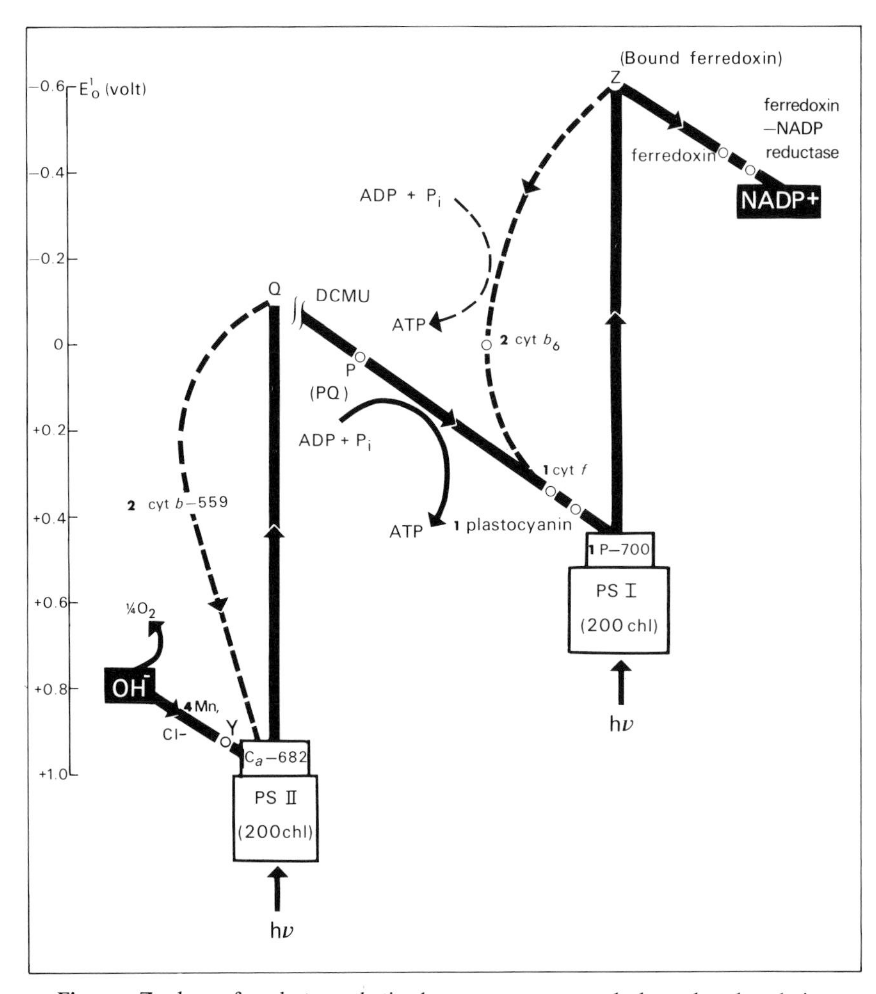

Fig. 4.4. Z-scheme for photosynthetic electron transport and photophosphorylation. The number beside a component indicates the number of molecules of that component per photosynthetic unit of 400 chlorophyll molecules. Two sites of ATP formation are located on the pathway between water and PS−1 (see text); one between water and plastoquinone (PQ) and the other between PQ and PS−1. A scale of redox potentials is shown on the left.

Photosynthetic electron transport is readily monitored by illuminating isolated chloroplasts in the presence of an electron acceptor, and measuring either the oxygen evolved or the amount of acceptor reduced (Hall & Rao, 1972). This reaction is known as the Hill reaction (Hill, 1939).

The herbicides 3(3,4-dichlorophenyl)1,1-dimethylurea (DCMU) and 3(*p*-chlorophenyl)-1,-1-dimethylurea (CMU) inhibit electron transport between

Q and PQ. Photoreduction of $NADP^+$ can be restored in the inhibited chloroplasts by the addition of an artificial electron donor such as reduced 2,6-dichlorophenolindophenol. Electrons from the artificial donor enter the electron transport chain between the light reactions, and photoreduction of $NADP^+$ is then driven by PS – 1.

PS – 1 is known as the far-red system because its absorption spectrum extends to longer wavelengths than that of PS – 2. At wavelengths beyond 700 nm, PS – 1 receives a high fraction of the quanta absorbed by chloroplasts. Chloroplasts contain one molecule of cytochrome *f* and one molecule of P-700 per 430 chlorophyll molecules, from which it is concluded that the photosynthetic unit contains about 400 chlorophyll molecules. As shown in Fig. 4.4 the chlorophyll molecules appear to be distributed about equally between the two photosystems.

4.4.3 PHOTOSYNTHETIC PHOSPHORLATION

Electron flow from water to $NADP^+$ is coupled to the formation of ATP. Until recently it was uncertain whether there is one or two energy conserving sites, where ATP is formed. This arose because of the conflicting experimental determinations of the amount of ATP formed during the transfer of 2 electrons from water to NADP (the P/e_2 ratio). Earlier measurements gave a P/e_2 ratio of one (Arnon *et al.*, 1958), suggesting one energy conserving site, but more recent determinations indicate P/e_2 ratios between 1.2 and 1.8 (Trebst, 1974). One energy conserving site is associated with the reoxidation of plastoquinone by cytochrome *f*, and the second (not shown on Fig. 4.4) is associated with the electron transfer from water to plastoquinone. However, the two sites do not necessarily yield two ATP per 2 electrons transferred from water to $NADP^+$, as indicated by the experimental P/e_2 ratios. A P/e_2 ratio of at least 1.5 is needed to provide enough ATP to drive the Calvin cycle, and more ATP is required for the C_4-pathway.

Another type of phosphorylation has been observed with isolated chloroplasts. Known as cyclic phosphorylation, it requires an exogenous cofactor such as phenazine methosulphate, pyocyanin or ferredoxin. Unlike non-cyclic phosphorylation, cyclic phosphorylation is not accompanied by any net change in oxidation or reduction and it is not inhibited by DCMU. The process is driven by light absorbed by PS – 1. Chloroplasts exhibit very low rates of cyclic phosphorylation in the absence of an exogenous electron carrier. It is uncertain, therefore, whether cyclic phosphorylation *in vivo* contributes a significant amount of ATP. Recent work suggests that oxygen may act as an alternative electron acceptor at PS – 1, when NADP is fully reduced, and provide extra ATP during electron flow from water to O_2 (Heber, 1975). The relative amounts of ATP formed by noncyclic electron flow to $NADP^+$ or O_2 could be regulated by the demands of the cell for NADPH and ATP.

4.4.4 PHOSPHORYLATION AND THYLAKOID STRUCTURE

Three main hypotheses have been proposed for the mechanism of ATP formation in mitochondria and chloroplasts (Slater, 1971), but here we will briefly consider only one of these, the Mitchell chemiosmotic hypothesis (Mitchell, 1966; see also chapter 5). The fundamental concept of the chemiosmotic hypothesis is that of vectorial flow of protons across the mitochondrial inner membrane or the thylakoid membrane. Mitchell proposed that the electron transport chains of mitochondria and chloroplasts consist of alternate electron and hydrogen carriers, organized in a vectorial fashion across the membrane. Transport of electrons through the chain of carriers results in the net transfer of protons across the membrane. For the thylakoid membrane, protons are transported from the outside (stroma or matrix) to the inside intrathylakoid space. This is depicted in Fig. 4.5 (Trebst, 1974). In this scheme, reduction of plastoquinone by Q

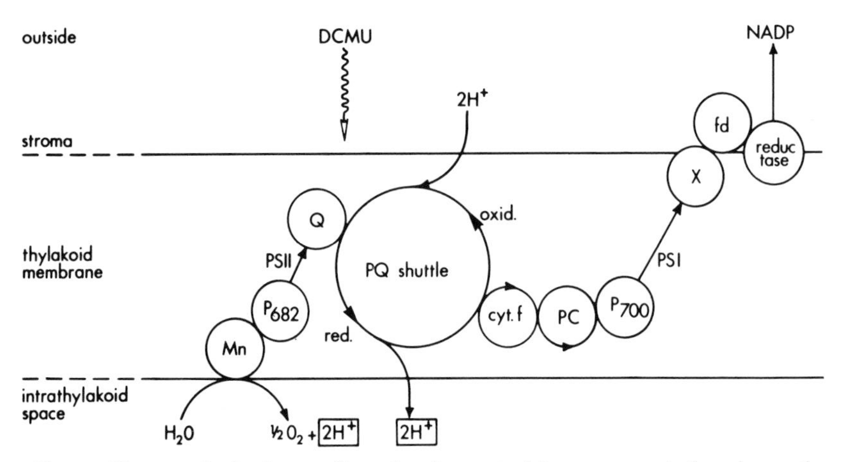

Fig. 4.5. Photosynthetic electron flow, showing vectorial arrangement of carriers and the movement of protons from outside to inside of the thylakoid. (From Trebst, 1974.)

occurs on the outside of the thylakoid and oxidation by cytochrome *f* on the inside, resulting in the transfer of 2 protons for each 2 electrons transported from water to $NADP^+$. Water oxidation is considered to take place on the inside to produce 2 protons per $\frac{1}{2}O_2$. According to the Mitchell hypothesis, the protons move back through the membrane through special channels which contain the coupling factor or ATPase, and result in formation of ATP from ADP and inorganic phosphate. As already discussed, part of the ATPase (CF_1) is on the outside of the thylakoid membrane, and part (CF_0) is embedded in the membrane. It is not yet established whether the proton gradient across the thylakoid membrane is an essential intermediate in the process of ATP formation during photosynthetic electron flow. However, with thylakoid membranes it has been shown that ATP formation can be driven by a pH gradient created artificially (Jagendorf & Uribe, 1966). Chloroplasts are suspended in a buffer of low pH (pH < 5) and then transferred to a buffer of pH 8, containing ADP and inorganic phosphate.

According to the Mitchell hypothesis, the energy-conserving sites are identified with the proton-releasing sites on the inside of the thylakoid. Substances such as ammonia which uncouple ATP formation from electron transport in the thylakoid membrane (termed uncouplers) are considered to destroy the proton gradient by transporting the protons across the membrane.

4.5 SUBCHLOROPLAST FRAGMENTS AND THE FRACTIONATION OF THE PHOTOSYSTEMS

4.5.1 DIGITONIN METHOD

Methods of isolation of chloroplasts were outlined in chapter 3. In this chapter, we will consider the fragmentation of the thylakoid membrane and the isolation of subchloroplast fragments enriched in PS–1 and PS–2 respectively. The thylakoid membrane is fragmented by a number of treatments including sonication, incubation with detergents, or mechanical shearing on passage through the valve of a French pressure cell (Boardman, 1970). Here, we will outline the digitonin method for obtaining subchloroplast fragments (Boardman, 1971).

Chloroplasts are isolated in a sucrose-phosphate buffer medium and resuspended in 50 mM phosphate buffer + 10 mM KCl. Digitonin, a nonionic detergent is added to a final concentration of 0.5% and the mixture incubated for 30 minutes at 0°C. The resulting subchloroplast fragments are sedimented by differential centrifugation. The first centrifugation is at 1,000 *g* for 10 minutes, followed by successive centrifugations at 10,000 *g* for 30 minutes, 50,000 *g* for 30 minutes and 144,000 *g* for 60 minutes. The larger fragments which sediment at 1,000 *g* (D–1) and 10,000 *g* (D–10) have a lower chl *a*/chl *b* ratio than chloroplasts, whereas the smaller fragments (D–50 and D–144) have much higher ratios (Table 4.3). The D–10 fractions accounts for about one-half of the chlorophyll of the chloroplast, while approximately one-third of the chlorophyll is divided about equally between D–50, D–144 and the 144,000 *g* supernatant.

The smaller fragments (D–144) are highly enriched in PS–1; for example, they have a chl/P–700 ratio of 205, compared with 440 for chloroplast, a low level of manganese and they lack cytochrome $b-559_{HP}$. They contain cytochromes *f*, b_6 and $b-559_{LP}$, and show the photochemical reactions of PS–1, including cyclic phosphorylation, but little or no PS–2 activity. The larger fragments (D–10) are enriched in cytochrome $b-559_{HP}$ and manganese and they have less P–700 and cytochrome *f* on a chlorophyll basis than do chloroplasts. Thus the small fragments have the photochemical properties of PS–1,

Table 4.3. Distribution of chlorophyll in subchloroplast fragments (from Boardman, 1971).

Fraction	Chl*a*+Chl*b* (%)	Chl*a*/Chl*b* (Ratio)
Chloroplasts	100	2.83
D-1	19.0	2.36
D-10	46.2	2.27
D-50	12.3	4.40
D-144	11.7	5.34
Supernatant	10.8	3.76

and the large fragments are enriched in PS−2. The non-ionic detergent Triton X−100, produces a more complete fractionation of the chloroplasts than does digitonin, but it has the disadvantage that the resulting PS−2 subchloroplast fragments are unable to evolve oxygen.

4.5.2 MECHANICAL DISRUPTION OF CHLOROPLASTS

Mechanical treatment of chloroplasts by passage through the French Press or by brief sonication disintegrates the chloroplasts into grana thylakoids and small vesicular structures derived from the stroma thylakoids (Park & Sane, 1971). The grana thylakoids are separated from the smaller vesicles by differential centrifugation. The composition and photochemical properties of the vesicles are similar to those of the D−144 subchloroplast fragments, i.e. they are highly enriched in PS−1. This led to the conclusion that stroma thylakoids contain only PS−1, whereas grana thylakoids contain both PS−1 and PS−2 (Park & Sane, 1971), gut an alternate explanation is that the PS−1 containing vesicles are derived from stroma thylakoids.

When chloroplasts are incubated with digitonin, PS−1 containing vesicles are first released from the stroma, but digitonin (and Triton X−100) also releases PS−1 fragments from grana thylakoids.

4.6 THYLAKOID STRUCTURE IN RELATION TO THE PHOTOSYSTEMS

Attempts have been made to isolate individual quantasome particles, active in electron transport from water to $NADP^+$ or ferricyanide. However, disruption of the thylakoid membrane into fragments comparable in size to the quantasome results in the complete loss of the ability to photo-oxidize water, although the small particles retain PS−1 activity, i.e. the small particles are capable of photo-reducing $NADP^+$ or methyl viologen with reduced indophenol dye as the electron donor. Biochemical evidence is lacking for the existence in the thylakoid membrane of a quantasome particle (containing one reaction centre of PS – 1 and one of PS – 2 and the associated light harvesting pigments and electron transfer components) which functions as a discrete unit independent of neighbouring reaction centres and electron transport chains. On the contrary, there is evidence for interaction between reaction centres and between electron transfer chains (Siggel *et al.*, 1972; Boardman *et al.*, 1974).

Studies with chlorophyll-deficient mutants of higher plants and with plants grown under different light intensities do not support the view that PS – 2 is confined to grana thylakoids in the higher plant chloroplast (Boardman *et al.*, 1974). Growth at low light results in greater development of grana, but the quantum efficiency for CO_2 fixation by leaves or for the reduction of 2:6 dichlorophenolindophenol by isolated chloroplasts is the same, irrespective of the light intensity during growth. If PS – 2 were confined to grana, then the plants grown at high light intensity should be less efficient than those grown at low intensity. Some chlorophyll-deficient mutants show poor development of grana, but they exhibit good PS – 2 activity. Studies on chloroplast development also support the conclusion that grana are not essential for PS – 2 activity (section 4.7).

We have already noted that the large particles seen in freeze-etching of mature chloroplasts are confined to the grana thylakoids. The large particles are not seen in developing chloroplasts prior to grana formation. The large particles seem to be related in some way to membrane stacking and to the presence of chlorophyll *b*, and do not correlate with functional photosystems, either PS – 1 or PS – 2.

4.7 ASSEMBLY OF THE THYLAKOID MEMBRANE

4.7.1 PROTOCHLOROPHYLLIDE

The structural events of chloroplast development were covered in chapter 3. In this chapter we will consider the net synthesis of chlorophyll and the formation of photosystems 1 and 2 on illumination of dark-grown seedlings. The small amount of protochlorophyllide (Mg-2-vinyl pheoporphyrin a_5) which is accumulated by the etioplasts of dark-grown seedlings is localized in the prolammelar bodies. Three spectoscopic forms of protochlorophyllide with slightly different absorption maxima (at 628 nm, 637 nm and 650 nm) can be distinguished. Two of these (Pchl-637 and Pchl-650) are converted rapidly to chlorophyllide *a* on illumination, while Pchl-628 is photo-inactive. The molecular basis for the difference in spectroscopic properties of the protochlorophyllides is not established. Different states of aggregation of the pigment or different modes of binding to protein have been suggested. In the photo-reduction of protochlorophyllide to chlorophyllide *a*, two hydrogen atoms are added to the 7,8 positions of the porphyrin ring and this requires that the protochlorophyllide is bound to a protein, termed holochrome (Boardman, 1966). The photoactive protochlorophyllide holochrome can be isolated from bean and barley seedlings. The origin of the hydrogen atoms for the photo-reduction is not known. In the membranes of the prolamellar body, the protochlorophyllide molecules appear to be organized into small units consisting of at least 4 protochlorophyllides, and there is some evidence that the units may contain as many as 20 chromophores.

4.7.2 CHLOROPHYLL ACCUMULATION

Chlorophyll accumulation on the illumination of 6-day-old dark-grown barley seedlings with white light is shown in Fig. 4.6. On turning on the light a rapid conversion of protochlorophyllide to chlorophyllide *a* occurs, followed by a lag of about 30 minutes before additional chlorophyll is formed. During the lag phase, the esterification of chlorophyllide *a* with phytyl alcohol to chlorophyll *a* takes place. Rapid chlorophyll synthesis occurs after about 2.5 hours. The ratio of chlorophyll *a* to chlorophyll *b* is shown by the broken line. Chlorophyll *b* cannot be detected immediately after the photo-conversion of protochlorophyllide, but it is formed during the lag phase and the ratio of chlorophyll *a*/chlorophyll *b* falls rapidly. It takes four hours of illumination, however, before the chlorophyll *a*/chlorophyll *b* ratio approaches that of the mature chloroplast. The enzymic steps in the formation of chlorophyll *b* are not known although radiotracer studies suggest that chlorophyll *b* is formed from chlorophyll *a* (Shlyk, 1971).

4.7.3 DEVELOPMENT OF THE PHOTOSYSTEMS

In the greening barley seedling, photosynthetic oxygen evolution is detected after 30 minutes of illumination. After 2 hours, the rate of oxygen evolution per gm fresh weight of leaf is as high as that in the greened leaf at 45 hours. However, when the photosynthetic rate of O_2 evolution is related to chlorophyll content of the leaf, it is 80-fold greater after 90 minutes than after 45 hours. The photosynthetic units of PS – 1 and PS – 2 are functional at an early stage of greening, but the size of the units are small compared with the photosynthetic units in the mature thylakoid membrane. During the greening process, chlorophyll *a* and chlorophyll *b* are synthesized and incorporated into the light harvesting units of the photosystems.

PS – 1 is active ahead of PS – 2. In isolated plastids from greening barley seedlings, appreciable PS – 1 activity, including cyclic phosphorylation is observed as early as 15 minutes after turning on the light (Henningsen & Boardman, 1973; Plesnicar & Bendall, 1972).

The cytochromes which are localized in photosystem 1 in the mature chloroplast (cytochrome *f*, cytochrome b_6 and cytochrome $b-559_{LP}$) are already present in the etioplasts of dark-grown seedlings but cytochrome $b-559_{HP}$ is

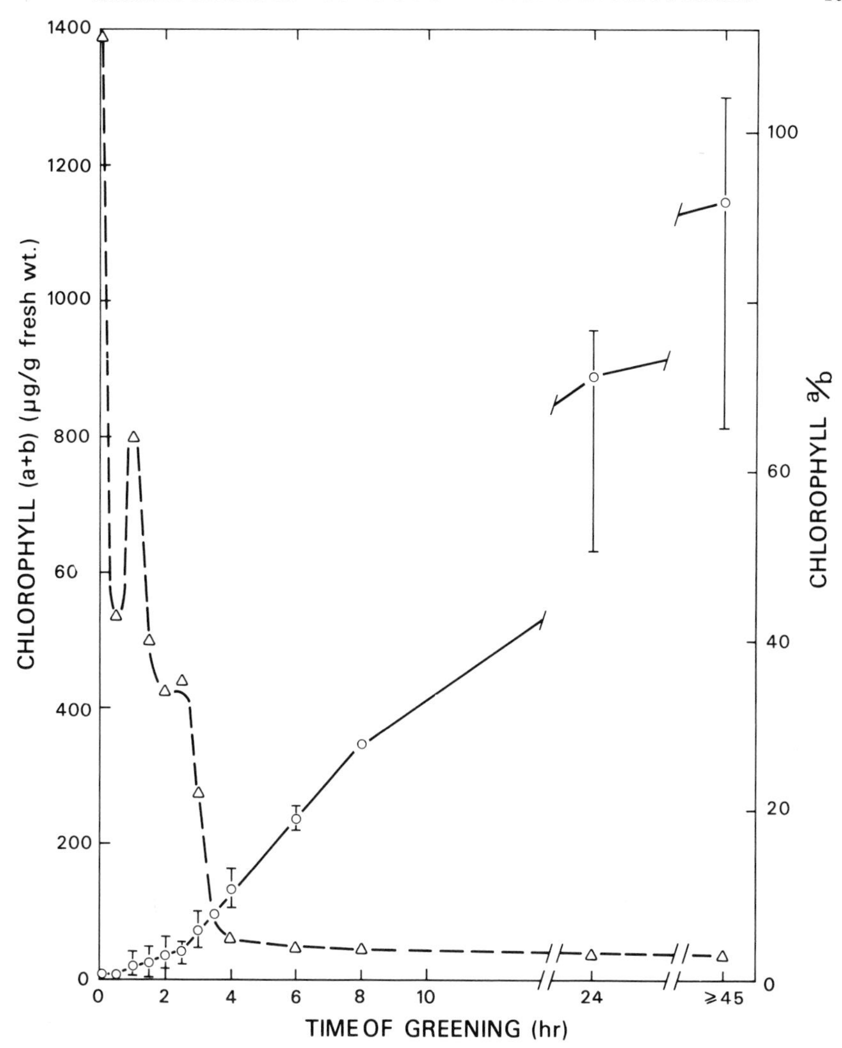

Fig. 4.6. Chlorophyll accumulation during the greening of dark-grown barley seedlings. The solid line indicates total chlorophyll $(a+b)$ and the broken line the ratio of chlorophyll a/chlorophyll b. (From Henningsen & Boardman, 1973.)

formed during greening (Table 4.4). However the synthesis of cytochrome $b-559_{HP}$ does not correlate with the onset of oxygen evolution and this cytochrome does not appear to be essential for photosystem 2 activity (Henningsen & Boardman, 1973). In greening pea seedlings, the photo-oxidation of cytochrome f, and the photo-reduction of cytochrome b_6 which require an active photosystem 1, could be detected after 20 minutes of greening.

There are no significant changes in the amounts of the phospholipids or galactolipids of greening pea leaves during the first 5 hours of greening, but the

Table 4.4. Cytochrome contents of greening barley leaves (from Henningsen & Boardman, 1973).

Time of greening (hr)	Cyt *f*	Cyt $b-559_{LP}$ + Cyt b_6 (male fresh weight)	Cyt $b-559_{HP}$
0	0.73	2.0	0
2	0.71	2.4	0.05
4	0.73	2.0	0.33
6	1.15	2.9	0.38
8	0.81	2.0	0.76
24	1.16	2.7	1.8

galactolipids increase after that time during grana formation. The fatty acid composition is also constant for 5 hours after illumination.

This constancy of the lipid and fatty acid compositions supports the electron microscope evidence that the membranes of the etioplast are the precursors of the first thylakoids. It is apparent that the etioplasts already contain many of the components which are required for functional photosystems (see also chapter 3).

4.7.4 THYLAKOID STRUCTURE DURING GREENING

Earlier studies with isolated plastids from bean and pea seedlings suggested that there was a correlation between the onset of grana formation and an active PS−2, but more recent work shows that grana are not essential for photosynthetic oxygen evolution. Grana formation occurs during the phase of rapid chlorophyll synthesis, which suggets that the degree of grana development may be related to the chlorophyll content of the plastid. This view is supported by work on plants grown at different intensities (section 4.6). If dark-grown seedlings are greened in intermittent illumination (2 minute light periods separated by 15 minute dark periods) grana are not formed, but the plastids exhibit good photochemical oxygen evolution.

4.8 RELATIONSHIP OF CHLOROPLAST TO CYTOPLASM

Chloroplasts are surrounded by an envelope which consists of two membranes, the outer and inner membranes. The inner membrane is a permeability barrier to the diffusion of many metabolites from the chloroplast to the cytoplasm and *vice versa*. For example, the inner membrane is practically impermeable to sucrose, which is a major product of photosynthesis. It is very slowly permeable to ATP and impermeable to pyridine nucleotides and hexosephosphates (Heber, 1974). How then are reduced carbon compounds transported into the cytoplasm and how is photosynthetic ATP made available for many other cell processes, such as protein synthesis?

Heldt and his colleagues (1975) have postulated that specific carriers or translocators are involved in the transfer of certain metabolites across the inner membrane of the chloroplast envelope (Heber, 1974). For example, the so-called phosphate translocator facilitates the transfer of 3-phosphoglycerate (PGA), dihydroxyacetone phosphate (DHAP), glyceraldehyde-3-phosphate (GAP) and inorganic phosphate across the inner membrane in a competitive fashion. Figure 4.7 depicts the transfer of carbon and phosphate from the

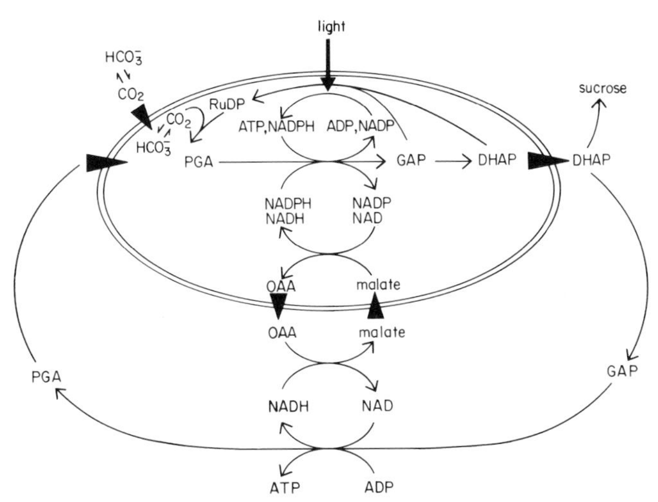

Fig. 4.7. Movement of carbon and phosphate energy from the chloroplasts to the cytoplasm. ATP is formed in the cytoplasm as the result of the transfer of dihydroxyacetone phosphate (DHAP) and oxaloacetate (OAA). Sucrose is formed from DHAP. (From Heber, 1974.)

chloroplast. It seems likely that sucrose is formed outside the chloroplast although this is not established. ATP and NADPH produced inside the chloroplast by photosynthetic electron flow are used in the conversion of PGA to DHAP. DHAP is exported from the chloroplast to the cytoplasm, where it is converted into sucrose or used for the generation of reducing power and ATP. The PGA can then re-enter the chloroplast.

Reducing power is transported across the inner membrane by a counter exchange of malate and oxaloacetate. NAD-dependent malate dehydrogenase occurs both in the cytoplasm and the chloroplast, while NADP-dependent dehydrogenase occurs in the chloroplast and is activated by light. Heldt *et al.* (1975) have postulated that the counter exchange of malate and oxaloacetate is facilitated by a dicarboxylate translocator.

Rapid transfer of metabolites between chloroplast and cytoplasm is crucial for the operation of the C_4-dicarboxylic pathway of photosynthesis. In plants with the C_4-pathway, CO_2 is fixed initially by reaction with phosphoenol-pyruvate in the mesophyll cells. The resulting oxaloacetate is reduced to malate or transaminated to aspartate. Depending on the species of C_4 plant, one or other of these acids is transported to the bundle sheath cells, where the C_4-acid is decarboxylated to give CO_2 and a 3-carbon compound. The CO_2 is then refixed by the Calvin carboxylation cycle (Hatch & Slack, 1970).

FURTHER READING

ARNTZEN C.J. & BRIANTAIS J.M. (1975) Chloroplast structure and function. In *Bioenergetics of Photosynthesis*, (ed. Govindjee). Academic Press, New York, pp. 52–113.

BOARDMAN N.K. (1968). The photochemical systems of photosynthesis *Adv. Enzymology*, **30**, 1–79.

BOARDMAN N.K. (1970) Physical separation of the photosynthetic photochemical systems. *Ann. Rev. Plant Physiol.* **21**, 115–40.

BRANTON D. (1968) Structure of the photosynthetic apparatus. In: *Photophysiology* Vol. III, (ed. A.C. Giese). Academic Press, New York. pp. 197–224.

CALVIN M. & BASSHAM J.A. (1962) *The Photosynthesis of Carbon Compounds*, Benjamin, New York.

HEBER U. (1974). Metabolite exchange between chloroplasts and cytoplasm. *Ann. Rev. Plant Physiol.* **25**, 393–421.

KIRK J.T.O. & TILNEY-BASSET R.A.E. (1967) *The Plastids*, Freeman, London and San Francisco.

MITCHELL P. (1966) Chemiosmotic coupling in oxidative and photosynthetic phosphorylation. *Biol. Rev. Cambridge Phil. Soc.* **41**, 445–502.

PARK R.B. & SANE P.V. (1971) Distribution of function and structure in chloropalst lamellae. *Ann. Rev. Plant Physiol.* **22**, 395–430.

TREBST A. (1974) Energy conservation in photosynthetic electron transport of chloroplasts. *Ann. Rev. Plant Physiol.* **25**, 423–58.

CHAPTER 5

PLANT MITOCHONDRIA

5.1 INTRODUCTION

Mitochondria are the sites of non-photosynthetic energy transduction in eukaryotic cells which carry out aerobic metabolism. Energy transduction includes those processes by which the chemical potential energy of organic substrates is transformed into a readily mobilized form, adenosine-5′-triphosphate (ATP). Organic substrates are oxidized and the free energy of oxidation is conserved by processes which are common to all mitochondria, regardless of source. Thus, with regard to the oxidative and phosphorylative processes, information obtained from studies in animal mitochondria is applicable to plant mitochondria.

Notable differences between plant mitochondria and animal mitochondria do occur, although these differences do not contradict the basic similarities in the mechanism of energy transduction. For example, plant mitochondria possess external reduced nicotinamide adenine dinucleotide (NADH) dehydrogenases which oxidize exogenous NADH; mitochondria from animal sources lack this capability. Mitochondria from many plant sources are relatively insensitive to cyanide inhibition, a feature not found in animal mitochondria. On the other hand, the β-oxidation pathway of fatty acids is located in animal mitochondria, whereas in plants, the enzymes of fatty acid oxidation occur in the glyoxysomes.

In this chapter, the morphology and function of plant mitochondria are discussed. In almost all cases, information is drawn from studies with mitochondria from higher plants. Emphasis is placed on the components of the plant mitochondria respiratory chain and their interactions with each other. Current ideas on oxidative phosphorylation are discussed with reference to knowledge gained from studies with animal and yeast mitochondria. Reversed electron flow and ion transport activities are considered with reference to studies in plant mitochondria. Structure and function relationships are sought, but in many instances, sufficient evidence is not available or available only from studies with mammalian or avian systems; it seems unwarranted, however, to draw exact parallels between animal and plant systems.

5.2 MORPHOLOGY

5.2.1 MORPHOLOGY *IN SITU*

Mitochondria in living cells are highly pleomorphic, as shown by phase contrast microcinematography by Hongladarom *et al.* (1965). Pleomorphism is reflected also in thin section electron microscopy, in which mitochondria appear as roughly circular profiles as well as highly elongated or irregular cross sections (Fig. 5.1a). The circular sections may represent transverse or oblique sections through an otherwise elongated organelle. The diameter of the elongated mitochondrion appears to be about 0·4 to 0·5 μm, while the length may be several micrometers long. Although rods or apparent spheres are the most common profiles seen, sections derived from branched or cup shaped organelles have also been discerned (Bagshaw *et al.*, 1964). The recent analysis of serial sections of yeast cells by Hoffman and Avers (1973), which showed that yeast contains a single, giant, branched mitochondrion, suggests that the irregular cross sections of mitochondria of other cells might also be sections of a single branched organelle.

The mitochondrion consists of a double membrane system with an inner convoluted membrane enclosing the matrix, and surrounded by a smooth outer membrane (Fig. 5.1a, 5.1b). High resolution electron micrographs of material

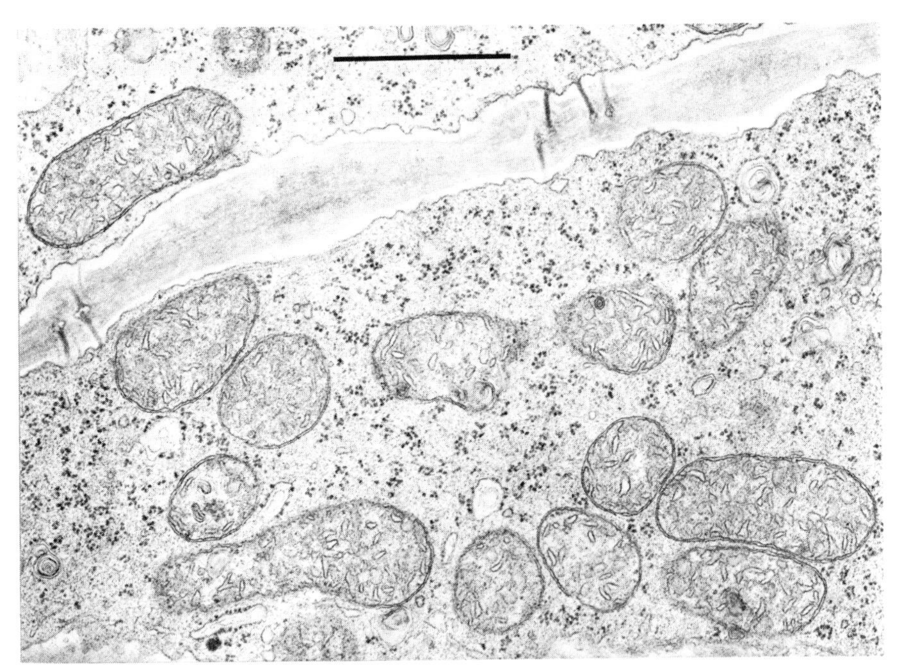

Fig. 5.1a. Mitochondria in phloem parenchyma cells of a maize leaf. Magnification bar = 1 μm. (Micrograph courtesy of O. E. Bradfute and Diane C. Robertson.)

fixed with glutaraldehyde and post-fixed with osmium tetroxide show the tripartite nature of both the inner and outer membranes. Each membrane has a thickness of approximately 9 nm (Baker *et al.*, 1968).

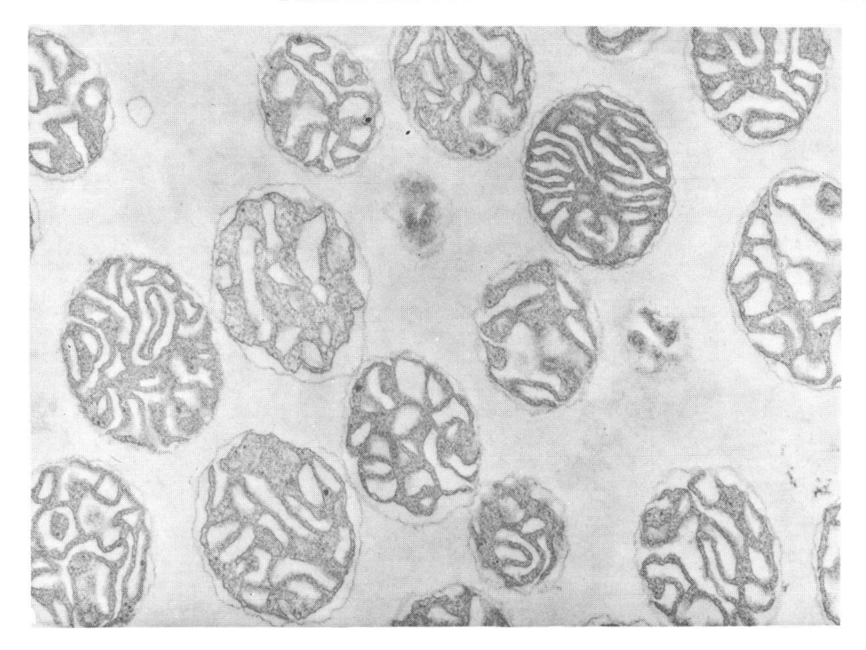

Fig. 5.1b. Isolated mitochondria from mung bean hypocotyls. Mitochdria have been suspended in 0·3 M mannitol prior to fixation. Magnification × 26,000. (Micrograph courtesy of W. D. Bonner, Jr.)

5.2.2 MORPHOLOGY OF ISOLATED MITOCHONDRIA

Electron micrographs of isolated mitochondria show circular cross sections, presumably reflecting a spherical shape when released from their cellular environment. The electron micrographs of the intact isolated mitochondrion show clearly the two membrane systems, as well as the tripartite organization of each membrane. The fine structure of isolated mitochondria is highly dependent upon the osmolarity of the suspending medium (Baker *et al.*, 1968). When mitochondria are suspended in 0·3 to 0·4 M sucrose or mannitol, the matrix appears contracted and electron dense (Fig. 5.1b), but when suspended in 0·2 M sucrose, the matrix appears more expanded and less electron dense, and resembles that of mitochondria seen *in situ*. The dense matrix of mitochondria suspended in 0·3 to 0·4 M sucrose or mannitol is due to the hypertonicity of the suspending medium. Since the inner membrane is generally regarded as the osmotic barrier, the dense nature of the matrix reflects a water loss, which is reversible when the organelles are suspended in 0·2 M sucrose.

Negatively-stained water-lysed mitochondria show that the inner membranes have the characteristic stalked particles similar to those reported for mammalian mitochondrial membranes (Fernandez-Moran, 1962; Parsons *et al.*, 1965). The particles have a headpiece with a diameter of 10 nm, attached to a stalk 3·5 to 4·5 nm wide and 4·5 nm long (Fig. 5.2). These resemble the particles identified with ATPase function in heart mitochondria (Racker *et al.*, 1969).

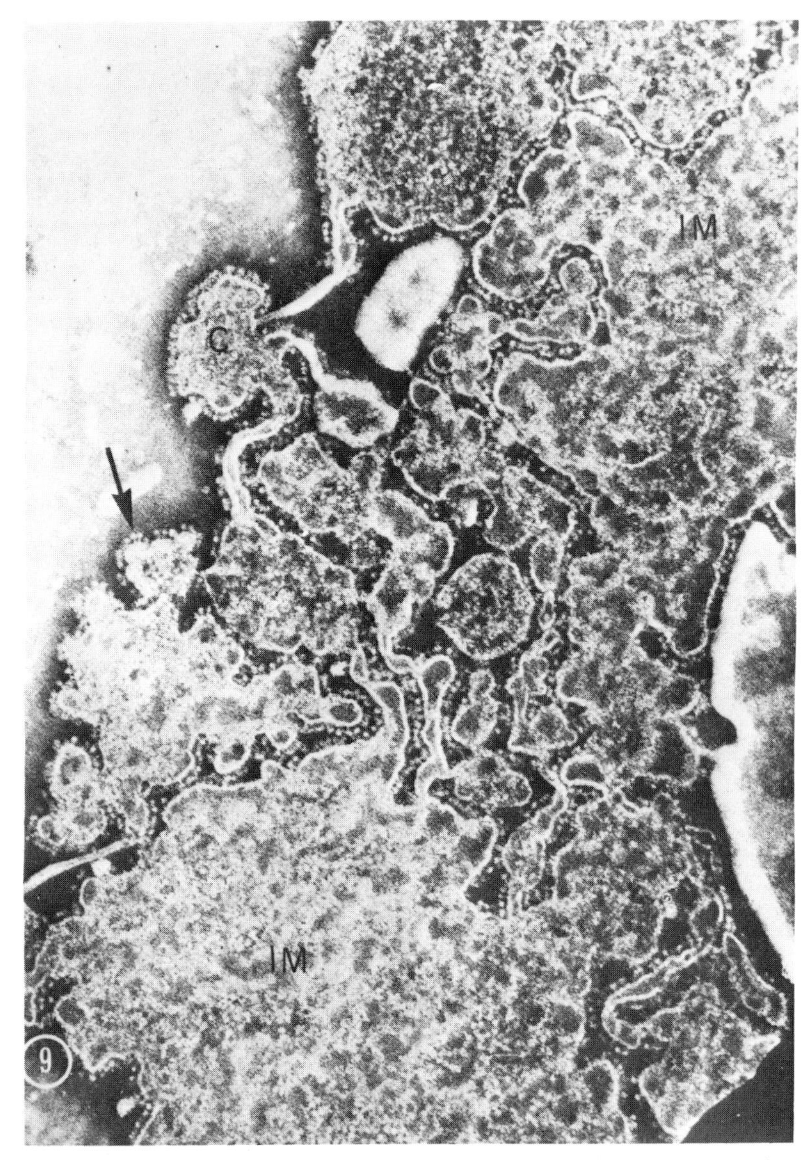

Fig. 5.2. Portion of a surface spread and negatively stained summer squash mitochondrion. The large areas of membrane (IM) are presumed to be part of the inner membrane forming the shell of the mitochondrion. The cristae (C) appear as smaller pieces of membrane of rounded shape connected together by narrower (possibly tubular) pieces. The membranes are coated with projecting knob-like subunits which are best seen lying in the plane of the object at the edge of the pieces of membrane (arrow). The dimension of the head of the subunit is 10 nm and the stem is 3·5–4·0 nm wide and 4·5 nm long. (Parsons *et al.*, 1965). (Reproduced by permission of the National Research Council of Canada from the Canadian Journal of Botany, Volume 43, 1965. pp. 647–55.)

5.3 ISOLATION AND PURIFICATION

5.3.1 TECHNIQUES OF ISOLATION AND PURIFICATION

Mitochondria have been obtained from a large number of plant sources including roots, storage tissue, stems and photosynthetic tissues. The usual problems of isolation, regardless of the source, are (a) the rupture of a rather rigid cell wall and (b) the prevention of damage to organelles through the release of intracellular, particularly vacuolar, contents. Ikuma (1970) listed a number of conditions for successful isolation of tightly-coupled mitochondria. These include (a) gentle tissue disruption, (b) rigorous exclusion of contaminating particles and (c) the use of a buffered grinding medium isotonic with mitochondria and containing a variety of protective reagents. Most investigators employ some device to reduce quickly the tissue to a coarse slurry, which is passed through a cloth filter to remove large debris. The fraction which sediments between 1,000 *g* and 10,000 *g* is collected as the mitochondrial fraction. This fraction will oxidize all the intermediates of the Krebs tricarboxylic acid cycle, exhibit respiratory control and yield ADP to O ratios approaching the theoretical value for the substrate used. The mitochondrial fraction can be further purified by density gradient centrifugation. This may be done in discontinuous sucrose gradients (Baker *et al.*, 1968; Douce *et al.*, 1972a) or Dextran-40 gradients (Solomos *et al.*, 1973). Mitochondria form a band at the interface between 1·2 and 1·5 M sucrose (Douce *et al.*, 1972a). This is recovered and diluted slowly to 0·3 M sucrose. This procedure yields mitochondria with intact outer and inner membranes as shown by electron microscopy. The integrity of the outer membrane is also shown enzymatically by the inability to reduce exogenous cytochrome *c* with NADH or succinate as substrates, unless the mitochondria have been subjected to mild osmotic shock which renders the outer membrane permeable to high molecular weight solutes.

During the disruption of cells and throughout the isolation procedures, a number of protective reagents must be present. Inclusion of sodium ethylenediamine-tetracetate (EDTA) in the isolation medium has been shown to give mitochondria with high respiration rates (Lieberman & Biale, 1955). EDTA probably removes cationic inhibitors although the specific cation complexed is unknown.

Many plant cells release phenolic compounds when ruptured. These phenolic compounds are oxidized in the presence of air and form polymers which are inhibitory to mitochondrial respiration and coupled phosphorylation. To prevent the damaging effects of phenolic compounds, a variety of reagents have been used successfully. Polyvinyl pyrrolidone is a competitive inhibitor of purified phenolase (Walker & Hulme, 1965), and has been used extensively as a protecting agent for mitochondrial isolation (Jones *et al.*, 1965; Hulme *et al.*, 1964; Wiskich, 1966). Other reagents include morpholinopropane sulphonate, cysteine, and sodium metabisulphite (Stokes *et al.*, 1968). Morpholinopropane sulphonate is thought to form complexes with phenolic compounds, while sulphydryl reagents inhibit phenoloxidases.

Free fatty acids are known to uncouple oxidative phosphorylation from electron transport (Borst *et al.*, 1962; Baddeley & Hanson, 1967). The uncoupling activity of fatty acids is reversed by the addition of bovine serum albumin. Bovine serum albumin also reverses the uncoupling activity of many other uncoupling agents, such as nitro- and halo-substituted phenols, dicumarol, and carbonyl-cyanide *m*-chloro-phenylhydrazone in rat liver mitochondria (Weinbach & Garbus, 1966). Dalgarno and Birt (1963) showed that free fatty acids were present in mitochondrial preparations from carrot root tissue. These included oleic, stearic, palmitic and some short chain fatty acids, as well as polyunsaturated C_{18} acids. Mitochondria isolated from such tissues were uncoupled as shown by a P/O ratio less than 0·1. When bovine serum albumin was included in the isolation medium, mitochondria became well coupled, with a P/O ratio greater than 1·6 with succinate as substrate. As a matter of routine, most isolation procedures include 0·1% (w/v) of bovine serum albumin (Cohn Fraction V, low in free fatty acids). Bovine serum albumin binds fatty acids, and other lipophilic uncoupling agents, but the nature of the binding is not clear.

5.3.2 ISOLATION FROM GREEN TISSUES

Mitochondria isolated from photosynthetic tissues are rarely free from chloroplasts or chloroplast fragments. Rocha and Ting (1970) subjected spinach leaf material to linear sucrose gradients (40 to 80% w/v) and obtained fractions after equilibrium. They found, nonetheless, that the mitochondrial fraction was contaminated with 13% intact chloroplasts and 6% broken chloroplasts. The degree of contamination was estimated from the activities of characteristic marker enzymes. Malate dehydrogenase and cytochrome *c* oxidase served as mitochondrial markers, while chlorophyll content and triose-phosphate dehydrogenase were chloroplast markers.

5.4 MITOCHONDRIAL MEMBRANES

5.4.1 STRUCTURE OF MEMBRANES

Electron micrographs of plant mitochondria show clearly the tripartite nature of both the outer and inner membranes. This may be interpreted as a lipid bilayer with the hydrophobic fatty acid chains oriented toward the interior of the bilayer (see chapter 2). The lipids of mitochondrial membranes are largely phospholipids. It is the current view that phospholipid bilayers are highly dynamic, with a high degree of fluidity in the fatty acid region, as well as high lateral mobility of the phospholipids in the plane of the membrane.

The membrane proteins may form loose interactions with the lipid bilayer, or may be very tightly associated with the membrane. The loosely associated proteins most likely have exposed hydrophilic side chains and are easily extracted from membranes. The proteins tightly associated with membranes presumably have exposed hydrophobic side chains and are pictured as partially or wholly embedded in the lipid bilayer. These proteins are extracted from membranes with difficulty. Indeed, they may require a lipid environment for optimal activity.

5.4.2 MEMBRANE LIPIDS

A survey of the lipid composition of mitochondrial membranes reveals great differences depending upon the source of mitochondria. McCarty *et al.* (1973) investigated the phospholipid composition of the inner and outer membranes of mung bean (*Phaseolus aureus*) and potato tuber (*Solanum tuberosum*) mitochondria which were prepared in such a way as to exclude contaminating particles. Phospholipids comprised 90% or more of the mitochondrial membrane lipids. The main phospholipids were phosphatidyl choline, phosphatidyl ethanolamine and phosphatidyl glycerol, except in the outer membrane which did not contain significant amounts of phosphatidyl glycerol. The phospholipid composition is shown in Table 5.1 together with that of beef heart mitochondria

Table 5.1. Phospholipid composition of mitochondrial membranes (McCarty *et al.*, 1973.)

	% total lipids			
Phospholipid	mung bean inner membrane	potato inner membrane	potato outer membrane	beef heart mitochondria
Phosphatidyl choline	33	33	36	38
Phosphatidyl ethanolamine	32	33	64	37
Phosphatidyl glycerol	23	19	none detected	16

for comparison. Minor amounts of lysophosphatidyl ethanolamine, phosphatidyl inositol, and phosphatidyl glycerol were also found. The fatty acids of the three principal phospholipids of mung bean mitochondria were palmitic, linoleic and linolenic acids. Stearic acid occurred in conjunction with phosphatidyl glycerol. Schwertner and Biale (1973), by contrast, found that phospholipids comprised only about 50% of the total lipids of mitochondria from avocado (*Persea* sp.), cauliflower (*Brassica oleracea*) and potato tubers, and found more phosphatidyl inositol than phosphatidyl glycerol. The principal fatty acids of the phospholipids of avocado, cauliflower and potato mitochondrial membranes were palmitic, linoleic and linolenic. The fatty acids of the neutral lipids of cauliflower mitochondria were C_{16} and shorter chain acids, while those of potato mitochondria included palmitic, oleic, linoleic and linolenic. The fatty acid composition of the membrane lipids can be highly variable and may reflect the growth conditions of the tissue. Mitochondria from cold grown mung beans (15°C) have a higher amount of unsaturated fatty acids than mitochondria isolated from mung beans grown at 25°C (see also chapter 2).

Sterols comprise about 2·6% of the total lipids of maize shoot mitochondria (Kemp & Mercer, 1968). These may be esterified, of which cholesterol and β-sitosterol are the principal compounds. Of the unesterified sterols, stigmasterol and β-sitosterol are the principal ones. The fatty acids esterified to the sterols include lauric, myristic, palmitic, and linolenic as the most abundant.

5·5 ENZYMES

5·5·1 ENZYMES OF THE TRICARBOXYLIC ACID CYCLE

Mitochondria contain all the enzymes of the tricarboxylic acid cycle. Isolated mitochondria oxidize all of the acids of the cycle, and chromatographic analysis shows that the products are those expected for the reactions of the tricarboxylic acid cycle (Lieberman & Biale, 1956; Avron & Biale, 1957; Bogin & Erickson, 1965).

5·5·1·1 *Citrate Synthetase*

Citrate synthetase (Citrate oxalacetate-lyase (CoA-acetylating)) activity is associated with a particulate fraction of leaf and root tissue from tobacco, bean, and soybean, which sediments at 10,000 *g*. This fraction consisted largely of mitochondria (Hiatt, 1962). Citrate was formed in the presence of acetyl CoA and oxalacetate. Citrate synthetase has been isolated from wheat scutellum mitochondria (Barbareschi *et al.*, 1974) with minimal contamination by glyoxysomes. The mitochondrial citrate synthetase was released by sonic disruption of the mitochondria and recovered in the supernatent fluid. The purified enzyme had a molecular weight of 96,000 daltons as determined by its elution volume in Sephadex G-100 gel filtration. The K_m for acetyl CoA and for oxalacetate were 4 μM and 34 μM respectively. The enzyme was inhibited competitively with respect to acetyl CoA by ATP. In these respects, the mitochondrial citrate synthetase from wheat scutellum mitochondria is similar to that from mammalian sources.

5·5·1·2 *Pyruvate Oxidase*

Pyruvate is oxidized by mitochondria with a requirement for catalytic amounts of one of the TCA cycle acids (Millerd, 1953; Walker & Beevers, 1956). TCA cycle acids were added to castor bean (*Ricinus communis*) mitochondria at a concentration of 0·001 M. Oxygen consumption ceased after 60 minutes. When pyruvate at a concentration of 0·01 M was then added, oxygen consumption continued at rapid rates and linearly up to three hours. Other cofactors required were NAD, coenzyme A, ATP, and thiamin pyrophosphate.

5.5.1.3 *Isocitrate Dehydrogenase*

A NAD-specific isocitrate dehydrogenase [L-iso-citrate:NAD oxidoreductase (decarboxylating)] has been purified from pea shoot (*Pisum sativum* var Alaska) mitochondria (Cox & Davies, 1967). The enzyme was released from pea shoot mitochondria ruptured by extrusion through a French pressure cell at 3,000 lbs in^{-2}. The enzyme, whose K_m for NAD was 0·22 μM, was activated by Mn^{2+} and Mg^{2+} and to a lesser extent by Zn^{2+}, and inhibited by NADH, with K_i =0·19 mM. The mitochondrial isocitrate dehydrogenase was specific for NAD which differentiates it from the cytosolic isocitrate dehydrogenase, which requires NADP as the electron acceptor.

5.5.1.4 *Malate Dehydrogenase*

Malate dehydrogenase (L-malate:NAD oxidoreductase) exists in both mitochondrial and cytosolic forms. Moreover, the mitochondrial malate dehydrogenase may occur as several isozymes. Ting *et al.* (1966) separated two isozymes from young maize (*Zea mays*) mitochondria after sonic disruption. Starch gel electrophoresis revealed a faint fast moving (toward anode) band and a major slow moving band. Grimwood and McDaniel (1970) also found a major slow moving band in polyacrylamide gel electrophoresis with several lighter fast moving bands. Boulter and Laycock (1966) attributed the minor bands to complexes of the mitochondrial malate dehydrogenase with other proteins, since re-electrophoresis of the eluted bands always gave a band in the main mitochondrial fraction as well as the original minor band. They determined the molecular weight of the main malate dehydrogenase to be 74,000 daltons.

Plant mitochondria oxidize malate readily, but glutamate must be included in the reaction mixture to remove the accumulated oxalacetate, due to the unfavourable equilibrium of the reaction. The oxidation of malate with endogenous NAD^+ is inhibited by rotenone and antimycin (Day & Wiskich, 1974), but is insensitive to these inhibitors when exogenous NAD^+ is supplied.

5.5.1.5 *Malic Enzyme*

Malate may be oxidized by a NAD^+-dependent malic enzyme [L-malate:NAD oxidoreductase (decarboxylating)] with the formation of pyruvate. Macrae and Moorhouse (1970) showed that pyruvate accumulated during malate oxidation by cauliflower bud mitochondria, unless thiamin pyrophosphate was included in the reaction medium. Under the latter conditions, malate was probably oxidized by both malic enzyme and malate dehydrogenase so that in the presence of cofactors for pyruvate oxidation and citrate formation, the latter accumulates. Malic enzyme is the main pathway for malate oxidation by wheat shoot mitochondria, since oxalacetate did not inhibit malate oxidation, although it did inhibit transiently the oxidation of citrate and pyruvate (Brunton & Palmer, 1973). The activities of malic enzyme and malate dehydrogenase differ in mitochondria from various sources (Macrae, 1971b). While the relative activity of malate dehydrogenase has in all cases been greater than the activity of malic enzyme, the accumulation of pyruvate *vs* oxalacetate may vary considerably. Pyruvate was accumulated in preference to oxalacetate by a ratio of 27·0 in cauliflower bud mitochondria, while the pyruvate/oxalacetate ratio for wheat shoot mitochondria was 0·13. When malic enzyme activity is high, mitochondrial NADH levels are raised, and thereby reduce oxalacetate accumulation by product inhibition of malate dehydrogenase. A strong pH dependence of the activities of the two enzymes was also observed (Macrae, 1971a). At pH 6·0 to 7·0, pyruvate accumulates, while at pH values between 7·0 and 8·0 pyruvate accumulation drops and oxalacetate accumulation rises. The pathway may reflect the pH profiles of malic enzyme and malate dehydrogenase. Below pH 7·0, the activity of malic enzyme would maintain a high internal concentration of NADH which would favour the conversion of oxalacetate to malate; above pH 7·0, the decreased activity of malic enzyme and the consequent drop in the NADH levels would favour the oxidation of malate by malate dehydrogenase.

5.5.2 ENZYMES OF FATTY ACID OXIDATION

Fatty acids were oxidized by a particulate preparation from peanut cotyledons (Stumpf & Barber, 1956). This fraction was identified as the mitochondrial fraction. Cooper and Beevers (1969a,b) have separated the particulate fraction from castor bean and have shown that the enzymes of the β-oxidation pathway as well as the enzymes of the glyoxylate pathway are associated instead with a heavy particle, the glyoxysome. Mitochondria from castor beans contained less than 5% of the glyoxylate cycle enzymes and virtually none of the β-oxidation enzymes.

5.5.3 ENZYMES OF FATTY ACID BIOSYNTHESIS

Isolated mitochondria from avocado mesocarp, flowerlets of cauliflower, and from white potato tubers were capable of incorporating radioactive acetate, acetyl-CoA, malonate, or malonyl-CoA into long-chain fatty acids (Yang & Stumpf, 1965; Mazliak *et al.*, 1972). Cofactor requirements included coenzyme A, NADPH, ATP, and Mg^{2+} or Mn^{2+}. The principal acids formed were palmitic and stearic acids by avocado mesocarp mitochondria, while mitochondria from cauliflower and white potato tubers synthesized some mono-unsaturated fatty acids as well, i.e., palmitoleic (9-hexadecanoic acid) and, with longer incubation times, oleic (9-octadecanoic) and cis-vaccenic (11-octadecanoic) acids. With the appearance of the C_{18} mono-unsaturated acids, stearic acid is found only in trace amounts, indicating that the unsaturated C_{18} acids were formed from stearic acid.

5.5.4 ENZYMES OF PHOSPHOLIPID BIOSYNTHESIS

The synthesis of phospholipids proceeds *via* the following reaction:

$$\text{Glycerol-3-phosphate} \xrightarrow[+\text{fatty acids}]{+\text{CoASH},\ +\text{ATP}} \text{Phosphatidic acid} \qquad (5.1)$$

or

$$\text{Diglyceride} + \text{ATP} \longrightarrow \text{Phosphatidic acid} \qquad (5.1a)$$

$$\text{Phosphatidic acid} + \text{CTP} \longrightarrow \text{CDP-diglyceride} + PP_i \qquad (5.2)$$

Mitochondria isolated from flowerlets of cauliflower contain all of the enzymes necessary for the formation of CDP-diglyceride from glycerol-3-phosphate, when coenzyme A, ATP, CTP and fatty acids are provided (Douce, 1971). Radioactivity from ^{32}P-ATP was found in phosphatidic acid in peanut cotyledon mitochondria (Bradbeer & Stumpf, 1960). The incorporation of ^{32}P into phosphatidic acid was stimulated by the presence of small amounts of α,β-diglyceride, indicating the presence of a mitochondrial diglyceride phosphokinase. ^{14}C-CTP was incorporated into CDP-diglyceride by cauliflower mitochondria (Sumida & Mudd, 1968). The radioactivity of CDP-diglyceride declines in the presence of α-glycerol phosphate or inositol, with the expected formation of phosphatidyl glycerol phosphate or phosphatidyl inositol. Using preparations carefully purified in sucrose density gradients from mung bean hypocotyl mitochondria, Douce *et al.* (1972b) showed that the CTP:phosphatidic acid cytidyl transferase activity was associated with the inner membrane fraction. Since the activity was not released upon sonication, it was assumed that the transferase was a membrane bound enzyme.

5.6 MITOCHONDRIAL ELECTRON TRANSPORT

5.6.1 COMPONENTS OF THE RESPIRATORY CHAIN

Reducing equivalents derived from the oxidation of the TCA cycle acids are oxidized in a stepwise manner in the mitochondrial electron transport chain. The electron transport chain, or respiratory chain, is a series of functionally linked electron carriers which undergo alternate reduction and oxidation, with molecular oxygen as the terminal electron acceptor. Electrons are donated by carriers of low redox potential to carriers of high redox potential. It is *via* these oxidation-reduction reactions that the main oxidative cellular energy transduction occurs, either through the formation of intermediate states which can be coupled to cellular work, (e.g., ion transport) or through the phosphorylation of ADP to form ATP, which can then mediate cellular endergonic reactions. The main components of the respiratory chain have been identified, both by characteristic reaction toward inhibitors and by spectral analysis. In its basic form, the respiratory chain in plant mitochondria is very similar to that of mitochondria from fungal or animal sources. The chain, (shown in Fig. 5.3) can be functionally separated into (a) NADH:coenzyme Q oxidoreductase; (b) succinate: coenzyme Q oxidoreductase; (c) reduced coenzyme Q:cytochrome *c* oxidoreductase; and (d) cytochrome oxidase. With refinements in detection systems, additional components have been identified. The concentration and molar ratios of the principal components have been determined by Lance and

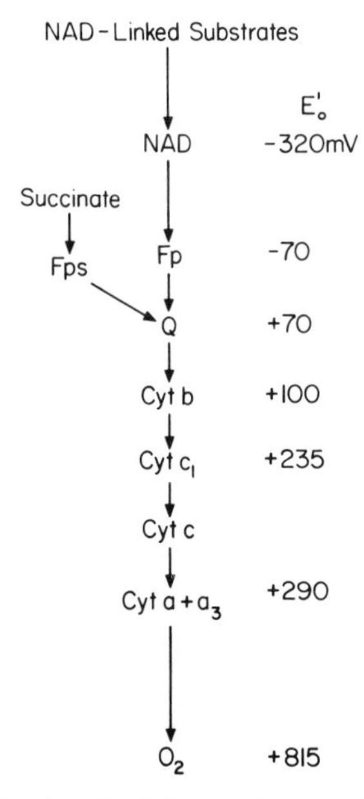

Fig. 5.3. The generalized mitochondrial electron transport chain. The components are NAD, nicotinamide adenine dinucleotide; Fp and Fp_s, the flavoproteins associated with NADH dehydrogenase and with succinate dehydrogenase respectively; Q, Coenzyme Q or the quinone containing carrier; cyt *b*, the complex of *b*-cytochromes; cyt c_1, the tightly bound *c*-cytochrome (cyt *c*-549 in plant mitochondria); cyt *c*, the salt extractable *c*-cytochrome; and cyt $a + a_3$, the *a*-cytochromes of cytochrome oxidase. Not shown are non-heme iron proteins which have been tentatively identified in plant mitochondria, and bound copper of cytochrome oxidase, which has been identified in cytochrome oxidase from animal sources only.

Bonner (1968) for mitochondria isolated from a number of sources and are given in Table 5.2. The concentrations of the cytochromes are quite similar to those of animal mitochondria.

Table 5.2. Concentration and stoichiometry of respiratory chain components in plant mitochondria (Lance & Bonner, 1968.)

Source	Conc. of carriers				
	cyt aa_3	cyt *b*	cyt *c*	Fp	PN
	nM per mg protein				
Helianthus tuberosus	0·10	0·10	0·15	0·38	0·92
Phaseolus aureus	0·11	0·12	0·17	0·58	4·10
Solanum tuberosus	0·17	0·20	0·27	0·89	3·50
Brassica oleracea	0·12	0·15	0·21	0·69	2·04
Symplocarpus foetidus	0·11	0·23	0·36	2·02	—

	Stoichiometry based on cyt *c* as unity				
Helianthus tuberosum	0·69	0·67	1·00	2·90	7·5
Phaseolus aureus	0·65	0·67	1·00	3·64	17·8
Solanum tuberosum	0·65	0·76	1·00	3·30	14·1
Brassica oleracea	0.60	0·75	1·00	3·40	11·5
Symplocarpus foetidus	0·32	0·65	1·00	5·80	—

5.6.1.1 *Nicotinamide Adenine Dinucleotide*

Malate dehydrogenase and isocitrate dehydrogenase are NAD^+ linked enzymes. The oxidation of substrates is accompanied by the reduction of endogenous NAD. This is shown by the strong fluorescence of reduced NAD. Carefully isolated mitochondria contain sufficient endogenous NAD to oxidize malate or isocitrate. The oxidation of malate or isocitrate by endogenous NAD is inhibited by rotenone or amytal. Malate oxidation is stimulated by the addition of NAD, but the oxidation then becomes insensitive to rotenone inhibition (Wiskich *et al.*, 1960; Day & Wiskich, 1974), just as the oxidation of exogenous NADH is insensitive to rotenone or amytal (Wilson & Hanson, 1969; Day & Wiskich, 1974) suggesting more than one pathway of NADH oxidation. Douce *et al.* (1973) and Day and Wiskich (1974) delineated three mitochondrial NADH dehydrogenases, one located on the outer membrane, a second located on the outer surface of the inner membrane, and a third on the inner surface of the inner membrane. Each dehydrogenase has a characteristic response to inhibitors. The outer membrane dehydrogenase is characterized by an antimycin insensitive NADH: cytochrome *c* reductase with added cytochrome *c*. In intact mitochondria, the NADH dehydrogenase of the outer surface of the inner membrane is coupled to cytochrome oxidase and goes through the antimycin sensitive site, but by-passes the rotenone sensitive site. This dehydrogenase shows NADH: cytochrome *c* reductase activity (antimycin sensitive) at low osmolarity only, due to the impermeability of the intact outer membrane to added cytochrome *c*. These first two dehydrogenases oxidize exogenous NADH. The NADH dehydrogenase of the inner surface of the inner membrane oxidizes the NADH linked to malate and isocitrate dehydrogenases. Electrons must go through the rotenone and the antimycin sensitive sites to cytochrome oxidase. In the intact mitochondrion, these various pathways interact, as demonstrated by the relief of antimycin or rotenone inhibition of malate: cytochrome *c* reductase activity by added NAD^+. Further evidence for the delineation of the NADH dehydrogenases is obtained through the P/O or ADP/O ratios of tightly-coupled mitochondria. Oxidation of malate or isocitrate gives ratios approaching 3·0, while the oxidation of NADH gives ratios approaching 2.0. The by-pass of the rotenone sensitive portion of the respiratory chain results in by-passing one of the phosphorylation sites as well (Wilson & Hanson, 1969).

5.6.1.2 *Flavoproteins of NADH Dehydrogenase*

Storey (1970c, 1971a) distinguished a flavoprotein, F_{PM}, which was rapidly reduced upon addition of malate. The reduction of this flavoprotein was inhibited by amytal. He tentatively assigned a midpoint potential, $E_{m\,7\cdot2} = -70$mV for this flavoprotein. F_{PM} is the flavoprotein involved in the first energy conservation site and hence is most likely the flavoprotein involved in the NADH dehydrogenase located in the inner surface of the inner membrane.

F_{Pha}, a high potential non-fluorescent flavoprotein is rapidly reduced upon the addition of exogenous NADH in mung bean mitochondria (Storey, 1970d). This reduction is insensitive to amytal (Storey, 1970c). Its midpoint potential is approximately $+110$mV (Storey, 1971a). Flavoprotein F_{Pha} could be the flavoprotein associated with the dehydrogenases which oxidize exogenous NADH, but it is not possible to determine if it is the flavoprotein of the outer membrane dehydrogenase or the inner membrane dehydrogenase from the information available.

5.6.1.3 *Flavoprotein of Succinate Dehydrogenase*

Isolated mitochondria from a number of plant sources oxidize succinate readily, with oxygen consumption rates of about 450 nM O_2 min^{-1} mg^{-1} protein (Douce *et al.*, 1972a). Activation by ATP is required to obtain maximal rates of respiration with succinate as substrate, as well as for rapid response to addition of ADP (Drury *et al.*, 1968). The activation by ATP is often attributed to the removal of inhibitory amounts of bound oxalacetate (Wiskich & Bonner, 1963) but the mechanism of the activation is not clear. The ATP effect is not due to phosphorylation mechanisms since neither oligomycin nor dinitrophenol affect the ATP activation (Singer *et al.*, 1973).

Hiatt (1961) reported the partial purification of succinate dehydrogenase from mitochondria from bean roots and tobacco (*Nicotiana tabacum*) leaves. The K_m of the enzyme for succinate was 1 mM. Malonate inhibits competitively, with $K_i = 0{\cdot}24$ mM. The apparent Michaelis constant of isolated mitochondria for succinate, in the presence of ADP and phosphate was 0·4 mM (Ikuma & Bonner, 1967). Singer *et al.* (1973) studied the succinate dehydrogenase of submitochondrial particles prepared by sonication of isolated mung bean and cauliflower mitochondria. The enzyme in submitochondrial particles was activated by a number of agents including substrate or Br^-. With Br^- activation, oxalacetate was removed, although it cannot be assumed that the oxalacetate was uniquely associated with the succinate dehydrogenase of the particles. Other activators included CoQ_{10}, NADH, NAD-linked substrate (i.e., malate plus pyruvate), and ADP (Oestreicher *et al.*, 1973). Succinate reducible flavoprotein was not detected spectrally (Storey, 1970a), but a flavoprotein associated with succinate dehydrogenase was determined chemically. Singer *et al.* (1973) found that succinate dehydrogenase contained covalently bound flavin as a histidyl-α-FAD. The flavin content was approximately 0·2 nM per mg protein. The molar ratio of flavin to enzyme was not determined.

5.4.1.4 *Ubiquinone*

Studies on the role of ubiquinone in the plant mitochondrial electron transport chain have not been extensive. Beyer *et al.* (1968) extracted ubiquinone from mung bean submitochondrial particles. A single ubiquinone was found which co-chromatographed with ubiquinone-10. The spectrum of the extracted ubiquinone-10 has an absorption peak at 275 nm in the oxidized form and at 290 in the reduced form. The ubiquinone was reduced by succinate and NADH; at an aerobic steady state, 38% of the ubiquinone was reduced by succinate, while 56% was reduced by NADH. At anaerobiosis induced by succinate or NADH, 88% and 84% respectively were reduced. Sodium hydrosulphite (dithionite) gave additional reduction (i.e., about 93% reduction). The quinones were virtually 100% oxidized in aerobic suspension in the absence of substrates. Ubiquinone is generally acknowledged as part of the mitochondrial respiratory chain and is placed at the juncture of succinate dehydrogenase and NADH dehydrogenase. This placement is based on the considerable work with animal mitochondria. Storey and Bahr (1972), basing their conclusions upon measurements of the half times of reduced to oxidized transitions, and the times for 50% reduced to the fully reduced state, suggested that ubiquinone is in the main respiratory chain of mung bean mitochondria. Ubiquinone is the link between the dehydrogenases and the cytochromes, as usually regarded in animal mitochondria, but Storey and Bahr (1972) in addition placed F_{Pha}, the high potential flavoprotein, between ubiquinone and the b-cytochromes. F_{Pha} has no counterpart in the animal mitochondrial respiratory chain.

5.6.1.5 *Cytochrome b*

The cytochromes are hemo-proteins. Three classes of cytochromes, distinguished by their spectral properties, as well as by the nature of their prosthetic groups are found in mitochondria (see Fig. 5.4). The *a*-cytochrome contains as its prosthetic group, heme *a*, while the *b*- and *c*-cytochromes contain a heme closely related to protoporphyrin IX. In the *c*-cytochromes, the heme is covalently linked to the protein *via* sulphur atoms in a thio-ether linkage. In the reduced state, the cytochromes exhibit strong absorption bands in the visible region of the spectrum which have been useful in their identification and in the analysis of their function. In addition, both the oxidized and reduced forms absorb strongly in the region around 400 nm, which is a characteristic of all heme compounds.

The *b*-cytochromes are best resolved when their spectra are determined at low temperatures, e.g., 77°K. Three *b*-cytochromes have been identified and two others are suggested. Their spectral properties are summarized in Table 5.3.

Table 5.3. The *b*-cytochromes of plant mitochondria.

Cytochrome	α-peak, 25°C nm	α-peak, 77°K nm	E_m mV	Reference
$b-556$	556	553-554	+75 to +100	(a)
		553	+75	(b)
$b-560$	560	557	+40 to +80	(a)
		557	+42	(b)
$b-558$	557-558	553-555	−70 to −105	(a)
$b-566$	566	561-563	−75	(a)
		562	−77	(b)
$S_2O_4^{2-}$ reducible	557-561	554, 560	−100	(a)

References: (a) Lambowitz & Bonner, 1974; (b) Dutton & Storey, 1971.

Considerable variability exists in the nomenclature of the *b*-cytochromes. In conformity to the International Union of Biochemistry, the *b*-cytochromes are designated according to the α-peak of their reduced spectrum at room temperature (25°C). It should be noted that there is a blue shift of about 3 nm in the spectrum at 77°K relative to the spectrum at room temperature. The use of the α-absorption peak is further complicated by the fact that some authors use the absorption maximum at 77°K to designate the various *b*-cytochromes. In older nomenclature, mammalian cytochrome $b-562$, as orginally described by Keilin, was designated cytochrome *b*. As other *b*-cytochromes were discovered with an α-absorption peak significantly different from 562 nm, these were designated with subscripts. More recently, cytochrome $b-566$ was thought to be directly involved in energy transduction and was designated cytochrome b_T, a transducing *b*-cytochrome, to differentiate it from cytochrome $b-562$, or b_K.

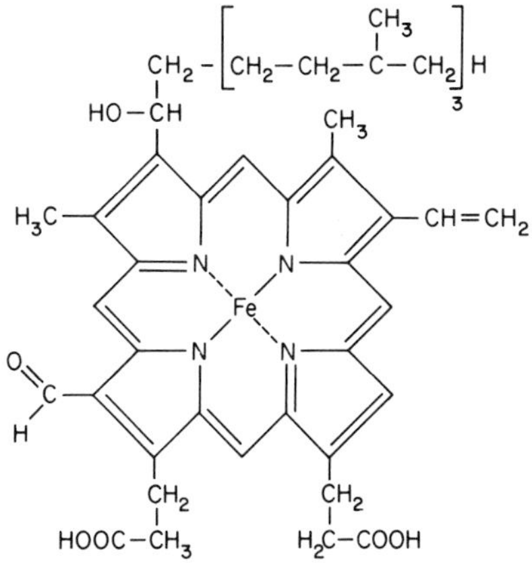

CH_3
CH_2 - [CH_2—CH_2—C—CH_2]$_3$ H
HO—CH
CH_3
H_3C—
—CH=CH_2
N
N
Fe
N
N
O=C—
H
CH_2
HOOC—CH_3
CH_2
H_2C—COOH

Heme a (a-Cytochromes)

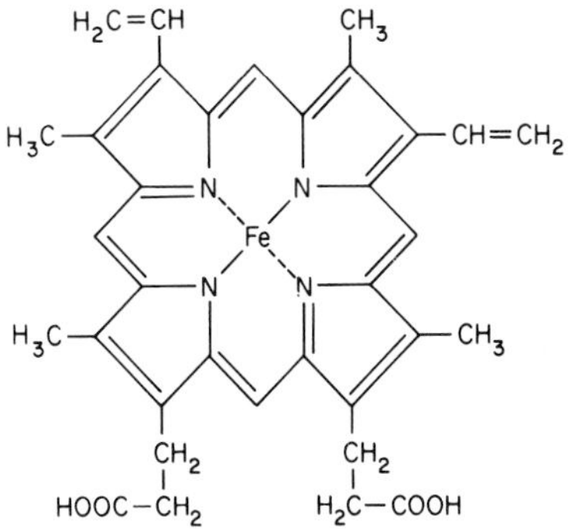

H_2C=CH
CH_3
H_3C—
—CH=CH_2
N
N
Fe
N
N
H_3C—
—CH_3
CH_2
HOOC—CH_2
CH_2
H_2C—COOH

Ferroprotoporphyrin IX (b-Cytochromes)

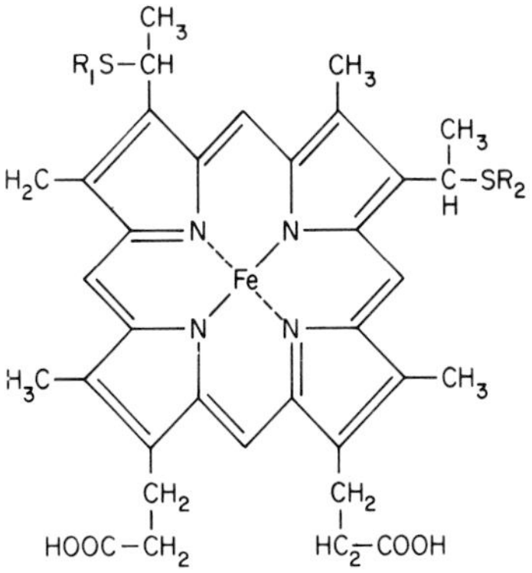

CH_3
R_1S—CH
CH_3
CH_3
—C—SR_2
H
H_2C—
N
N
Fe
N
N
H_3C—
—CH_3
CH_2
HOOC—CH_2
CH_2
HC_2—COOH

Heme c (c-Cytochromes)

Fig. 5.4. Prosthetic groups of the cytochromes.

The multiplicity of *b*-cytochromes is due to different *b*-cytochromes in mitochondria rather than a splitting of the absorption bands at low temperature, since the peak heights do not change in synchrony in the presence of reducing agents, inhibitors or uncouplers (Lance & Bonner, 1968). The *b*-cytochromes are placed in the respiratory chain according to the following sequence (Storey, 1973):

$$\text{Cyt } b-556 \rightarrow \text{cyt } b-560 \rightarrow \text{cyt } c-551$$

Cytochrome $b-560$ (557) was placed on the oxygen side of cytochrome $b-556$ (553) as a result of the determination of the rates of oxidation of the 556 and 560 components by an oxygen pulse of an anaerobic suspension of mitochondria. The 560 component was oxidized with a half-time of oxidation of 6 to 8 msec while the 556 component was oxidized with a half-time of 150 to 200 msec. The reduction by succinate of these two components in anaerobiosis showed, however, that $b-560$ was reduced more slowly than $b-556$ which was contrary to the expected rates in view of the rates of oxidation (Storey & Bahr, 1972; Storey, 1973). The slow reduction was ascribed to the more negative redox potential of $b-560$. The midpoint potentials of $b-560$ and $b-556$ would predict that $b-560$ would be on the substrate side of $b-556$. Further resolution of the sequence of the *b*-cytochromes is necessary.

Cytochrome $b-566$ was thought to be analogous to the $b-566$ (b_T) of mammalian mitochondria. Cytochrome $b-566$ from animal mitochondria was found to undergo a midpoint potential shift as well as an enhanced reduction in anaerobic suspension when the respiratory chain was energized (Chance *et al.*, 1970). This was interpreted as the formation of a high energy intermediate of phosphorylation directly involving cytochrome $b-566$. In plant mitochondria, the midpoint potential shift of $b-566$ was not observed (Dutton & Storey, 1971; Lambowitz *et al.*, 1974). Although enhanced reduction of $b-566$ by ATP or by energization of the respiratory chain could be demonstrated in plant mitochondria, it could be explicable by reverse electron flow through the *b*-cytochromes (Lambowitz *et al.*, 1974; Lambowitz & Bonner, 1974). Thus the status of a transducing *b*-cytochrome in plant mitochondria is in question. In fact, cytochrome $b-566$ was excluded from the main sequence of the respiratory chain, since it remains oxidized in anaerobic suspensions (succinate reduced) while other *b*-components, pyridine nucleotides and fluorescent flavoproteins are reduced (Storey, 1969, 1974). There was a lack of equilibration between the low redox potential carriers with cytochrome $b-566$. The function of cytochrome $b-566$ is left uncertain.

5.6.1.6 *Cytochrome c*

Two *c*-type cytochromes have been detected in plant mitochondria. They have the same relationship as cytochrome *c* and c_1 in animal mitochondria (Lance & Bonner, 1968). The room temperature spectrum shows a large peak at 550 mn which shifts to 547 nm at liquid nitrogen temperature (77°K). As with cytochrome *c* in animal mitochondria, the cytochrome *c* (cyt $c-547$*) is easily extracted by salt solutions. A second component with a low temperature absorption peak at 549 nm remains after extensive washing with phosphate buffer. The 547 nm absorbing component is recovered in the phosphate buffer extract while the 549 nm absorbing component remains in the pellet. Both are reducible by ascorbate, which differentiates them from the *b*-cytochromes. The spectral properties of the 549 absorbing *c* component and its strong binding to mitochondrial membranes relate this *c* component to cytochrome c_1 of animal and yeast mitochondria. The midpoint potentials of the two *c*-cytochromes of mung bean mitochondria have been determined to be +235 mV in both cases (Dutton & Storey, 1971). The half-time of oxidation of cytochrome $c-547$ and $c-549$ are 3·0 and 3·1 msec respectively when KCN treated anaerobic mitochondria were pulsed with 14 μM O_2. The electron transfer sequence of the *c*-cytochromes was given as cyt $c-549$ to cyt $c-547$ (Storey & Bahr, 1972).

Cytochrome *c* (cyt $c-547$) is essentially identical to cytochrome *c* from all eukaryotic sources. Cytochrome *c* from one source will react with the reductase and oxidase from quite distantly related sources. The amino acid sequence of cytochrome *c* from a large number of sources is shown to have a high degree of homology (Nolan & Margoliash, 1968; Dickerson *et al.*, 1971). This homology is all the more striking when the tertiary structure is considered. For example, the amino acids about the heme show a high degree of conservatism among the cytochrome *c* proteins examined. Those amino acid residues important to the structure and function of the protein have suffered few substitutions in the course of evolutionary history.

5.6.1.7 *Cytochrome oxidase*

The cytochrome oxidase of plant mitochondria contains cytochromes *a* and a_3, as does that from animal mitochondria. These are two spectroscopically differentiated components, although two separate chemically different entities have not been isolated. The optical properties are well differentiated in the presence of cyanide or azide, which binds to cytochrome a_3. The α-band of cytochrome oxidase at room temperature is located at 602 nm; at 77°K, there is a blue shift to 598 nm. The reduced spectrum of cytochrome *a* is revealed in the difference spectrum of an azide treated aerobic suspension minus an aerobic suspension. All components are oxidized except for cytochrome *a*, which shows a symmetrical reduced α-band at 598 nm. The a_3 spectrum is shown in a difference spectrum of an anaerobic (succinate reduced) plus azide suspension, minus an aerobic plus azide suspension. The reduced α-peak of cytochrome *a* cancels and the reduced α-peak of cytochrome a_3 is shown with its maximum at 603 nm (Lance & Bonner, 1968). The midpoint potentials of cytochromes *a* and a_3 are

*The 77°K absorption peak.

+190 and +380 mV respectively (Dutton & Storey, 1971). Half-times of oxidation in oxygen pulsed anaerobic suspension are 2·0 msec and 0·8 msec respectively for cytochromes *a* and a_3 (Storey, 1970b).

5.6.1.8 *Non-heme Iron Proteins*

Few investigations have been carried out on the occurrence and nature of non-heme iron proteins in higher plant mitochondria. Electron paramagnetic resonance (epr) signals characteristic of iron sulphur centres (non-heme iron proteins) were observed by Cammack and Palmer (1973) in mitochondria from *Helianthus tuberosus* and *Arum maculatum* with components at g= 2·02 and 1·93, 2·05 and 1·92, and at 2·10 and 1·87. Schonbaum *et al.* (1971) obtained an epr signal in NADH-reduced skunk cabbage (*Symplocarpus foetidus*) mitochondria for a component at g= 1·94 and some complex components near g= 2·00. These iron sulphur centres no doubt participate in electron transport, since they undergo oxidation and reduction. At present, their precise functions are not known. They may be analogous to the iron sulphur centres identified in the NADH dehydrogenase segment of yeast and pigeon heart submitochondrial particles, which are believed to be closely involved in energy coupling (Ohnishi, 1973).

5.6.2 CYANIDE RESISTANT RESPIRATION

An unusual characteristic of respiration in plant mitochondria is a partial insensitivity to cyanide inhibition. Partial insensitivity is exhibited to inhibition by azide, antimycin and 2-heptyl hydroxy-quinolin-N-oxide, all of which are potent inhibitors of oxygen uptake in animal mitochondria. This cyanide insensitivity may be almost 100% as in the spadix mitochondria from some aroid species, notably of *Arum maculatum* and *Symplocarpus foetidus*, partial as in mung bean mitochondria, or completely lacking as in the mitochondria from fresh, dormant white potato tubers (Bahr & Bonner, 1973a). In the latter, a cyanide insensitivity, and indeed a cyanide stimulation of respiration, may be induced upon vigorously aerating slices of potato tuber tissue in water for 24 hours. Mitochondria isolated from such aged potato tuber slices are much less inhibited by antimycin or cyanide (Hackett *et al.*, 1960).

Because of the insensitivity to cyanide at concentrations which completely inhibit the respiration of animal mitochondria, it is unlikely that the cyanide insensitive respiration is due to an incomplete inhibition of cytochrome oxidase. This 'excess' oxidase hypothesis has been considered by various workers and received strong support from the finding that cytochromes *a* and *c* were incompletely reduced in the presence of cyanide (Chance & Hackett, 1959). The insensitivity to antimycin, which inhibits electron transport between the cytochrome *b* to cytochrome *c* region favours the argument that the cytochrome system is by-passed entirely and that there exists an alternate oxidase in mitochondria showing cyanide insensitivity, although mitochondria of *A. maculatum* and *S. foetidus* have the conventional cytochrome complement (Bendall & Hill, 1956; Chance & Hackett, 1959). By and large, cytochrome *c* and cytochrome oxidase are reduced in the presence of cyanide (Bendall & Bonner, 1971). The incomplete reduction of cytochromes found by Chance and Hackett could be attributed to (a) the fact that the cytochromes are not always reduced by substrates relative to the reduction by dithionite; (b) possible spectral interference in the measurement of cytochrome *c* due to an oxidized *b*-cytochrome; or (c) in coupled mitochondria, significant reverse electron transport may cause the carriers on the oxygen side of the respiratory chain coupling site to become partially oxidized (Bonner & Bendall, 1968; Bendall & Bonner, 1971). Chance and Hackett (1959) and Bendall and Hill (1956) reported, however, that a *b*-type cytochrome becomes oxidized in the presence of cyanide and oxygen. Bendall and Hill called their *b*-component from *A. maculatum* cytochrome b_7 (α_{max} 560 nm), while Chance and Hackett identified an oxidizable *b*-component with an α_{max} at 558·5 nm. It was hypothesized that a *b*-cytochrome functioned as the shunt to the alternate oxidase. Such a role for a *b*-cytochrome is favoured, since Bahr and Bonner (1973b) reported that the known flavoproteins and ubiquinone had equal access to oxygen by either pathway, based on observations of their oxidation in the presence or absence of cyanide, and Storey and Bahr (1969a) found no identifiable carriers among flavoprotein, ubiquinone, the known *b*-cytochromes or *c*-cytochromes which could mediate electron transfer to the alternate oxidase. Cytochrome b_7 was identified with cytochrome $b-557$ (77°K) (Chance *et al.*, 1968), but since the oxidation rate of cytochrome $b-557$ was unaffected by m-chlorobenzhydroxamic acid, an inhibitor of the alternate oxidase (Schonbaum *et al.*, 1971) it was felt that cytochrome $b-557$ does not play a part in the alternate pathway, and should not be equated with Bendall and Hill's cytochrome b_7 (Erecinska & Storey, 1970). Bendall and Bonner (1971) showed that thiocyanate and other metal binding agents also inhibit the alternate pathway, suggesting a role for non-heme iron proteins or other metalloproteins. Efforts to identify non-heme iron proteins which may be the alternate oxidase have not been successful (Cammack & Palmer, 1973). Strong epr signals characteristic of iron-sulphur proteins in mitochondria of aged Jerusalem artichoke (*H. tuberosus*) and in *A. maculatum* mitochondria were found. However, these are most likely the iron sulphur proteins of NADH:ubiquinone reductase since they were not reducible by succinate, and were, moreover, unaffected by hydroxamic acids. Schonbaum *et al.* (1971) found that the epr signals in NADH reduced skunk cabbage submitochondrial particles were enhanced by treatment with m-iodobenzhydroxamic acid. In addition to a signal at $g = 1·94$, which is characteristic of NADH dehydrogenase, a set of complex signals near $g = 2·0$ was detected at 77°K. It was thought that the $g = 2·0$ signal may originate from the alternate pathway.

The possibility that the alternate oxidase may involve a flavoprotein has been explored. Flavoprotein oxidases which reduce oxygen with the formation of hydrogen peroxide are known, but such oxidases have not been identified for cyanide resistant mitochondria (Bendall & Bonner, 1971).

The function of cyanide resistant respiration is unclear. In the spadix tissue of maturing flowers of *Arum* and *Symplocarpus*, it may serve the function of thermogenesis. A rise in temperature of the spadix tissue ten degrees above the ambient has been recorded. This thermogenesis may aid these plants in pollination, since flowering occurs in early spring.

5.7 ENERGY LINKED REACTIONS OF MITOCHONDRIA

5.7.1 OXIDATIVE PHOSPHORYLATION

5.7.1.1 *Coupling sites*

Oxidative phosphorylation, the process whereby the reaction

$$ADP + P_i \rightarrow ATP + H_2O$$

is coupled to the oxidation of reduced electron carriers of the respiratory chain occurs in plant mitochondria in the same manner as does the reaction in animal mitochondria. The sites of phosphorylation are identical (see Baltscheffsky & Baltscheffsky, 1974). These are the sites I in the NADH: ubiquinone reductase segment, II in the cytochrome *b*:cytochrome *c* reductase segment and III in the cytochrome oxidase segment of the respiratory chain. It has not been possible to identify precisely those components of the respiratory chain which are the energy transducers in each of the three sites, although these have been approximated from the change in redox potentials ($\Delta E_0'$) between adjacent carriers (Lehninger, 1965), by application of the crossover theorem (Chance & Williams, 1955), and from shifts in the midpoint potential of certain electron carriers upon the addition of ATP to uncoupled anaerobic suspensions of mitochondria (Wilson & Dutton, 1970a,b; Lindsay & Wilson, 1972; Chance *et al.*, 1970; Ohnishi, 1973; Devault, 1971). The $\Delta E_0'$ indicates the thermodynamic feasibility of coupling between two adjacent components. Thus, coupling sites were thought to be located between endogenous NAD and the flavoprotein of NADH: ubiquinone reductase; between cytochrome *b* and cytochrome *c* of the cytochrome *b*:cytochrome *c* reductase; and between cytochrome *a* and oxygen of cytochrome oxidase. The application of the crossover theorem by and large confirms the general location based on thermodynamic grounds. According to the crossover theorem, electron transport through the coupling site is the rate limiting step in tightly coupled mitochondria. The carrier on the substrate side of the coupling site should become reduced, while the carrier on the oxygen side would become oxidized. Upon the addition of the phosphate acceptor, ADP, there would be observed a rapid transient oxidation of the carrier on the substrate side, and a reduction of the carrier on the oxygen side of the coupling site, concomitant with the release of controlled respiration in the state 4-state 3 transition (see Table 5.4).

Table 5.4. Definition and values of steady states in mitochondria (Chance & Williams, 1955.)

State	(O_2)	ADP levels	Substrate	Respiratory rates	Rate-limiting substance
1	>0	low	low	slow	ADP
2	>0	high	~0	slow	substrate
3	>0	high	high	fast	respiratory chain
4	>0	low	high	slow	ADP
5	0	high	high	0	oxygen

Wilson and Dutton (1970a) reported that the midpoint potential (E_m) of cytochrome a_3 becomes more negative upon energization of an anaerobic suspension of rat liver mitochondria by ATP. Similarly they report that the E_m of one of the *b* cytochromes (α_{max} 564 nm) increases upon energization by ATP (Wilson & Dutton, 1970b; Chance *et al.*, 1970). These shifts of the midpoint potentials were attributed to changes in ligand interaction energy of the heme iron upon energization, and that the iron atoms of these cytochromes were directly involved in energy coupling.

Such a change in the midpoint potential was postulated by Wang (1970). In his model, the phosphoimidazole group becomes a much weaker coordinating ligand for the Fe(II) than imidazole, and hence should lower the midpoint reduction potential of the corresponding electron carrier. Similar experiments in the iron-sulphur centres of the site I region of the respiratory chain showed that of the five iron sulphur centres identified, ATP affected the reduction potential of only one of these centres, designated as centre I (Ohnishi, 1973). Addition of ATP caused a partial oxidation of centre I when the reduction potential was poised at a value where the iron was almost completely reduced. These observations suggested that the reduction potential of centre I was dependent upon the phosphate potential, and that the addition of ATP caused a lowering of the midpoint potential of the centre I iron sulphur protein. The significance of the changes in the midpoint potential was questioned, at least for the site II region, since none of the *b*-cytochromes showed an ATP-induced potential shift when investigated in plant mitochondria (Dutton & Storey, 1971; Lambowitz *et al.*, 1974). One must conclude that there is a fundamental difference in the function of the *b*-cytochromes in plant and animal mitochondria or that the changes in the midpoint potential do not reflect the formation of an energy transducing species. Lambowitz *et al.* took the latter view and attributed the midpoint potential changes to a reverse electron flow with cytochrome $b-566$ and cytochrome a_3 equilibrating with the redox mediators by way of cytochrome *c*, while the iron sulphur centre of site I equilibrated through the nicotinamide adenine nucleotide pool.

5.7.1.2 *ADP:O Ratios*

The general location of the three coupling sites means that for tightly coupled mitochondria, predictable ratios of the moles of ADP phosphorylated to the gram-atoms of oxygen consumed (ADP/O or P/O) may be obtained for each substrate. Thus succinate should give a ratio of 2·0, NAD-linked substrates a ratio of 3·0 and α-ketoglutarate a ratio of 4·0 (one phosphorylation *via* succinyl-CoA, three phosphorylations *via* the respiratory chain NADH). These ratios are largely confirmed in plant mitochondria. Early attempts gave results much below the predicted ratios, but with refinements in the procedure for the isolation of plant mitochondria, the inclusion of required cofactors, and the exclusion of competing reactions, predicted ratios have been approached, as shown in Table 5.5. Exogenous NADH does not interact with respiratory chain NAD,

Table 5.5. Oxidative phosphorylation by isolated mitochondria (Douce *et al.*, 1972a).

Source	Substrate	P:O	Respiratory control index
Phaseolus aureus	succinate	1·68	3·2
	malate	2·56	5·4
	NADH	1·55	4·5
Solanum tuberosum	succinate	1·64	3·4
	malate	2·59	4·2
	NADH	1·48	2·4

but with flavoproteins of the external dehydrogenase. These ultimately enter the respiratory chain at the level of cytochrome *b* and hence electrons go through two coupling sites only (Douce *et al.*, 1973).

5.7.1.3 *Energy Coupling in Cyanide Resistant Respiration*

Mitochondria of *Symplocarpus* are fully capable of energy conservation in the uninhibited state. Hackett and Haas (1958) found P/O ratios greater than 3 for α-ketoglutarate oxidation by skunk cabbage mitochondria. In the presence of cyanide, the energy coupling sites associated with the cytochromes are by-passed, and energy coupling involves only the NADH:ubiquinone reductase portion of the respiratory chain. Storey and Bahr (1969b) obtained ADP/O ratios of 1·3 with succinate as substrate in skunk cabbage mitochondria in the absence of cynide; the ratio was zero in the presence of cyanide. With malate as the substrate, the ADP/O ratios were 1·9 and 0·7 in the absence and presence of 0·3 mM KCN respectively, showing that coupling sites II and III were effectively by-passed while coupling site I, which could be activated by malate but not by succinate, was still functioning in the presence of cyanide.

Wilson (1970a), however, has proposed that energy coupling occurs in the alternate pathway as well. He found that cyanide treated mitochondria of *A. maculatum* and *Acer pseudoplatanus* produced ATP, with an ATP/O ratio of 0·5 at high O_2 concentration (greater than 100 μM) with succinate as the substrate. Submitochondrial particles of *A. maculatum* which lacked malate dehydrogenase activity oxidized succinate with P/O ratios of 0·2 to 0·6 (Wilson 1970b). Since malate dehydrogenase-depleted particles were still capable of phosphorylation, the possibility that the phosphorylation was due to the oxidation of a product of succinate oxidation, i.e., malate, could be eliminated, therefore suggesting that a phosphorylation site exists in the alternate pathway which functions only at high O_2 concentration. Cyanide inhibited mitochondria from spadices of *A. maculatum* and from mung bean hypocotyls still retain an energy related function through the cytochromes. Bonner and Bebdall (1968) showed that a substrate linked reverse electron flow is active in spadix mitochondria. Cytochrome *c* and cytochrome oxidase of cyanide treated mitochondria were reduced by ascorbate plus *N,N,N′,N′*-tetramethyl-*p*-phenylenediamine (TMPD), an electron donor to cytochrome *c*. Subsequent addition of succinate caused a reoxidation of cytochrome *c* and cytochrome *a* with a partial reduction of cytochrome *b*. The reoxidation of cytochromes *c* and *a* was prevented by uncoupling concentrations of carbonylcyanide-phenyl-hydrazone or 2,4-dinitrophenol. Wilson and Moore (1973) also showed that cyanide inhibited mung bean mitochondria could carry out oxidation of ascorbate plus TMPD. The oxidation was thought to involve a reverse electron flow through coupling site II and energized by ATP or the high energy intermediate.

5.7.1·4 *Mechanism of Coupling*

The mechanism of energy coupling remains an intractable problem in spite of the impressive array of workers in this area. Three outstanding hypotheses are under active consideration: (a) the chemical intermediate hypothesis; (b) the chemiosmotic hypothesis; and (c) the conformation coupling hypothesis.

The operation of the chemical intermediate hypothesis may be summarized as follows:

$$A_{red} + B_{ox} + I \rightarrow A_{ox}{\sim}I + B_{red}$$

$$A_{ox}{\sim}I + X \rightarrow A_{ox} + X{\sim}I$$

$$X{\sim}I + P_i \rightarrow X{\sim}P + I$$

$$X{\sim}P + ADP \rightarrow ATP + X$$

where A and B are adjacent electron carries at the coupling site, X and I are unknown couplers common to all coupling sites (see Greville, 1969). The scheme postulates the existence of both non-phosphorylated and phosphorylated intermediates. A non-phosphorylated intermediate accounts for the action of uncouplers, which is independent of the presence of phosphate. The non-phosphorylated intermediate can also be coupled to work functions such as ion transport, which is sensitive to uncouplers, but not to the phosphorylation inhibitor, oligomycin. The phosphorylated intermediate transfers a phosphoryl group to ADP. Hill and Boyer (1967) showed that the bridge oxygen between the β and γ phosphorus of ATP is furnished by ADP. Hence, the mechanism of phosphorylation involves an activation of phosphate.

The chemiosmotic hypothesis in its simplest form as first proposed by Mitchell (1961) involves a vectorial metabolism in which the elements of water are transported to opposite sides of the mitochondrial membrane (Fig. 5.5).

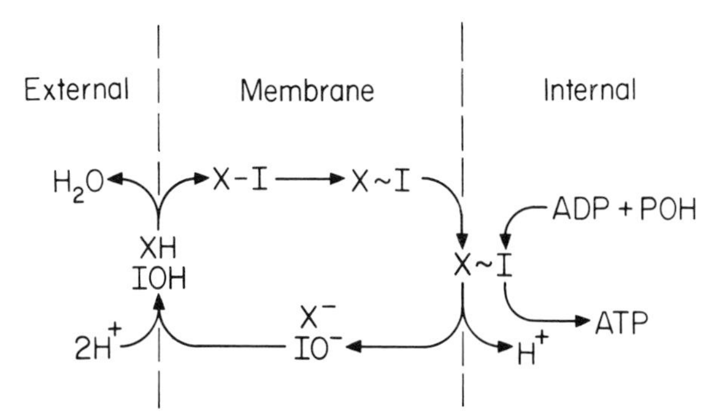

Fig. 5.5. Representation of the Mitchell chemiosmotic scheme for oxidative phosphorylation. (Redrawn from Mitchell, 1966.)

Thus oxido-reduction reactions create a pH gradient as well as a potential difference (outside positive) which tends to drive H^+ back across the membrane into the inner compartment. This force is the proton-motive force, and it is this flow of protons through the coupling site (a reversible ATPase) which drives ATP synthesis. The phosphorylation reaction is represented as follows:

$$ADP + P_i + H^+_{out} \rightarrow ATP + H_2O + H^+_{in}$$

It can then be shown that in the absence of a transmembrane electrical potential, a pH differential of 3·5 units is required to poise the ratio of ATP/ADP = 1 while a potential difference of 210 mV would be required in the absence of pH gradients (Mitchell, 1966). The translocation of protons and equivalent OH^- is effected by ionizable groups which are designated XH and IOH, corresponding to components of an ATPase. The proposed intermediate X-I of the ATPase must have a sufficiently low hydrolysis constant at the high electrochemical potential of H^+ in the outer phase to come to reverse equilibrium with water according to the reaction

$$X-I + H_2O \rightarrow XH + IOH$$

On the other hand, the hydrolysis constant of the intermediate X~I must be in the order of 10^5 M, and the intermediate X~I must be in equilibrium with the ATP/(ADP+P_i) couple in the inner phase, so that

$$X{\sim}I + Pi + ADP \rightarrow X^- + Io^- + 2H^+ + ATP$$

The system vibrates between states in which the intermediate X–I is alternately accessible to the outer and inner phases, being X–I when in contact with the outer phase and X~I when in contact with the inner phase. The transition of X–I to X~I is due not to the pumping of energy into X–I, but to a lowering of the ground state energy for X–I hydrolysis by some 10,000 calories per mole on translocation through an anisotropic membrane.

While experimental verification of the chemiosmotic mechanism has been realized only in chloroplasts by acid-base trasitions (Jagendorf & Uribe, 1966; see also chapter 4), similar experiments have not been successful in mitochondria. Cockrell *et al.* (1967) have shown ATP synthesis after establishing a K^+ gradient by valinomycin-treated mitochondria. However, Glynn (1967) argued that ATP synthesis *via* a K^+ gradient could be explained equally well on the basis of a membrane potential rather than cation transport down a chemical gradient through an ATPase.

In their general aspects, the chemical intermediate hypothesis and the chemiosmotic hypothesis differ primarily in the nature of the initial driving force for coupled phosphorylation. The two hypotheses converge at the level of the ATPase in that both call for an unknown intermediate, X~I (Greville, 1969).

The conformational coupling hypothesis of Boyer (1967) differs in that the ATPase has a high affinity for P_i and ADP. Tightly bound ATP is formed *via* a nucleophilic attack by ADP on orthophosphate in a S_N2 reaction with the displacement of water (Korman & McLick, 1970). According to the model, there exists an energy requirement for the release of bound ATP from the complex, by changing the complex from one having a high affinity for ATP to one having low affinity (Boyer *et al.*, 1973). The advantages of this hypothesis are that it explains most of the exchange reactions observed in coupled phosphorylation, and the coupling of ATP to energy yielding reactions of mitochondria.

5.7.2 REVERSE ELECTRON FLOW

Electron flow and energy transduction through the coupling sites of the respiratory chain are reversible processes. These are shown by the reduction of endogenous NAD and cytochrome *b* when substrates of higher reduction potentials are oxidized as well as the ATP induced oxidation of reduced cytochrome $a+a_3$ when the terminal oxidase is inhibited by sulphide. The properties of energy linked NAD reduction by plant mitochondria are similar to those reported in animal mitochondria (Chance, 1961; Chance & Hollunger, 1963). These include the requirement for succinate oxidation and for ATP (Storey, 1971b). Although ATP is required, the reduction of NAD is not sensitive to oligomycin. Hence NAD is reduced by reverse electron flow through the coupling site but without the participation of the ATPase as such. Uncouplers either inhibit the reduction when added before succinate, or reverse the reduction when added after a steady state reduction is attained. This can be interpreted as an effect upon the coupling site in preventing reverse electron flow, as well as a general release of controlled respiration so that the endogenous NADH is now oxidized rapidly through the coupling site. It is not possible to distinguish between the two alternatives.

Reverse electron flow through coupling sites II and III has been demonstrated by Storey (1972) and Lambowitz *et al.* (1974). When ATP is added to sulphide-inhibited or to anaerobic suspensions of mung bean mitochondria, the *b*-cytochromes become reduced while cytochrome *c* and cytochrome $a+a_3$ become oxidized. The effect is reversed by uncouplers or by phenazine methosulphate (PMS) which mediates electron flow between cytochromes *b* and *c*. The reverse electron flow through coupling site II involves an ATPase (Lambowitz *et al.*, 1974). Hydrolysis of ATP is observed in an anaerobic suspension of mung bean mitochondria supplemented with ascorbate plus TMPD and ATP. Addition of PMS caused an increase in the rate of ATP hydrolysis which is to be expected if PMS formed a shunt of electrons from reduced cytochrome *b* to cytochrome *c*.

5.7.3 ION TRANSPORT

Mitochondria contain two compartments, one of which is readily accessible to low molecular weight solutes such as sucrose or mannitol, and a second inaccessible to sucrose or mannitol. The former is identified with the intermembrane space, and the latter, the mitochondrial matrix. The volume enclosed by the inner membrane behaves as an osmometer (Lorimer & Miller, 1969; Overman *et al.*, 1970). Selective permeability to solutes and active transport are properties of the inner mitochondrial membrane. The movement of solutes across the inner membrane may be determined by assay for net increases in intramitochondrial contents. If the movement results in the net increase or decrease in the osmolarity of the sucrose-inaccessible volume, such solute movements will be reflected in volume changes of the inner compartment which can be monitored by changes in the light scattering properties of a mitochondrial suspension. Thus mitochondrial swelling is accompanied by a decrease in light scattering, while a contraction is reflected by an increase in light scattering.

5.7.3.1 *Monovalent Cations*

Permeability of the inner membrane of plant mitochondria to potassium and chloride ions is restricted. Slow, passive permeability to K^+ and Cl^- may be observed on transfer of mitochondria to isosmotic KCl, but when mitochondria are energized by the addition of substrate (NADH), rapid loss of both K^+ and Cl^- occurs, accompanied by mitochondrial contraction (Wilson *et al.*, 1969; Kirk & Hanson, 1973). In contrast to the energized extrusion of K^+ and Cl^-, mitochondria undergo an NADH induced swelling in potassium acetate solution, and K^+ and acetate ions are taken up. This has been interpreted as an active transport of acetate, while K^+ penetrates along an electrochemical potential and chloride is not transported. The NADH-induced swelling in potassium acetate is inhibited by uncouplers of oxidative phosphorylation, by ADP plus P_i, or by respiratory chain inhibitors (Wilson *et al.*, 1969; Lee & Wilson, 1972), but is restored when oligomycin is present with ADP plus P_i. An intermediate of oxidative phosphorylation possibly mediates active transport of acetate, but not of chloride. This is strengthened by the observation that mitochondria respire with expected respiratory control and ADP/O ratio when suspended in isotonic KCl, but lose respiratory control in potassium acetate.

Ionophorous antibiotics, valinomycin and gramicidin D, increase the permeability of the mitochondrial membrane to potassium ion as shown by the increased rate of passive swelling in potassium chloride solutions (Kirk & Hanson, 1973; Miller *et al.*, 1970a). Valinomycin also increases the rate of NADH-induced swelling in potassium acetate, as well as in chloride, phosphate, sulphate and nitrate salts of potassium (Kirk & Hanson, 1973; Wilson *et al.*, 1972) although the swelling due to acetate and phosphate reverses on exhaustion of NADH. This has been interpreted to mean that the swelling is due to the enhanced permeability of the inner membrane to potassium, as well as an active transport of acetate or phosphate. In the case of the latter anions, swelling is thought to be due to an enhanced permeability toward K^+ while the anions diffuse along an electric potential. On cessation of NADH-supported respiration, acetate and phosphate leak out according to their chemical potential, followed by potassium ions, while no leakage of chloride, nitrate or sulphate occurrs as they are distributed along their chemical potential (Wilson *et al.*, 1972). The picture which emerges is that the movement of potassium ion is controlled by the movement of anions, but K^+ flux can be modified by ionophores such as valinomycin or gramicidin D. Valinomycin may stimulate K^+ uptake *via* a H^+ exchange as found by Rossi *et al.* (1967) in liver mitochondria, but such K^+ uptake due to H^+ exchange is not accompanied by swelling of the mitochondria.

5.7.3.2 *Divalent Cations*

Ca^{2+} does not cause marked stimulation of respiration in plant mitochondria as it does in animal mitochondria, except with exogenous NADH as substrate

(Miller *et al.*, 1970b; Miller & Koeppe, 1971). A slight release from controlled respiration is detected with malate plus pyruvate or with succinate as substrates. This is associated with the accumulation of calcium and inorganic phosphate by mitochondria (Miller *et al.*, 1970b). Extensive Ca^{2+} uptake by plant mitochondria occurs only in the presence of inorganic phosphate (Hodges & Hanson, 1965; Elzam & Hodges, 1968; Earnshaw *et al.*, 1973; Chen & Lehninger, 1973). The phosphate dependent Ca^{2+} transport is energy dependent and may be supported by substrate oxidation or by ATP (Elzam & Hodges, 1968). The energy dependence seems to be at the level of an intermediate of oxidative phosphorylation. Substrate-supported Ca^{2+} transport is sensitive to uncouplers and respiratory inhibitors (Chen & Lehninger, 1973) as well as to ADP (Elzam & Hodges, 1968). The ATP supported Ca^{2+} transport is inhibited by oligomycin. Similar observations have been reported for Sr^{2+} transport by mitochondria (Johnson & Wilson, 1972). Other anions, e.g. acetate, arsenate, sulphate, chloride or nitrate promote neither Ca^{2+} nor Sr^{2+} uptake (Hodges & Hanson, 1965; Johnson & Wilson, 1972) although all will produce a metabolically independent swelling, while arsenate and acetate produce an active swelling as well, indicating that mitochondrial membranes are permeable to these anions, and will actively transport arsenate or acetate, as well as phosphate (Hanson & Miller, 1967; Johnson & Wilson, 1972; Lee & Wilson, 1972). Uptake of Ca^{2+} and inorganic phosphate results in the deposition of electron dense material in mitochondria which is dependent upon the concentrations of both Ca^{2+} and phosphate, and the time of incubation (Peverly *et al.*, 1974). The composition of the deposits has not been ascertained, but is believed to be a form of calcium phosphate. The deposition of the phosphate salt of divalent alkaline earth metal ions may account for the contraction of mitochondria induced by Ca^{2+}. Since the volume changes of mitochondria are measured by light scattering changes, the formation of crystals within the mitochondria may increase the ight scattering properties of the suspension, and be interpreted as a contraction.

5.7.3.3 *Anion Transport*

The importance of anion transporters in the movement of solutes across the mitochondrial inner membrane has been studied extensively in mammalian mitochondria (Chappell & Haarhoff, 1967; Harris, 1969). In these studies, the role of the phosphate transporter and the malate transporter is emphasized. Similar investigations were carried out by Phillips and Williams (1973b) and by Wiskich (1974). The presence of anion transporters was demonstrated by the spontaneous swelling of mitochondria in solutions of the ammonium salts of phosphate or malate. Ammonium salts were used because the mitochondrial membrane is readily permeable to ammonium ion. Mitochondrial swelling under these conditions is indicative of an osmotic adjustment due to the net influx of solutes into the mitochondrial matrix (Overman *et al.*, 1970; Wilson *et al.*, 1973). The swelling in ammonium phosphate was inhibited by

N-ethylmaleimide which inhibits the phosphate-hydroxyl antiporter while the swelling in ammonium malate as well as the malate supported respiration was inhibited by 2-butylmalonate, 2-phenylsuccinate, benzylmalonate or *p*-iodobenzylmalonate, all inhibitors of the dicarboxylate carrier (Phillips & Williams, 1973a,b). The anion transport system is interpreted as follows: (a) a phosphate-hydroxyl antiporter which transports phosphate in exchange for hydroxyl ion; (b) a malate-phosphate antiporter which transports malate in exchange for phosphate; (c) a tricarboxylate-malate antiporter which transports tricarboxylate anions in exchange for malate; (d) other dicarboxylate anions enter by exchange with malate. It is the prevailing view that anions are actively transported, and that cations follow the anion transport along an electric gradient (Hanson & Miller, 1967). The essential role of anion transport in determining cation movement, however, is modified by the presence of cation ionophores.

FURTHER READING

Bonner W.D.,Jr. (1965) Mitochondria and electron transport. In *Plant Biochemistry* (eds. J.F. Bonner & J.E. Varner). Academic Press.

Bonner W.D.,Jr. (1973) Mitochondria and plant respiration. In *Phytochemistry*, vol. 3 (ed. L.P. Miller). Van Nostrand Reinhold.

Dawson A.P. & Selwyn M.J. (1974) Mitochondrial oxidative phosphorylation. In *Companion to Biochemistry* (eds. A.T. Bull, J.R. Lagnado, J.O. Thomas & K.F. Tipton). Longman.

Goddard D.R. & Bonner W.D.,Jr. (1960) Cellular respiration. In *Plant Physiology*, vol. 1A (ed. F.C. Steward). Academic Press.

Greenberg D.M. (ed.) (1967) *Metabolic Pathways*, vol. 1. Energetics, Tricarboxylic Acid Cycle and Carbohydrates. Academic Press.

Hanson J.B. & Hodges T.K. Energy linked reactions of plant mitochondria. In *Current Topics in Bioenergetics*, vol. 2 (ed. D. Rao Sanadi). Academic Press.

Lehninger A.L. (1964) *The Mitochondrion: Molecular Basis of Structure and Function*. W.A. Benjamin, Inc.

Munn E.A. (1974) *The Structure of Mitochondria*. Academic Press.

Öpik H. (1974) Mitochondria. In *Dynamic Aspects of Plant Ultrastructure* (ed. A.W. Robards). McGraw-Hill.

Sato S. (ed.) (1972) Mitochondria. In *Selected Papers in Biochemistry*, vol. 10. University Park Press.

Slater E.C., Zaniuga Z. & Wojtczak L. (eds.) (1967) *Biochemistry of Mitochondria*. Academic Press.

Wainio W.W. (1970) *The Mammalian Mitochondrial Respiratory Chain*. Academic Press.

CHAPTER 6

MICROBODIES

6.1 INTRODUCTION

Most of the organelles of the plant cell were detected by the classical techniques of light microscopy and were described by the turn of this century. Microbodies however were discovered relatively recently with the advent of the electron microscope. They were first recognized in the early 1950's as small spherical bodies in electron micrographs of mammalian kidney and liver tissue and similar organelles were reported in plant tissues in the early 1960's. Later in that decade they were isolated from plant tissues and from studies of their biochemical functions were recognized as being of major importance in plant cell metabolism. It is perhaps interesting to point out that the physiological significance to plant cell metabolism of major organelles, such as mitochondria and chloroplasts, was recognized long before their component biochemical reactions had been studied in detail, while the discovery and isolation of microbodies allowed some already well known metabolic pathways to be ascribed to a specific organelle.

Two types of microbody with identical structures but with distinct physiological functions have been isolated from plant tissues. They have been called *peroxisomes* and *glyoxysomes* to distinguish their separate functions in cell metabolism. Peroxisomes occur in the leaves of higher plants and are closely associated with chloroplasts. They are the sites for the oxidation of glycollic acid, a product of carbon dioxide fixation. The oxidation of this compound, results in a release of carbon dioxide and oxygen uptake which is called photorespiration. Glyoxysomes on the other hand, occur abundantly during the germination of those seeds which store fats as a reserve material, and contain the enzymes necessary for the breakdown of fatty acids to acetyl-CoA and the synthesis of succinate from acetyl-CoA.

All microbodies appear to contain flavin-dependent oxidases and catalase. The oxidation of a substrate by a flavin-linked oxidase is accompanied by the uptake of oxygen and the production of hydrogen peroxide, which is broken down by catalase to oxygen and water.

$$\left.\begin{matrix}\text{Reduced substrate}\\ \text{Oxidized substrate}\end{matrix}\right) \left(\begin{matrix}\text{FAD}\\ \text{FADH}_2\end{matrix}\right) \left(\begin{matrix}H_2O_2 \rightarrow H_2O + \frac{1}{2}\,O_2\\ O_2\end{matrix}\right.$$

Microbodies, therefore, contribute to the uptake of molecular oxygen by the cell, but unlike mitochondria they do not contain any electron transport system which would be needed for the recovery of energy as ATP.

6.2 STRUCTURE AND OCCURRENCE

Microbodies are usually spherical, but can be ellipsoidal or dumbell shaped, and range from 0·2 to 1·5 μm in cross sectional diameter. Because of the low contrast between the matrix of the microbody and the cytosol they are not detectable by light microscopy but in electron micrographs they can be seen to have a single limiting membrane enclosing a granular matrix of moderate electron density (Fig. 6.1). The core of the microbody commonly contains

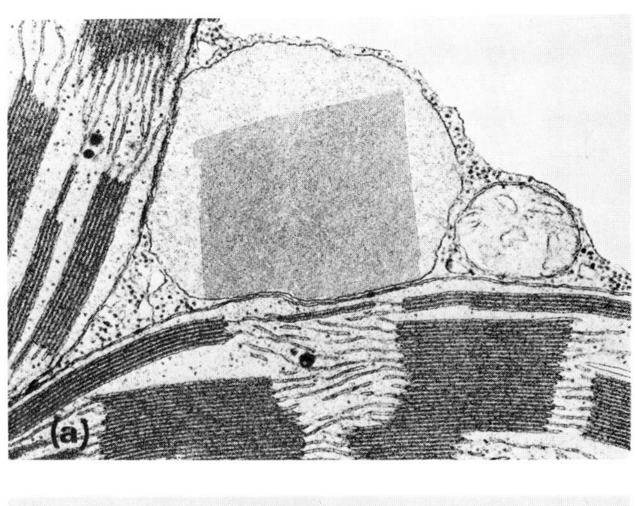

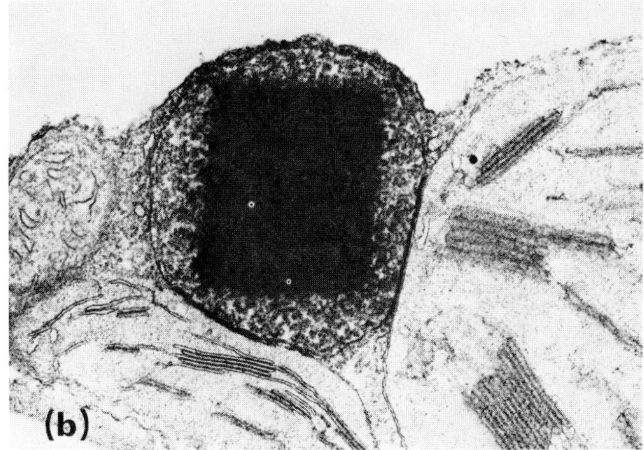

Fig. 6.1.a. A portion of a tobacco leaf cell showing a microbody with a crystalline inclusion appressed to two chloroplasts. A mitochondrion lies to the right of the microbody. (Magnification × 33,000).

b. A portion of a tobacco leaf cell, incubated in DAB medium, showing a heavy deposition of osmium throughout the crystalline inclusion of the microbody. (Magnification × 30,500). (Reproduced with permission from Frederick and Newcomb (1969). Original prints supplied by Professor E. H. Newcomb.)

fibrillar inclusions or a single large crystalline inclusion which may be formed by a reorganization of the fibrillar material. These crystalline bodies have been shown to have catalase activity (Frederick & Newcomb, 1969). No ribosomes or any form of nucleic acids have been detected in the microbody and hence they are not considered to be capable of self-replication. Since they are found in many tissues in association with the endoplasmic reticulum they are thought to be formed by this structure. Microbodies are therefore structurally very simple and cannot be confused in electron micrographs with any organelle except perhaps lysosomes from which they can be distinguished by cytochemical techniques.

The presence of catalase is a distinguishing feature of microbodies and this can be detected cytochemically using the dye 3,3′-diaminobenzidine (DAB). In fixed sections, catalase remains active and in the presence of hydrogen peroxide the dye is oxidized to give an osmiophyllic electron-dense product in the core of the microbody.

Microbodies have been seen in electron micrographs of many plant tissues. They have been found in the leaves of angiosperms, gymnosperms and bryophytes where they are numerous and are invariably located near, or appressed to, chloroplasts. Microbodies have also been found in yeast and hyphal fungi and in species representing a wide range of algal phyla. In the fat-storing seeds of higher plants they are numerous at certain stages of germination and are found in the cells in close association with *spherosomes* which are large lipid storage bodies. In the roots of higher plants a catalase-containing microbody has been found in association with the endoplasmic reticulum but since it does not contain either glycollate oxidase or enzymes of the glyoxyllate cycle, its function in the cell is not clear.

6.3 ISOLATION

Microbodies appear to be very fragile, since homogenization of plant tissue damages them to such an extent that their constituent enzymes appear in soluble fractions after removal of chloroplasts and mitochondria by centrifugation. More gentle techniques of tissue breakage however allow microbodies to be isolated at a yield of usually about 10% of the organelles present in the whole tissue as judged by the solubilization of enzymes presumed to be present in the microbody. The method of breakage depends upon the tissue. Leaves are ground for a brief time with a mortar and pestle at 0–4°C in a buffer containing an osmoticum such as sucrose at a concentration of 0·4 to 0·8 M. Other tissues may be chopped or finely sliced with razor blades in a similar medium. The resulting homogenate is squeezed through cheesecloth and the brei centrifuged at a low speed to remove cell debris. The supernatant is then centrifuged at 6,000 to 10,000 *g* to obtain a pellet containing broken chloroplasts, microbodies and some mitochondria.

Microbodies are separated from the other organelles by layering this resuspended pellet fraction on a discontinuous or continuous density gradient of sucrose ranging in concentration from 1·3 to 2·5 M, and centrifuging in an ultracentrifuge for 3–4 hr at 40,000 *g*. Microbodies increase in density during this process by a loss of water and ultimately form a band in the gradient at a density of 1·24 to 1·26 g cm^{-3}. This density is higher than that of other organelles and they are clearly separated from mitochondria which sediment at a desntiy of 1·16 to 1·19 g cm^{-3} (Fig. 6.2). On the separation of the gradient into fractions

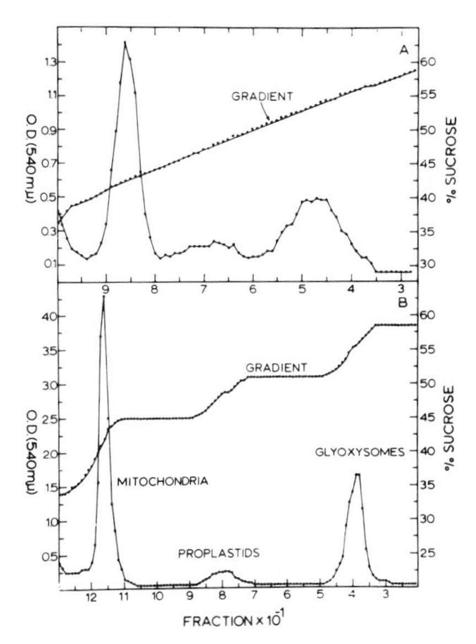

Fig. 6.2. The distribution of protein in continuous (A) and discontinuous (B) sucrose density gradients after centrifugation of crude particles from castor bean endosperm. (Reproduced with permission from Cooper & Beevers, 1969a.)

the microbodies are detected by assaying for specific 'marker' enzymes. Typical marker enzymes for peroxisomes are catalase, glycollate oxidase and hydroxypyruvate reductase, while those for glyoxysomes are catalse, malate synthetase and isocitrate lyase. Mitochondria are characterized by the presence of cytochrome *c* or succinic dehydrogenase, and chloroplasts by their chlorophyll pigments.

Much of the activity of the microbody marker enzymes is found in the soluble non-particulate fraction of gradients, indicating that the yield of microbodies is low. Despite these low yields, sufficient amounts of intact microbodies have been isolated from several plant tissues to be able to determine their metabolic functions. This has resulted in the distinguishing of at least two types of microbody, peroxisomes and glyoxysomes, with distinctly different enzyme complements and physiological functions.

6.4 GLYOXYSOMES

There are many plant species in which lipid is the main storage material in the cotyledons or endosperm of the seed. During the first few days of seed germination there is a dramatic decrease in the lipid content of the seed and sugars, principally sucrose are formed. These sugars are subsequently translocated to the growing embryo or embryonic axis. This lipid to carbohydrate conversion has been correlated with an increase in the activity of the glyoxysomal enzymes, malate synthetase and isocitrate lyase, and as germination proceeds and the lipid reserves are depleted, the number of lipid storage bodies or spherosomes decrease and there is a drop in the activities of malate synthetase and isocitrate lyase. The elucidation of the metabolic processes involved in this process of *gluconeogenesis*, ultimately led to the isolation of particles in which were localized the crucial enzymes of this pathway. These particles were found to be morphologically similar to animal peroxisomes and were called glyoxysomes.

The breakdown of lipids is initiated by their hydrolysis to fatty acids. Triglycerides are hydrolized to glycerol and fatty acids by the enzyme lipase while phospholipids are hydrolized by phospholipase. The resultant long-chain fatty acids are subsequently degraded by the successive removal of 2-carbon fragments in the process of β-oxidation.

6.4.1 β-OXIDATION

In the process of β-oxidation the removal of each 2-carbon fragment from a long chain fatty acid involves a succession of five reactions (Fig. 6.3). The sequence is initiated by the activation of the substrate by coenzyme-A, catalyzed by the enzyme fatty acid thiokinase in the presence of ATP. The fatty acyl-CoA is then oxidized by the removal of hydrogen from carbons 2 and 3 of the chain and a double bond between 2 and 3 is formed. In this reaction the hydrogen is transferred to FAD. This reacts with molecular oxygen to produce peroxide which is broken down by catalase, and results in the uptake of one mole of oxygen for every two moles of fatty acyl-CoA oxidized. The unsaturated acyl-CoA produced is hydrated to form 3-hydroxy acyl-CoA, the reaction being catalysed by enoyl hydratase or crotonase, and this product then oxidized, with a concomitant reduction of NAD^+, to form a 3-keto acyl-CoA by the action of hydroacyl-CoA dehydrogenase, In the final reaction the 3-keto acyl-CoA is cleaved by the enzyme thiolase into acetyl-CoA and a new fatty acyl-CoA. The fatty acyl-CoA re-enters the reaction sequence and successive acetyl-CoA units are generated.

Fatty acids with an even number of carbons yield only acetyl-CoA units but those with an odd number of carbons result in the formation of acetyl-CoA and propionyl-CoA. In plant tissues propionyl-CoA is degraded by a modified β-oxidation sequence to yield acetyl-CoA and carbon dioxide.

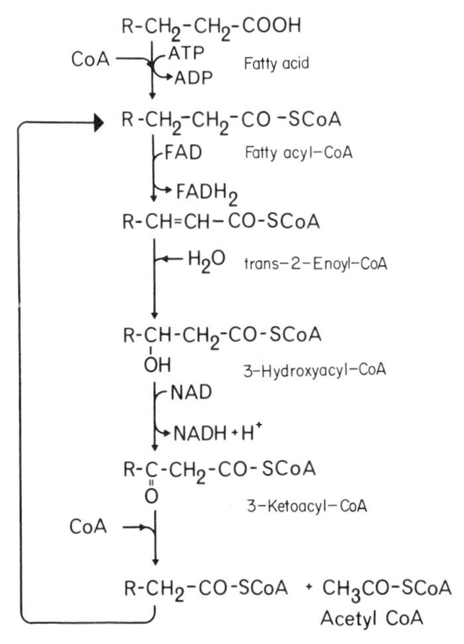

Fig. 6.3. The β-oxidation pathway.

The β-oxidation pathway was unequivocally demonstrated to occur in plant tissues by Stumpf and Barber (1956) who showed that long chain aliphatic acids were oxidized to carbon dioxide by mitochondrial preparations from germinating peanut cotyledons, when these preparations were supplemented with a number of cofactors including ATP, CoA, and NAD^+. The rate of release of $^{14}CO_2$ from specifiically labelled butyric and palmitic acids was consistent with their degradation by β-oxidation and subsequent oxidation by the TCA cycle.

6.4.2 THE GLYOXYLLATE CYCLE

The acetyl-CoA derived from fatty acid breakdown in germinating seeds could be consumed by the TCA cycle, as indicated by the experiments of Stumpf and Barber (1956). However this would not result in the net formation of a glucose precursor since for each molecule of acetyl-CoA consumed two molecules of carbon dioxide would be produced. Furthermore, it was known at the time that little of the fatty acid was oxidized to CO_2 but instead contributed to a net synthesis of sugars.

The problem of how acetyl-CoA was converted to a glucose precursor was solved by the discovery of the glyoxyllate cycle, by Kornberg and Krebs in 1957. This cycle represents a modification of the tricarboxylic acid cycle in which two molecules of acetyl-CoA are consumed and a molecule of succinic acid is formed (Fig. 6.4). Five enzymes are involved in this process three of which,

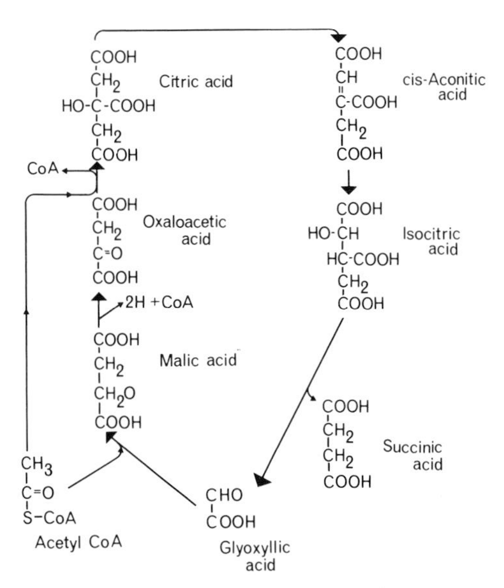

Fig. 6.4. The glyoxyllate cycle.

citrate synthetase, aconitase, and malic dehydrogenase are components of the TCA cycle. The remaining two enzymes, the key enzymes of the glyoxyllate cycle, are isocitric lyase (isocitratase) and malate synthetase. The first reaction of the glyoxyllate cycle, catalyzed by citrate synthetase, is the condensation of oxaloacetate and acetyl-CoA to form citrate, which is then converted to isocitrate by the action of aconitase. The next reaction, unique to this cycle, is the cleavage of isocitrate into succinate and glyoxyllate catalyzed by isocitrate lyase. One of the products of this reaction, glyoxyllate, is then condensed with a second molecule of acetyl-CoA under the catalytic action of malate synthetase, to produce one molecule of malate. Malate is then oxidized by malate dehydrogenase to oxaloacetate with the concomitant reduction of NAD^+. The overall equation for the cycle is therefore:

$$2 \text{ acetyl-CoA} + NAD^+ \rightarrow \text{succinate} + NADH + H^+$$

Succinate produced by the glyoxyllate cycle can then be converted to hexose by conversion to oxaloacetate, by the action of succinic dehydrogenase and fumarase. The oxaloacetate is converted to phosphoenolpyruvate, a reaction catalised by phosphoenolpyruvate carboxykinase,

$$\text{oxaloacetate} + ATP \rightleftharpoons \text{phosphoenolpyruvate} + CO_2 + ADP$$

and the phosphoenolpyruvate converted to glucose by a reversal of the reactions of the Embden-Meyerhof-Parnas pathway. Thus four molecules of acetyl-CoA will give rise to two molecules of succinate and this in turn will result in the formation of one molecule of glucose and the loss of two molecules of carbon dioxide (Fig. 6.5).

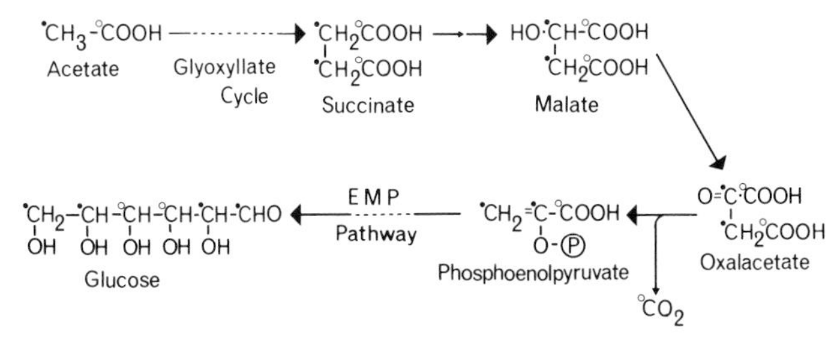

Fig. 6.5. The pathway of incorporation of [^{14}C] from [1-^{14}C]-acetate (O) or [2-^{14}C]-acetate (●) into glucose during gluconeogenesis in plant tissue.

The glyoxyllate cycle was first demonstrated in the bacterium *Pseudomonas* (Kornberg & Krebs, 1957) grown on two-carbon compounds, and has since been shown to operate in many micro-organisms and plant tissues. The evidence for the cycle is based on the presence of the two key enzymes, malate synthetase and isocitrate lyase, and on the distribution of ^{14}C in organic acids, sugars and carbon dioxide when the tissue is supplied with specifically labelled [^{14}C] acetate. Malate synthetase and isocitrate lyase are found in a wide variety of plant tissues (Carpenter & Beevers, 1958), particularly in fatty seedlings where they increase in activity during germination. Similarly the activities of aconitase and citrate synthetase increase in these tissues during germination. Incubation of castor bean endosperm tissue with [1-^{14}C] acetate or [2-^{14}C] acetate results initially in the formation of [^{14}C] malate, and subsequently the radioactivity from [1-^{14}C] acetate results in about an equal labelling of CO_2 and sucrose. In the cotyledons of germinating peanut and sunflower seedlings (Bradbeer & Stumpf, 1959) and castor bean endosperm (Canvin & Beevers, 1961), [1-^{14}C]-acetate was converted to carboxyl-labelled malate and to sucrose in which the glucose moiety was labelled in the 3 and 4 carbons, while [2-^{14}]C acetate gave rise to malate labelled in the methylene carbons and to sucrose where the glucose moiety was labelled in the 1, 2, 5 and 6 carbons. These patterns or labelling of the products of acetate metabolism are consistent with the operation of a glyoxyllate cycle in these tissues (Fig. 6.5).

6.4.3 METABOLIC FUNCTIONS OF THE GLYOXYSOME

The reactions of the glyoxyllate cycle and the pathway of β-oxidation were generally thought to be associated with the mitochondria since the enzymes of these pathways were usually present in the mitochondrial fraction isolated from cell homogenates. Elegant studies by Beevers' group at Purdue University however showed that sucrose density centrifugation of crude particulate fractions of castor bean endosperm resulted in the separation of three distinct bands of particles which were identified as proplastids, mitochondria and a new particle sedimenting at a high density. Enzymes of the glyoxyllate cycle, isocitrate lyase and malate synthetase were found exclusively in this dense particle together with catalase and a large proportion of the glycollate oxidase of the gradient. On the other hand citrate synthetase and malate dehydrogenase were associated with both the mitochondrial band and the band containing the new particle, while succinic dehydrogenase, fumarase and NADH oxidase were located exclusively in the mitochondrial band with cytochrome oxidase (Fig. 6.6).

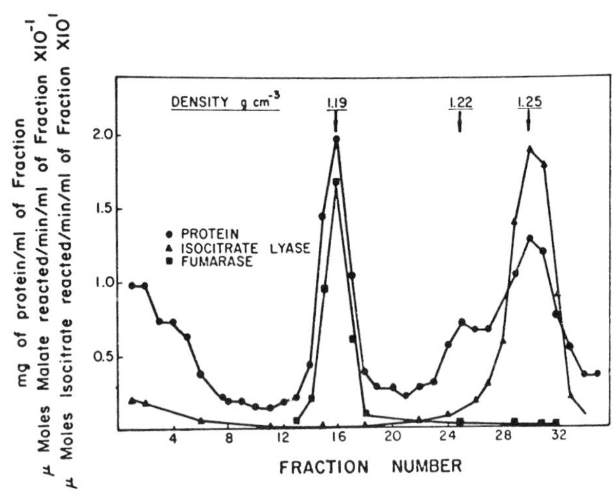

Fig. 6.6. The distribution of protein, fumarase and isocitric lyase after sucrose density gradient separation of the components of a crude particulate fraction of the endosperm of germinating castor bean. (Reproduced with permission from Breidenbach & Beevers, 1967.)

The enzyme distribution indicated that the TCA cycle enzymes were located in the mitochondria while the enzymes of the glyoxyllate cycle were compartmentalized in the denser particle. These particles were therefore called glyoxysomes (Breidenbach & Beevers, 1967; Breidenbach *et al.*, 1968). The isolated glyoxysomes were found to be organelles with a single unit membrane bounding a finely granular matrix. Similar structures were recognized in electron micrographs of intact castor bean endosperm tissue indicating that the isolated organelles were not artefacts of the isolation and centrifugation processes. Since the finding of glyoxysomes in endosperm tissue, they have been reported to be the site of the glyoxyllate cycle in the cotyledons of a number of fat-storing seeds including those of watermelon, sunflower, peanut and cucumber. Microbodies containing catalase and enzymes of the glyoxyllate cycle have also been found in yeast (Szabo & Avers, 1969) and although these have been called peroxisomes they clearly have the enzyme complements of glyoxysomes.

In addition to glyoxyllate cycle activity, the glyoxysomes were shown to be the site of β-oxidation in castor bean endosperm (Cooper & Beevers, 1969b; Hutton & Stumpf, 1969). Glyoxysomes isolated from this tissue oxidized palmitoyl-CoA to acetyl-CoA with a concomitant reduction of NAD^+ and uptake of oxygen. Addition of [^{14}C] oxaloacetate during this reaction resulted in the formation of [^{14}C] citrate and [^{14}C] malate from palmityl-CoA indicating that the acetyl-CoA produced by the β-oxidation process was consumed in the glyoxyllate cycle also located in the organelle (Cooper & Beevers, 1969b). Similarly, ricinoleate, linoleate and palmitate were oxidized by glyoxysomes of castor bean with the formation of acetyl-CoA, and three enzymes of the β-oxidation complex, crotonase, β-ketothiolase, and β-hydroxyacyl dehydrogenase were found to be located specifically in the organelle (Hutton & Stumpf, 1969). The activity of the β-oxidation complex in this tissue during the germination of the seed was found to parallel the increase in activity of the glyoxyllate cycle enzymes indicating that an integrated system for lipid utilization develops together with the formation of the glyoxysome.

The principal pathways of gluconeogenesis are therefore compartmentalized in a single organelle, the glyoxysome. The complete pathway of gluconeogenesis however involves at least three organelles, the spherosome, the glyoxysome and the mitochondrion. The reactions are initiated by a hydrolysis of lipids in the spherosome by the action of lipase, and the glycerol and fatty acids produced diffuse out of the organelle. Glycerol is utilized directly by the EMP pathway in the cytosol and contributes to sucrose synthesis (Beevers, 1956) while the fatty acids diffuse into glyoxysomes which are located near the spherosome. In the glyoxysome the fatty acid is oxidized in the β-oxidation pathway and the acetyl-CoA released is converted to succinate by the action of the glyoxyllate cycle enzymes located in the organelle. Although the β-oxidation pathway is reversible, the equilibrium is presumably maintained in a catabolic direction by the removal of peroxide, produced in the oxidation step of the pathway, by the catalase present in the glyoxysome. Reduced NAD^+ produced in these reactions is probably oxidized in the mitochondria. Succinate produced in the glyoxysome is further metabolized to oxalacetate and finally to phosphoenolpyruvate in the mitochondria since succinic dehydrogenase and fumarate are not component enzymes of the glyoxyllate cycle and are absent from glyoxysomes.

This compartmentalization of the β-oxidation complex and the glyoxyllate cycle together in an organelle discrete from the mitochondrion is probably the reason why acetyl-CoA is utilized in gluconeogenesis in plant tissues rather than being oxidized to carbon dioxide and water as in animal tissues. Free acetyl-CoA is presumably not released from the glyoxysome and made available for oxidation in the mitochondrion, while consumption of succinate, the final product of the glyoxyllate cycle, by the mitochondrion, would not stimulate the rate of oxidation in the TCA cycle since this can only occur by increasing the supply of acetyl-CoA. Some measure of control of competing metabolic pathways is therefore achieved by separation of these reactions within different organelles.

6.5 PEROXISOMES

Glycollic acid is produced in large amounts in the chloroplast, as a by-product of the reactions of carbon dioxide fixation. The formation of glycollate has been proposed to occur by the oxidation of ribulose-1,5-bisphosphate (RBP) by molecular oxygen, a reaction which would produce a two carbon fragment of phosphoglycollic acid and a three carbon fragment of phospholgyceric acid (PGA) instead of two molecules of PGA resulting from the normal carboxylation of RBP by CO_2. It has been found that RBP carboxylase acts as an oxygenase in the presence of molecular oxygen and that phosphoglycollate and PGA are produced in this reaction *in vitro* (Andrews *et al.*, 1973). Thus the enzyme RBP carboxylase can act as an oxygenase or as a carboxylase and the formation of phosphoglycollic acid is favoured by high partial pressures of oxygen. Glycollate is produced from phosphoglycollate by the action of phosphoglycollate phosphatase, a chloroplast enzyme. Glycollate is also produced, *in vitro*, from fructose-6-phosphate by the action of the chloroplast enzyme, transketolase, which may be due to the oxidation, by hydrogen peroxide, of the glycolaldehyde-thiamine pyrophosphate, an intermediate complex in this enzyme reaction (Bradbeer & Racker, 1961). Glycollate is released from the chloroplast into the cytosol, where it is further metabolized in a specific metabolic pathway, the initial reactions of this pathway being located in the peroxisome.

6.5.1 THE GLYCOLLATE PATHWAY

The pathway of glycollate metabolism in leaves has been elucidated by infiltrating excised leaves or leaf discs with ^{14}C-labelled glycollate or other intermediates of the pathway. Glycollate is converted rapidly to glyoxyllate which may be oxidized non-enzymatically to carbon dioxide and formic acid, in the presence of hydrogen peroxide. In this reaction the carbon dioxide is derived from the carboxyl carbon of glycollate and the formate from the methyl carbon (Tolbert & Burris, 1950). However the presence of catalase, which breaks down peroxide, is thought to preclude such a total degradation of glyoxyllate in the peroxisome and the supply of ^{14}C-labelled glycollate or glycoxyllate to leaf tissue in the light has been found to give rise initially to labelled glycine and serine and subsequently to labelled glyceric acid hexoses and sucrose (Tolbert, 1963; see Fig. 6.7).

Infiltration of [2-^{14}C] glycollate into leaves in the light was found to give [2-^{14}C] glycine but serine was found to be labelled in the 2 and 3 carbons (Tolbert & Cohan, 1953; see Fig. 6.7). From this evidence it was concluded that two molecules of glycine give rise to one molecule of serine with a loss of one molecule of carbon dioxide; the 1 and 2 carbons of serine are derived from carbons 1 and 2 of glycine respectively, while the 3 carbon of serine is derived from the 2 carbon of glycine and carbon dioxide arises from carbon 1 of glycine. In wheat leaves in light [3-^{14}C] serine was converted to [3-^{14}C] glycerate

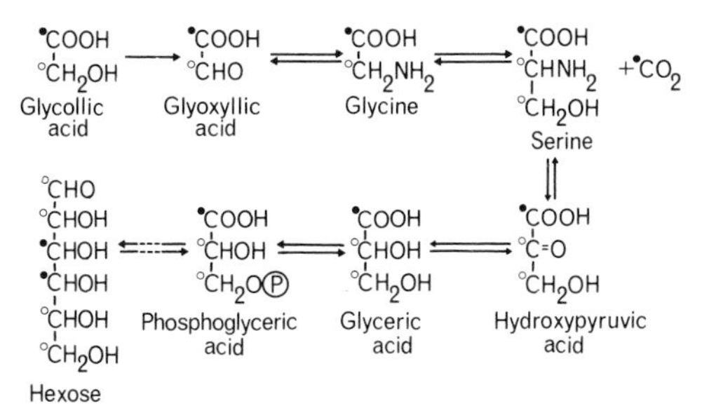

F[g. 6.7. The glycollate pathway. The labelling pattern of intermediates of the pathway ane hexose is indicated for when they are derived from [1-^{14}C]-glycollate (●) or [2-^{14}C]-glycollate (○).

and this was incorporated into hexose presumably by the reactions of the Embden-Meyerhof-Parnas pathway (Rabson *et al.*, 1962). This pathway of hexose formation from glycollate has been confirmed by thc finding that [2-^{14}C] glycollate gives rise to glycose labelled in the 1, 2, 5 and 6 carbons and [3-^{14}C] serine to glucose labelled in the 1 and 6 carbons (Jiminez *et al.*, 1962).

The glucollate pathway is therefore gluconeogenic in light. The conversion of glyceric acid derived from glycollate, to glucose and sucrose is inhibited by DCMU an inhibitor of Photosystem II (Miflin *et al.*, 1966). In the dark, supplied [^{14}C]-glycollate is converted into TCA cycle acids rather than sugars which may be due to an oxidation of pyruvate derived from glyceric acid, or may result from a direct conversion of glyoxyllate to malate.

Glycollate produced during photosynthesis in the presence of $^{14}CO_2$ is usually found to be uniformly labelled i.e. both carbons have the same specific activity. This distribution of radioactivity would be expected if the glycollate was derived from carbons 1 and 2 of ribulose-1,5-bisphosphate. Consequently all the compounds arising from glycollate are uniformly labelled. Serine, in particular, has been found to be uniformly labelled while phosphoglyceric acid was predominately carboxyl labelled, indicating that serine was produced from glycollate rather than directly from phosphoglyceric acid formed in CO_2-fixation (Rabson *et al.*, 1962). The uniformly labelled glyceric acid formed in this pathway in turn produces uniformly labelled hexoses instead of the 3,4-^{14}C-hexoses resulting from incorporation of ^{14}C from the photosynthetic carbon cycle.

The operation of the glycollate pathway in leaves has been shown by tracer experiments but it has been confirmed by the detection of enzymes in leaves which are necessary for some reactions of the pathway. The initial reaction, the oxidation of glycollate to glyoxyllate is catalysed by glycollate oxidase, an enzyme first isolated by Zelitch and Ochoa (1953). It has FMN as the prosthetic group and utilizes molecular oxygen as the electron acceptor. The enzyme is competitively inhibited by bisulphite addition compounds of aldehydes, α-hydroxy-sulphonates, which have the general structure R-CHOH-SO_3H and are therefore structural analogues of glycollate. The most commonly used α-hydroxysulphonate is α-hydroxypyridylmethane sulphonate (HPMS) and treatment of leaf discs or infiltration of excised leaves with this compound results in the accumulation of glycollic acid while having no effect upon the rate of CO_2-fixation. Experiments of this type have shown that 40 to 70% of the carbon fixed in photosynthesis will accumulate as glycollate in tissues treated with HPMS and these results have been interpreted as indicating that a large fraction of the carbon fixed in photosynthesis is metabolized by this pathway.

Transaminases are present in leaves which catalyse two steps of the pathway: a glutamate-glyoxyllate transaminase catalysing the formation of glycine is widespread in plant tissues, and a serine-pyruvate aminotransferase catalysing the formation of serine to hydroxypyruvate is found in leaves. The conversion of glycine to serine is catalysed by serine hydroxymethyltransferase which has been found in pea and wheat leaves (Cossins & Sinha, 1966). This reaction is inhibited by isonicotinyl hydrazide, and treatment of leaf tissue with this compound during photosynthesis in $^{14}CO_2$ causes an accumulation of [^{14}C] glycine and [^{14}C] glycollate and a decrease in the incorporation of ^{14}C into glucose (Miflin *et al.*, 1966). Use of this inhibitor thus provides additional evidence for the operation of the pathway.

Two types of glyoxyllate reductase have been found to occur in leaves. One, an NADP-linked enzyme, is thought to be located in the chloroplast, while a second NAD-linked enzyme is present in the cytoplasm and is referred to as hydroxypyruvate reductase since the enzyme isolated from some sources has a higher activity to hydroxypyruvate than to glyoxyllate. The enzyme appears to catalyse the reduction of hydroxypyruvate to glycerate in the glycollate pathway.

6.5.2 METABOLIC REACTIONS OF THE PEROXISOME

It was generally accepted that the enzymes of the glycollate pathway were soluble proteins of the cytosol and the failure of several attempts to localize these enzymes, particularly glycollate oxidase, in discrete organelles confirmed this idea. The first successful localization of enzymes of glycollate metabolism in a discrete organelle was achieved by Tolbert and coworkers at Michigan State University (Tolbert *et al.*, 1968). They demonstrated that sucrose density gradient centrifugation of spinach homogenates separated three bands of particles: broken chloroplasts, mitochondria, and small bodies distinctly separated from, and denser than the other organelles. Electron microscopic examination of these bodies showed them to be organelles bounded by a single unit membrane and since they closely resembled peroxisomes from animal cells in size and morphology, they were referred to as leaf peroxisomes. The most important finding was that these peroxisomes contained the bulk of the activity of the glycollate oxidase, catalase and hydroxypyruvate reductase (NAD-glyoxyllate reductase) of the gradient, while cytochrome *c* oxidase was specifically located in the mitochondrial fraction. Peroxisomes were later isolated from the leaves of nine other plant species including tobacco, maize, and sugarcane, and all contained the same enzyme complement (Tolbert *et al.*, 1969). These studies therefore demonstrated that processes of glycollate oxidation and peroxide breakdown were localized together in an organelle discrete from the chloroplast. Electron microscope studies, particularly by Newcomb, have since revealed the presence of peroxisomes in leaves of many plant species (e.g. Frederick & Newcomb, 1969).

Further studies in Tolbert's laboratory have shown that other enzymes of the glycollate pathway are also localized in the peroxisome. Two aminotransferases, glutamate-glyoxyllate aminotransferase catalysing the conversion of glyoxyllate to glycine, and serine-pyruvate aminotransferase catalysing the conversion of serine to hydroxypyruvate, were found in peroxisomes isolated from leaves of various species. Serine hydroxymethyltransferase is the only enzyme of the glycollate pathway not found in the peroxisome, and is probably located in the mitochondrion. Supply of [^{14}C] glycollate and [^{14}C] glyoxyllate to isolated peroxisomes gave rise only to [^{14}C] glycine and while oxygen uptake occurred, no $^{14}CO_2$ release was detected (Kisaki & Tolbert, 1969).

The glycollate pathway appears to require enzymic steps located in three subcellular organelles (Fig. 6.8). Glycollate, formed in the chloroplast is

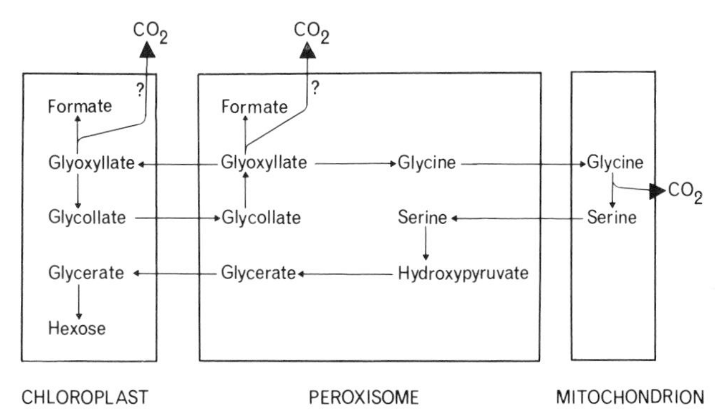

Fig. 6.8. The distribution of the reactions of the glycollate pathway among organelles of the leaf cell.

oxidized to glyoxyllate in the peroxisome. The glyoxyllate may then be exported to the chloroplast where it could be reduced to glycollate by the action of NADP-dependent glyoxyllate reductase which is located specifically in the chloroplast (Tolbert *et al.*, 1970). Such a coupling of alternate oxidation and reduction reactions. a 'glycollate-glyoxyllate shuttle', has been proposed as a mechanism for controlling the levels of reduced NADP in the chloroplast, but no unequivocal evidence for such a reaction *in vivo* has been found. Glycine is formed from glyoxyllate in the peroxisome and then transferred to the mitochondrion where it is converted to serine, with a concomitant loss of carbon dioxide. The conversion of serine to glycerate can then be accomplished by enzymes localized in the peroxisome. Further metabolism of glycerate to hexose appears to be confined to the chloroplast since these reactions would be initiated by the formation of phosphoglyceric acid, a reaction catalysed by phosphoglycerate phosphatase which is located in the chloroplast.

The metabolism of glycollate by leaf tissue results in the release of carbon dioxide, but the exact site of this CO_2 release in the cell is still a controversial question. It has been proposed that CO_2 is evolved by the mitochondria during the conversion of glycine to serine, since [^{14}C] glycine is as good a precursor as [^{14}C] glycollate for $^{14}CO_2$ evolution in leaves and the $^{14}CO_2$ evolved is derived from the carboxyl groups of these compounds. The presence of catalase in the peroxisomes is thought to minimize the non-enzymatic oxidation of glyoxyllate by H_2O_2 to formate and carbon dioxide and no evolution of CO_2 by isolated peroxisomes from glycollate or glyoxyllate could be detected by some workers (Kisaki & Tolbert, 1970). This loss of CO_2 from glycine, which amounts to only 25% of the total carbon passing through the glycollate pathway, would not account for the large losses of CO_2 which occur as a result of the photorespiration of glycollate in leaves. However the amount of catalase present in the peroxisome may not *preclude* the non-enzymatic oxidation of glyoxyllate and it has been shown that both [^{14}C] glycollate and [1-^{14}C] glyoxyllate can be decarboxylated by peroxisomal fractions at pH 8·0 (Halliwell & Butt, 1974). An enzyme is also present in chloroplasts which catalyses the decarboxylation of glyoxyllate to formic acid and CO_2 (Zelitch, 1972). Present evidence indicates therefore that three subcellular organelles have the capacity to decarboxylate components of the glycollate pathway and each may contribute to the production of the carbon dioxide in photorespiration (Fig. 6.8).

6.5.3 PHOTORESPIRATION

The oxidation of glycollate in the peroxisome is accompanied by a consumption of oxygen and ultimately results in the release of carbon dioxide. The net result is a respiratory gas exchange where the substrate of respiration is glycollate rather than glucose. Since glycollate is only formed in light this respiration is light-dependent and is called *photorespiration*.

The direct measurement of oxygen uptake during a net photosynthetic release of oxygen, or CO_2 release during net photosynthetic CO_2 fixation is impossible. However, indirect methods have demonstrated that photorespiration occurs in plants and that the rate of CO_2 loss in light is higher than that in the dark. Measurements of photorespiration rates vary considerably with the assay method used and can only be considered as approximations of the magnitude of the process. Nevertheless, it is becoming clear that the process of photorespiration has great importance in decreasing the rate of net photosynthesis in many plants.

Photorespiration can be detected by measuring the flux, that is the simultaneous uptake and release, of oxygen or carbon dioxide of photosynthesizing tissue by using isotopic tracer methods. The uptake of $^{18}O_2$ by leaves during photosynthesis has been detected but experiments using $^{18}O_2$ have generally given equivocal results which have been difficult to interpret.

The measurement of carbon dioxide flux during photosynthesis is much more convenient and depends upon the accurate measurement of carbon dioxide concentration in the atmosphere using an infra-red gas analyser and a simultaneous measurement of the total activity of supplied $^{14}CO_2$ with an ion counter or Geiger-Müller counter. When a plant is placed in a closed system in light, there is a rapid uptake of carbon dioxide and the CO_2 concentration of the atmosphere around the plant decreases to a concentration at which the uptake of CO_2 exactly balances the output of CO_2 by the plant (Fig. 6.9). This concentration of carbon dioxide is called the *CO_2 compensation point* of the plant and is usually measured in parts per million (ppm) of CO_2 in air. Plants such as tobacco, sunflower and wheat have compensation points ranging from 35 to 100 ppm indicating a marked loss of CO_2 (i.e. photorespiration) during photosynthesis, but others such as maize and sugarcane have compensation points of 3 to 10 ppm indicating a low loss of CO_2 or a low photorespiration rate. If photosynthetic carbon fixation is similarly measured in a closed-system but in an atmosphere containing $^{14}CO_2$, the CO_2 arising from the plant by photorespiration will, over short time periods, be $^{12}CO_2$. Thus, while a decrease in the CO_2 *concentration* of the atmosphere around the plant will occur, the *radioactivity* of $^{14}CO_2$ in the atmosphere will appear to decrease at a faster rate because of the efflux of unlabelled CO_2 from the plant, and continues to decrease even after the compensation point is reached (Fig. 6.9a). In a plant which

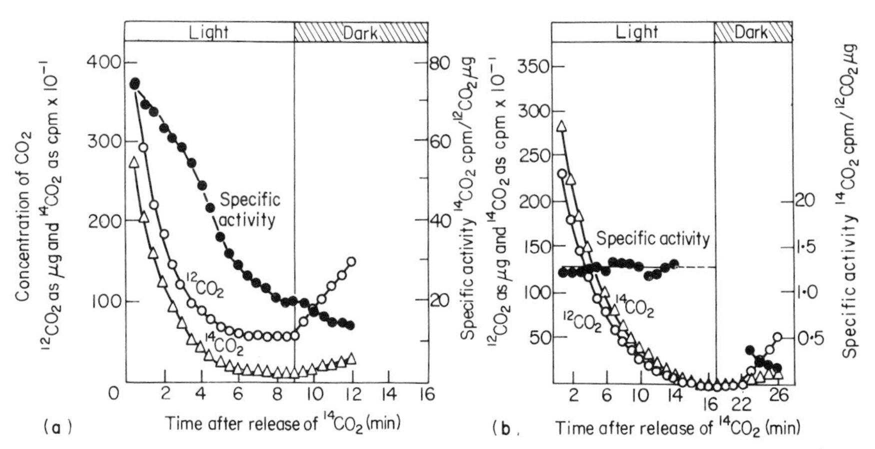

Fig. 6.9. The concentration of CO_2 and $^{14}CO_2$ and the specific activity of $^{14}CO_2$ around a detached sunflower leaf (a) and a detached maize leaf (b) during illumination and in subsequent darkness in an atmosphere of 21% oxygen and at 21°. (Reproduced with permission from Hew *et al.*, 1969.)

is photorespiring, therefore, a decrease in the *specific radioactivity* of $^{14}CO_2$ will occur, while in a plant with no photorespiration the uptake of $^{14}CO_2$ will not be accompanied by an efflux of $^{12}CO_2$ and the specific radioactivity of the $^{14}CO_2$ in the atmosphere will remain constant (Fig. 6.9b). In the dark the specific radioactivity of the $^{14}CO_2$ decreases with the efflux of $^{12}CO_2$ and the rate of decrease of this activity is a measure of dark respiration. The rate of photorespiration as measured by these methods has been shown to be 1·5 to 2·5 times that of dark respiration, while in plants with low photorespiration, such as maize, loss of CO_2 in light is only a fraction of the rate of CO_2 loss in the dark. This method clearly indicates that photorespiration occurs in some species while it is absent in others.

Another method which has been used to detect differences in the rate of photorespiration and dark respiration was devised by Zelitch (1968). In this method leaf discs are allowed to fix $^{14}CO_2$ for a period of 45 to 60 minutes. The remaining $^{14}CO_2$ is then quickly flushed out of the closed system, and the release of $^{14}CO_2$ from the tissue into CO_2-free air is measured over short periods in the light and the dark. The method is based on the assumptions that the rate of $^{14}CO_2$ loss is a measure of total CO_2 loss i.e. that the specific radioactivity of the CO_2 evolved remains constant over short time periods and that low CO_2 tensions have no effect on the loss of CO_2. These assumptions may not be valid for all photosynthetic tissues but within these limitations the method is a very rapid and sensitive means for detecting photorespiration. Photorespiratory loss of CO_2 in tobacco has been shown by this method to be 2 to 5 times higher than in the dark while CO_2 loss from maize is not detectable (Fig. 6.10a). This method has been similarly used to detect photorespiration in other species.

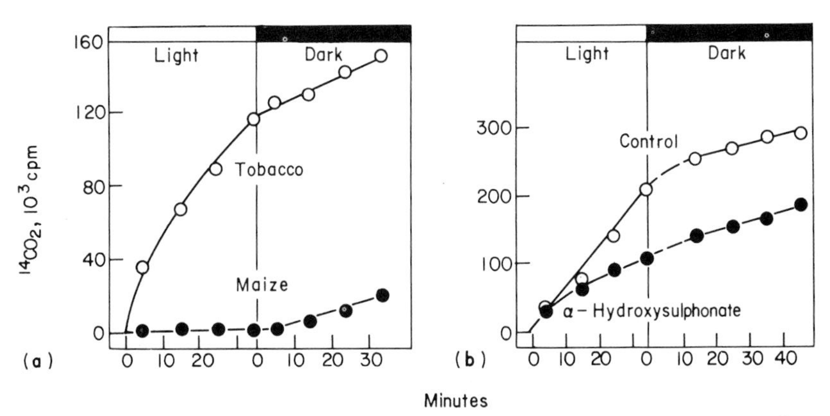

Fig. 6.10.a. A comparison between the release of $^{14}CO_2$ by tobacco and maize discs in the light and the dark after previously being supplied $^{14}CO_2$ in light.

(b). The effect of α-hydroxysulponate on the releaseof $^{14}CO_2$ from tobacco leaf discs in the light and dark. (Reproduced with permission from Zelitch, 1968.)

Photorespiration is not simply a stimulation of dark respiration since the two processes of photorespiration and dark respiration respond differently to changes in oxygen concentration. Dark respiration of leaves has an optimal rate at about 2% oxygen in air and any increase in the O_2 concentration up to 100% does not increase the rate (Fig. 6.11). Net photosynthesis however is inhibited

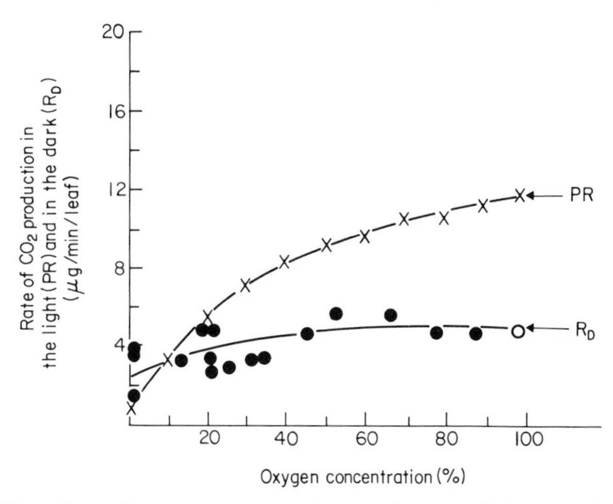

Fig. 6.11. The effect of oxygen concentration on the rate of photorespiration (PR) and dark respiration (R_D) of detached soybean leaves. (Reproduced with permission from Forrester *et al.*, 1966.)

by oxygen, a phenomenon called the *Warburg effect*, and this is attributable to an increase in the rate of photorespiration rather than to an inhibition of photosynthesis *per se*. A reduction in the oxygen concentration in the atmosphere around a leaf from the ambient 21% O_2 of air has been shown to lower the compensation point and increase the rate of net photosynthesis, while an increase in concentration raises the compensation point and lowers net photosynthesis (Forrester *et al.*, 1966). Plants with photorespiration have been found to evolve CO_2 at a high rate when they are transferred to darkness after a period of photosynthesis. This post-illumination CO_2 burst lasts for a few minutes before a steady dark respiration is established and the magnitude of the burst has been found to depend on the light intensity and oxygen concentration during the previous period of photosynthesis: the CO_2-burst increases with an increase in light intensity and decreases with a decrease in O_2 concentration (Tregunna *et al.* 1961). This burst has been explained as being due to the continued slow oxidation, in the dark, of a product produced in photosynthesis after the uptake of CO_2 by photosynthesis has stopped. Photorespiration therefore appears to have different characteristics than dark respiration and to have a different substrate.

The substrate for photorespiration is thought to be glycollic acid. Glycollic acid production in leaves and in chloroplasts is stimulated by an increase in oxygen concentration as is photorespiration, while the inhibition of glycollate oxidation by HPMS has been found to lower the rate of photorespiration in leaf discs to that of dark respiration (Fig. 6.10b). Photorespiratory loss of CO_2 is also stimulated by the infiltration of leaves with glycollic acid while acetate has no effect on this rate. Thus the oxidation of glycollate, mediated by the leaf peroxisomes results in a photorespiratory loss of carbon dioxide by the leaf.

Plants with low compensation points are said to lack photorespiration. These plants are C_4 plants, that is, the primary carboxylation step is catalysed by phosphoenolpyruvate carboxylase rather than by RBP carboxylase. This efficient fixation of CO_2 has been postulated to preclude the formation of glycollate, but maize for example, has been shown to possess a glycollate pathway (Osmond, 1969) and peroxisomes occur in the mesophyll cells of the leaf. Presumably in such plants glycollate may be formed and oxidized, but the efficiency of refixation of the CO_2 resulting from glycollate oxidation is such that no CO_2 is released from the plant and hence no photorespiration is detected.

Estimates from various plant species suggest that from 15 to 40% of the carbon fixed in photosynthesis is lost by photorespiration. There is no unequivocal evidence to show that energy in the form of ATP is recovered during the oxidation of glycollate and the process appears to be a wasteful one in terms of energy conservation. An important question therefore arises: does photorespiration serve a useful function or is it simply an inevitable consequence of photosynthesis being carried on in an atmosphere of 21% oxygen? It has been suggested that photorespiration might act as a 'safety valve' for the plant, in that, under conditions of high light intensity and low carbon dioxide concentration, the oxidation of glycollate would consume both excess reduced $NADP^+$ and excess oxygen, which would serve to protect the chloroplast from photo-oxidative damage. If the process of photorespiration imposes limitations on the growth of plants, it would be expected that there would have been some selective pressure to eliminate it by natural selection during the evolution of the higher plants. The occurrence of C_4 plants, which have low photorespiration, may represent such an evolutionary step to correct for the presence of photorespiration by the development of an efficient CO_2-fixation mechanism. Since variations in the rate of photorespiration occur within a single species (Zelitch, 1971) it may be possible to select artificially for low photorespiration in crop plants and consequently increase net photosynthetic productivity and crop yield.

6.6 ONTOGENY AND TURNOVER OF MICROBODIES

The development of microbodies in plant cells is a difficult process to follow since they usually occur in small numbers and are difficult to characterize enzymatically. Small, membrane-bound organelles occur in close proximity to the endoplasmic reticulum but these do not always give a positive reaction to DAB, the usual criterion for the presence of catalase. Nevertheless catalase is invariably the first enzyme to be detected in microbodies and it appears to be the most active enzyme in the developing organelle. Studies of microbody development have been carried out principally on tissues having large populations of microbodies such as the endosperm and cotyledons of fatty seedlings.

In the endosperm of fat-storing seeds, particularly of the castor bean, the process of germination results in the development of large numbers of microbodies which reach numbers per cell twice that of mitochondria (Vigil, 1970). This increase in number is correlated with increases in the activities of isocitrate lyase and malate synthetase in the whole tissue. It has been suggested by several investigators that microbodies develop from the endoplasmic reticulum by a budding process. In castor bean the formation of microbodies has been linked directly to the endoplasmic reticulum and connections between the cisternae of the rough endoplasmic reticulum and developing microbodies have been clearly demonstrated (Vigil, 1970). This connection does not appear to persist during the whole course of organelle development although they increase seven-fold in size and increase in density suggesting a continuous addition of newly synthesized protein. The depletion of lipid in the endosperm or cotyledons of fatty seedlings is accompanied by a decrease in glyoxysomal enzymes which is attributable to a decrease in the number of glyoxysomes as measured by enzyme activity and total protein (Gerhardt & Beevers, 1970). It is not clear how these microbodies are destroyed but some at least become enclosed in autophagic vacuoles and are digested.

Increases in peroxisomal enzymes have been detected in germinating seedlings and etiolated seedlings on exposure to light and some of these changes have been correlated with the size and structure of microbodies. The development of microbodies in leaves of wheat seedlings in light involves distinct changes in the biochemical and physical characteristics of the organelle. During the first few days of germination an increase occurred in the density of catalase-containing particles from 1·17 to 1·24 g cm^{-3} on separation on sucrose gradients, which was paralleled by an increase in catalase, hydroxypyruvate reductase and glycollate oxidase (Fierabend & Beevers, 1972).

In etiolated leaves small microbodies occur which are rich in catalase but low in other peroxisomal enzymes. These bodies have not been unequivocally characterized as peroxisomes because of the difficulty of isolation. Upon exposure to light there is an activation of glycollate oxidase and hydroxypyruvate reductase, but these changes in enzyme activity have not been correlated with an increase in size and number of peroxisomes. The activity of glycollate oxidase is stimulated by exposure of etiolated tissue to red light and this activation is reversed by far-red light. These effects probably reflect a stimulation of leaf development rather than a specific phytochrome involvement in peroxisome development. However an increase in glycollate oxidase occurs under continuous far-red light in mustard seedlings where chloroplast development is inhibited, suggesting that substrate activation is not involved in this increase of enzyme activity. Similar increases in the activities of glycollate oxidase, and hydroxypyruvate reductase have been found in cotyledons of germinating seedlings on exposure to light (Trelease *et al.*, 1971; Kagawa *et al.*, 1973).

In germinating seeds and in greening leaves the microbodies have easily identifiable functions, those of glyoxysomes or peroxisomes, and only one type of microbody occurs in each tissue. However an interesting situation occurs in some seeds of the *Cucurbitacea*. In cucumber and watermelon for example, lipid is the main storage material in the seed and during germination there is an increase in activity of glyoxysomal enzymes in the cotyledon. During normal development, the cotyledons become exposed to light and turn green. The glyoxysomal activity of the microbodies, as measured by the activity of malate synthetase and isocitrate lyase, decreases while the peroxisomal activity, as measured by glycollate oxidase activity, increases. In one organ therefore in response to environmental stimulus there is a transition from lipid breakdown and gluconeogenesis, to glycollate metabolism, which is reflected in a change in the microbody population from a glyoxysomal to a peroxisomal function. Since the two microbodies are morphologically indistinguishable the question arises as to whether the structure of the glyoxysome is retained and there is a selective replacement of enzymes to give it a peroxisomal function or whether there is an autolysis of the glyoxysome and a replacement of it by a newly synthesized peroxisome.

The changes in microbody enzyme complements which occur during seedling development have been correlated with changes in the fine structure of the microbodies. It has been reported that in cucumber seedlings the decrease in glyoxysomal enzyme activity is not accompanied by a decrease in the number of microbodies as determined by electron microscopy. The subsequent exposure of the cotyledons to light caused a great increase in the activity of peroxisomal enzymes, particularly that of glycollate oxidase, while there was not a concomitant increase in the number of microbodies (Trelease *et al.*, 1971). No evidence was found for the degradation of microbodies during the loss of glyoxysomal enzyme activity as has been reported in castor bean (Vigil, 1970). These studies suggest that there is a continuity of microbody structure in the transition from a glyoxysomal to a peroxisomal function. On the other hand, in developing watermelon seedlings, the decrease in the total activity of glyoxysomal enzymes in isolated microbodies which occurred after the depletion of lipid, was accompanied by a decrease in the total protein of these microbodies. This was interpreted as indicating a destruction of glyoxysomes in the tissue (Gerhardt & Beevers, 1970). Normally the peroxisomal activities of the microbody increase as glyoxysomal activity decreases, but in this tissue the decrease in glyoxysomal activity of the microbodies could be separated from the increase in the peroxisomal activity. Brief exposure of dark-grown cotyledons to light at an early stage of development stimulated the activities of peroxisomal enzymes, while this did not affect the rate of decline of the activity of glyoxysomal enzymes (Kagawa *et al.*, 1973). These results suggest that the formation of peroxisomes can occur while active glyoxysomes are present and that two populations of microbody can occur in the tissue at the same time. The apparent contradiction in the results from these two tissues has still to be resolved, and it is still not clear whether the transition from a glyoxysomal function to a peroxisomal function in the microbodies of these tissues is due to enzyme replacement in a single microbody or to synthesis of a new microbody.

6.7 ALGAL MICROBODIES

The microbodies of photosynthetic higher plants have distinct physiological roles either as peroxisomes or glyoxysomes which are active at different stages of the life cycle of the plant. In the algae however their role is not quite as clear since both functions of the microbody, i.e. glycollate oxidation and gluconeogenesis, may be carried on in a single cell. Microbody-like organelles have been found in a wide range of algae, and they are most numerous in algal cells grown on two-carbon compounds. Catalase has been detected cytochemically by DAB staining *in situ* but has not been shown to be universally present in algal microbodies.

Algae grown autotrophically, that is, in light on carbon dioxide, produce glycollic acid which is metabolized by a pathway similar to that of higher plant leaves (Bruin *et al.*, 1970; Lord & Merrett, 1970). Glycollate oxidase however is not present in algae, with the possible exception of some members of the Zygnematales, Ulotrichales and Charophyceae (Fredrick *et al.*, 1973) but the oxidation of glycollate is catalysed by glycollate dehydrogenase. This enzyme is not FMN dependent and does not couple to molecular oxygen so that hydrogen peroxide is not formed during glycollate oxidation (Nelson & Tolbert, 1970). This may to some extent account for the low activities of catalase reported in many algae. The natural electron acceptor for glycollate dehydrogenase is as yet unknown.

While the presence of a glycollate pathway has been established in green algae, the extent to which glycollate is metabolized by this pathway during photosynthesis is not known. In contrast to higher plant cells, serine formed during photosynthesis in $^{14}CO_2$ in several algae is carboxyl labelled indicating its formation directly from PGA rather than by the glycollate pathway (Bruin *et al.*, 1970). There is also a lower rate of CO_2 loss by photorespiration in algae than in higher plants (Cheng & Colman, 1974) suggesting perhaps that little glycollate is formed in algal photosynthesis. Glycollate is excreted by algal cells and this has been interpreted to indicate that the algae have a low rate of glycollate metabolism. This can now be discounted however since little glycollate is excreted during steady-state photosynthesis (Watt & Fogg, 1966; Colman *et al.*, 1974) and all the glycollate formed appears to be metabolized in the algal cell.

Many unicellular algae can be grown in the dark on two-carbon compounds such as acetate or ethanol as sources of both energy and carbon for growth. *Chlorella* grown in the dark on acetate, incorporates [^{14}C]-acetate into protein and carbohydrate without prior degradation while simultaneously about half of the acetate taken up is oxidized to CO_2 in a manner consistent with the operation of the TCA cycle (Syrrett *et al.*, 1964). These cells have high levels of malate synthetase and isocitrate lyase, thereby allowing the formation, from acetate, of four-carbon compounds which can form precursors of amino acids and hexoses (Syrrett *et al.*, 1963). Autotrophically-grown cells however have little isocitrate lyase activity and [^{14}C]-acetate supplied to these cells in the dark is oxidized to CO_2. Furthermore the activity of isocitrate lyase increases when autotrophically-grown cells are supplied with acetate in the dark and no cell division takes place during this time. Similar increases in malate synthetase and isocitrate lyase in response to growth on two-carbon compounds have been reported in *Euglena gracilis*.

The enzymes of both the glycollate pathway and the glyoxyllate cycle have therefore been found in several algae and the question arises as to whether these enzymes are compartmentalized in a microbody. The limited number of studies which have been done suggest, on the basis of their enzyme activities, that microbodies isolated from algae have either a glyoxysomal function or are of the non-specialized type often present in non-green plant tissues (Huang & Beevers, 1971). In *Euglena gracilis* grown on ethanol the clyoxylate cycle enzymes malate synthetase and isocitrate lyase together with glycollate dehydrogenase and glyoxyllate reductase were detected in a particulate fraction distinct from mitochondria and, while no catalase activity was detected, the enzyme complement indicates a microbody with a glyoxysomal function (Graves *et al.*, 1972). However in another flagellate, *Chlorogonium elongatum*, grown either on acetate in the dark or autotrophically in the light, a catalase-containing microbody was found which had neither glyoxysomal nor peroxisomal enzymes associated with it. In the acetate-grown cells isocitrate lyase and malate synthetase were found to be located in the cytosol (Stabenau & Beevers, 1974), while in autotrophically-grown cells, glycollate dehydrogenase, hydroxypyruvate reductase and glyoxyllate-glutamate aminotransferase were$_2$found to be localized in a mitochondrial fraction together with cytochrome oxidase, and malate and isocitrate dehydrogenases (Stabenau, 1974).

It is apparent that algal microbodies do not have a peroxisomal function and that this is correlated with the oxidation of glycollate in these cells by glycollate dehydrogenase, thus eliminating the requirement for the compartmentalization of glycollate oxidation with catalase. However, microbodies have been isolated from only a few algae primarily because vigorous methods are required to break algal cell walls and microbody isolation is therefore difficult to achieve. It remains to be seen whether peroxisomes occur in algae such as *Nitella* and *Spirogyra* which have relatively high levels of catalase and also have glycollate oxidase rather than glycollate dehydrogenase.

FURTHER READING

Beevers H. (1969) Glyoxysomes of castor bean endosperm and their relation to gluconeogenesis. *Ann. N.Y. Acad. Sci.* **168**, 313–24.

Gibbs M. (1969) Photorespiration, Warburg effect and glycolate. *Ann. N.Y. Acad. Sci.* **168**, 356–68.

Jackson W.A. & Volk R.J. (1970) Photorespiration. *Ann. Rev. Plant Physiol.* **21**, 385–432.

Tolbert N.E. (1971) Microbodies—peroxisomes and glyoxysomes. *Ann. Rev. Plant Physiol.* **22**, 45–74.

Vigil E.L. (1973) Structure and function of plant microbodies. *Sub-Cell. Biochem.* **2**, 237–85.

Zelitch I. (1971) *Photosynthesis, Photorespiration and Plant Productivity*. pp. 352. Academic Press, New York.

CHAPTER 7

MICROTUBULES

7.1 INTRODUCTION

Microtubules were first described in spermatozoids of the moss, *Sphagnum*, by Irene Manton in 1957. Since the introduction, six years later, of glutaraldehyde as a fixative for electron microscopy, their presence has been revealed in a very wide variety of plant and animal cells. They have not been observed in prokaryotes.

These organelles are associated with several processes in the repertoire of movements exhibited by the eukaryotic cell, including; movements of individual cells by flagellar motion or axopod extension; movement of components within cells, as in mitosis, axoplasmic flow or transport of certain pigment granules; and morphogenetic movements involving the generation and maintenance of cell shape. In the case of plant cells, in addition to chromosome movements during cell division, the presence of microtubules has been correlated with definition of the plane of cell cleavage, formation of the cell plate and determination of cell wall architecture.

In certain cell types, microtubules are seen only during mitosis and meiosis, suggesting that they were involved originally in these processes alone, and only later did they come to be used for extranuclear functions in advanced eukaryotic organisms. Indeed, it has been suggested (Margulis, 1970) that acquisition of the microtubule was a major factor in permitting the evolution of the eukaryote cell.

7.1.1 DESCRIPTION

Under the electron microscope, microtubules appear as long, unbranched cylindrical structures with a diameter of 22–25 nm (Fig. 7.1 and Fig. 7.4). In transection, they show an electron-lucent core, *ca.* 12 nm in diameter, and a wall of, usually, 13 electron-dense subunits. The subunits are 4–5 nm wide and are organized along the long axis of the microtubule into protofilaments. There is an axial displacement of subunits between adjacent protofilaments, resulting in a visible pitch relative to the long axis of the tubule (Fig 7.2).

Microtubules are usually separated from each other or from adjacent cellular components by a clear space of 10–40 nm. Also, in longitudinal view, an electron-lucent zone is often observed along the sides of the microtubule. These observations suggest that each tubule may be surrounded by a specialized region or layer of material. In certain systems, microtubules show 'arms' projecting from the walls of the tubules at regular intervals along their length.

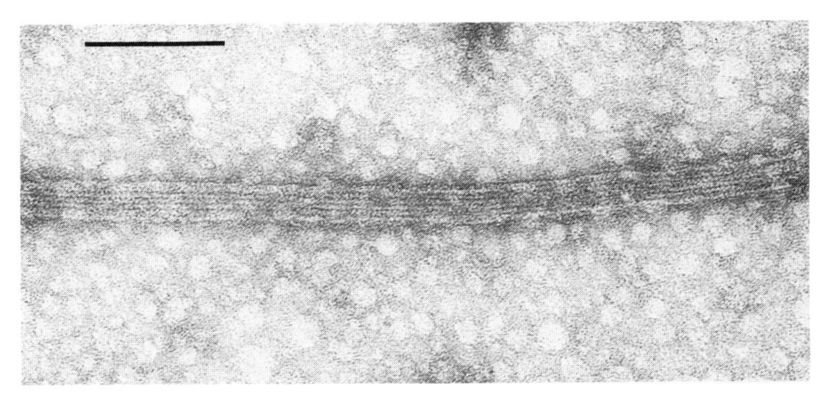

Fig. 7.1. Negatively stained microtubules prepared from brain, showing axially aligned protofilaments and their substructural periodicity. Scale marker 0·1 μm.

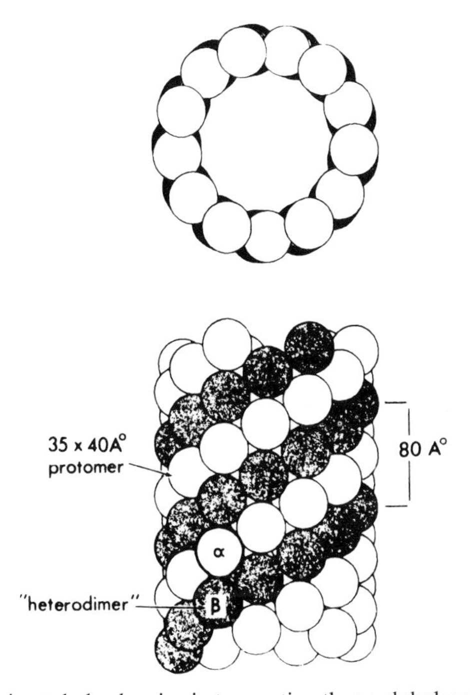

Fig. 7.2. Model microtubule, showing in transection the 13 globular subunits comprising the wall, and in longitudinal view, the axial displacement of subunits in adjacent protofilaments. From Bryan (1974) *Fedn. Proc. Fedn. Am. Socs. exp. Biol.* **33**, 152–57, reproduced by permission of the author and Fedn. Am. Socs. exp. Biol.

Such lateral projections act, in some instances, as cross-bridges between adjacent microtubules. Alternatively, projections can extend from microtubules to adjacent membranes such as the plasmalemma, the nuclear envelope, the endoplasmic reticulum or associated vesicles.

7.1.2 BACKGROUND

Eukaryote mitotic spindles are a heterogeneous group of intracellular structures characterized by the presence of anisotropically arranged microtubules. The number of microtubules within the spindle of a given species appears to reflect the mass of the chromosomes, and can vary from a few to several thousand. The highly oriented microtubules are responsible for the weak form-birefringence of the spindle under polarized light and there are natural fluctuations in spindle birefringence during the different phases of mitosis.

Mitotic arrest can be achieved with a variety of physical and chemical agents. Spindle birefringence is also affected by such agents. For example, birefringence in dividing *Lilium* pollen mother cells is abolished within seconds of exposure to low temperature. Upon return to normal temperatures, spindle birefringence reappears in a few minutes, after which mitosis proceeds normally. Exposure to high hydrostatic pressure causes a similar effect. Chemical agents with anti-mitotic activity, such as the alkaloids colchicine, podophyllotoxin or vinblastine, also abolish spindle birefringence. The important feature of all these effects is that they are reversible upon removal of the inhibitory agent.

Based on: (a) natural fluctuations in spindle birefringence, (b) reversible effects of inhibitors on both birefringence and mitosis and (c) the findings of many earlier studies that the major components of the spindle are synthesized prior to prophase, Inoué and Sato in 1967 proposed a model in which the spindle is envisaged as a labile structure in dynamic equilibrium with a pool of subunits. Mitosis is thus a process which reflects the sequential assembly and disassembly of different structures to perform different functions. Inhibitory agents may act by disrupting the dynamic equilibrium. This model had a profound effect in engendering the concept of *certain* microtubules being labile structures capable of being polymerized or depolymerized under cellular control (Tilney, 1971).

However, several lines of evidence suggest that there are different classes of microtubules. Morphological variations in arrangement and in associated components have already been briefly described. Perhaps more importantly, microtubules of cilia or flagella are not depolymerized by treatments which reversibly destroy the birefringence of the spindle. Similarly, certain cytoplasmic microtubules of higher plant cells appear to be resistant to anti-mitotic chemicals. The generalization has been made that microtubules be classified as *stable*, i.e. those in, for example, cilia and flagella, or *labile*, i.e. cytoplasmic microtubules in many types of animal cells (Margulis, 1973).

7.2 BIOCHEMICAL STUDIES

7.2.1 DRUG INTERACTION

The responses of certain types of microtubule to anti-mitotic agents, together with their apparent repeating substructure, thus gave rise to the concept of the microtubule as an assembled polymer of soluble subunits. Early investigations into the nature of the subunit were hampered by the absence of a method or an assay for the recognition of a soluble microtubule component, even when such enriched sources as flagella or the mitotic apparatus were examined. This problem was overcome in the mid-1960's when it was postulated that the microtubule system was the direct target of the anti-mitotic drug, colchicine, and that the subunit might therefore be recognizable as a drug receptor.

Preliminary experiments with human carcinoma cells in culture established that, at the low levels of colchicine which depolymerized microtubules, there were no effects on other areas of metabolism such as respiration or protein synthesis (Taylor, 1965). Consideration of the kinetics of drug uptake suggested the presence of a single type of binding site. Work in two laboratories confirmed, using radiolabelled drug, that colchicine itself was indeed bound non-covalently to a single, soluble, protein species (Borisy & Taylor, 1967; Wilson & Friedkin, 1967). Since colchicine-binding could be assayed quantitatively, using gel filtration or ion exchange procedures, the reaction assumed great importance in the subsequent isolation and characterization of the subunit (Adelman *et al.*, 1968).

It was quickly established that colchicine is bound in a stoichiometric 1:1 relationship to a dimeric protein with a molecular weight of 120,000 (see Wilson & Bryan, 1974). Denaturation to the protomers results in loss of binding. The binding affinity is high, of the order of 2×10^6 litres $mole^{-1}$, and it is important to note that there is no binding at all of the colchicine isomer, lumicolchicine. Although the protein can be stored in the frozen state, an outstanding feature of the binding site is its instability in solution, where it shows first order decay kinetics and a half-life of the order of hours even under optimal conditions of pH, temperature and ionic environment. Inclusion in the medium of the nucleotide, GTP, or the alkaloid, vinblastine, stabilizes and binding activity to some extent.

Evidence that the colchicine binding component is the microtubule subunit was initially circumstantial. High levels of colchicine binding protein were obtained from material rich in microtubules, e.g. dividing cells, cilia and brain cells. The physical characteristics of the binding protein with regard to its molecular weight and dimeric nature were similar to those of the major protein species in flagella and cilia. Evidence of a more direct nature was obtained through ultrastructural observations of the solubilization of the central pair nucleotide, GTP, or the alkaloid, vinblastine, stabilizes the binding activity to binding activity.

The binding site for colchicine seems to be masked when the subunits are assembled into microtubules. This, together with the instability of the site, was responsible for initial lack of success in obtaining colchicine-binding activity from stable outer doublet microtubules of flagella. However, the presence of the colchicine site in such subunits has been demonstrated using appropriate solubilization procedures (Wilson & Meza, 1973). Thus, the colchicine binding protein has been obtained from a variety of types and sources of microtubule. It is now generally accepted that labile microtubules are depolymerized by the binding of colchicine to subunits in the soluble pool, with consequent disruption of the equilibrium.

The colchicine binding moieties isolated from diverse sources of microtubules display similar physical and chemical properties (see Bryan, 1974). The name 'tubulin' has been applied to this dimeric, functional subunit of microtubules. Tubulin is an acidic, globular protein with a sedimentation coefficient of 6s and an agreed molecular weight of 110–120,000, whose behaviour approximates that of a prolate ellipsoid with an axial ratio of 5:7. The amino acid composition prompted initial comparison with actin, but subsequent comparable peptide maps of both proteins argued against any homology between tubulin and actin (Stephens, 1970). Treatment of tubulin with chaotropic agents such as urea or guanidine. HCl results in a reduction of the apparent molecular weight to 50–60,000. The two protomeric polypeptides can be resolved by appropriate electrophoretic analysis and have been named α and β tubulin, the more mobile electrophoretic component being β tubulin. A number of chemically reactive sites have been distinguished on the tubulin heterodimer. Two sites are known to be involved in drug interaction, a site for colchine and podophyllotoxin, and a separate site for vinblastine. There are also sites which bind endogenous

Table 7.1. Colchicine binding activity from various animal and plant sources*.

		Bound ^{3}H-colchicine	
Author	Source	cpm mg^{-1} protein in extract supernatant	cpm g^{-1} fresh weight
Borisy & Taylor, 1967	squid axoplasm	180,000	—
	pig brain	50,000	—
	rabbit muscle	4,800	—
	HeLa cells	54,000	—
	sperm tail	60–90,000	—
	mitotic apparatus	13,000	—
	slime mould	100	—
Wilson & Friedkin, 1967	grasshopper embryo	40,000	—
Haber *et al.*, 1972	yeast	7,000 (colcemid)	—
Hart & Sabnis, 1973	pea internode	(80)**	220
(and unpubl.)	hogweed petiole	(135)	480
	hogweed vascular tissue	(195)	970
	barley coleoptile	(90)	310
	mustard root tips	(540)	1,390

*Original data, uncorrected for colchicine concentration or radiocounting efficiency.

**Calculated values for comparison with animal sources; colchicine binding activity not detectable in 100,000 × *g* supernatants of higher plant extracts, without pre-concentration of protein by ammonium sulphate precipitation.

ligands, including a site where GTP (guanosine triphosphate) is tightly and non-exchangeably bound and another site to which GTP is bound in such a way as to allow exchange with exogenously added GTP. Dephosphorylation of the terminal phosphate of GTP and transphosphorylation between the two sites have been suggested to play a role in microtubule assembly. In addition, other sites may be involved in interactions with Ca^{2+} and Mg^{2+} ions. The relationship of all these sites to the protein-protein contact sites necessarily involved in the vertical and lateral bonding of the subunits in the assembled tubule is unclear.

Recent examinations of plant preparations for a tubulin-like moiety with characteristic colchicine-binding activity have met with limited success. Such activity has either been undetectable or present only to a low degree (Haber *et al.*, 1972; Hart & Sabnis, 1973; Burns, 1973; Heath, 1975a). The stability and specificity of the binding render it unlikely that all such activity is attributable to a plant tubulin. Table 7.1 shows the levels of colchicine binding activity from a variety of tissues, including plants.

7.2.2 POLYMERIZATION

Since the behaviour of microtubules *in vivo* obviously involves their strictly regulated assembly-disassembly, a primary aim throughout the biochemical investigations was to develop a system in which polymerization could be studied *in vitro* (see Borisy *et al.*, 1974). Early studies in this area did indeed demonstrate the presence of ordered structures in tubulin preparations. However, the relevance of such systems to the situation *in vivo* was considered doubtful, either because the process did not show characteristic kinetic responses to colchicine and temperature, or because the structures did not look like microtubules. It was therefore a notable advanace when authentic microtubules from rat brain, sensitive to colchicine and temperature, were first assembled *in vitro*, the crucial factors being inclusion in the medium of GTP and the calcium-chelating agent, EGTA (ethylene-glycol-amino ethylether tetracetic acid) (Weisenberg, 1972). It was subsequently reported that polymerization, using pig brain tubulin, also requires nucleation sites which are in the form of discs (29 nm diameter, 17 nm lumen). These successful polymerization studies finally confirmed that the colchicine binding protein is the microtubule subunit. In addition, they have provided a means of demonstrating the probable universality of the tubulin moiety, via co-polymerization of tubulin from such disparate sources as *Chlamydomonas* flagella and pig brain.

The ability to polymerize microtubules *in vitro* has initiated several exciting new avenues of research. In the first place, it has provided a method for isolating *microtubule* protein, rather than tubulin, the drug receptor. Cycles of centrifugation at 37°C and 0°C (i.e. with subunits alternately in the polymerized and depolymerized states) allow microtubules to be isolated with a high degree of purity. Both the ultrastructure and the chemical composition of such preparations have been examined. Electrophoretic analyses of purified microtubules by several laboratories has confirmed the presence of more than ten other proteins, including some with a molecular weight greater than 200,000. The roles of these proteins are the subjects of intense investigation. Some seem to be involved in the regulation of assembly. Others, with a molecular weight close to that of dynein (the ATPase 'arms' of flagellar microtubules) may be involved in the functioning of microtubules.

Secondly, *in vitro* polymerization has made the assembly process itself amenable to experimental study. The initial test of biological relevance seems settled, with the endothermic, reversible, polymerization system responding to many factors known to regulate *in vivo* assembly of microtubules. The questions of nucleation centres and morphopoeitic proteins have already been mentioned. In addition, microtubules have been successfully polymerized *in vitro* on cellular components such as asters, basal bodies and kinetochores. The pathway of polymerization is being deduced from combined biochemical and ultrastructural examination of experimentally induced intermediates. A commonly observed intermediate is a ring structure of tubulin, varying in dimensions and complexity with different investigators. Whether the ring structure serves as a nucleation centre for stack-wise growth of microtubules, or whether it uncurls to form protofilaments which then associate laterally, is unclear.

7.3 BIOLOGICAL STUDIES

Microtubules appear to be involved in several functional roles within the plant cell (see Newcomb, 1969). For convenience, these can be grouped under five headings: (1) Cytoskeletal role; generation and maintenance of cell shape (2) Cell wall architecture; cell differentiation (3) Intracellular transport (4) Chromosome movements; cell plate formation (5) Cell motility; cilia and flagella. In this section, we shall consider the involvement of microtubules in these phenomena, with emphasis on possible operative mechanisms, and drawing, where necessary, on information obtained from studies of animal cells.

7.3.1 CYTOSKELETAL ROLE

Despite the forces of surface tension, animal and some lower plant cells can maintain a non-spherical shape in the absence of a rigid cell wall. Numerous electron microscope studies have revealed that microtubules are present in large numbers and with distinctive orientations in such asymmetric cells or cell extensions. Correlations between cell asymmetry and microtubules have been extensively studied in animal cells, for example in erythrocytes, explanted nerve ganglia or neuroblastoma cells, lens epithelial cells and mesenchyme cells of echinoderm embryos. When these cells are treated with microtubule depolymerizing agents (high hydrostatic pressure, low temperature or colchicine), they assume spherical shapes.

The unicellular alga, *Ochromonas*, lacks a cell wall or pellicle, yet has a characteristic pear-shape, with a narrow tail or rhizoplast, and a bulbous anterior bearing a small projection, the beak. Two sets of microtubules are involved in maintaining this shape. The plasma membrane of the main cell body is underlain by a series of curved microtubules which extend into the rhizoplast. The other set of microtubules is associated with the beak complex. Disassembly of the microtubules with colchicine or hydrostatic pressure leads to loss of the characteristic cell shape; removal of the depolymerizing agent allows the normal shape to regenerate even in the absence of protein synthesis (Brown & Bouck, 1973).

The two sets of microtubules in *Ochromonas* are differentially sensitive to colchicine and pressure. At relatively high concentrations of colchicine (10 mg ml^{-1}) or at high pressure (8,000 psi), both sets of microtubules are affected but at different rates: first, disassembly of rhizoplast tubules is correlated with disappearance of the posterior tail; this is followed by loss of beak tubules and beak asymmetry. Lower concentrations of drug (5 mg ml^{-1}) or lower pressure (6,000 psi) selectively disassemble only the rhizoplast microtubules. These results possibly indicate separate roles for each set of microtubules in determining the shape of this cell. Assuming that different tubulins are not involved, the beak and rhizoplast microtubules may represent two equilibrium systems competing for a common pool of subunits. Alternatively, the differential sensitivity of the two systems, and their sequential reassembly, may reflect characteristics of the sites which initiate microtubule polymerization.

The protozoan, *Actinosphaerium*, has been extensively used in similar studies on the morphopoietic function of microtubules. The organism is a spherical cell bearing numerous, long, relatively stiff, protoplasmic extensions or axopods, containing a central core, the axoneme (Fig. 7.3a). Each axoneme is composed

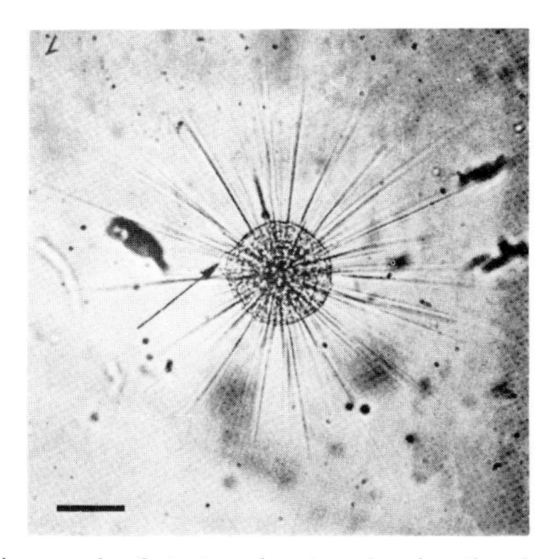

Fig. 7.3a. Photomicrograph of *Actinosphaerium* showing the slender cell extensions or axopods radiating from the cell body. Scale marker 100 μm.

of several hundred axially aligned microtubules, arranged into two rows that coil about a central axis (Fig. 7.3b). This strict pattern does not seem to be determined by a template of initiating sites, since, during axopod growth, the microtubule array is initially disorganized and becomes progressively more geometrically perfect.

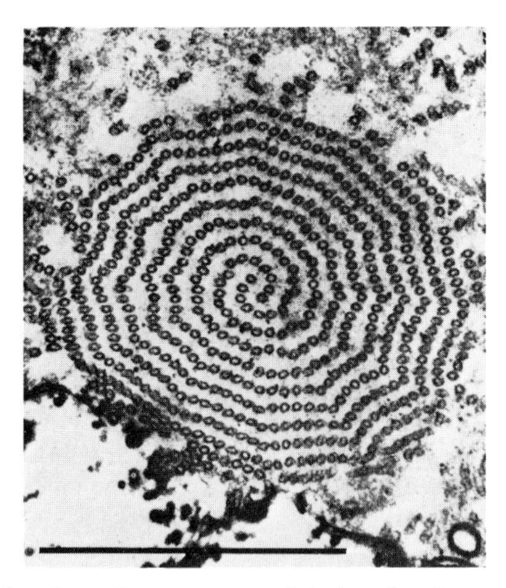

Fig. 7.3b. Cross section through an axoneme of *Actinosphaerium* near the axopod base. Two spiral rows of microtubules are coiled about a central axis in a precise geometrical arrangement. Scale marker 1 μm. From Tilney & Porter (1965) *Protoplasma* **60**, 317–43, reproduced by permission of the authors and Springer-Verlag.

When low temperature, pressure or drugs are employed to depolymerize microtubules of *Actinosphaerium*, the axopods undergo immediate retraction. In contrast, D_2O, which stabilizes microtubules, prevents the breakdown of axopods by cold or pressure. Retractions and extensions are a normal feature of individual axopods and serve to move the cell across its substratum. Thus, the cell must exert a fine control over the behaviour of microtubules in each axopod separately (see Gunning & Steer, 1975).

In flowering plants, the male sperm is formed from the generative cell which usually has no cell wall and lies within the pollen grain. Growth of the pollen tube is accomplished by the vegetative cell, while the generative cell adopts a spindle-shape before entering and travelling down the pollen tube. Development of the spindle shape is accompanied by the appearance of microtubules aligned parallel to the long axis of the generative cell. Treatment of the generative cell with drugs leads to a loss of microtubules and the cell adopts a spherical form (Sanger & Jackson, 1971).

Thus, microtubules are apparently involved in both generating and maintaining changes in cell shape. The mechanisms regulating orientation and assembly of microtubules in this role are still unknown. Microtubules in the axonemal cores of *Actinosphaerium* axopods are connected with their neighbours by cross-bridges. Whether these lateral links function in stabilizing the extended structure or in orienting the microtubules has yet to be determined.

7.3.2 CELL WALL ARCHITECTURE

During interphase in higher plant cells, cytoplasmic microtubules lie close to the cell membrane and are arranged circumferentially along the lateral walls of the cells but are randomly disposed underlying the transverse walls (Fig. 7.5a). In the earliest paper reporting the presence of microtubules in higher plant cells, Ledbetter and Porter (1963) noted the parallel alignment of cellulosic microfibrils of the wall and the subjacent cytoplasmic microtubules (Fig. 7.4). Numerous subsequent studies have confirmed this observation, in algal cells and in higher plant cells undergoing both primary and secondary wall formation (see Hepler & Palevitz, 1974).

In the long xylem fibres of *Salix*, wall microfibrils are deposited in two different orientations simultaneously in the middle and at the extremities of the cell. Even in this situation, microtubule alignment mirrors that of the overlying microfibrils, indicating the ability of the cell to maintain two sets of microtubule-microfibril associations. In the lorica stalk of the alga, *Poteriochromonas*, the primary wall microfibrils are arranged helically and tend to fasciate into ribbon-like fibrils, 20 nm in width. Every such wall fibril coincides precisely with a microtubule, the plasmalemma separating the two sets of linear structures (Schnepf *et al.*, 1975).

Other examples of the association between microtubules and microfibrils are afforded by cells undergoing irregular or sculptured deposition of secondary wall material. During xylogenesis, microtubules are specifically grouped under the developing wall thickenings in vessels, and, once again, are oriented parallel to the microfibrils. Similar situations prevail during the development of wall thickenings in differentiating guard cells and during the deposition of the nacreous walls of sieve elements.

Attempts have been made to confirm the association between wall microfibrils and microtubules by treating cells with colchicine. However, plant tissues seem relatively resistant to a wide variety of drugs (Heath, 1975b) and, even at high concentrations of colchicine, not all cytoplasmic microtubules in plant cells are depolymerized (Pickett-Heaps, 1967; Wooding, 1969). Also, such levels of colchicine are known to affect other cellular processes, in particular, membrane-related phenomena (see Wilson & Bryan, 1974). Therefore, although wall microfibril orientation is often distorted in the presence of colchicine, such results should be interpreted with caution.

7.3.3 INTRACELLULAR TRANSPORT

There is much evidence to indicate that wall precursors are transported to the wall in dictyosome-derived vesicles. On the other hand, there is no evidence to

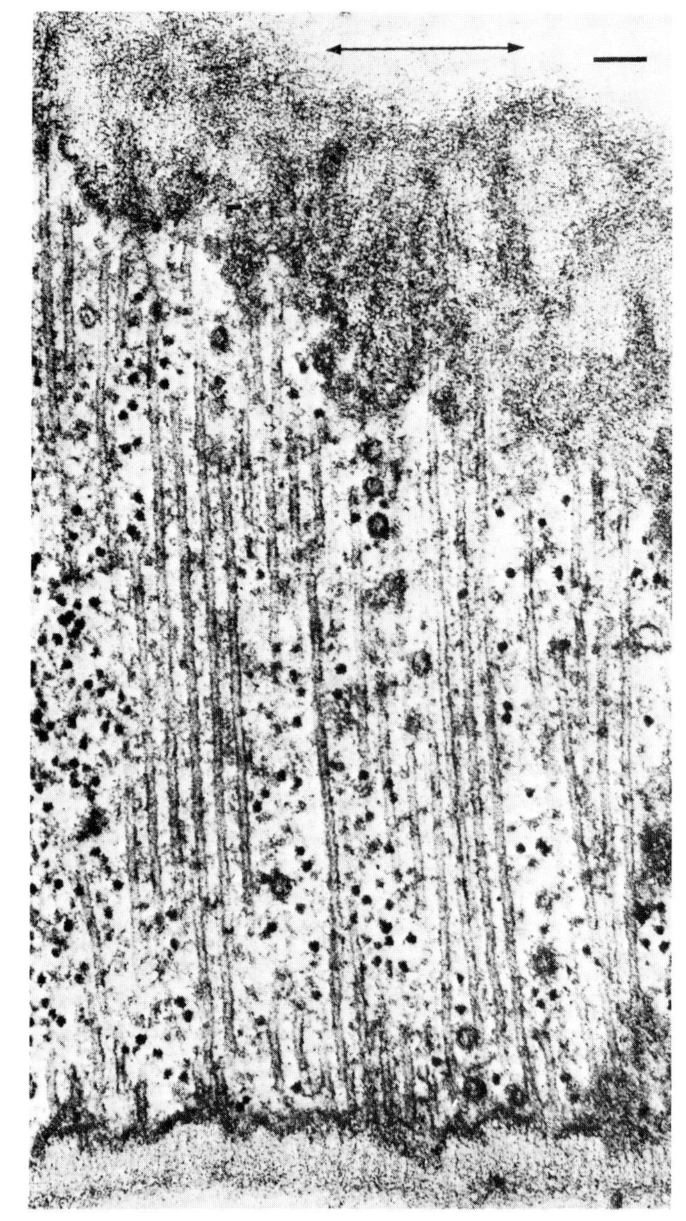

Fig. 7.4. Microtubules running parallel to one another immediately beneath a primary cell wall in root tips of bean (*Phaseolus vulgaris*). The microfibrils of the wall are oriented parallel to the microtubules. The axis of cell elongation (indicated by the arrow) is at right angles to the orientation of the microtubules and microfibrils. Vesicles can be seen among the microtubules. Scale marker 0·1 μm. From Newcomb (1969) *A. Rev. Pl. Physiol.* **20**, 253–88, reproduced by permission of the author and Annual Reviews, Inc.

suggest that microtubules provide the motive force for such transport. However, it has been proposed that the parallel microtubules girdling the cells of higher plants peripherally provide a framework for the guidance and alignment of these vesicles. In some cases the microtubules lie so close to one another that it has been argued that they could function only by excluding vesicles or elements of the endoplasmic reticulum, preventing their fusion with the plasmalemma at certain sites. In either case, a vectorial role for microtubules is indicated.

Ledbetter and Porter suggested that microtubules may be involved in orienting and even driving cyclosis of cytoplasm, a process which might indirectly be responsible for the alignment of wall microfibrils. However, several lines of evidence have recently implicated cytoplasmic microfilaments in this role (Hepler & Palevitz, 1974). These filaments, 5–8 nm in width, are more consistently located and aligned within the zones of streaming cytoplasm. The microfilaments are morphologically identical to F-actin, one of the major contractile proteins of muscle. Recently, microfilaments in cells of both *Nitella* and higher plants have been decorated with heavy meromyosin, a characteristic test for actin-like filaments. In primitive organisms, such as the slime mould *Physarum*, there is overwhelming evidence to indicate that similar microfilaments provide the motive force for rapid cytoplasmic streaming.

However, around the axonemes of Heliozoan axopods described earlier, cytoplasmic particles stream to and from the cell body. Microfilaments have not been seen in this structure. By contrast, in the giant coenocytic alga, *Caulerpa*, microtubules are located within the zones of, and parallel to, the numerous sluggish cytoplasmic streams. Mitochondria, plastids and other cytoplasmic components are constantly circulated to and from the growing tips of this large, asymmetric cell. In the chromatophores of animal cells, microtubules have been shown to be essential to the movements of pigment granules. Anti-mitotic agents depolymerize microtubules and simultaneously disrupt the alignment and arrest the movement of the pigment granules (Murphy, 1975). Thus, it remains possible that certain types of cytoplasmic transport require a microtubule framework in both a vectorial and an active role.

7.3.4 CELL DIVISION

At least two types of microtubule can be distinguished in most spindles, those that traverse the spindle from pole to pole (continuous fibres), and those that connect the kinetochores of the chromosomes to the poles (chromosomal or kinetochore fibres). During separation of chromatids in anaphase, the continuous fibres lengthen, increasing the distance between the poles, while the chromosomal fibres usually—but not always—shorten, further separating the chromatids. After telophase, the continuous spindle may persist and be added to in the interzonal region by numerous additional microtubules, leading to formation of the phragmoplast and eventually, the cell plate.

Considerable changes in microtubule organization (Fig. 7.5, A–F) take

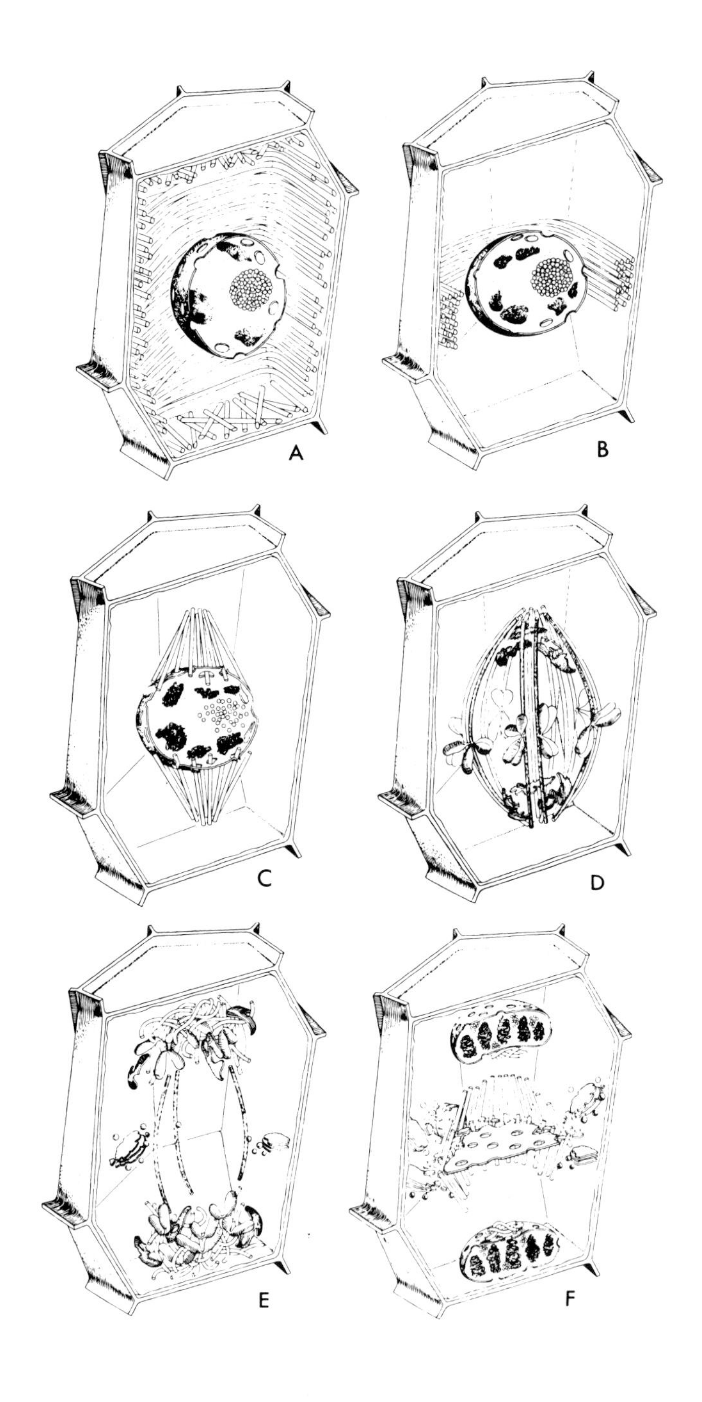
A
B
C
D
E
F

place during the events of a typical mitotic cycle (Fuge, 1974). With the onset of division, the peripheral cytoplasmic microtubules of the interphase cell disappear. Concomitantly, a band of microtubules is formed, several layers deep, lying close to the cell wall and girdling the nucleus perpendicularly to the prospective spindle axis. This preprophase band of microtubules appears to predict the plane of the future cell plate (Pickett-Heaps & Northcote, 1966). This has been demonstrated in divisions destined to give daughter cells of different size, for example in the asymmetric divisions of guard mother cells of wheat.

At the beginning of prophase, the number of microtubules in the preprophase band decreases, and new microtubules appear close to the nuclear membrane, this time running parallel to the prospective spindle axis. The microtubules in this so-called 'clear zone' in turn appear to furnish protein material for the spindle. It is possible that a total transformation of these extranuclear microtubules into spindle microtubules takes place.

Prior to spindle formation, the nuclear envelope either disintegrates or becomes perforated at the prospective poles, and cytoplasmic (clear zone) microtubules enter the karyoplasm at these sites. Some of these microtubules appear to establish connections with the chromosomal kinetochores. Since the number of kinetochore microtubules (kMts) increases as prometaphase proceeds, and since clear zone microtubules can be utilized only immediately after breakdown of the nuclear envelope, it is likely that *de novo* microtubule assembly also occurs at the kinetochores. Thus, it appears that the kinetochore, which varies in appearance in different organisms from amorphous to highly structured, acts as an initiating site or microtubule morganizing centre (MTOC).

In prometaphase, continuous fibres also begin to form and to increase in number. There is some controversy as to whether in higher organisms the continuous microtubules pass from pole to pole. A large proportion appear to overlap in the interzone between the half-spindles (Fig. 7.6), and have been termed non-kinetochore microtubules (nkMts). In lower plants, and all animal cells, with centrioles, the continuous or nkMts originate near the centrioles. These loci are represented by amorphous, electron-dense aggregates of material, and behave as another set of MTOCs. The essential feature of the centriole in mitosis seems to be in determining the spindle axis. In higher plants that lack centrioles, neither the location nor form of polar MTOCs is clear.

During metaphase, engagement of sister chromatids to opposite poles takes place *via* microtubules, followed by movement of the chromatids to the equatorial plate. At this stage, the maximum number of spindle microtubules is attained, the numbers often trebling between prometaphase and metaphase. Estimates from microtubule counts in serial sections of the metaphase spindle of the sea urchin, *Arbacia*, indicate a total microtubule length of 50,000 μm. The very large

Fig. 7.5. Diagram of changes in microtubule orientation and distribution during the phases of mitosis in a higher plant cell. A—interphase; B—preprophase; C—prophase; D—metaphase; E—anaphase; F—telophase. From Ledbetter & Porter (1970) *Introduction to the Fine Structure of Plant Cells*, p. 44, Springer-Verlag, Berlin, reproduced by permission of the authors and Springer-Verlag.

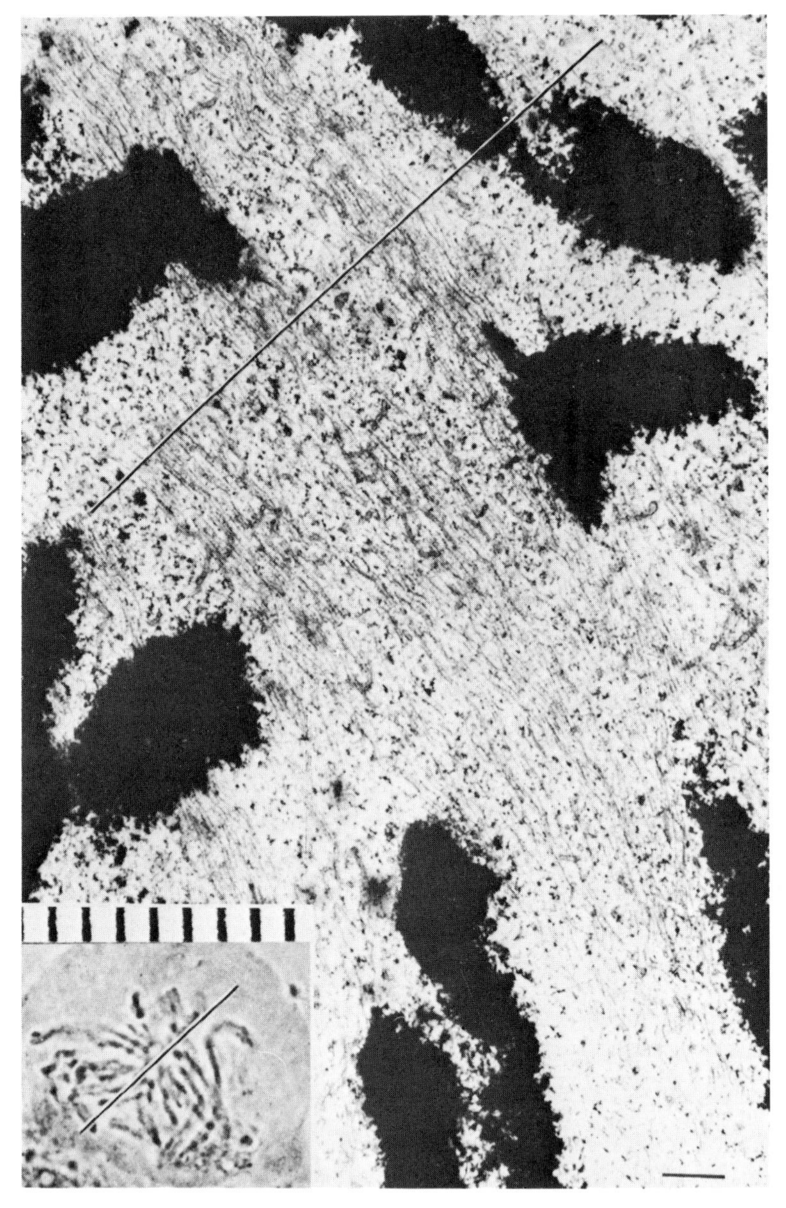

Fig. 7.6. Electron and light micrographs of a dividing cell in prometaphase showing kinetochore and nonkinetochore microtubules. The line indicates the equatorial plate. Scale marker 1 μm. From Bajer (1968) *Symp. Soc. exp. Biol.* **22**, 285–310, reproduced by permission of the author and University Press, Cambridge.

spindles of *Haemanthus* endosperm cells contain considerably more microtubules than those of *Arbacia*.

Anaphase separation of the chromatids involves direct translation of the kinetochores at almost uniform velocity (0·2–4 μm min^{-1}) towards their respective poles. Two experimentally separable processes are involved: first, a movement of the chromatids towards their poles, during which the chromosomal fibres shorten to as little as 20% of their metaphase length; secondly, an elongation of the spindle itself, during which the distance between the poles increases, thereby further separating the two groups of chromatids.

Time-lapse films of dividing *Haemanthus* endosperm cells show that many other objects, such as vesicles and granules, are commonly transported polewards during prometaphase-anaphase. Conversely, in onion root tip cells, the poleward movement of the chromosomes is accompanied by a reciprocal movement of other materials from the poles towards the mid-plate. Elements of the endoplasmic reticulum accumulate at the poles during metaphase, enter the spindle at anaphase, pass between the chromosomes moving in the opposite direction, and eventually aggregate and coalesce at the mid-plate to give rise to the transverse cell wall at telophase.

Dissolution of kMts and nkMts during telophase is accompanied by the formation of new microtubules in the interzone, leading to the formation of the phragmoplast in higher plant cells (Fig. 7.7). Phragmoplast microtubules appear to be involved in the movement, alignment and fusion of ER- and

Fig. 7.7. Phragmoplast microtubules at the mid-plate during telophase. Vesicles contributing to the forming cell plate can be seen aligned among the microtubules. Scale marker 1 μm. From Hepler & Jackson (1974) *J. Cell. Biol.* **38**, 437–46, reproduced by permission of the authors and The Rockefeller University Press.

dictyosome-derived vesicles to form the cell plate—yet another example of microtubule-associated intracellular transport.

Thus, distinct changes in the polarity and distribution of spindle microtubules take place during the various phases of the mitotic cycle. It should be borne strongly in mind that such changes occur under conditions of very low protein synthesis. Consequently, both assembly and functioning of the mitotic apparatus must involve regulation of microtubule polymerization at a physico-chemical level. Interaction of microtubules with other components of the mitotic apparatus must also be regulated at this level. Several hypotheses have been proposed to explain the mechanism of mitotic movements, although little direct experimental evidence to support the models has yet been obtained.

One prevailing hypothesis to account for spindle elongation (Brinkley & Cartwright, 1971) is that the poles are pushed apart by tip growth of continuous microtubules in the course of which subunits from disassembled kMTs may be utilized. Shortening of the chromosomal fibres is thought to occur by depolymerization at the poles.

Another hypothesis based on the sliding of adjacent spindle microtubules was proposed by McIntosh *et al.* (1969). Earlier, several workers had reported the existence of arms or cross-bridges between spindle microtubules. The basis of the sliding tubule hypothesis is that the intertubular bridges in the spindle serve as mechano-chemical elements. Sliding of two microtubules against each other is thought to be generated by the successive breaking and reforming of cross-bridge linkages, a process that is energy-consuming and mediated *via* ATPase activity of the bridges. It has been estimated that the energy released through the hydrolysis of approximately 20 ATP molecules would suffice to move an average chromosome from the metaphase plate to the pole.

A prerequisite for the McIntosh model is an ordered system of long microtubules, emanating from the poles and reaching far into the opposite half-spindles in metaphase. Microtubules from opposite poles are suggested to have antiparallel polarity with regard to the direction in which their lateral arms could produce force. The kMts in each half-spindle would also show polarity; their lateral arms would only produce force directed towards the spindle equator. If one of the long interpolar microtubules and a kMt of antiparallel polarity came into contact at anaphase, the two tubules, due to the directed activity of their arms, would push themselves in opposite directions by sliding. The model is analogous to the generally accepted model for the interaction of actin and myosin in skeletal muscle. The mechanism could cause chromosome separation, as well as spindle elongation when anti-parallel nkMts interact where they overlap.

A third hypothesis involving actin-microtubule interactions (Forer & Behnke, 1972) has derived from observations of actin-like thin filaments in the mitotic or meiotic spindles of both animal and plant cells. Fluorescein-labelled heavy meromyosin binds to isolated sea urchin spindles. At the ultrastructural level, microfilaments within the spindle have been decorated with heavy meromyosin, a test for actin-like proteins (Gawadi, 1971; Forer & Behnke, 1972). Several other recent publications have confirmed the presence of actin-like filaments in dividing cells. In kangaroo rat cells, actin was only demonstrable in association with chromosomal fibres (Sanger, 1975), suggesting an actin-myosin *type* interaction as the force-producing mechanism for chromosome movements, that is, for the autonomous movements of chromosomes in anaphase, as opposed to spindle elongation which may be controlled solely by microtubule assembly-disassembly.

Myosin-like proteins have not yet been located in spindles. Several workers have suggested that microtubules might serve as rigid structures against which the actin filaments could exert a contractile force by the formation and breakage of cross-links. A Ca^{2+}-dependent ATPase (similar in this respect to both myosin and dynein) is present in the isolated spindle in a concentration three times that in the cytoplasm (Mazia *et al.*, 1972).

The theories on the mechanisms involved in chromosome separation have arisen from observations of living or fixed, intact cells. Such studies are hampered by the inability to experiment directly with the mitotic apparatus, except as it exists in the complexity of its cellular environment. Exciting possibilities have been raised by recent successful attempts to isolate and experiment with the mitotic apparatus *in vitro*. Intact spindles from the eggs of the surf clam show normal sensitivity to temperature and colchicine. These isolated spindles can incorporate brain tubulin during their reassembly and such hybrid spindles are also cold labile, Ca^{2+}-sensitive and capable of considerable increase in overall length (Rebhun *et al.*, 1974). Moreover, early anaphase spindles isolated from kangaroo rat cells will continue chromosome motion, in the *absence* of exogenous spindle subunits, when ATP is added (Cande *et al.*, 1974). These results already suggest that while spindle growth requires microtubule polymerization, anaphase movements do not.

7.3.5 CELL MOTILITY; CILIA AND FLAGELLA

Microtubules, arranged in a characteristic 9+2 configuration (Fig. 7.8a), are the major structural components of flagella, cilia and sperm tails (Warner, 1974). The flagellar apparatus consists of a membrane-bounded, slender, cylindrical cell extension subtended by its basal body, the intracellular organelle which appears to be its origin and kinetic centre.

Basal bodies are identical in structure and apparently homologous with the centrioles of animal and lower plants cells. During mitosis, centrioles appear to determine the poles of the spindle axis and to organize the assembly and alignment of spindle microtubules. Following mitosis, in flagellated cells, the centrioles migrate to the cell periphery and organise the assembly of the flagellar apparatus. The mechanism whereby centrioles act as microtubule initiating centres, the reasons for their stability during the life cycle of the cell, and the factors controlling basal body replication in multiflagellated cells, are all unknown.

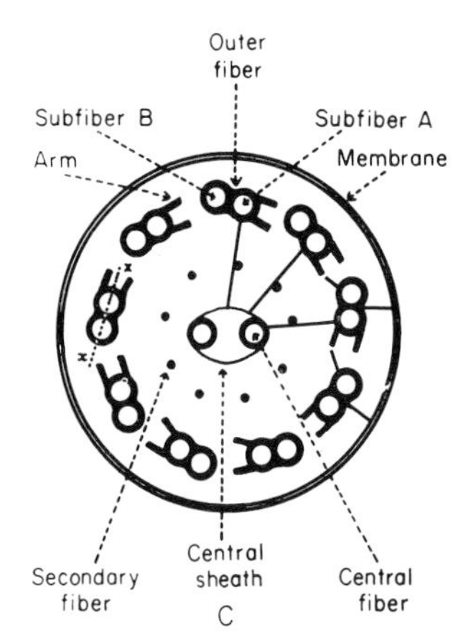

Fig. 7.8a. Diagram of flagella cross section, showing arrangement of microtubules, arms and cross-links.

Basal bodies are composed of nine sets of triplet microtubules (A, B and C) with lateral connections between the A and C tubules of adjacent triplets. At the proximal end of each basal body, a thin filament runs from the A tubule of each triplet to a central hub, forming a cartwheel pattern. At the distal end is the complex transitional region between the microtubules of the basal body and those of the cilium.

The shaft of the cilium is characterized by a central pair of axially oriented microtubules, surrounded by a ring of nine doublets (Fig. 7.8a). The A subfibre of a doublet possesses a complete wall of 13 protofilaments. However, the B subfibre has only 10 protofilaments in its wall, and shares the three protofilaments of the A tubule forming the common partition between them.

From the A tubule of a doublet, two short arms project towards the B tubule of the adjacent doublet. These arms have been isolated and shown to possess Mg^{2+}-dependent ATPase activity (Gibbons, 1965). The arms are approximately 30 nm long, 9 nm wide, and spaced at intervals of 16–22 nm along the length of the A tubule. The enzyme has been called dynein after its postulated role in converting chemical energy into mechanical force. In addition to the dynein arms, cross-bridges have been seen extending between the outer doublets and the flagellar membrane. Radial links, connecting the central pair of tubules with the outer doublets, have also been described. Further, the two microtubules of the central pair differ from one another, one member exhibits two rows of short projections, 18 nm long, and spaced at intervals of 16 nm. Thus, a complex of radially or longitudinally arranged arms, bridges or filaments is present in the flagellar matrix in association with microtubules (Fig. 7.8b).

The most widely accepted hypothesis to account for flagellar motion is the sliding microtubule model (Satir, 1974). Applied to flagella, the hypothesis states that the force responsible for motion is produced when the axonemal

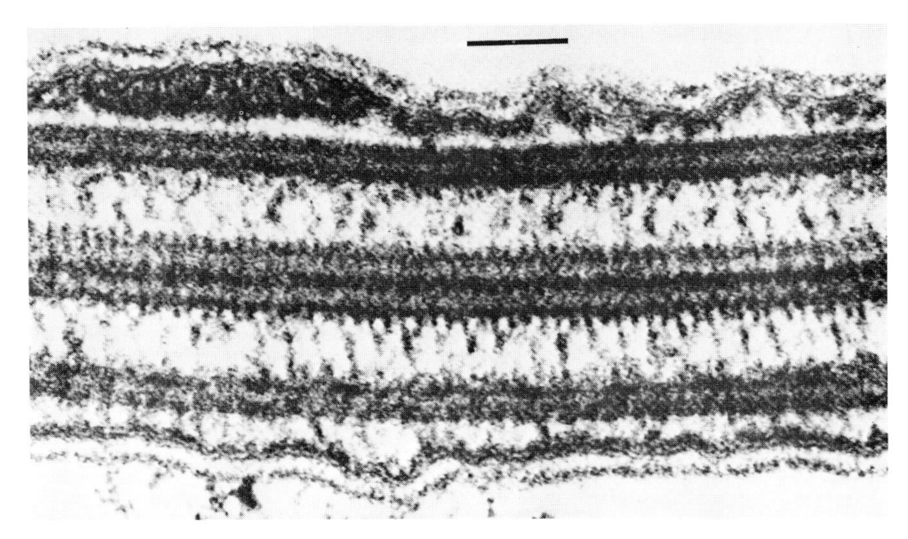

Fig. 7.8b. Longitudinal section of a flagellum showing connections between the central and outer doublet microtubules, and bridges between the peripheral tubules and the cell membrane. Scale marker 0·1 μm. Fig. 8.8b. from Ringo (1967 *J. Cell Biol.* **33**, 543–71, reproduced by permission of The Rockefeller University Press.

microtubules, which do not change in length, tend to slide with respect to one another. Accordingly, the model predicts that in different stroke positions, the morphological relationships of the microtubules will change in a systematic fashion, so that the geometry of the bent flagellum will be reflected in the displacement of the microtubules. This has, indeed, been demonstrated using serial sections of the lateral gill cilia of the freshwater mussel.

Other evidence has also accumulated to support this hypothesis. Glycerinated or demembranated flagella beat normally upon the addition of ATP and Mg^{2+} or Ca^{2+} ions. When such naked axonemes are briefly treated with trypsin, the circumferential and radial links holding the axoneme together are interrupted, whereas the dynein arms are trypsin-resistant. Addition of ATP now causes a sliding of the microtubules, so that the axoneme grows very much longer and thinner as groups of doublets crawl over one another. This sliding must normally be converted to bending by the series of intermicrotubule connections, within the axoneme.

Thus, microtubules are implicated, in both plant and animal cells, in a variety of processes involving the active movement of cells or cellular components. However, the possibility that a single basic mechanism of action underlies their seemingly diverse roles is still in doubt. The dynamic equilibrium model involving assembly and disassembly of microtubules cannot be applied to the stable microtubules of flagella or the cytoplasmic microtubules of higher plant cells. Neither the assembly-disassembly hypothesis, nor the sliding filament model *alone* appears adequate to explain the complex functioning of microtubules in the mitotic apparatus. Even within a process such as spindle elongation, microtubules may function in different ways. For example, the spindles of HeLa cells show an initial rapid phase of elongation, followed by a subsequent slow phase: colchicine only inhibits the slow phase (presumably by interfering with microtubule assembly) but does not affect the rapid phase which may be based on a sliding mechanism. Furthermore, dividing plant and animal cells are remarkably different in their sensitivity to drugs: the spindles of many plant cells require treatment with approximately 1,000-fold higher concentrations of a wide variety of anti-mitotic drugs to achieve mitotic arrest.

7.4 CONCLUDING REMARKS

It is noteworthy that close collaboration between the electron microscopist and the biochemist has been demanded at almost every stage in the development of knowledge concerning microtubules. After the discovery of the microtubule as an ubiquitous organelle of the eukaryote cell, the physiological manipulation of cells in ablation experiments, together with continuing ultrastructural work, confirmed the interaction of microtubules in a variety of cellular activities. Such studies also stimulated development of the concept of the microtubule as a macrostructure in dynamic equilibrium with its constituent subunits. Biochemical investigation into the nature of the subunit initially depended heavily on the electron microscope for evidence that the moiety was relevant to the microtubule. The phase of study culminating in the development of an *in vitro* polymerization system also relied on the electron microscope to provide evidence of the biological relevance of both the process and final structure.

Knowledge concerning microtubules has been derived mainly from studies involving animal tissues or cells of lower plants. Elucidation of the roles of microtubules in higher plant cells has not proceeded so rapidly. Ultrastructural investigations have confirmed the ubiquity of the organelle and have shown its oriented presence to be correlated with a variety of processes. While it is clear that microtubules are involved in various stages of cell division, their precise role in the earlier events which predict the plane of cleavage is unclear. Similarly, their functioning in other aspects of plant development, including cell wall growth and differentiation, remains conjectural. There are, of course, practical difficulties in studying microtubules in the cells of higher plants. Such problems vary from the general, i.e., lack of tissue specialization and presence of a vacuole and cell wall, to the more specific paucity of microtubules in the meagre protoplasm of the differentiated higher plant cell. In addition, there would seem to be a problem in the resistance of certain plant cytoplasmic microtubules to depolymerizing agents. Since this includes, in some cases, stability to such physical agents as temperature and pressure, it would seem that it is a real property of the microtubule rather than an indirect effect of, say, cellular impermeability towards a drug. This property may be related to the stability of flagellar microtubules. It remains to be seen whether the stability is due merely to a lack of an equilibrating subunit pool or to some other feature of this type of microtubule.

Much remains to be learned about plant microtubules, their subunits and associated components, not only to gain further understanding of their roles in specific processes of plant growth and development, but also for comparison of their properties with those of equivalent components from animal sources. The latter aspect, in addition to answering questions concerning the evolutionary origins of this important group of proteins, would also seem to be a prerequisite to determining any common modes of functioning. How microtubules function, what factors regulate their activity in various cellular processes, and what relationships exist among these diverse activities are some of the questions for research.

FURTHER READING

BAJER A.S. & MOLE-BAJER J. (1972) Spindle dynamics and chromosome movements. *Int. Rev. Cytol.* Suppl. 3.

GUNNING B.E.S. & STEER M.W. (1975) *Ultrastructure and the Biology of Plant Cells*, pp. 312. Edward Arnold, London.

HEPLER P.K. & PALEVITZ B.A. (1974) Microtubules and microfilaments. *A Rev. Pl. Physiol.* **25**, 309–62.

INOUÉ S. & SATO H. (1967) Cell motility by labile association of molecules. The nature of mitotic spindle fibres and their role in chromosome movement. *J. gen. Physiol.* (Suppl.) **50**, 259–92.

MARGULIS L. (1973) Colchicine sensitive microtubules. *Int. Rev. Cytol.* **34**, 333–61.

SLEIGH M.A. (1974) *Cilia and Flagella*, pp. 500. Academic Press, London and New York.

WILSON L. & BRYAN J. (1974) Biochemical and pharmacological properties of microtubules. *Adv. Cell Molec. Biol.* **3**, 21–72.

CHAPTER 8

THE ENDOMEMBRANE SYSTEM AND THE INTEGRATION OF CELLULAR ACTIVITIES

8.1 INTRODUCTION

The invention of ultra-thin sectioning gave electron microscopists the first sight of a new level of organization within the cell: a system of membranes forming microstructures and sub-compartments throughout the protoplasm. Some components of this system are small and discrete: the mitochondria, plastids and microbodies. Because they therefore survive cell breakage intact, biochemists have been able to characterize the specialized functions of each type. Others, the nuclear membrane, the plasma membrane (plasmalemma) and the tonoplast, bound discrete compartments so large that, though they do not survive cell breakage intact, many of their specialized functions have been worked out by cytologists and physiologists using whole cells.

Subtract these organelles from the membrane system of the cell and the remainder, a considerable bulk of material, is arranged as parallel, paired sheets fused at the edges to enclose a narrow and empty-looking lumen. The membranes are also fused to each other around holes in the twin sheet, and these fenestrations are occasionally so frequent as to reduce the paired membrane to a layer of tubules, or even isolated vesicles. A small proportion of this double-membrane system, the Golgi apparatus, is organized into units of recognizable morphology, the Golgi bodies. The rest, the endoplasmic reticulum (ER), is relatively amorphous and, more often than not, distributed apparently haphazardly about the cytoplasm. Offering few clues for the microscopist and a host of problems for the biochemist, this material and its function in the cell is only just beginning to be understood.

8.2 TECHNIQUES

8.2.1 ELECTRON MICROSCOPY

Thin (*ca* 50 nm) sections provide a very limited image of the membrane system. Seen in profile, not much can be told about the way its parts are, or are not, connected. What is described from an isolated thin section as a collection of vesicles may in fact be a network of tubules, anastomosing in three dimensions, or a stack of paired lamellae, extensively fenestrated, or even a single tubule, coiled back on itself many times (Fig. 8.1). There are methods which provide information on structure in three dimensions. Thanks to the tremendous depth

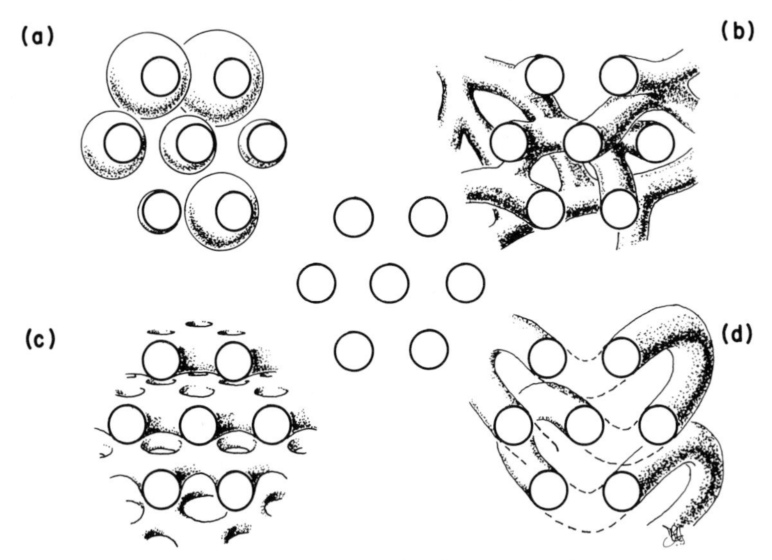

Fig. 8.1. Identical thin sections can be produced from quite different structures:
(a) a collection of vesicles
(b) a three-dimensional network of tubules
(c) a stack of fenestrated lamellae
(d) a single, twisted tubule

of focus of the electron-microscope, using high-voltage electrons it is possible to view all the organelles within a thick (2 μm) section. The specimen can then be tilted slightly to provide stereo-pairs of micrographs.

The freeze-etch technique often provides information on membrane topography. In frozen tissues the hydrophobic bonds holding the two halves of each membrane together are virtually non-existent. The activity of cell water is so low, the lipids of the two halves of the bimolecular leaflet are no longer forced together. The freeze-fracture, which follows planes of weakness, may run through large areas of interconnected membrane. Freeze-etch studies often allow a direct comparison to be made between views of the internal surface and cross-fracture of the same membrane system, as in Fig. 8.2 where the ER is seen as continuous sheets arranged in concentric layers. Fenestrated ER can only be shown by freeze fracture (Fineran, 1973b).

To obtain a truly comprehensive picture of membrane interconnections requires serial sectioning. So lengthy and difficult it is rarely attempted, this technique has nevertheless proved highly rewarding. Not until 1973 was it discovered by serial sectioning that there is only *one*, highly branched mitochondrion in a yeast cell, and the same is true for the unicellular green alga, *Chlorella* (Atkinson *et al.*, 1974; Gunning & Steer, 1975). Though freeze-etch pictures of plant cells have already provided much information on the shape of the endomembrane complex, thick section stereomicrography and serial sectioning are as yet techniques of the future.

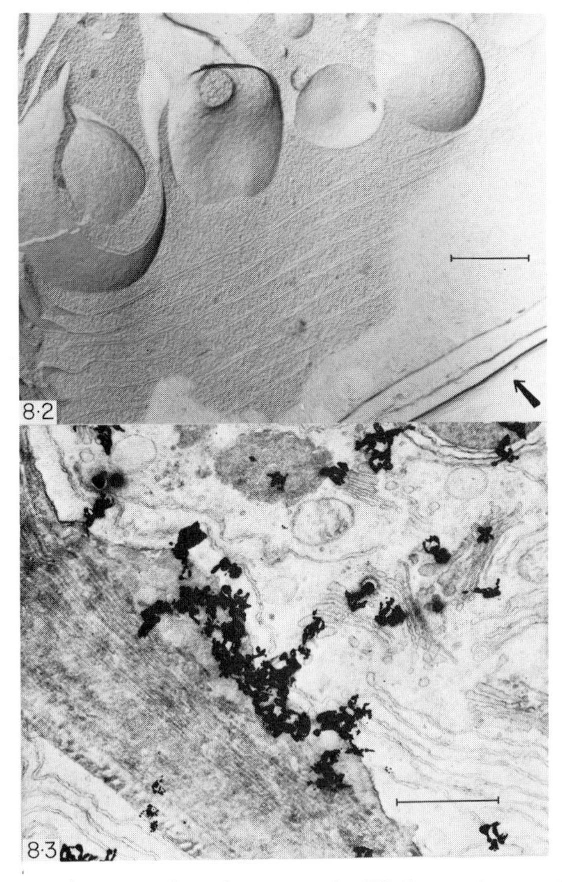

Fig. 8.2. Freeze-etch preparation of pea root tip. ER is seen in cross-fracture and surface view, and forms almost continuous sheets arranged in concentric layers. The arrow indicates the direction from which heavy metal was evaporated and the scale line represents 1 μm. (From Northcote, 1968.)

Fig. 8.3. Autoradiograph of a thin section of a wheat root cap cell exposed to [6-^{3}H] glucose for 30 minutes. Silver grains are located over slime-polysaccharide deposited between the wall and the plasmalemma, and over Golgi bodies in the cytoplasm. The slime deposit is sited under a gap in the peripheral ER. The scale line represents 1 μm. (From Northcote & Pickett-Heaps, 1966.)

High resolution autoradiography is a powerful tool in the investigation of the synthetic activities of endomembranes. First, the living tissue is given a pulse of tritium-labelled precursor. At fixed intervals of time afterwards, portions of tissue are prepared for electron-microscopy, sectioned and the stained sections are coated with a very thin layer of radiosensitive, silver halide emulsion. To prevent the development of this emulsion by chemicals in the section, an inert layer of carbon is deposited on the sections before they are coated. From the changing distribution of silver grains over the organelles of the cell, the pulse of radioactive precursor can be followed from the site of synthesis of the first insoluble product to the final resting place of the completed polymer (Fig. 8.3). Because the β-particle may have moved in the plane of the section before hitting the emulsion, the practical limit of resolution of the technique is 100 nm and since the organelles in the cell are fairly crowded it may not be possible to decide unequivocally which of them contained the labelled material.

Proper controls for autoradiographic experiments would have to include some method of identifying the radioactive product in the section. For example, pre-treating the section with solutions of degradative enzymes specific for the polymer in question should eliminate the radioactivity associated with the organelles involved in its synthesis. Also, a precursor specific for the product under investigation should be used, e.g. [^{3}H]-orotic acid for RNA. Non-specific precursor, such as [^{3}H]-glucose, is only useful with cells producing a single polymer-type, e.g. slime polysaccharide in root-cap cells or cellulose in differentiating xylem vessels. Always important, but especially necessary in this case, is a back-up biochemical analysis showing the distribution of label from the precursor in all polymeric fractions at the times that the tissue portions were fixed for electron-microscopy. It must be admitted that proper controls of this nature are hardly ever reported.

8.2.2 BIOCHEMISTRY

To isolate the various parts of the endomembrane complex, the biochemist breaks open the cells of a lump of tissue, in a pH-buffered solution to neutralize vacuolar acids, and separates membrane fractions from the rest of the cytoplasm by centrifuging. To homogenize is to use high shear forces to fragment all membranes. The little pieces spontaneously re-anneal forming small vesicles or 'microsomes'. Small quantities are homogenized in some sort of pestle and mortar arrangement, large quantities in a blender. Any compartmentalization of soluble substances in organelles is lost by this method, and the in/out orientation of the membranes of the vesicles does not necessarily represent that *in vivo*.

To prepare intact organelles, shear forces are minimized by crushing or chopping the tissue in a solution which is also buffered osmotically to prevent bursting. A few organelles will inevitably be broken and many of them will be left behind in pieces of more or less intact tissue. So, unlike homogenization, quantitative recovery of membrane from the tissue does not occur and thus it is impossible to relate directly, say, enzyme activities in isolated organelles, to the total activity in the tissue.

Differential centrifugation involves centrifuging at successively higher speeds for longer times to effect a separation on the basis of size. Thus debris and cell wall are spun out at 4,000 *g* for about 15–20 minutes, organelles at 10,000 *g* for about 30 minutes, and total membrane by 100,000 *g* for 1 hour. The membrane which pellets between 10,000 *g* and 100,000 *g* is referred to as a microsomal fraction and this material is often assumed to be ER, which probably does make up the bulk of it. Differential centrifugation is now only used to prepare organelle or microsomal fractions for subsequent analysis by density gradient centrifugation or, since resuspension of pelleted organelles causes some breakage, to remove debris before the total membrane material plus soluble cytoplasm is layered onto a gradient (Lord *et al.*, 1973).

Density gradients are prepared in the centrifuge tube by pouring in a progressively less concentrated solution of sucrose in the buffered medium. The membrane preparation is added on top and centrifuged into the gradient. In an isopycnic separation, centrifugation is continued until the components of the sample reach their equilibrium positions in the gradient and so separation is effected on the basis of density alone. In a rate zonal separation, centrifugation is stopped before equilibrium is reached. Smaller particles sediment faster and so the ordering of components in the gradient is determined by size as well as density. Continuous gradients in which the density changes linearly with distance over a range selected from 1·06 g cm^{-3} (16% w/w sucrose) to 1·23 g cm^{-3} (50% w/w sucrose) are conventionally used to separate the components of membrane fractions. When the density of a particular component is known, discontinuous or 'step' gradients are sometimes used for bulk preparation. However this method does not produce fractions as pure as those from continuous gradients. A heavy particle arriving late at an already crowded interface between two densities in a step gradient may be trapped there or, as appears to happen more often, plummet through carrying light membranes down with it. (Note that results from sucrose gradients cannot be compared with those from experiments involving gradients of 'ficoll', a synthetic sucrose polymer sometimes used, since sucrose is osmotically active and organelles and vesicles lose water progressively as they fall through a gradient of it.) Centrifuging organelles and microsomes through linear density gradients produces distinctive patterns in the distribution of protein in the gradient, so it seems the endomembrane complex can be resolved into different types of membrane. The next step is to identify them.

If the gradient fractions are fixed and embedded for electronmicroscopy, stained thin sections can reveal the presence of organelles with a characteristic morphology, such as attached ribosomes. It would involve a lot of work to establish that a particular organelle was *not* present. Negative-staining is easier and faster, but less reliable since phosphotungstic acid solution is a powerful protein extractant and so, unless the membranes are glutaraldehyde-fixed, alters the appearance of organelles (Mollenhauer *et al.*, 1973). Many membrane fractions, after conventional post-fixation with osmium and uranyl acetate staining of sections, appear as simple vesicles with no recognizable morphology. Efforts have been made to develop electron-dense stains which are specific for individual membranes of the cell, but so far only the plasmalemma has been stained selectively with any success using periodic acid and various heavy metal anions, e.g. chromate and phosphotungstate (Leonard & Van Der Woude, 1976). However, the chemical basis of the staining reaction is by no means certain (it is likely that the carbohydrate moiety of plasmalemma glycoproteins is involved) and its specificity for this membrane cannot always be repeated.

Gradient fractions can be characterized by various enzyme activities associated with them. The enzyme proteins may be freely soluble and enclosed by the membrane, or loosely associated with the membrane surface (extrinsic), or firmly bound in the membrane (intrinsic). Obviously only intrinsic enzymes can be trusted as markers. For organelles with unique and well-characterized functions, enzymes provide reliable identification, e.g. succinate dehydrogenase, an enzyme unique to the tricarboxylic acid cycle and very firmly bound, is used to indicate the presence of inner mitochondrial membrane. Otherwise, to determine the distribution in the cell of an enzyme activity localized in a gradient is a major problem.

Sometimes enzymes can be located in thin sections of cells by cytochemical techniques for electron-microscopy, e.g. specific phosphatases can be tracked down if inorganic phosphate released from supplied substrate is precipitated *in situ* on thin sections by heavy metal cations. Thiamin pyrophosphatase (TPPase) was shown to be a marker for the Golgi body in vertebrate cells in this way. The specificity of the reaction must be checked by comparison of results obtained using a range of alternative substrates. An enzyme localized by cytochemistry on thin sections cannot with certainty be used to establish the identity of membrane fractions which have not been treated with glutaraldehyde or heavy metal cations, both of which are powerful inhibitors of many enzymes.

An association of enzymes with particular membranes of the cell has sometimes been built up as a result of some knowledge of cell physiology, e.g. Na^+/K^+ stimulated ATPase and callose synthetase are thought to be markers for plasmalemma since salt uptake and callose synthesis take place at the cell surface. However, the limitation of callose synthesis to the outer surface of the plasmalemma *in vivo* may be due to *substrate* availability and it would seem equally likely that the tonoplast, for which at present there is no marker enzyme, is also involved in salt uptake.

Alternatively, enzyme markers become accepted by long association. Thus, because the activity of NADH-specific cytochrome *c* reductase correlates well with the proportion of ribosome-bearing fragments in membrane fractions, this enzyme is widely accepted as a marker for ER. (Note that the inner mitochondrial membrane also shows NADH-specific cytochrome *c* reductase, which differs from that of the ER in being sensitive to antimycin A. mitochondria must be shown to be absent from the fraction and the activity demonstrably antimycin-insensitive when this marker is used for ER.) This sort of association between an enzyme and a specific membrane is only of any use for tissues and organisms in which it has been established as there is no reason to believe that markers are universal. Thus, glucose-6-phosphatase is an ER-marker in certain cells only, e.g. vertebrate liver and kidney, and together with 5′-adenosine monophosphatase (AMPase, 5′nucleotidase), a marker for plasmalemma in animals, does not occur in the equivalent membranes of plant cells.

Note that while the detection of marker enzyme activity can indicate which types of membrane are present, the absence of activity may be due to failure of the assay for a number of reasons other than absence of the enzyme. Also, no membrane fraction can be established as 'pure' on the basis of marker enzymes while there are still types, like the tonoplast, for which a marker does not exist.

8.3 COMPOSITION AND CHARACTERISTICS OF THE MEMBRANE TYPES

There is considerable evidence for broad homology between the nuclear membrane and the membranes of the ER, and some indications that the outer mitochondrial membrane and the plastid envelope, while essentially similar to the ER, are intermediate in composition between it and the specialized inner membranes of these organelles. From the little that is known of their composition, the membranes of the Golgi apparatus are closely allied to those of the ER, but the plasmalemma is markedly different from this membrane.

8.3.1 NUCLEAR MEMBRANE

Nuclear membrane is prepared from isolated nuclei by sonication to fragment the membrane, high salt treatment to remove associated nucleic acid, and density gradient centrifugation to collect the fragments. In a careful study, taking precautions to minimize phospholipid loss by enzyme degradation during isolation, Philipp *et al.* (1976) found no difference between the lipid and fatty acid composition and the nature and level of enzyme activities in nuclear membrane and ER from plants. Both membranes, for example, had NADH- and NADPH-specific cytochrome *c* reductases at similar levels, and both possessed a cytochrome of the b_5 type. They could be distinguished by the higher mean buoyant density of nuclear membrane (1·20 g cm^{-3}), attributed to a higher content of nucleic acid and protein. (Values between 1·08 and 1·12 g cm^{-3} are recorded for the density of ER without attached ribosomes in sucrose gradients, and between 1·16 and 1·18 for the mean buoyant density of 'rough' ER.) In animal cells ER and nuclear membrane appear to be identical except for the higher density of the latter (Franke, 1974). They also resembled their animal counterparts in having a higher proportion (60%) of their phospholipid as phosphatidyl choline ('lecithin') than the other membranes of the cell. Phosphatidyl ethanolamine was the second most abundant phospholipid at 20–24%. They differed from animal nuclear membrane and ER, and resembled other plant membranes in their high sterol content (*ca* 30%), and high content of unsaturated fatty acids as esters in the lipid. Oleate (18:1) and linoleate (18:2) accounted for 80% of these and palmitate (16:0) made up most of the rest.

In thin sections, the outer nuclear membrane is often seen to bear ribosomes on its cytoplasmic face, just like both membranes of the ER, and when nuclei were isolated from pea shoots in high concentrations of Mg^{2+}, the outer membrane was densely covered with ribosomes (Stavy *et al.*, 1973).

On the basis of lipid composition, enzymic activities and ability to bind ribosomes, nuclear membrane cannot be distinguished from ER. It differs morphologically in possessing nuclear pore complexes which probably contribute to its higher buoyant density.

8.3.2 OUTER MITOCHONDRIAL MEMBRANE

The outer membrane can be released from isolated mitochondria either by a slightly hypotonic medium (osmotic lysis) or digitonin to which the inner membrane is relatively resistant. The heavy, intact inner membrane vesicles and their enzymic contents are then easily separated from the light fragments of outer membrane.

Turnip outer mitochondrial membrane was found to have NADH-specific cytochrome *c* reductase and a b_5-type cytochrome, just like the microsomal fraction, from which it differed in lacking firmly bound, NADPH-specific cytochrome *c* reductase (Day & Wiskich, 1975). Compared with the bulk microsomes, the inner membrane of cauliflower mitochondria had a low (40%) proportion of its phospholipid as phosphatidyl choline due to replacement by the specialized mitochondrial phospholipid, diphosphatidyl glycerol ('cardiolipin'). The outer membrane resembled the inner membrane in that only 40% of its phospholipid is phosphatidyl choline, but was intermediate between this membrane and the microsomes in its content of diphosphatidyl glycerol (Moreau *et al.*, 1974). Isolated plastids and mitochondria have been shown to contain less of the plant sterols sitosterol and stigmasterol than the other membranes, but about the same concentration of cholesterol in $\mu g\ mg^{-1}$ protein. As a result, the sterol content of these organelles is only one-tenth that of the bulk membranes of the cell (Hartmann *et al.*, 1973). Mannella & Bonner (1975) estimate the ratio of sterol:phospholipid in the outer mitochondrial membrane as 0·17 for mung bean hypocotyl and 0·06 for potato tuber. These authors report that the degree of unsaturation of the fatty acids in outer membrane lipids was intermediate between that of light microsomes and the value for the highly unsaturated lipids of the inner membrane. For mung bean hypocotyl, the average number of double bonds per fatty acid was 1·3 for light microsomes, 1·7 for the outer and 2·0 for the inner membrane. The same pattern was seen in experiments on potato tuber, although in cauliflower Moreau *et al.* (1974) reported that the outer membrane was more saturated than the microsomal membranes.

In summary, the plant outer mitochondrial membrane has enzyme activity which shows it to be ER-like, but in its lipid composition it has characteristics intermediate between the ER and the very specialized inner membrane.

8.3.3 CHLOROPLAST ENVELOPE

This membrane is prepared from isolated chloroplasts by osmotic rupture. Chloroplast lamellae are chiefly composed of galactolipid (92% of the lipid), especially monogalactosyl diglyceride and the low content of phospholipid is largely made up of phosphatidyl glycerol. The fatty acids are highly unsaturated, with linolenate (18:3) as the major type at 83 mole %. Trans-Δ^3-hexadecenoate is a minor fatty acid component unique to the lamellae. The envelope was found to differ from the microsomes and resembled the lamellae in its high galactolipid content, in the complete absence of phosphatidyl ethanolamine, and relatively high proportion of linolenate (18:3). It also shared a number of features with the microsomal fraction in respect of which the envelope differed from the lamellae. These were: phosphatidyl choline as major phospholipid, digalactosyl diglyceride as the major galactolipid, and a relatively high proportion of palmitate (16:0) amongst the esterified fatty acids (Mackender & Leech, 1974).

Like the outer mitochondrial membrane, the chloroplast envelope appears to be a curious chimaera, intermediate between the ER and the specialized membrane system it encloses.

8.3.4 GOLGI MEMBRANES

Membrane fractions enriched in Golgi bodies are prepared by density-gradient centrifugation of organelles obtained after chopping plant tissues. Glutaraldehyde is routinely added to the extraction medium to 'stabilize' the structure of the Golgi bodies though it now seems that the organelle survives preparation intact and only requires this fixative for electron microscopy (Bowles & Kauss, 1976). Golgi bodies from plant sources sediment with the fraction of buoyant density 1·12 to 1·15 g cm^{-3} in sucrose gradients and are identified in gradient fractions by their distinctive morphology. Thin section studies of whole cells in higher plants show the Golgi body as a stack of 5 to 8 closed membrane sacks. Each sack, or 'cisterna', is flattened to a disc, 0·5 to 1 μm in diameter. Around the perimeter of the disc there is progressive fenestration through, and localized swelling of, the cisternal membranes, grading into isolated vesicles beyond the edge of the disc (Fig. 8.4). In some cases a single layer of parallel fibrils has been seen in the 10–15 nm wide, electronlucent region between the cisternae. Freeze-etch pictures (Fineran, 1973a) confirm this structure for the organelle. Electron-microscope sections and negative-stain images of glutaraldehyde-fixed material show that this structure is maintained in isolated organelles. The cisternae remain joined in a stack as they are in the cell, but separate in media of high ionic strength (Mollenhauer *et al.*, 1973).

Inosine diphosphatase (IDPase) has been used as a marker for Golgi bodies and, for some plant tissues, a single peak of activity is observed on sucrose gradients, coinciding with the Golgi body-rich fraction. However, IDPase activity

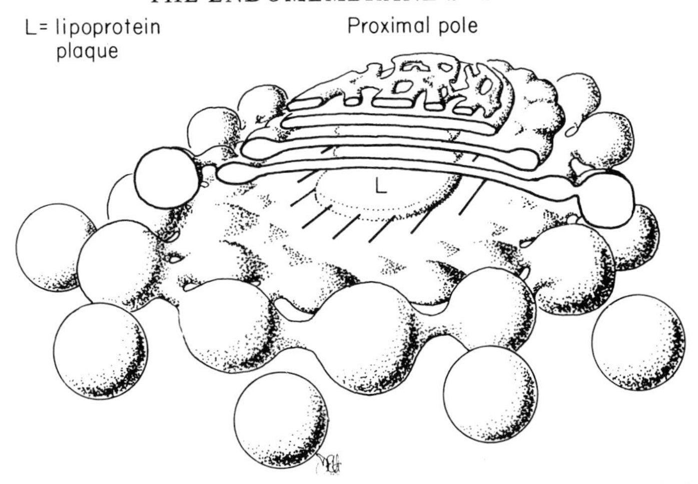

Fig. 8.4. The structure of the Golgi body. Based on the freeze-etch pictures of Fineran (1973a) which are probably close to the real shape of this organelle in the cell. Conventional electron microscope methods involving fixation and dehydration are more likely to cause distortion.

was found in purified nuclear membrane (Philipp *et al.*, 1976), and using mung bean hypocotyl a second peak of activity was observed in light membrane characterized as ER (Bowles & Kauss, 1976).

Thin sections of Golgi body fractions from sucrose gradients show that a variety of less organized vesicles is also present, and the possibility that these preparations are contaminated with other membrane types cannot be excluded. The purest Golgi body fractions to date have been obtained from rat liver. Giving a rat ethanol induces the liver Golgi bodies to fill up with light lipoprotein particles which cause a distinctive change in the buoyant density of the organelle and serve as internal structural markers for electron-microscopy. The purified Golgi body membranes contained the cytochrome b_5 electron-transport chain (NADH-specific cytochrome *c* reductase and cytochrome b_5) characteristic of ER, but lacked NADPH-specific cytochrome *c* reductase and some other ER markers. UDP galactose:N-acetyl-glucosamine galactosyltransferase was restricted to the Golgi body fractions (Bergeron *et al.*, 1973). Other evidence from animal cells indicates that fucosyl and sialyl transferases are primarily associated with the Golgi apparatus. 'Pure' Golgi fractions have not yet been prepared from plant material, but it has been shown that maximum activities of a UDP galactose:N-acetyl-glucosamine galactosyltransferase involved in glycoprotein biosynthesis (Powell & Brew, 1974), and UDP glucose:sterol glucosyltransferase (Bowles *et al.*, 1976) were found in the Golgi body-enriched fraction of sucrose density gradients. The evidence then, such as it is, suggests that Golgi membranes have some affinity with the ER, but are specially enriched in glycosyltransferases involved in the synthesis of glycoprotein and glycolipid.

8.3.5 PLASMA MEMBRANE

Plasmalemma is obtained from tissue homogenates by sucrose density gradient centrifugation after removing mitochondria whose density range overlaps that of this membrane. Reported mean buoyant densities range from 1·16 to 1·18 g cm^{-3}. This fraction has been identified as plasmalemma on the basis of specific staining, and characterized by the presence of Na^+/K^+ stimulated ATPase activity, and glucan synthetase activity at high UDP glucose concentrations (callose synthetase), and the absence of ER and Golgi body markers (NADH-specific cytochrome *c* reductase and IDPase) (Leonard & Van Der Woude, 1976). Since none of the markers can be considered exclusive to the plasmalemma and non-staining membrane is always seen in the fraction, firm characterization of this membrane in plants will depend on better means of separation or identification. Investigators have tried to label the external surface of this membrane in intact cells, e.g. using radioactive iodine, hoping to trace the membrane in a gradient by its label. The most successful attempt has involved the carbohydrate-binding protein, concanavalin A (conA), which has been used to prepare plasmalemma from wall-less mutants of *Neurospora* and yeast protoplasts. Treatments with conA just before lysis stabilizes the membrane as large sheets and in this form it is easily separated from small vesicles. Removing the conA causes the sheets to vesiculate and now the plasmalemma can be separated from large contaminants which had co-sedimented with it (Scarborough, 1975).

Plasmalemma fractions from animal cells have a much higher content of glycoprotein and glycolipid than the other membranes of the cell and the carbohydrate moieties of these compounds form a glycocalyx over the external surface. Scarborough's work with conA suggests a similar layer is present on the surface of the fungal plasmalemma, and the specific staining of the plant plasmalemma after treatment with periodic acid, which creates reactive aldehyde groups on carbohydrate residues, implies that this is also true for the plant plasmalemma. When thin sections of protoplasts in plasmolysed onion tissue were stained for carbohydrate, only the outer surface of the plasmalemma reacted. Since this response was unaffected by treatment with concentrated cellulase and pectinase, it is unlikely that it was due to wall polysaccharide (Roland & Vian, 1971).

Bailey and Northcote (1976) have obtained a pure preparation of plasmalemma from the green alga, *Hydrodictyon*, by syringing out the contents of these huge cells, leaving the plasmalemma firmly attached to the wall. Since there was insufficient material for direct chemical analysis, the composition of the membrane was estimated from the radioactivity in each phospholipid after feeding [$^{32}P_i$] to algae for 20 days, sufficient to saturate the membranes with label. Phosphatidyl choline was very much the major phospholipid, accounting for 58% of the radioactivity in identified phospholipids. Phosphatidyl serine, though only traces were found in whole cells, was a substantial component of the plasmalemma at 8%. Sterol:phospholipid molar ratios between 1·0 and 1·2 have been recorded for plant membranes isolated on a density gradient and identified as plasmalemma-rich by staining. This represents a considerable enrichment in sterol over the total (0·4) and light membrane (0·2) fractions (Hodges *et al.*, 1972). High sterol content will tend to reduce the fluidity of the membrane and so stabilize the cell boundary. In animal cells also phosphatidyl serine is predominantly associated with the plasmalemma and, to a lesser extent, the Golgi apparatus. Again, in animal cells, there is an increase in the proportion of sterols from ER to Golgi apparatus to plasmalemma, but they differ from *Hydrodictyon* in that this sequence is associated with a marked decline in the contribution of phosphatidyl choline to the phospholipids.

In summary, plasmalemma-enriched membrane fractions have a set of enzyme activities very different from that of the ER. There is some evidence that this membrane, like its animal counterpart, is more glycosylated than other membranes, from which it also differs in lipid composition.

8.4 FUNCTIONAL RELATIONSHIPS BETWEEN MEMBRANES

8.4.1 MEMBRANE SYNTHESIS

Since mitochondria and chloroplasts appear to make only a small fraction of their protein and lipid, and the plasmalemma none at all, there seems little doubt that most of this material is ultimately derived from the ER which has been shown to be the only site in the cell competent to synthesize the major membrane constituents.

Microsomal fractions from onion stem and spinach leaves synthesized phosphatidyl choline from CDPcholine and phosphatidic acid and phosphatidyl ethanolamine from CDPethanolamine and phosphatidic acid (Marshall & Kates, 1973). Beevers' group have shown that the terminal enzymes for the synthesis of phosphatidyl choline, phosphatidyl serine and phosphatidyl inositol are found only in a light membrane fraction (buoyant density 1·12 g cm^{-3}) prepared from chopped endosperm of germinating castor beans (Beevers, 1975). The whole cytoplasm, soluble enzymes as well as membranes, was centrifuged into a linear gradient containing 1 mM EDTA. Electron-microscope sections showed that the membranes carried no ribosomes and they had sedimented separately. In gradients containing 3 mM Mg^{2+}, activity at 1·12 g cm^{-3} largely disappeared and the membrane, NADPH-specific cytochrome *c* reductase, and phosphorylcholine glyceride transferase (the terminal enzyme in phosphatidyl choline synthesis) activities were all spread around 1·16 g cm^{-3}. This membrane was now coated with ribosomes. It was concluded that the final steps in the synthesis of these phospholipids are located exclusively in the ER (Lord *et al.*, 1973). A key enzyme of sterol biosynthesis, transfarnesyl pyrophosphate-squalene synthetase, is almost all associated with fractions rich in ER (Hartmann *et al.*, 1973).

Polyribosomes, the groups of ribosomes linked by mRNA and active in protein synthesis, occur free in the cytoplasm and attached to the surface of the

ER and outer nuclear membrane. They can be seen in glancing sections of these membranes as folded or spiral chains of ribosomes. The potential of the ER to synthesize membrane proteins is therefore not in doubt. There is good evidence from mammalian cells that secretory proteins are synthesized by ribosomes attached to ER and pass rapidly through this membrane into the lumen (Jamieson, 1975). Liver cells synthesize protein mainly for secretion and 70% of the polysomes are bound. Puromycin released most of the ribosomes from rough ER isolated from liver, indicating that it is the attachment of the growing peptide to the membrane which holds the ribosomes on. Now, this does not mean that proteins synthesized by rough ER are inevitably incorporated into or through the membrane. In non-secretory animal tissue-culture cells the ribosomes were bound to ER only by their attachment to mRNA. Interestingly, for rough ER from bean hypocotyls, 80% of the bound ribosomes required puromycin for release and only 10% were released by RNase alone (the others could be removed by high salt) (Dobberstein *et al.*, 1974). Since only a few plant cell types secrete large amounts of protein, e.g. cells of the aleurone layer of cereal grains, it appears that most of the protein made by the rough ER is being incorporated into the membrane rather than through it. However, only 20 to 25% of the cytoplasmic ribosomes were recovered in the rough ER fraction isolated from bean hypocotyls (*cf* liver cells) and it is quite possible that membrane proteins are made on free polyribosomes and subsequently incorporated into the appropriate membrane.

To find out more about the synthesis and movement of membrane in the cell, a number of investigators have fed radioactive precursors of lipid and protein and traced the pattern of distribution of labelled product in various cell fractions.

In rat liver cells, nuclear membrane and ER incorporate a variety of [^{14}C]-amino acids into membrane protein with identical kinetics, and the labelling of the phosphatidyl choline of these membranes by [^{14}C]-choline rules out either synthesis in the nuclear membrane and transfer to the ER, or synthesis in the ER and transfer to the nuclear membrane (Franke, 1974). These results confirm the close similarity of these two membranes. No comparable findings have been reported for plants.

When [^{14}C]-choline was supplied to castor bean endosperm, phosphatidyl choline in the ER was labelled before that of any other membrane and part of this labelled phosphatidyl choline appears in mitochondrial membrane within a short time (Beevers, 1975). As the enzyme studies suggested, mitochondria are dependent on the ER to synthesize the bulk phospholipids of their membranes. In similar experiments with animal tissues, the specific activity of phosphatidyl choline and phosphatidyl ethanolamine in the outer membrane was intermediate between that of the ER and that of the inner membrane. This is consistent with the movement of these phospholipids from the ER to the outer membrane to the inner membrane (not a very startling idea since access to the inner membrane from outside the mitochondrion must be *via* the outer). The dependency of this organelle on the ER for phospholipids is not total. The terminal enzyme for the synthesis of phosphatidyl glycerol was found in both ER and mitochondria of castor bean endosperm. This phospholipid is a precursor for diphosphatidyl clycerol (DPG), in which the inner membrane is especially rich, and isolated animal mitochondria have been shown to synthesize phosphatidyl glycerol *and* DPG.

Whether the lipids of the chloroplast lamellae are made 'on site', or imported from somewhere else in the cell, is still a matter of controversy (Morré, 1975).

Another mystery is the site of synthesis of phospholipid precursors. Although fatty acid synthetase is soluble, recent work has indicated that all the activity is membrane-enclosed in the cell (Weaire & Kekwick, 1975; Donaldson, 1976), and at least part of it was found in 'heavy' organelles, proplastids and mitochondria. Certainly, when [^{14}C]-acetate was supplied to castor bean endosperm, lipid in organelles was labelled at the earliest measurement (5 minutes) making it unlikely that fatty acid synthesis was restricted to the ER. If mitochondria and plastids are able to synthesize some of their own fatty acids, this could explain how they maintain a high level of unsaturation against the influx of lipid from the ER, since phospholipid molecules in membranes are dynamic structures in which acyl moieties can be replaced or exchanged. In addition it is possible that the fatty acids of ER-derived phospholipid could be desaturated in the plastid or mitochondrion. Desaturases are always intimately associated with membrane systems since the participation of a special electron transport chain is required, and phospholipid appears to be the preferred substrate for the production of linoleate (18:2) from oleate (18:1) in *Chlorella* (Sedgwick, 1972).

Independent, internal synthesis of special components, e.g. phosphatidyl glycerol and unsaturated fatty acids, is one way the distinctive composition of organelle membranes may be maintained. Another way would be to use the intermembrane space, e.g. between inner and outer mitochondrial membranes, to control the movement of individual lipid types. The differences in lipid composition between these two membranes support the idea and there are reports of 'exchange' proteins in the soluble cytoplasm which can ferry specific phospholipids in solution between membranes (McMurray & Dawson, 1969; Morré, 1975). The fact that the ER and the outer membrane differ in composition suggests that lipid must also be 'filtered' as it passes between these two membranes. For this reason, if no other, it is unlikely that the ER is structurally continuous with these outer membranes in the cell. Although movement between the two halves of the bimolecular leaflet is restricted (see chapter 2), lipids and many membrane proteins are freely mobile in the plane of the membrane and any differences in composition would soon disappear. Conventionally prepared, electron microscope sections have provided occasional observations of interconnections between all the membrane types of the cell, but phenomena seen only infrequently by electron microscopy should be treated with suspicion. Membrane conformations are stabilized by weak bonds, and the covalent cross-linking of proteins by glutaraldehyde could force lipidic structures to re-organize to create continuities, for example where two membranes were closely appressed. For certain species, e.g. the fern *Pteris*, an unusually large number of conjoinings between membranous organelles has been reported (Crotty & Ledbetter, 1973) which makes it unlikely that these observations have universal significance. The close similarity, biochemically, between nuclear membrane and ER would support observations that these two membranes are in clear luminal continuity, but can also be explained as the result of the total re-structuring of the nuclear membrane from ER at every mitosis. It has been suggested that membrane could be transferred from the ER to the outer organelle membranes by intermittent direct connections, and lipid components might also move in solution on 'carrier' proteins (Morré, 1975).

From thin section studies, electron microscopists suggested that the ER and the plasmalemma are linked functionally *via* the Golgi apparatus, the cell's complement of Golgi bodies. Golgi bodies can be seen adjacent to a sheet of ER or the outer nuclear membrane, aligned with the cisternae parallel to the plane of these membranes. The portion of membrane apposed to the Golgi body bears no ribosomes but projects as small tubules into the cytoplasm between. It is hypothesized that the tubules vesiculate and the 'primary' vesicles formed then re-fuse to form a new Golgi body cisterna (Morré & Mollenhauer, 1974; Gunning & Steer, 1975). Also, circular membrane profiles, similar to those in the vicinity of the vesiculated rim of the cisternae, are found close by and apparently fused with the plasmalemma. The membrane of these 'secretory' vesicles is stained heavily after treatment of sections with barium permanganate and the plasmalemma stains in the same way. ER does not stain heavily. The membranes of the Golgi cisternae show a progressive increase in staining intensity from ER-like at one pole to plasmalemma/'secretory' vesicle-like at the other (Grove *et al.*, 1968). There is, then, ultrastructural evidence to suggest that the Golgi apparatus is linked to both ER and plasmalemma by alternating membrane vesiculation and fusion, and some aspect(s) of the transition between ER and plasmalemma membranes takes place within the Golgi apparatus. Of course, equivalent fiffinity for heavy metal is no real indication of biochemical similarity, no more than the intermediate density of the Golgi body can be considered as evidence for its function as an intermediate between ER and plasmalemma.

After feeding labelled precursors, the specific radioactivity of phospholipid in membrane fractions was reported to be highest in ER, intermediate in Golgi bodies and lowest in plasmalemma (Morré, 1975), a pattern consistent with the proposed function of the Golgi body.

Studies of the incorporation of a pulse of radioactive amino acid, [guanido-^{14}C]-arginine, into membrane protein in rat liver have shown that membrane protein in a 'rough microsomal' fraction was labelled very early on. The kinetics of labelling suggested that some membrane protein passed from ER to Golgi apparatus to plasmalemma. Thus, the rough ER fraction reached maximum specific activity within 10 minutes of feeding. By 20 minutes the specific activity had fallen by one-fifth, as if part of the newly labelled protein had left the fraction. At the same time the specific activity of a Golgi body-enriched fraction, already higher than that of the ER, continued to rise reaching a maximum at 30 minutes and declining until 60 minutes to five-sixths the maximum value. Now, between 10 and 60 minutes, the membrane protein of a plasmalemma fraction accumulated label in two phases, firstly as label was lost from the rough ER and then as label disappeared from the Golgi fraction (Fig. 8.5, Franke *et al.*, 1971). These results indicate that:

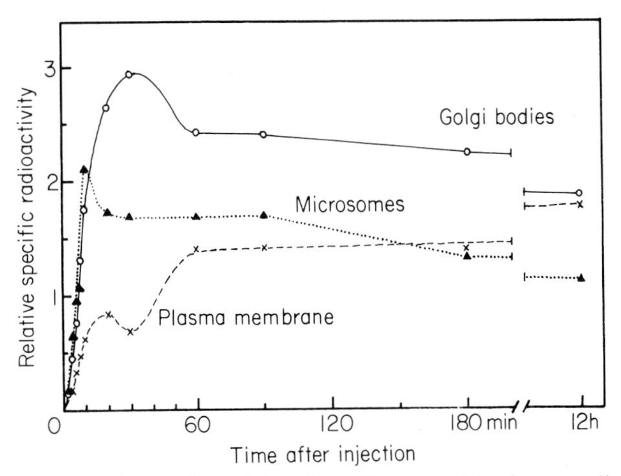

Fig. 8.5. Changes in the specific activity of membrane proteins in organelle fractions from rat liver after feeding a pulse of [guanido-^{14}C] arginine.
○——○ Golgi body-enriched fraction
▲ . . . . . ▲ microsomes
× - - - - × plasmalemma-enriched fraction
(from Franke *et al.*, 1971.)

(a) proteins of the ER may contribute to the plasmalemma *without* the mediation of the Golgi apparatus,

(b) only a part of the newly labelled ER and Golgi apparatus membrane protein turns over sufficiently rapidly to be able to give rise to the observed increase in plasmalemma labelling,

(c) if the newly labelled material of the Golgi apparatus is derived from the ER (and this is by no means obvious from the data since there is no lag between incorporation into ER and into Golgi apparatus fractions) then it must be transferred from a region of synthesis and rapid turnover within the bulk ER since the specific activity of the Golgi apparatus was higher than that of the ER fraction.

Similar experiments have been performed with onion stem tissue, feeding [U-^{14}C]-leucine (Morré & Van Der Woude, 1974). In this case a lag of 30 to 60 minutes was observed after incorporation into microsomal and nuclear membranes had begun and before Golgi apparatus and plasmalemma fractions began to be labelled. The authors discern a second phase of incorporation into plasmalemma, lagging 30 minutes behind the initial simultaneous labelling of Golgi apparatus and plasmalemma. Their results are consistent with transfer of newly synthesized protein from microsomal to Golgi membranes in plant cells and, as in rat liver, the ER may contribute material directly as well as *via* the Golgi apparatus. Morré and Van Woude assumed the label in membrane fractions represented incorporation into membrane protein only. Though the cells of onion stem do not secrete protein on the same scale as rat liver, this assumption is not entirely justified. A hydroxyproline-rich glycoprotein is a substantial component of cell walls and its synthesis probably takes place in the ER and the Golgi apparatus (Gardiner & Chrispeels, 1975).

If, as this work suggests, the bulk of the membrane protein in the ER and the Golgi bodies is not transferred, this would explain how these membranes and the plasmalemma maintain distinctive enzyme activities.

However, there is considerable evidence that the membranes in the Golgi body stack are moving rapidly through it. Certain unicellular algae, e.g. *Prymnesium* and *Chrysochromulina*, have only one Golgi body per cell. By way of compensation, there may be as many as 30 or more cisternae per stack. The organelle elaborates complex scales which coat the outer surface of the alga and the polarity of the stack is obvious from the gradation in complexity of scale precursors seen inside the cisternae in thin sections (Manton, 1966). This is the most direct evidence that whole cisternae are shuttled along the stack in one direction, from the ER, which apposes the immature ('proximal') end of the stack, to the plasmalemma with which the cisternae full of mature scales fuses. The Golgi body in higher plants and animals differs from that of these algae in that the cisternae vesiculate before fusing with the plasmalemma, but evidence from higher organisms also suggests the whole stack turns over. The number of cisternae per stack decreases in starvation or when 'secretory' vesicle production is stimulated and increases when vesicle production is inhibited (Morré *et al.*, 1971). The alteration in staining properties across the Golgi stack affects the whole cisternal membrane and not just the peripheral vesiculations (Grove *et al.*, 1968), and in isolated Golgi bodies this property spreads within a short time to the ER-like cisternae of the stack so that *all* the membrane resembles plasmalemma (Frantz *et al.*, 1973).

Golgi-specific enzyme proteins *could* be retained in the organelle while the lipidic sack and plasmalemma-specific proteins pass on if there is movement of membrane protein between cisternae. The centre of the cisterna where it is closely appressed to its neighbours is the most likely site for this activity and the intercisternal space here is filled by a lipoprotein plaque (Mollenhauer *et al.*, 1973), a hydrophobic environment in which lipophilic membrane proteins would be mobile. It is apparent from the staining evidence and the intermediary status of the Golgi body between ER and plasmalemma that there is a gradient in the concentration of substrate, protein and lipid, for the membrane-modifying glycosyltransferases of the Golgi, and the cisternal membrane at the proximal pole of the stack has the highest concentration. Substrate binding alone could ensure that the net movement of Golgi enzymes is in the direction of the proximal pole. The glycosylation of the products, Golgi body-modified plasmalemma-specific proteins and lipids, would make them too hydrophilic to move rapidly between cisternae in this hypothetical scheme.

8.4.2 SYNTHESIS AND SECRETION OF EXTRACELLULAR MATERIAL

While investigations into the function of the Golgi apparatus in animals have concentrated on protein glycosylation and other membrane modifications, plant biochemists have been more concerned with a role for this organelle in polysaccharide biosynthesis.

Gardiner and Chrispeels (1975) have presented evidence that the glycosylation of the hydroxyproline-rich glycoprotein takes place in the Golgi apparatus. The completed polymer is a cell wall rather than a plasmalemma component and carries arabinosyl side-chains on its hydroxyproline residues. Organelle fractions were prepared on density gradients from carrot root tissue pulse-labelled with [^{14}C]-proline. Most of the label in hydroxyproline was associated with a Golgi body-enriched fraction. This fraction coincided with peak activity in the gradient of an enzyme which transferred arabinosyl residues from UDP-arabinose onto endogenous protein acceptors. (It should be noted, though, that the gradients used to prepare organelles contained glutaraldehyde and the distribution of enzyme activities on glutaraldehyde-free gradients is not strictly comparable).

From studies of thin sections, electron-microscopists proposed that the Golgi apparatus was involved in the synthesis and secretion of cell wall polysaccharides. The increase in luminal volume at the cisternal periphery was thought to represent the polymerization or decantation of a batch of wall material which travelled to the plasmalemma in a secretory vesicle and was discharged into the wall as the vesicle membrane fused with the plamalemma. This evidence has been critically reviewed by O'Brien (1972).

Much of the biochemical evidence has come from work on the root cap. It seems that to follow polysaccharide biosynthesis by autoradiography requires the extremely high rates of uptake and incorporation found only in certain differentiating or differentiated cells. The root cap is a very rapidly growing and metabolizing tissue—the entire cap of *ca* 10,000 cells in maize is replaced every 24 hours. The cap cells secrete large quantities of slime to ease the passage of the cap through the soil. This material is a hydrated polysaccharide with a chemical composition similar to the matrix components of primary cell walls, pectin, but the structure of the polymer has been modified to reduce gelation, e.g. maize root slime-polysaccharide has a substantial content of residues of fucose, a sugar absent from maize root pectin. Slime producing root cap cells each have several hundred Golgi bodies (root parenchyma cells have only *ca* 30) with a characteristic 'hypertrophied' morphology: the vesiculations of the cisternal periphery are very swollen.

Northcote and Pickett-Heaps (1966) fed [6-^{3}H]-glucose to wheat roots and analysed the pattern of incorporation of this label into polymeric material of root cap organelles by high resolution autoradiography. After 5 minutes of feeding, silver grains were confined mainly to the immediate vicinity of the Golgi bodies. By 10 minutes, both Golgi bodies and secretory vesicles were labelled, but very little radioactivity was associated with the wall. Subsequent incubation of labelled roots in non-radioactive glucose solution for 10, 30 and 60 minutes provided a time-series of sections showing progressive loss of radioactivity from the Golgi bodies and secretory vesicles and its accumulation outside the plasmalemma (Fig. 8.3). Chemical analysis of the high molecular weight polysaccharide in the root cap after 15 minutes exposure to radioactive glucose showed that more than 70% of the label in this material was in galactosyl residues. Galactose occurs only in pectin-type polymers in angiosperm cell walls and is the only major unit of these polysaccharides to retain a hydrogen atom on carbon 6. Since less than 3% of the polysaccharide label was in glucose units of α-cellulose at this time, the labelled material represents slime-polysaccharide or matrix components of the wall.

This is strong evidence that the Golgi apparatus is responsible for the secretion of pectin-type polymers. Though high molecular weight material appeared first of all in the Golgi apparatus, this does not establish that it was synthesized here. The early stages in polymerization must involve oligosaccharides which may be lost from the sections.

The differing functions of ER and Golgi apparatus in the synthesis and secretion of slime-polysaccharide in the root cap have been investigated by Bowles and Northcote (1972), and compared with the synthesis and secretion of the ordinary matrix components of the wall in the rest of the root. They fed [U-^{14}C]-glucose to maize roots and, taking different parts of the root, prepared organelle fractions by differential and 'step' density gradient centrifugation. These were characterized from thin sections. Radioactive polysaccharide was found in a rough ER fraction and a fraction rich in Golgi bodies. Chemically, the labelled material in membranes from the root tip resembled slime polysaccharide and that in membranes from mature root tissue had the composition of the cell wall matrix. It could have been slime or pectin from the wall which was mixed with and bound to, or enclosed in, membranes when the tissue was chopped. To test this, membrane fractions were prepared from unlabelled tissue chopped in an extraction medium with radioactively labelled, soluble slime and wall components. The membrane fractions were essentially unlabelled.

When [U-^{14}C]-glucose was fed to maize roots over times ranging from 5 to 90 minutes, radioactivity was incorporated steadily into slime, wall and membrane polysaccharide up to 30 minutes. After this time the slime and wall continued to accumulate label at the same rate but there was no further increase in the radioactivity of each of the sugar residues of the polysaccharide in membranes, confirming that these polymers are precursors of the slime and wall matrix (Bowles & Northcote, 1974).

Bowles and Northcote (1976) investigated this precursor material after labelling it to saturation from supplied [U-^{14}C]-glucose. Some of the polysaccharide was freely soluble with a high molecular weight (>40,000), most of the rest was so firmly membrane-bound it required protease digestion for release. These types were found in both membrane fractions, but the Golgi body-rich fraction had more of the former, whereas the ER had much more of the latter. Most of the membrane bound polysaccharide of the ER was as short chains (MW < 4,000), but *all* the polysaccharide segments bound to Golgi body membranes had molecular weight greater than 4,000. Interestingly, though all the *other* sugars of slime-polysaccharide were found in labelled residues of the short chains, fucose, which always occupies a terminal position on pectin side-chains, was absent.

The earliest polymeric precursors of polysaccharides will have the lowest molecular weights and it seems from this work that they are membrane-bound. The glycosyltransferases involved in chain-extension during pectin synthesis are also all bound to membranes (Villemez *et al.*, 1965; McNab *et al.*, 1968; Odzuck & Kauss, 1972). This early precursor is associated chiefly with the ER. Chain extension, addition of terminal fucosyl groups to slime polysaccharide and release from the membrane occur progressively as this material leaves the ER and passes through the Golgi apparatus and there is no evidence that synthesis is restricted to a particular section of the endomembrane complex.

ER fractions for both cap and mature root contained much more radioactive polysaccharide than the equivalent Golgi body fraction. However, since a whole-root ER fraction had 40 times as much lipid, the radioactivity in polysaccharide *per unit quantity of membrane* (i.e. weight of lipid) for the Golgi body-rich fraction was twice the value for ER. This could be because a smaller fraction of the ER is devoted to polysaccharide synthesis, or because the polymer chains are longer in the Golgi body, or both.

With the biosynthetic machinery saturated from [^{14}C]-glucose, Bowles and Northcote (1974) compared the rate of production of radioactive wall and slime with the steady-state levels in membrane polysaccharide. This enabled them to calculate the rates of turnover of polysaccharide in the membrane compartments expected for different models of secretion. For example, labelled wall polysaccharide is produced at 2,000 cpm per minute and the steady state level of radioactivity in wall polysaccharides in the Golgi bodies in 5,000 cpm. Therefore the entire polysaccharide content of the Golgi body will be replaced every 2·5 minutes *if* all the wall material has to pass through the Golgi apparatus.

Supposing polysaccharide and membrane sack move through the stack as a single entity, then for the average stack of 5 to 6 cisternae, one cisterna is released roughly every 0·5 minutes. This seems very fast, but it is remarkably close to values determined microscopically in other plants. Working with the Chrysophycean alga *Pleurochrysis*, Brown (1969) showed by time-lapse cinephotomicrography that the single Golgi body and all the other organelles rotate inside the cell wall, 360° in 15 to 20 minutes. Now the lateral displacement observed in thin sections between successive cisternae released from the Golgi body was 15°, indicating that the time between the release of successive cisternae is about 0·75 minutes. Schnepf (1961) measured the volume of slime produced by glands on the leaves of the carnivorous plant *Drosophyllum lusitanicum* over set time periods. From the size and number of Golgi vesicles in the gland, he estimated that the rate of production of vesicles by each Golgi body necessary to maintain the observed rate of secretion was 3 per minute at 28°C.

Now there is good reason to suspect that 2 cisternae per min is an overestimate of the speed required to shift all the matrix to the wall *via* the Golgi apparatus in maize roots. Firstly, while the recovery of slime and wall components is probably close to 100%, the recovery of Golgi bodies is certainly much lower and so the steady-state level of radioactivity in this organelle is underestimated. Secondly, the wall fraction from mature root tissue was shown to contain a large amount of labelled glucose polymer which was only a minor component in the membrane fractions. It could be cellulose or contaminant starch, but it means that the measured rate of increase in polysaccharide label of the wall fraction is probably an overestimate of the rate of production of wall matrix by the endomembranes. Nevertheless, observed rates of turnover of cisternal stacks can account even for this overestimate of the rate at which polysaccharide would have to pass through the Golgi body and there is no need, therefore, to invoke secretion *via* the vesiculating cisternal periphery independent of the turnover of the stack, or transfer direct from ER to the wall.

Similar calculations indicate that if slime-polysaccharide is secreted only *via* Golgi bodies, the entire polysaccharide content of these organelles is displaced every 20 seconds, a figure corrected for cellulose. The changing pattern of label in polysaccharide monomers after feeding [^{14}C]-glucose for different times confirmed that slime-polysaccharide was synthesized faster than wall matrix polymers. Some of it had been secreted from the cells within 2 minutes of supplying the labelled precursor. These results pay tribute to the furious synthetic activity of root cap tissue. Even using conservative estimates of the rate of turnover of cisternae, Morré *et al.* (1971) calculate that secretory vesicles contribute enough new membrane to the plasmalemma of a maize root cap cell to replace it entirely every 4 to 8 hours. How the excess is recycled is not known. Careful studies of Golgi bodies and numbers of secretory vesicles in thin sections of maize root caps fixed at different times have shown the existence of rhythmic fluctuations in secretory activity. The organelles reach a peak of activity every 3 hours, synchronized over the whole cap. Whole batches of roots can be synchronized by transfer to fresh solution, which induces a peak of activity at around 1 hour later (Morré *et al.*, 1967). Since the results obtained by Bowles and Northcote refer to the first three quarters of an hour after transfer of roots to [^{14}C]-glucose solution, they probably represent peak activity for root cap Golgi bodies.

Though still in debate, it now seems unlikely that the cellulose microfibrils of the wall are synthesized in the ER or Golgi apparatus in higher plant cells (see also chapters 1 and 7). Low incorporation into glucose relative to other wall monomers in the polysaccharides of ER and Golgi body-rich fractions of maize root confirms the earlier autoradiographic evidence from developing xylem vessels. In this tissue only the plasmalemma was labelled from [^{3}H]-glucose at the time it was being incorporated exclusively into cellulose (Wooding, 1968). However, Golgi body-enriched membrane fractions synthesized radioactive β-1,4 glucan when supplied with UDP-[^{14}C]-glucose (Van Der Woude *et al.*, 1974; Shore & MacLachan 1975). The β-1,4 glucans synthesized from UDP-glucose have been shown to be cellulose and not β-1,4 glucan sections in glucomannan, a matrix component synthesized from GDP sugars (Villemez, 1974). Golgi body cellulose synthetase showed distinctive kinetics when compared with the activity in a plasmalemma-enriched fraction, and was much less active than the plasmalemma enzyme at high (1 mM) substrate concentration. It seems, then, that the activity in Golgi bodies was not due to contamination by plasmalemma and that the Golgi apparatus may be ferrying the enzyme to the plasmalemma in a less active form. A similar mechanism has been shown for chitin synthetase in yeast and *Mucor*. Chitin is structurally similar to cellulose and is produced as a microfibril on the outer surface of the plasmalemma by an enzyme particle which spans this membrane. Chitin synthetase is made in an inactive zymogen form, found in microsomal membranes, which can be converted to an active enzyme by protease digestion. An inhibitor of protease found in the soluble cytoplasm is thought to control this transition, preventing the activation of the enzyme before it reaches its operational site in the plasmalemma (McMurrough & Bartnicki-Garcia, 1973; Durán *et al.*, 1975). The cellulose component of the scales of Chrysophycean algae appears to be made in Golgi cisternae, but this difference between them and higher plants may well amount to nothing more fundamental than the timing of cellulose synthetase activation.

8.4.3 DIFFERENTIATION

Inevitably the elements of the endomembrane complex are heavily involved in the differentiation of plant cells. Normal cell growth and development require the biosynthesis of intracellular membrane and extracellular wall material, both of which involve the endomembranes. When a cell differentiates the specialized nature of these products must be the result of altered biosynthetic activity of the endomembrane complex in response to information received, ultimately, from the nucleus. For example, by comparison with a parenchyma cell of the same root, slime-producing root cap cells synthesize more Golgi body membrane, more polysaccharide, and the biosynthetic machinery includes different enzymes, e.g. fucosyltransferases.

As the source of bulk membrane components the ER plays a unique role in the development of specialized organelles in differentiating cells. Undoubtedly the best characterized example is the development of glyoxysomes in castor bean. These organelles are found only in endosperm tissue and develop between

2 and 4 days after germination, by *de novo* synthesis of protein and lipid. Glyoxysomes do not contain the enzymes for phospholipid synthesis. When [^{14}C]-choline is supplied to endosperm tissue, the first membrane fraction to be labelled is the ER and this label subsequently appears as phosphatidyl choline in the glyoxysome membrane. In the early stages of germination, glyoxysomal enzymes can be detected in the ER fraction (Beevers, 1975). Because there is no evidence that glyoxysomes can synthesize protein, the kinetics of incorporation of radioactive amino acid into these organelles was interpreted as the result of synthesis in the ER followed by transfer to the glyoxysome. Thus the incorporation of [^{35}S]-methionine into glyoxysomes lagged 30 minutes behind the labelling of the ER. During a 'chase' with unlabelled methionine, the radioactivity in ER membranes decreased at the same time as the glyoxysome membrane continued to accumulate label (Bowden & Lord, 1976). In thin sections, the coincidence of a sheet of ER abutting end-on to a developing microbody is reported as frequent, both for glyoxysomes and for leaf peroxysomes. Although their membranes are seen to touch, the lumina of the two organelles appear quite separate. When segments of juvenile leaves are incubated in a solution of 3,3′-diaminobenzidine and H_2O_2 (to localize catalase activity) and prepared for electron microscopy, the developing microbodies show a strong electron-opaque deposit but no such product can be detected in the contiguous ER (Gruber *et al.*, 1973).

From electron-microscope sections, the vacuole of plant cells is thought to begin with the distension of portions of ER to form numerous 'pro-vacuolar' vesicles which then coalesce and further expand. A combined thin section and freeze-etch study of later stages in vacuole formation showed that each expanding pro-vacuole had a sheet of fenestrated ER lying close to its bounding membrane over the whole external surface (Fineran, 1973b).

These observations suggest that the microbodies and the vacuole develop in two phases. A primary event is the formation of a separate compartment from part of the ER. This pro-organelle must already differ from the bulk ER because its subsequent development is so different, but it is not known to what extent any of the components for its specialized function are present at this stage. Further development of this structure appears to be by transfer of lipid and protein from apparently undifferentiated ER close to or touching its bounding membrane.

Some correlations between the distribution of the ER and spatial patterns of wall biosynthesis in differentiating cells have suggested a morphogenetic role for this membrane. For example, in the differentiation of sieve-tubes, the siting of the sieve-pores in the end walls is pre-patterned by plates of ER, and in the differentiation of xylem vessels sheets of ER cover the internal surface of the plasmalemma between wall thickenings (Northcote, 1968). In both cases, the result of this pattern is a reduced deposition of secondary wall against the section of plasmalemma persistently overlain by ER. In many thin section electron micrographs of plant cells, large areas of plasmalemma appear to be covered closely by a continuous sheet of ER and it is difficult to see how secretory vacuoles could get to the plasmalemma. Figure 8.3 shows secretory vacuoles apparently pouring through a gap in such a barrier. Juniper and Pask (1973) have shown that the deposition of slime polysaccharide is restricted to the outer tangential wall of maize root cap cells and their micrographs show that fenestration of the peripheral ER is most apparent on this side of the cell. The ER, therefore, may not only produce the polysaccharide of the wall but also control the site of its deposition. How the ER is organized in the correct pattern is a central problem of cell morphogenesis. Certainly the membrane is well placed to receive and is competent to interpret information from the nucleus and the possibility that it also 'responds' to chemical information from other organelles or outside the cell cannot be discounted.

The role of endomembrane complex in differentiation is undoubtedly not restricted to the 'motor' function of assembling new structures in response to a changing supply of nucleus-derived information. The effects of plant growth regulators in the tissue environment of the cell on nuclear activity must be mediated by the cytoplasm and, because of its position in the front line, the plasmalemma is strongly implicated. We know virtually nothing about the mechanisms involved in this 'sensory' function of the endomembrane complex. The first indication of the nature of this interaction may be the discovery that RNA polymerase in plants is activated by a factor released from plasmalemma by auxin (Cherry, 1974, see also chapter 13).

8.5 CONCLUSIONS

The working definition of the endoplasmic reticulum is essentially negative. All the membrane of the cell not obviously specialized is relegated to this category. Biochemical investigation of its activities, stimulated and supervised by electron microscopy, has gone some way toward explaining the low level of its organization and replacing the negative definition with something more positive. The ER is the *meristem* of the cell. It is membrane in the process of synthesizing membranes, creating a pool of components and distributing them about the cell for the maintenance and enlargement of the specialized organelles. For this purpose the outer membranes of plastids, mitochondria and the nucleus can be considered as rather more specialized 'limbs' of the ER, the two former membranes acting as placentas for the very specialized membranes they enclose. The new protein and lipid components must somehow cross the placental gap in a solubilized form. Material from the ER apparently contributes to the expansion of the tonoplast, microbodies and the plasmalemma in the same way (Fig. 8.6).

However, the glycoproteins, glycolipids and polysaccharides intended for the plasmalemma and wall are not transferred in this way. Probably they are too hydrophilic to be shuttled in and out of different sides of lipid bilayers. Once synthesized, they never leave the membrane (and its lumen) which is incorporated as a unit into the plasmalemma (and the wall). This vesicular transfer cannot

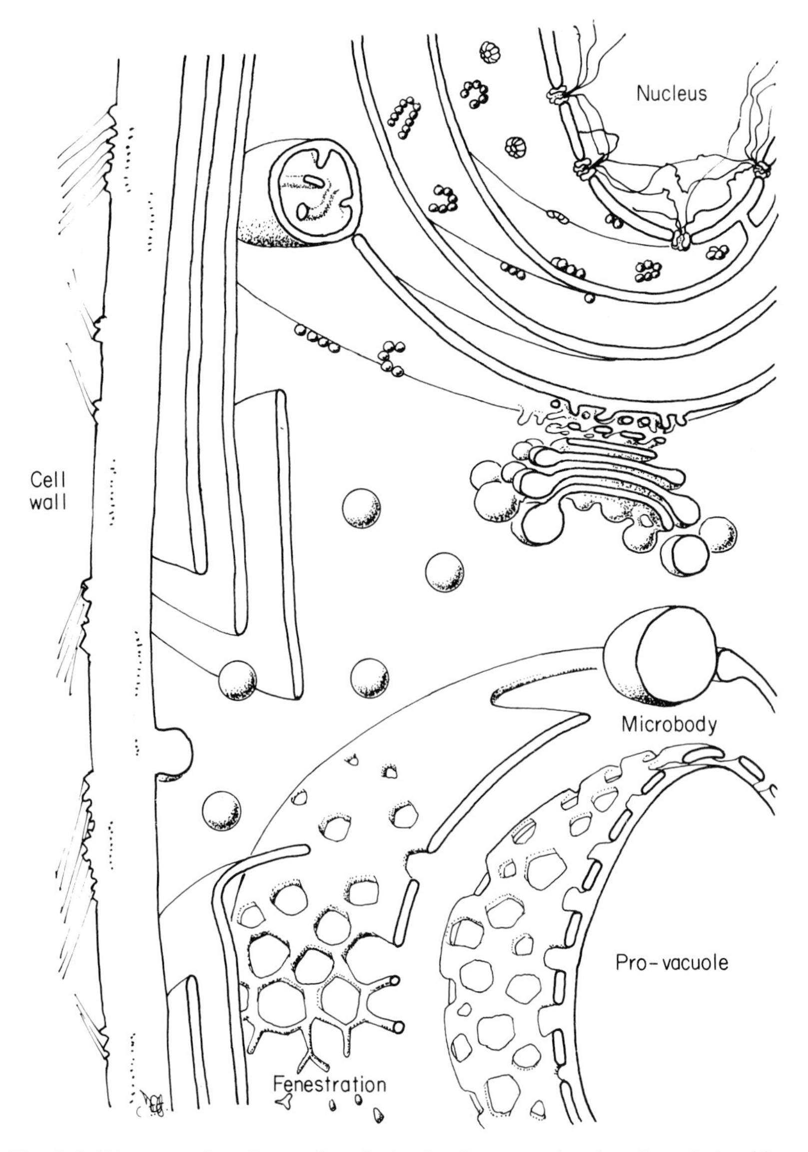

Fig. 8.6. Diagram of a plant cell early in development showing the relationships between the ER and the other membranes of the cell proposed from electron microscope and biochemical evidence.

provide the selectivity of carrier proteins. Therefore all the membrane which is subsequently incorporated into the plasmalemma by fusion passes through a phase of close appression (the Golgi body) to allow the selective redistribution of membrane components.

In that the endomembrane complex is the cell's machinery for organelle and wall synthesis, the specialized nature of these ER products in differentiated cells can only come about as a result of reprogramming the synthetic machinery. Essentially it is the endomembrane complex which builds the plant cell and the nucleus which stores, selects and supplies designs for the appropriate tools.

FURTHER READING

BEEVERS H. (1975) Organelles from castor bean seedlings: biochemical roles in gluconeogenesis and phospholipid biosynthesis. In *Recent Advances in Chemistry and Biochemistry of Plant Lipids*, pp. 287–299 (eds. T. Galliard & E. I. Mercer). Academic Press, New York.

BOWDEN L. & LORD J.M. (1976) The cellular origin of glyoxysomal proteins in germinating castor bean endosperm. *Biochem. J.* **154**, 501–6.

BOWLES D.J. & NORTHCOTE D.H. (1976) The size and distribution of polysaccharides during their synthesis within the membrane system of maize root cells. *Planta (Berl.)* **128**, 101–6.

DONALDSON R.P. (1976) Membrane lipid metabolism in germinating castor bean endosperm. *Plant Physiol.* **57**, 510–15.

MORRÉ D.J. & MOLLENHAUER H.H. (1974) The endomembrane concept: a functional integration of endoplasmic reticulum and Golgi apparatus. In *Dynamic Aspects of Plant Ultrastructure*, pp. 84–137 (ed. A.W. Robards). McGraw-Hill, Maidenhead.

MORRÉ D.J. (1975) Membrane biogenesis. *Ann. Rev. Plant Physiol.* **26**, 441–81.

NORTHCOTE D.H. (1968) The organisation of the endoplasmic reticulum, the Golgi bodies and microtubules during cell division and subsequent growth. In *Plant Cell Organelles*, pp. 179–197 (ed. J.B. Pridham). Academic Press, London.

O'BRIEN T.P. (1972) The cytology of cell-wall formation in some eukaryotic cells. *Bot. Rev.* **38**, 87–118.

PHILLIP E.I., FRANKE W.W., KEENAN T.W., STADLER J. & JARASCH E.D. (1976) Characterization of nuclear membranes and endoplasmic reticulum isolated from plant tissue. *J. Cell Biol.* **68**, 11–29.

SECTION TWO

GENE EXPRESSION AND ITS REGULATION IN PLANT CELLS

INTRODUCTION

The activities, potentialities, and developmental fate of all cells are determined by the hereditary genetic material stored within the nucleus, and also to a lesser but still important extent, by that stored within the plastids and mitochondria. Hereditary information is coded within the nucleotide sequences of the DNA, and the generalized concept of the expression of genetic information is usually portrayed as follows:

$$\text{replication} \circlearrowleft \text{DNA} \xrightarrow{\text{transcription}} \text{mRNA} \xrightarrow{\text{translation}} \text{protein}$$

This 'central dogma' states that the information contained within the DNA is only expressed, in terms of cellular processes, through the action of the ultimate gene products, the proteins, most of which are enzymatic in nature. The central dogma also points to there being two principal points at which gene expression may be regulated; i.e. gene transcription and m-RNA translation. In fact, a further control point should be added, since the catalytic activity of enzymes may be regulated in several ways in addition to the regulation of the synthesis of the enzyme. Thus the regulation of gene expression may be more adequately described by the following scheme:

$$\text{DNA} \xrightarrow[(1)]{} \text{mRNA} \xrightarrow[(2)]{} \text{protein} \underset{(3)}{\rightleftharpoons} \text{enzyme}$$

in which the potential control points at transcription, translation and enzyme activation/inactivation are numbered.

Section II considers the biochemical and structural basis of gene expression and its regulation in higher plants and places it in context with what is known from the more extensive studies carried out with other organisms. The principal hypothesis of the regulation of gene expression stems from the discoveries relating to the control of enzyme levels in bacteria. In 1961, to account for the known effects of metabolites on the synthesis of specific enzymes in bacteria, Jacob and Monod proposed that the activity of the genes which code for specific enzyme proteins (*structural genes*) is controlled by other genes, not necessarily linked to the structural genes, and known as *regulator genes.* A regulator gene codes for the synthesis of a specific cytoplasmic protein, known as a *repressor*, which has the capacity to interact with a region of chromosomal DNA (the *operator*) adjacent to the initiation end of the structural gene. The binding of a repressor to a specific operator, which may control a sequence of structural genes (known collectively with the operator as an *operon*) inhibits transcription of m-RNA on those structural genes. It is also held that the repressor protein has the capacity to bind one or more low molecular weight substances, known as *effectors*, which may alter its capacity to bind to the operator. Effectors which combine with the repressor and prevent its binding to the operator are termed *inducers*, since they allow enzyme synthesis from a previous repressed state; inducers are commonly substrates of the first enzyme in a pathway. Effectors which combine with the repressor and enhance binding with the operator, inhibit or repress enzyme synthesis; such effectors are known as *co-repressors* and are commonly products of a reaction pathway. In repressible systems of this nature the structural genes are normally fully expressed in the absence of the co-repressor and the enzymes are thus known as *constitutive*. As shown in Fig. 11.1 this hypothesis elegantly accounts for the formation of enzyme in response to the substrate (*enzyme induction*) and for the cessation of enzyme formation in response to the end product (*enzyme repression*).

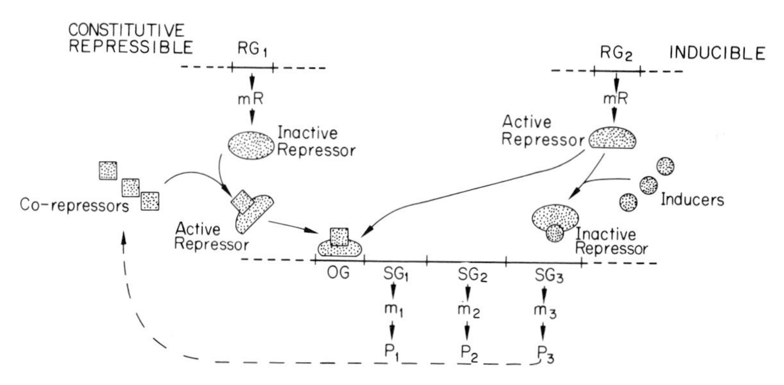

Fig. 11.1. A generalized scheme of the Jacob–Monod model for constitutive repressible and inducible systems of gene regulation. The regulator genes (RG) are shown as coding for the formation of either active repressors (Inducible system) or inactive repressors (Constitutive repressible system) which bind to the operator gene (OG) and block the transcription of the structural genes of that operon (SG_1 to SG_3). P=protein and m=mRNA; P_3 is considered to be an enzyme whose catalytic activity leads to the formation of a co-repressor for the constitutive repressible system. (Modified with permission from F.A.L. Clowes and B.E. Juniper (1968) *Plant Cells*. Blackwell Scientific Publications Ltd., Oxford.)

It is not yet clear how fully these concepts account for gene expression and its regulation in eucaryotic organisms. As will be seen from Chapter 9, the structure, organization and complexity of nuclear DNA are wholly different from the rather simple arrangements found in bacteria. There seems to be far too much DNA than would be necessary to code for the predicted number of structural genes and attendant regulator genes as estimated from the Jacob–Monod model. Several suggestions have been put forward to account for this unexpected genetic largesse, but the most likely ideas involve the extra DNA in the regulation of the structural genes (*see* Britten & Davidson, 1969).

Additionally, the separation of the site of transcription (in the nucleus) from that of translation (in the cytoplasm) creates further opportunities for regulation. It will be seen that the processes of RNA and protein synthesis in the nuclei and cytoplasm of plants are similar in principle, but different in detail, from those in higher animals (chapters 9 and 10).

A very important aspect of plant molecular biology, and one which is unique to plants, is the presence of a substantial store of genetic information in the chloroplasts (chapter 11). The coding potential of chloroplast DNA is vastly greater than that of mitochondrial DNA, and many of the activities of higher plant cells, and indeed a large part of their total protein, is coded for in the chloroplast genome. The regulatory processes involved here must be highly complex, since it is known that some chloroplast proteins are coded for and synthesized within the plastids, and others coded in the nucleus, and synthesized in the cytoplasm; Fraction I protein, the most abundant protein in green leaves, is composed of two types of sub-units, one of nuclear, and one of plastid, origin. Clearly, here the regulation of gene expression must involve the integration of transcription and translation within three separate cell compartments, the nucleus, the cytoplasm and the chloroplast.

As stated briefly above, gene expression may also be regulated after the final gene product, the protein, has been formed, since a gene can only be considered to be expressed when the protein it codes for performs its specific function. Thus the activation and inactivation of enzymes, through the several known mechanisms, becomes properly a facet of gene expression (chapter 12).

The integration of the activities of different cells, particularly with regard to development, inevitably requires on the one hand, the generation and transmission of intra- and inter-cellular stimuli and, on the other, specific responses to these stimuli. In higher animals, both electrical and chemical stimuli appear to be important; in higher plants inter-cellular integration seems to be almost completely chemical, through a narrow range of hormonal substances. Although understanding of the mechanism of hormone action in plants is still incomplete, the regulation of gene expression is probably involved (chapter 13).

All organisms need to be able to perceive their environment and adapt to environmental changes. Evolution, having favoured non-motility for plants, has inevitably endowed them with extremely subtle mechanisms for detecting and responding to certain environmental factors. The best example here is the response of plants to light, in which specific photoreceptors react to signals from the radiation environment by initiating changes in metabolism and development. As an example of external regulation of gene expression in higher plants, chapter 14 deals with the mechanism of action of phytochrome, an important photoreceptor found in all green plants.

Section II therefore, is concerned with the genetic material of higher plants, its organization, complexity and regulation, with particular reference to the control of cell metabolism, to cell-cell communication in hormone action and to responses to the environment via photoreceptor action.

CHAPTER 9

THE NUCLEUS AND THE ORGANIZATION AND TRANSCRIPTION OF NUCLEAR DNA

9.1 INTRODUCTION

The importance of the cell nucleus has been recognized for some considerable time and a great deal of literature concerned with its general organization and composition and the structure and behaviour of chromosomes has been derived from cytological investigations. This present chapter is mainly concerned with recent developments in research which provide a more detailed understanding of the organization and function of nuclear DNA, the mechanism of RNA synthesis and the question of the control of gene expression in higher plants.

9.2 NUCLEAR STRUCTURE AND COMPOSITION

In general, nuclei are spherical or disc-shaped, although occasionally they may appear lobed. They vary in size in different species but are usually within the range 1–10 μm in diameter (Fig. 9.1). Nuclei may be isolated by gently breaking cells in a suitable osmoticum, filtering the homogenate, for example through muslin, followed by low-speed centrifugation. Attempts are usually made to free the nuclei of contaminating cytoplasm, cell wall fragments and other organelles by centrifugation through sucrose. However, methods for obtaining high yields of intact, purified nuclei from higher plant cells are at present not very satisfactory. For certain purposes it is often easier to work with chromatin preparations which consist essentially of chromosomal and nuclear material obtained from broken nuclei.

The major part of the cellular DNA is present in the nucleus, arranged in a highly organized fashion within the chromosomes which at different times may show characteristic bands or loops (see Fig. 9.1). Chromosomes are composed of approximately equal amounts of DNA and basic proteins known as histones (see Table 9.7). In addition, much smaller quantities of RNA and acidic proteins are present which are probably of great importance in controlling gene expression. There is wide variation in the number of chromosomes observed within the nuclei of different species although in diploid cells of a single species the number is normally constant. During interphase the chromosomes are less highly condensed than at mitosis and are more difficult to detect. At this stage at least part of the DNA is being transcribed into RNA. It is possible to identify chromosomes on the basis of morphological features such as size, position of the centromere and the presence or absence of constrictions. In addition, certain

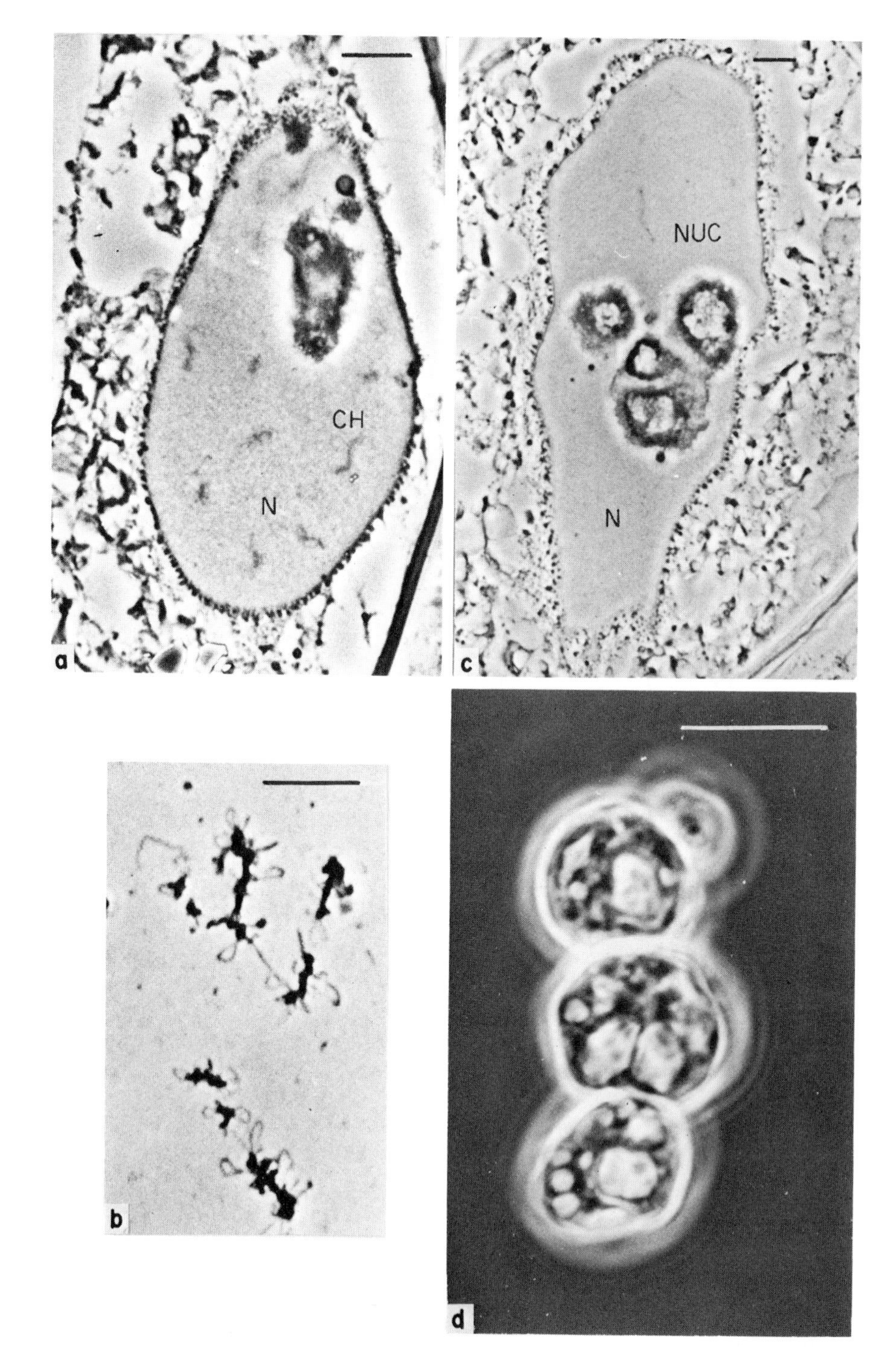
CH
N
a
NUC
N
c
b
d

areas of heterochromatin, which appear more highly condensed and therefore take up more stain, may be distinguished from the remainder of the chromosome, termed euchromatin. In certain cell types, such as embryo suspensor cells, giant polytene chromosomes occur. In such situations it is possible to study the structural organisation of chromosomes in much greater detail.

Recent developments have made it possible to identify individually each of the 46 chromosomes in man on the basis of characteristic banding patterns produced by certain staining procedures. These techniques and their application to the study of plant chromosomes have been reviewed by Vosa (1975).

Organelles known as nucleoli become visible within the nucleus during interphase. These are very prominent spherical bodies with different staining properties to the remaining part of the nucleus. They arise in association with specific regions, the nucleolar organizers, of certain chromosomes. Nucleoli are not surrounded by a membrane but they do have a highly ordered structure (Fig. 9.2). The central core consists of a region of fibrils 8–10 nm across surrounded by a region of dense granules 15–20 nm in diameter. Mutants which lack nucleoli are unable to synthesise ribosomes and it is now known that the genes coding for ribosomal RNA normally occur at the nucleolar organizer region of chromosomes. Here the DNA is transcribed into ribosomal-RNA (rRNA) which is assembled to form ribosomes in the outer region of the nucleolus by combination with ribosomal proteins. The nucleolus is most prominent, therefore, when synthesising large numbers of ribosomes. This explains the old observation that there is a general correlation between nucleolar volume and the rate of protein synthesis of cells.

The fact that nuclei are surrounded by a membrane was deduced from the observation that they are selectively permeable to small molecules. The nature of the double membrane may be readily observed with the aid of the electron microscope (Roberts & Northcote, 1971). It is often interconnected with the cytoplasmic membrane system, the endoplasmic reticulum, and is perforated by a large number of pores. This is demonstrated quite clearly in Fig. 9.3 which shows a surface view of a portion of the nuclear membrane of *Acetabularia*. The apparent diameter of a nuclear pore is of the order of 100 nm. However, the observation that colloidal particles of much smaller size do not readily pass through these pores suggests that their effective size is substantially less, probably due to the presence of a protein-containing matrix within the pore.

Fig. 9.1. Phase contrast light micrograph of nuclei and isolated nuclear contents from *Acetabularia*. These giant nuclei are unusually large; the scale line represents 10 μm. Chromosomes (CH) and nucleoli (NUC) can be seen within the nuclei (N) fixed *in situ* (a and c) and the appearance of isolated chromosomes and nucleoli are shown in b and d. Note the large aggregates of nucleolar material with an internal cavern system (c and d) and the 'lampbrush-chromosome-like' morphology of the chromosomes (b). (Fig. 9.1a, c and d from Spring *et al.*, 1974; 9.1b, previously unpublished micrograph from the same authors.)

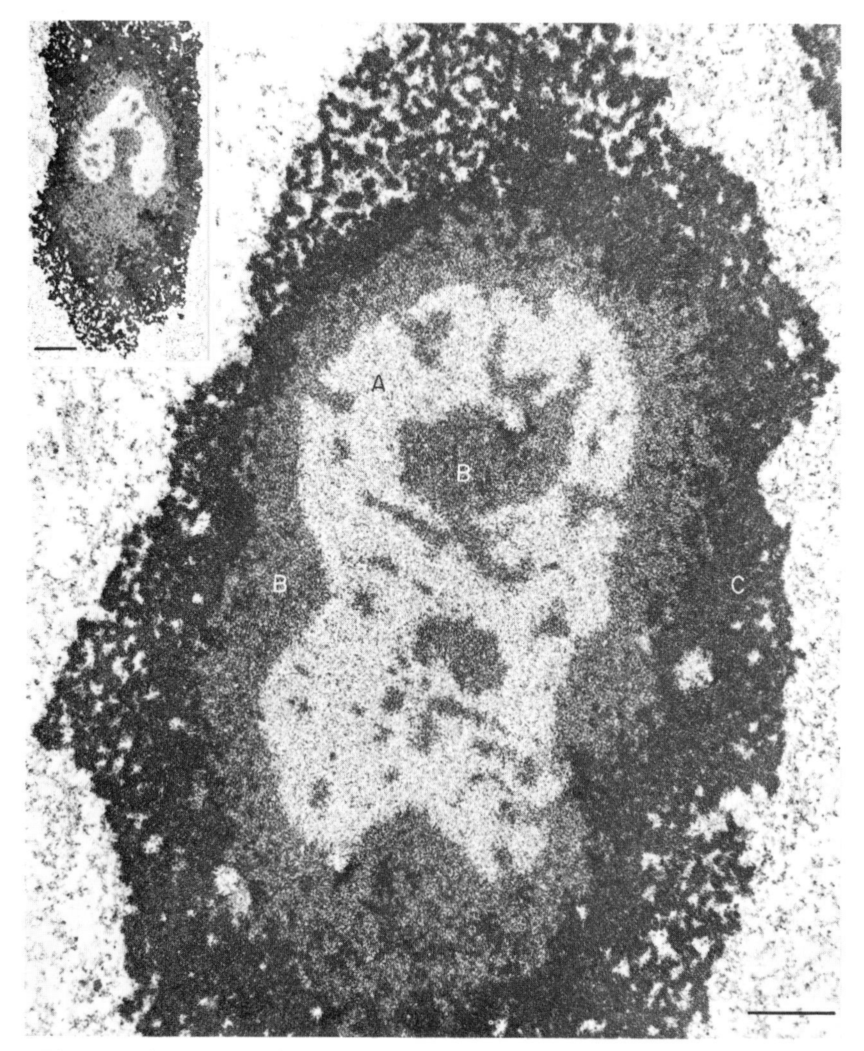

Fig. 9.2. Ultrastructural details of the composition of the nucleolar subunits of *Acetabularia* revealed by electron microscopy. Note the finely fibrillar texture in the internal zone A and the dense packing of small granulofibrillar structures in zone B, extensions from which deeply penetrate zone A (see also the *insert*) and the very dense packing of granules in the cortical zone C. The scale indicates 1 μm, the scale in the *insert* denotes 2·5 μm. (From Spring *et al.*, 1974.)

9.3 NUCLEAR DNA CONTENT

The amount of DNA present in nuclei may be estimated by nuclear isolation followed by extraction and direct chemical measurement of DNA content. A more elegant approach is to carry out quantitative microspectrophotometry of

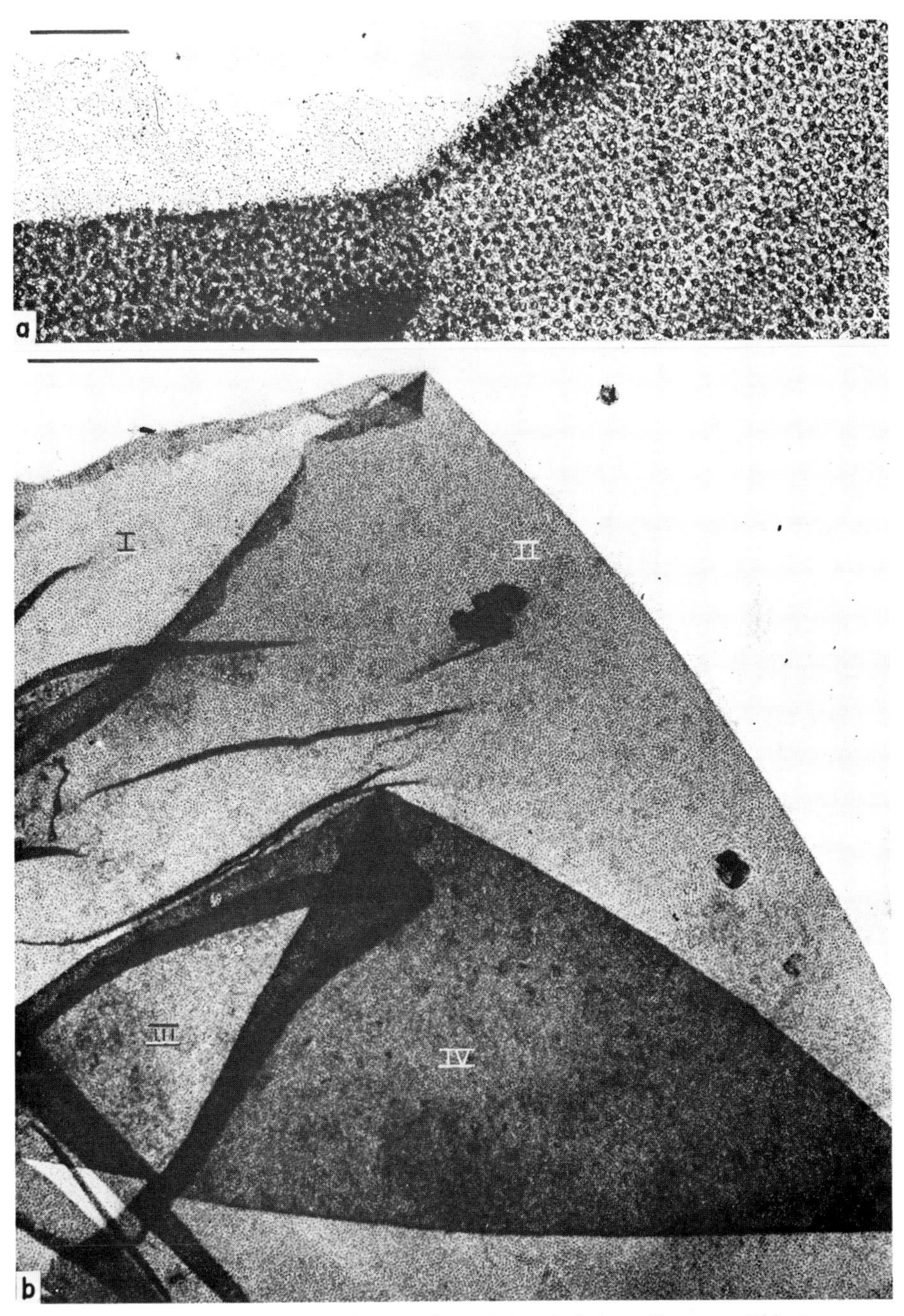

Fig. 9.3. Manually isolated nuclear envelope of *Acetabularia mediterranea*. This shows a sequence of overfoldings (I to IV in b). Note the high density of nuclear pore complexes and the association with some fibrillar material as identified in the upper region of a. The scale in (a) denotes 1 μm, the scale in (b) denotes 10 μm. (From Spring *et al.*, 1974.)

individual nuclei following Feulgen staining. Table 9.1 compares the nuclear DNA content of a range of higher plants with the amounts of DNA found in lower organisms. Two conclusions can be drawn from such a comparison. Firstly, and perhaps not surprisingly, higher plants contain considerably more DNA than bacteria and fungi. For example, *Pisum sativum* contains 1,000 times as much DNA as *Escherischia coli*. Secondly, however, the amount of DNA present in different higher plants may vary by as much as 100 fold. Further comparison shows that there is little correlation between chromosome number and DNA content.

Table 9.1. DNA content of different organisms.

Genus and species	DNA amount (picograms per 2*C* nucleus)
Flowering plants	
Tradescantia ohioensis (4×)	89·0
Allium globosum	75·8
Tradescantia virginiana (4×)	70·0
Tulipa gesneriana	51·5
Scilla campanulata	46·0, 40·0
Tradescantia ohioensis	45·9
Allium karataviense	45·4
Scilla sibirica	44·5
Allium angulosum (4×)	41·2
Allium cepa	33·5
Vicia faba	28·7, 25·8, 26·8, 23·9
Chrysanthemum nipponicum	26·8
Allium jesolianum	24·1
Secale cereale	18·9, 17·7
Allium fuscum	18·4
Phalaris minor (4×)	17·1
Allium sibiricum	15·2
Lolium perenne	9·9
Helianthus annuus	9·8
Pisum sativum	9·1
Kalanchoë daigremontiana	6·7
Vicia angustifolia	6·1
Lycopersicum esculentum	3·9
Raphanus sativus	3·1
Cucurbita pepo	2·6
Lupinus albus	2·3
Linum usitatissimum	1·4

Microorganisms

Fungus	Haploid amount of DNA (picograms per cell)
Aspergillus sojae	0·088
Aspergillus nidulans	0·044
Neurospora crassa	0·017
Bacterium	**DNA per cell (picograms per cell)**
Bacillus megatherium	0·070
Salmonella typhimurium	0·011
Escherichia coli	0·009
Bacteriophage	**DNA per particle (picograms)**
T_4	0·000025
T_5	0·000018
T_1	0·000007

Modified from Rees and Jones (1972).

One explanation of the increased DNA content of higher plants is that more complex organisms require a greater store of genetic material than bacteria and viruses. Although it is difficult to estimate precisely the extent of the increased coding requirement it is quite clear that this is in itself not a satisfactory explanation for the presence of such large amounts of DNA in some plants. There often exists a wide variation in the *2C* DNA content of plant species within one genus and it is unreasonable to conclude that small differences between related plants are a direct consequence of gross differences in nuclear DNA. Furthermore, some plants contain 20 times as much DNA as mammals and yet the latter are normally accepted as comprising a more complex group of organisms. It therefore appears that the amount of nuclear DNA is not directly related to the complexity of the organism. It is also possible to argue on other grounds that the total amount of nuclear DNA is not required in order to code for additional proteins. Firstly, it is difficult to envisage functions to account for a 1,000 fold increase in the number of enzymes. Secondly, if each higher plant contained such a large number of essential genes it would, on numerical grounds, be prone to an unrealistically high incidence of mutation.

The size of the genome of plants with relatively low nuclear DNA contents is of the order of 2×10^{12} daltons. Given that the molecular weight of a gene required to code for a messenger-RNA (mRNA) molecule for an average-sized polypeptide is 10^6 daltons, this represents, in principle, 2×10^6 genes. A more realistic estimate would be that 1% of this DNA actually codes for proteins. Experimental evidence for this view and a discussion of the organization, properties and possible function of the remaining DNA is given later.

It used to be widely proclaimed that when differences in ploidy are taken into account the amount of DNA in each nucleus of a single species is constant. The observation that the gene number remains constant, and the fact that many plant cells are totipotent, supports the view that differentiation involves the selective control of gene expression. However, the rule of DNA constancy is not inviolate. It is now known that specific regions of the genome may become selectively amplified at certain stages of development. This undoubtedly occurs to the genes for ribosomal RNA during oocyte development in Amphibia. Although there exist no established cases of ribosomal RNA gene amplification in higher plants, the satellite DNA present in melon may be under- or over-replicated in different tissues (Ingle & Timmis, 1975).

One very intriguing observation is that the amount of nuclear DNA in flax may become altered if the plants are subjected to certain regimes of nitrogen and phosphorus fertilizer. The alteration in nuclear DNA content is stable and inherited (Evans, 1968). Although this phenomenon is well established it is not known how widespread it may be and there is no satisfactory explanation to account for it.

9.4 FRACTIONATION AND PROPERTIES OF DNA

9.4.1 GENERAL PROPERTIES

Although single-stranded DNA does occur naturally most DNA molecules consist of two anti-parallel chains of nucleotides. Each nucleotide consists of deoxyribose, a phosphate group and a base. The four bases that occur in all types of DNA are the purines adenine and guanine and the pyrimidines thymine and cytosine. Adenine is capable of forming hydrogen bonds with thymine, and guanine with cytosine. Sometimes modified bases are present in small amounts and in plant nuclear DNA a substantial proportion of the cytosine residues are replaced by 5-methyl-cytosine. In double stranded DNA there is an equivalence between the amount of adenine and thymine and guanine and cytosine plus 5-methyl-cytosine. The two strands are joined together by hydrogen bonds between complementary pairs of bases to form a right handed helix. The bases are at right angles to the sugar residues, and project inwards, and the phosphate groups are on the outside. In the B form of DNA there are ten base pairs in one complete turn of the helix, which extend for 3·4 nm. The width of the helix is 1·8 nm. An alternative configuration is possible, the A form, where the bases are tilted and 11 pairs are accommodated per turn of the helix. Double stranded RNA and DNA-RNA hybrids adopt this latter configuration.

Many DNA molecules are known to be several millimetres in length. It is very difficult to determine the actual length of very large DNA molecules because they are extremely sensitive to shearing forces produced by various isolation procedures. It is known that the complete genome of *E. coli*, with a molecular weight of $2{\cdot}9 \times 10^9$ daltons, consists of a single circular DNA molecule of length approximately 1 mm. The total amount of DNA in a higher plant cell is equivalent to a length of approximately 1 m, but it is not known how many molecules this comprises.

The base composition of DNA is normally expressed as the per cent G+C content. This value, which is characteristic of a given species, varies widely in different organisms and in bacteria may be anything from 30% to 70%. The base compositions of a number of plant DNAs are given in Table 9.2.

9.4.2 BUOYANT DENSITY CENTRIFUGATION

One of the most common methods of fractionating DNA is to exploit differences in density between different types of DNA in concentrated salt solutions. Caesium chloride is the most commonly used. The method involves prolonged high speed centrifugation of purified DNA in concentrated caesium chloride. The centrifugation produces a gradient of salt concentration in the centrifuge tube and the DNA migrates to occupy a position in the gradient which corresponds to its own buoyant density. A number of factors affect the buoyant density, such as the nature of the caesium salt, the presence of heavy metals or DNA-binding dyes, the pH and the temperature. Under constant conditions (usually 25°C in caesium chloride at neutral pH) the buoyant density of DNA is related to the GC content:

$$\%\text{G+C content} = \frac{\text{buoyant density (g cm}^{-3}) - 1{\cdot}660}{0{\cdot}098} \times 100$$

DNAs with different base compositions can therefore be separated by this method. Figure 9.4 shows the fractionation of *Phaseolus aureus* DNA (buoyant density $= 1{\cdot}695$) and *E. coli* DNA (buoyant density $= 1{\cdot}710$) in caesium chloride.

Most nuclear DNAs from higher plants have buoyant densities within the range 1·69–1·71 g cm^{-3}. However, the presence of 5-methyl-cytosine serves to reduce the density slightly, thereby giving rise to an under-estimate of the GC content. In general, 1% methylation decreases the buoyant density by 1 mg cm^{-3}. Certain sequences of bases may also distort the relationship between base composition and buoyant density. Furthermore, single-stranded DNA is denser than double-stranded DNA of similar base composition by approximately 0·015 g cm^{-3} and under alkaline conditions the density is increased by 0·06 g cm^{-3}. This is due partly to the fact that DNA becomes single-stranded under these conditions and also partly because the deprotonated adenine and thymine residues are neutralized by binding caesium ions.

9.4.3 SATELLITE DNA

Figure 9.4 shows that DNA from a particular organism often forms an apparently homogenous band in caesium chloride. The width of the band depends

Table 9.2. Base composition of plant DNA. (Moles per 100 moles.)

	Adenine	Thymine	Guanine	Cytosine	5-methyl cytosine	G+C
Ascomycetes						
Saccharomyces cerevisiae	31·7	32·6	18·3	17·4	—	35·7
Neurospora crassa	23·0	23·3	27·1	26·6	—	53·7
Green algae						
Chlamydomonas globosa	19·0	19·5	30·3	28·2	2·75	61·25
Spirogyra sp.	30·7	30·4	19·2	19·8	—	39·0
Gymnosperms						
Pinus sibirica	29·2	30·5	20·8	14·6	4·9	40·4
Ginkgo biloba	31·6	33·5	17·2	17·7	—	34·9
Angiosperms						
Daucus carota (carrot)	26·7	26·9	23·1	17·3	5·9	46·5
Zea mays (maize)	26·8	27·2	22·8	17·0	6·2	46·0
Triticum aestivum (wheat)	27·3	27·1	22·7	16·8	6·0	45·5
Cucurbita pepo (pumpkin)	30·2	29·0	21·0	16·1	3·7	40·8
Phaseolus vulgaris (French bean)	29·7	29·6	20·6	14·9	5·2	40·7
Allium cepa (onion)	31·8	31·3	18·4	12·8	5·4	36·3
Arachis hypogaea (peanut)	32·1	32·3	17·6	12·3	5·7	35·6
Gossypium hirsutum (cotton)	32·8	32·9	16·9	12·7	4·6	34·2

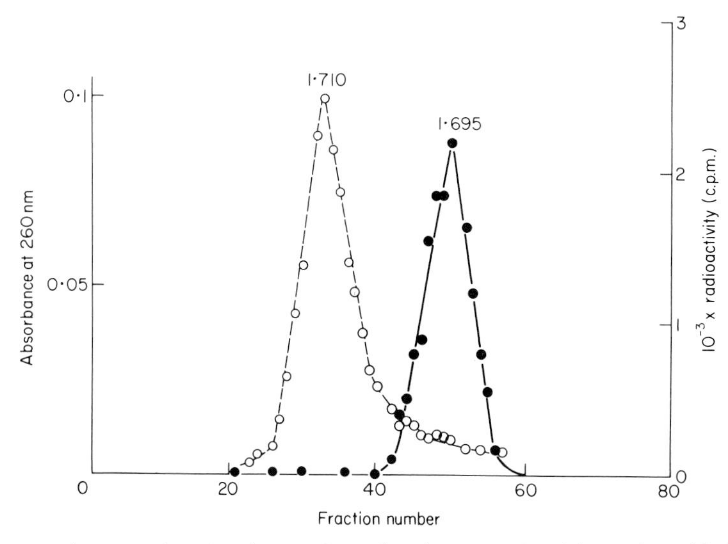

Fig. 9.4. The separation of a mixture of *E. coli* and *P. aureus* DNA in caesium chloride. A trace of *E. coli* DNA labelled with ^{3}H thymidine was mixed with *P. aureus* DNA and centrifuged at 30,000 rpm at 25°C in caesium chloride for 3·5 d using a fixed-angle centrifuge rotor. Three-drop fractions were collected and the *P. aureus* DNA located by absorbance at 260 nm (solid circles, continuous curve) and *E. coli* DNA detected by measurement of radioactivity (open circles, dotted curve). (Unpublished result of D. Grierson.)

upon the slope of the density gradient, the molecular weight of the DNA and the difference in base composition between different regions of the DNA. In most plants certain regions of the genome are sufficiently different in base composition for them to occupy a position distinct from the main-band DNA. If they represent only a small percentage of the total DNA such sequences are difficult to detect but when present in large amounts they produce one or more peaks of satellite DNA. Melon DNA provides an example where the satellite is denser than the main-band (Fig. 9.5) but light satellites have also been observed, for example in flax. It has been assumed that the majority of satellite DNAs are nuclear in origin but recent studies with cucumber suggest that they may sometimes represent organelle DNA (Kadouri *et al.*, 1975). Although they are widely distributed in plants no satellites have so far been found in monocotyledonous plants using caesium chloride fractionation (Table 9.3). However similar components can be detected using other techniques. For example, substances such as heavy metals and dyes often selectively bind to certain sequences within the DNA. This alters their buoyant density and thus generates satellites. One common method is to fractionate DNA in caesium sulphate gradients in the presence of silver ions. Using this procedure, Ingle (unpublished) has demonstrated the presence of a satellite DNA in *Scilla* (a monocotyledon) although in caesium chloride no such satellite is observed.

Table 9.3. Species distribution of satellite DNAs.

Family	Species	Common name	Buoyant density ($g\ cm^{-3}$) Main	Sat.	% Sat.
Dicotyledons					
Ranunculaceae	*Ranunculus acris*	Meadow buttercup	1·699	—	0
	Ranunculus ficaria	Lesser celandine	1·696	—	0
	Helleborus niger	Christmas rose	1·694	—	0
	Anemone coronaria	de Caen	1·696	—	0
	Pulsatilla vulgaris	Pasque flower	1·699	—	0
	Clematis montana		1·700	—	0
	Trollius europaeus	Globe flower	1·698	—	0
	Aconitum napellus	Monkshood	1·697	—	0
Cucurbitaceae	*Cucumis melo*	Melon	1·692	1·706	25
	Cucumis sativus	Cucumber	1·694	1·702	28
				1·706	16
	Cucurbita pepo	Marrow	1·696	1·706	18
		Pumpkin	1·695	1·707	16
		Squash	1·695	1·706	17
	Citrullus vulgaris	Watermelon	1·693	1·708	3
	Bryonia dioica	White bryony	1·696	1·706	5
	Momordica charantia	Balsam pear	1·695	—	0
	Lagenaria vulgaris	Bottle gourd	1·692	1·707	9
	Luffa cylindrica		1·696	1·707	6
Leguminosae	*Vicia faba*	Broad bean	1·694	—	0
	Vicia benghalensis	Purple vetch	1·694	—	0
	Pisum sativum	Pea	1·695	—	0
	Phaseolus coccineus	Runner bean	1·693	1·702	24
	Phaseolus vulgaris	French bean	1·693	1·703	19
	Phaseolus aureus	Mung bean	1·692	1·705	5
Linaceae	*Linum usitatissimum*	Flax	1·699	1·689	15
	Linum grandiflorum rubrum	Red flax	1·698	—	0

Table 9.3. (*continued*)

Family	Species	Common name	Buoyant density ($g\ cm^{-3}$) Main	Sat.	% Sat.
Solanaceae	*Solanum tuberosum*	Potato	1·695	1·707	4
	Solanum crispsum		1·698	1·710	6
	Solanum capsicastrum	Christmas orange	1·693	—	0
	Lysopersicon esculentum	Tomato	1·694	1·705	8
	Atropa belladonna	Deadly nightshade	1·694	—	0
	Nicotiana tabacum	Tobacco	1·697	—	0
	Petunia hybrida	Garden petunia	1·696	—	0
Monocotyledons					
Liliaceae	*Lilium regale*	Regal lily	1·698	—	0
	Hyacinthus orientalis	Hyacinth	1·700	—	0
	Puschkinia libanotica		1·699	—	0
	Chlorophytum elatum variegatum	Spider plant	1·693	—	0
Amaryllidaceae	*Allium cepa*	Onion	1·691	—	0
Commelinaceae	*Tradescantia virginiana*		1·695	—	0
Gramineae	*Secale cereale*	Rye	1·702	—	0
	Zea mays	Maize	1·701	—	0
	Triticum aestivum	Wheat	1·703	—	0
	Hordeum vulgare	Barley	1·701	—	0
Lemnaceae	*Lemna minor*	Duckweed	1·703	—	0

Modified from Ingle *et al.* (1973).

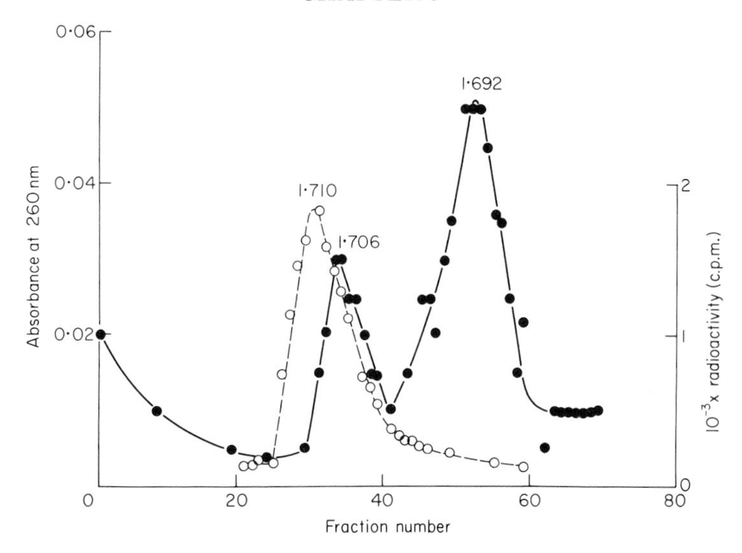

Fig. 9.5. The separation of a mixture of *E. coli* and melon DNA in caesium chloride. The experimental details were as described in the legend to Fig. 9.4. Melon DNA was detected by absorbance at 260 nm (solid circles, continuous curve) and *E. coli* DNA by measurement of radioactivity (open circles, dotted curve). (Unpublished result of D. Grierson.)

9.4.4 DISSOCIATION AND REASSOCIATION OF DNA

When double-stranded DNA in solution is heated above a certain temperature the hydrogen bonds between complementary base pairs are broken and the two strands dissociate. The process, which can also be brought about by treatment of DNA with alkali, is commonly referred to as melting or denaturation. Strand separation is accompanied by an increase of approximately 35% in the absorbance of ultra-violet light. This hyperchromic effect provides a means of monitoring the dissociation of double-stranded DNA by measuring the increase in absorbance of a DNA solution at 260 nm as the temperature is raised. The melting curves of DNA from a variety of sources are compared in Fig. 9.6. The temperature at which 50% of the DNA is melted, the Tm, depends upon a number of factors such as the salt concentration and pH. Under standard conditions, i.e. when the DNA is dissolved in standard saline citrate (S.S.C. being 0·15 M sodium chloride, 0·015 M trisodium citrate, pH 7) the base composition and the Tm are related in the following way:

$$\text{GC content } (\%) = (\text{Tm} - 69{\cdot}3) \times 2{\cdot}44$$

In viruses, the majority of sequences within the genome are similar in base composition and for a given DNA a sharp melting profile is observed with a characteristic Tm. In contrast, higher organism DNA often consists of distinct subfractions, each with a slightly different Tm. This produces a broad melting

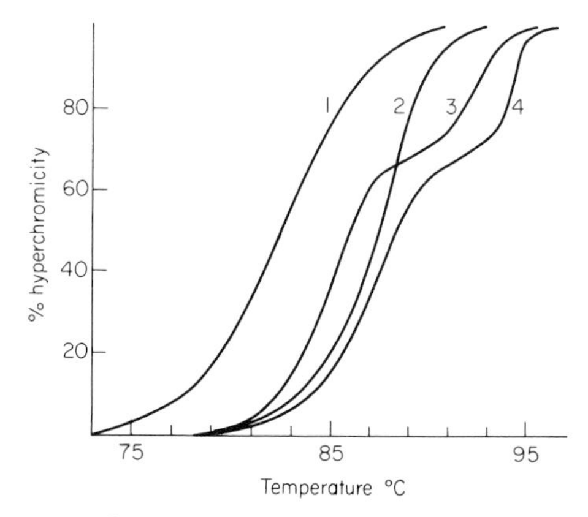

Fig. 9.6. Melting curves of various DNA samples. (1) Musk melon main-band and (4) satellite DNA. (3) a 2:1 mixture of DNA from bacterium C and phage P58. (2) *B. subtilis* DNA. Melting was carried out in S.S.C. (From Bendich & Anderson, 1974.)

profile which represents a number of overlapping melting curves. With DNA samples from organisms that contain prominent satellite bands in caesium chloride the melting curve often has two or more phases. Even where no large satellites are present heterogeneity within the DNA can readily be detected by plotting the *rate of change* of absorbance at 260 nm of a DNA solution as a function of temperature.

If, following denaturation of DNA by alkali or by heating, the pH of the solution is restored to neutrality or the temperature reduced below the Tm, complementary strands of DNA show a marked tendency to recombine to form double-stranded molecules. This process of duplex formation is termed renaturation or reassociation. The rate of the reaction is governed by the concentration of DNA sequences, pH, temperature and salt concentration. Renaturation is normally carried out at moderately high salt concentrations at a pH close to neutrality and at a temperature approximately 25°C below the Tm. Under these conditions, single-stranded DNA molecules collide at random. Most collisions occur in such a way that complementary sequences of bases are not in register and the strands do not reassociate. If, however, two strands approach in an anti-parallel configuration and potential regions of complementarity are in close proximity a nucleation event will take place and hydrogen bonds will form. Whether or not the strands remain together depends upon the nature of the neighbouring bases on either side of the nucleation site. If complementary sequences are arranged in the correct register along the length of the molecule a rapid zippering effect takes place to produce a stable double-stranded DNA molecule.

DNA renaturation can be followed by measuring any property that distinguishes single-stranded from double-stranded DNA. The decrease in absorbance at 260 nm that accompanies duplex formation is often exploited for this purpose but fractionation by hydroxyapatite is also commonly used. Under appropriate conditions only double-stranded DNA binds to hydroxyapatite (a hydrated form of calcium phosphate); single strands pass straight through a column and double-stranded material may subsequently be eluted either by increasing the salt concentration or by raising the temperature above the Tm. This procedure has the advantage that the properties of the single-stranded and renatured fractions may subsequently be studied separately.

9.4.5 REPEATED SEQUENCE DNA

Information about the sequence content of DNA can be obtained by studying the renaturation rate. Double-stranded DNA is first sheared to lengths of a few hundred nucleotides by ultrasonication or by forcing through a narrow gauge needle. This is designed to liberate short sequences of DNA and to allow them subsequently to react independently of neighbouring sequences. The DNA is then denatured and allowed to renature at the desired concentration under controlled conditions. Renaturation is followed by monitoring the absorbance of the solution at 260 nm in a spectrophotometer cell maintained at the appropriate temperature, or alternatively by taking samples at intervals and fractionating them by hydroxyapatite chromatography. In some situations partially double-stranded segments of DNA are produced, which have single-stranded 'tails'. These can be removed by treating the reassociated DNA with SI nuclease, which is specific for single-stranded DNA, before hydroxyapatite fractionation.

The reassociation reaction follows second order kinetics, which means that it is governed by the concentration of the *reacting sequences*. This is determined by the DNA concentration and the sequence content. For example, synthetic poly-dA. dT renatures almost instantaneously, whereas at the sme concentration naturally occurring DNA takes much longer to renature. In general the renaturation rate is inversely proportional to the number of base pairs in the genome (providing that no repetitive sequences are present). In principle, therefore, the number of 'genes' or sequences may be estimated by comparing the renaturation rate of an unknown DNA with that from an organism with a well characterized genome. The results of renaturation experiments are often expressed as a 'Cot' plot. In this form the extent of DNA renaturation is plotted as a function of the initial concentration of the DNA (in moles of nucleotides per litre) multiplied by the renaturation time (in seconds) plotted on a log scale. One advantage of this is that different types of DNA can easily be compared by their Cot$\frac{1}{2}$ values (the value of Cot at which 50% of the DNA becomes renatured) (Fig. 9.7a). For a variety of microorganisms and viruses the Cot$\frac{1}{2}$ is related to the genome size. However, this is not strictly true for eukaryotic DNA which behaves in a more complex way.

In bacteria and viruses most DNA sequences are represented only once in the genome but in higher organisms many sequences occur in multiple copies. For this reason a plot of the renaturation of higher plant DNA deviates markedly

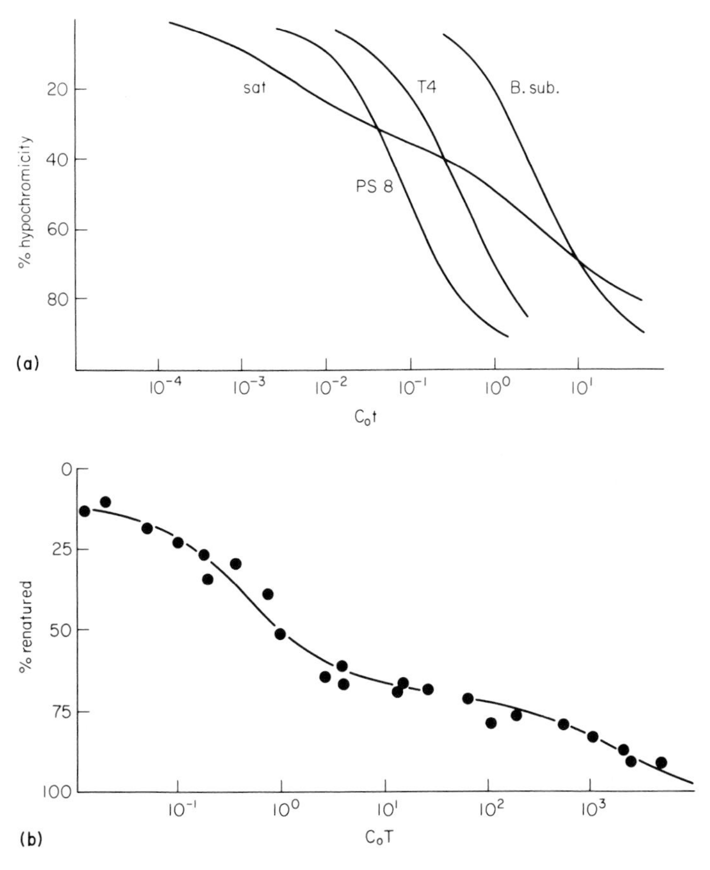

Fig. 9.7. Reassociation kinetics of various DNA samples. (a) Musk melon satellite (sat.), phage PS8 (PS8), *Bacillus subtilis* (*B. sub.*), phage T4 (T4). The data are corrected for differences in GC content and to take account of different solvent concentrations. (b) *Helianthus annuus* DNA. The slowest reassociating fraction behaves approximately as expected for sequences present in one copy per haploid genome. (a) redrawn from Bendich and Anderson, 1974; (b) from Flavell *et al.*, 1974.)

from a second order curve (Fig. 9.7b). A substantial proportion of the genome is so highly repetitive that it renatures even more rapidly than bacterial DNA. A second, intermediate, fraction contains families of repetitive sequences present in a lower frequency. Finally, there exists a unique fraction of DNA which essentially comprises sequences present in only one or a small number of copies in the genome. Flavell *et al.* (1974) have studied the reassociation kinetics of the DNA from 23 higher plant species, which vary in *2C* nuclear DNA content between 1·7 and 98×10^{-12} g. The results, summarized in Table 9.4, show that in plants with relatively large genomes an average of 80% of the DNA consists of sequences present in 100 copies or more. For plants with smaller genomes

Table 9.4. Proportion of repeated-sequence DNA* in species with a nuclear DNA mass between 5 and 98 pg.

Species	Ploidy	2*C* nuclear DNA content (pg)	Proportion of repeated sequences** (%)
Poa trivialis	2×	6·9	82
Tropaeolum majus	2×	7·3	70
Pisum sativum	2×	9·9	75
Helianthus annuus	2×	10·7	69
Zea mays	2×	11·0	78
Hordeum vulgare	2×	13·4	76
Poa annua	4×	13·8	87
Triticum monococcum	2×	14·0	80
Secale cereale	2×	18·9	92
Vicia faba	2×	29·3	85
Allium cepa	2×	33·5	95
Triticum aestivum	6×	36·2	83
Avena sativa	6×	43·0	83
Tulipa kaufmanniana	2×	62·5	73
Hyacinth orientalis	4× − 1	98·1	75
Mean			80 ± 2.0

Proportion of repeated-sequence DNA* in species with nuclear DNA mass below 4 pg.

Species	Ploidy	2*C* nuclear DNA content (pg)	Proportion of repeated sequences** (%)
Linum usitatissimum	2×	1·5	59
Capsella bursa-pastoris	4×	1·7	46
Veronica persica	4×	1·9	63
Stellaria media	7×	2·5	69
Lamium purpureum	2×	2·7	60
Senecio vulgaris	8×	3·5	74
Daucus carota	2×	2·1	62
Beta vulgaris	2×	2·7	63
Mean			62 ± 2.9

*Proportion of DNA sequences present in an excess of ten copies as indicated by hydroxylapatite chromatography.

**Corrected for 20% hypochromicity of reannealed DNA.

From Flavell *et al.* (1974).

the average amount of repetitve DNA is 62%. These results indicate that part of the variation in nuclear DNA mass can be accounted for by variation in the amount of repeated sequence DNA.

9.5 THE GENES FOR RIBOSOMAL-RNA

More is known about the properties, organization and function of the genes for rRNA than any other part of the genome. These DNA segments direct the synthesis of rRNA molecules which form the structural 'backbone' of ribosomes, without which cells cannot function. It has been known for some time that they are associated with the nucleolus because certain mutants of the toad *Xenopus*, which lack nucleoli, are incapable of synthesizing rRNA. During development these mutants die because they are unable to synthezise ribosomes.

9.5.1 DNA-RNA HYBRIDIZATION

The genetic origin of any RNA molecule can be studied by DNA-RNA hybridization. This technique is similar in principle to DNA-DNA reassociation. It depends upon the fact that any RNA molecule synthesized on a DNA template has a base sequence complementary to the DNA strand from which it was transcribed. Under suitable conditions, therefore, DNA can be induced to combine with complementary-RNA (cRNA) to produce an RNA-DNA double helix. This provides a means of locating and measuring the genes that code for a particular type of RNA.

One of the commonest procedures is to prepare pure DNA, denature it by heating or treatment with alkali, and to attach the single strands to a solid matrix such as a nitrocellulose filter. During subsequent treatments the immobilized DNA strands are prevented from reassociating but they are capable of forming hybrids when they are placed in a solution of nucleic acid containing complementary base sequences. Using radioactive RNA the formation of DNA-RNA hybrids can be detected by measuring the amount of radioactivity that becomes associated with the filter. The reaction is affected by a number of factors such as RNA concentration, salt concentration, temperature, the presence of compounds such as formamide, which lowers the Tm, and the reaction time. With rRNA at a concentration of 2 μg cm^{-3} at 70°C in 6 × S.S.C. the reaction is virtually complete within a couple of hours (Fig. 9.8). After a suitable time the filters are washed, given a mild treatment with ribonuclease to remove any unhybridized RNA, and the amount of radioactive RNA remaining is measured in a liquid scintillation counter. If the amount of DNA initially applied to the filter is known then the extent of hybridization can be calculated from the specific radioactivity of the RNA. The amount of DNA that hybridizes to rRNA can either be calculated directly from the saturation plateau (Fig. 9.8) or, more accurately, from a double reciprocal plot of the data (Scott & Ingle, 1973). In most plants between 0·05 and 1 % of the DNA hybridizes to rRNA. This represents from a few hundred to several thousand genes per diploid nucleus, depending on the species (Table 9.5). The 25s and 18s rRNA molecules hybridize to separate sites on the DNA in accordance with the ratio of their molecular weights (1·3:0·7; approximately 2:1 by mass but 1:1 molar ratio).

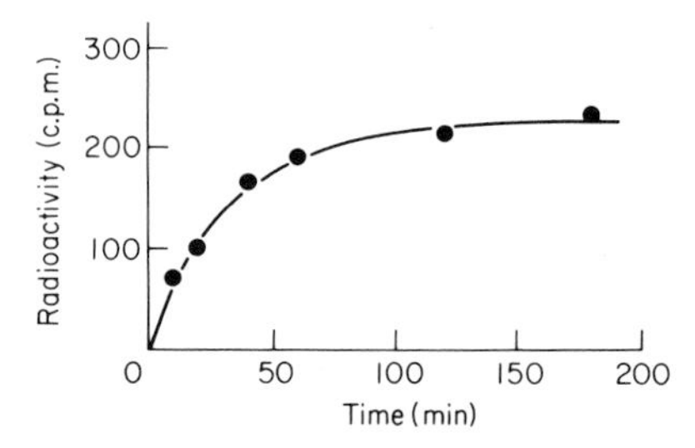

Fig. 9.8. Saturation hybridization of nuclear DNA from *Phaseolus aureus* with radioactive 25s rRNA.

25s rRNA (molecular weight $1 \cdot 3 \times 10^6$ daltons) labelled with [^{3}H]-uridine at a specific activity of $1 \cdot 4 \times 10^4$ c.p.m. μg^{-1} was hybridized to nuclear DNA (2 μg per filter) for increasing periods of time in $6 \times$ S.S.C. at 70°C. (From Grierson, 1975).

Table 9.5. Number of ribosomal-RNA genes in different plants.

Plant	% DNA hybridized	Number of genes per haploid complement
Swisschard (*Beta vulgaris, v. cicla*)	0·20	1,150
Maize (*Zea mays*)	0·18	3,100
Wheat (*Triticum vulgare*)	0·092	2,100
Cucumber (*Cucumis sativus*)	0·96	4,400
Onion (*Allium cepa*)	0·090	6,650
Pea (*Pisum sativum*)	0·17	3,900
Artichoke (*Helianthus tuberosus*)	0·022	260

From Ingle & Sinclair (1972).

This indicates that there are equal numbers of genes for each type of rRNA. In cases where they do not hybridize in equimolar amounts this may be due to sequence homology between the two RNAs, or alternatively may arise by contamination of the RNA samples with other types of RNA.

Multiple genes for rRNA are a physiological necessity in order to provide sufficient sites of synthesis for rRNA molecules to meet the high metabolic demand for ribosomes. Plants often contain thousands of these genes although the number varies between varieties of the same species and between different species. Most of the available evidence suggests that all the copies are identical and this provides a clear example of repetitive DNA with a known function.

Their localization within the nucleolus has been confirmed by *in situ* hybridization of rRNA to cytological preparations (Brady & Clutter, 1972).

In many eukaryotes ribosomal DNA occupies a separate position in caesium chloride gradients relative to main band DNA. However, since they represent 1%, or less, of the genome these genes are not easily detected by studying the distribution of the DNA. Nevertheless, they can readily be pinpointed by hybridizing rRNA with separate fractions of DNA from the gradient. The fractionation of main band DNA from the rRNA genes of *Phaseolus aureus* is shown in Fig. 9.9. In general, ribosomal DNA from plants has a higher density

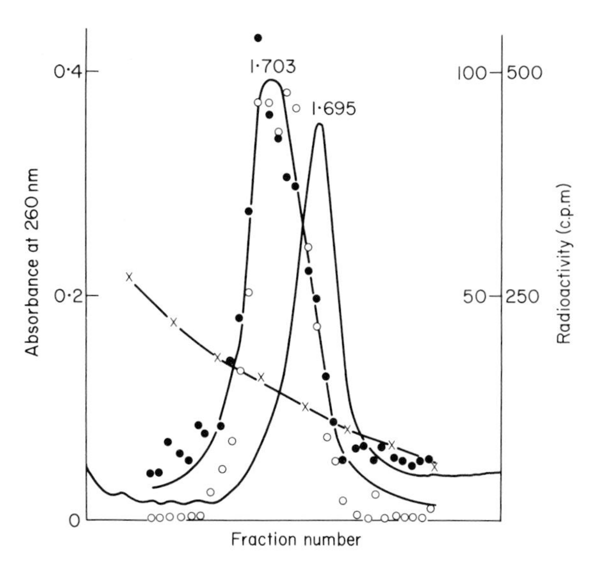

Fig. 9.9. Separation and detection of the DNA sequences coding for ribosomal RNA. Nuclear DNA from *Phaseolus aureus* was fractionated in caesium chloride (see Figs. 9.4 and 9.5) and the fractions from the gradient were divided into two and hybridized to either 25s (cytoplasmic) or 23s (chloroplast) rRNA labelled with [^{3}H]-uridine. Smooth, curve, absorbance of DNA at 260 nm; solid circles, hybridization with 25s rRNA; open circles, hybridization with 23s rRNA. The gradient of CsCl is shown and the calculated densities of the rRNA genes and the main band DNA are indicated. (Unpublished result of D. Grierson.)

than main band DNA. This is consistent with the fact that in higher plants rRNA has a base composition in excess of 50% GC, whereas the bulk of nuclear DNA often has a lower GC content. In this sense rRNA genes represent a type of satellite DNA. However, in different plants the rRNA genes are often present in similar numbers and with a similar density irrespective of the presence or absence of large amounts of satellite DNA. It therefore appears that ribosomal DNA and other satellites may occasionally band together in caesium chloride but this is entirely fortuitous (Ingle *et al.*, 1975).

9.5.2 ELECTRON MICROSCOPY

The rRNA genes can also be studied by electron microscopy of preparations from isolated nucleoli. This approach provides a clear indication of their organization. Figure 9.10 shows an electron micrograph of the partially

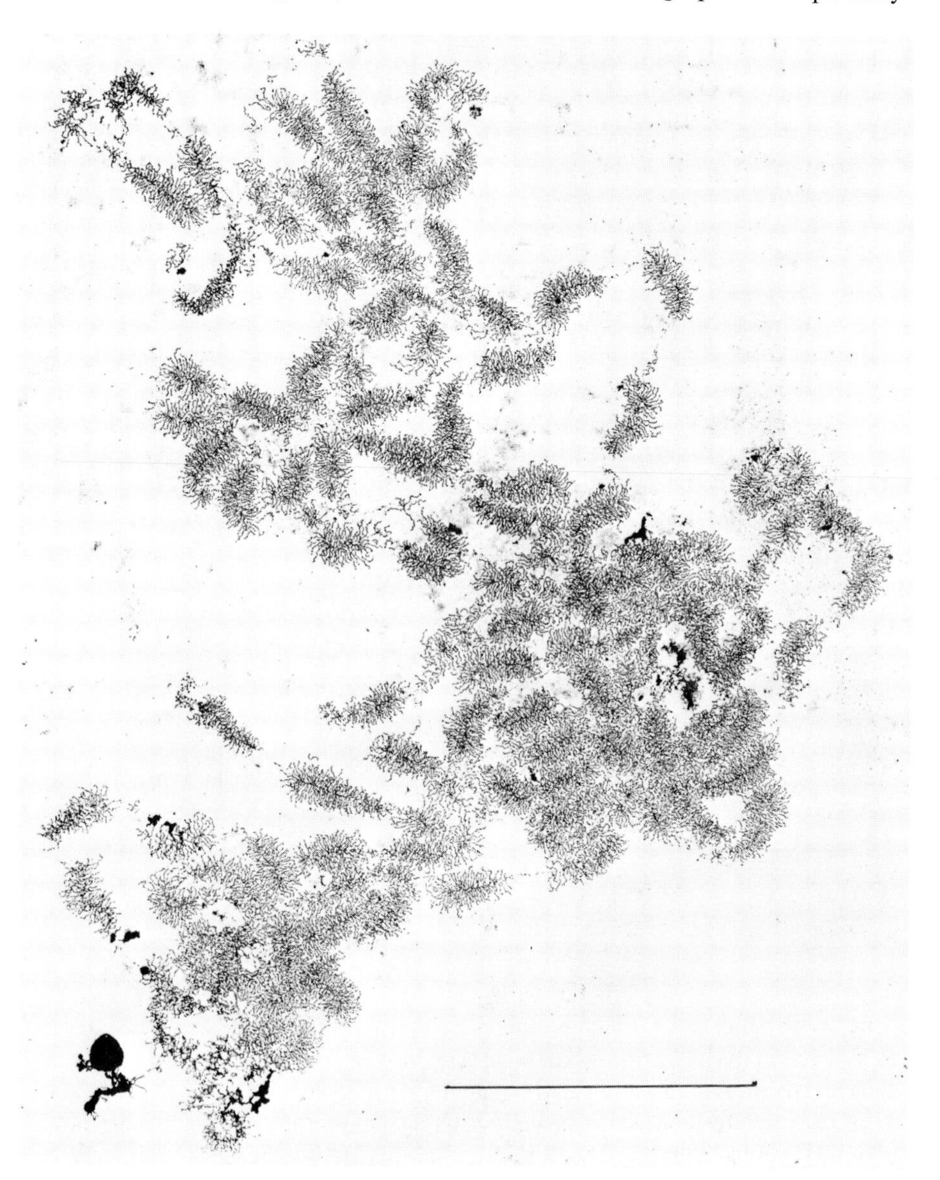

Fig. 9.10. Spread contents of a complete nucleolar aggregate subunit from *Acetabularia mediterranea*. Note the numerous 'matrix' units. The scale indicates 5 μm. (From Spring *et al.*, 1974.)

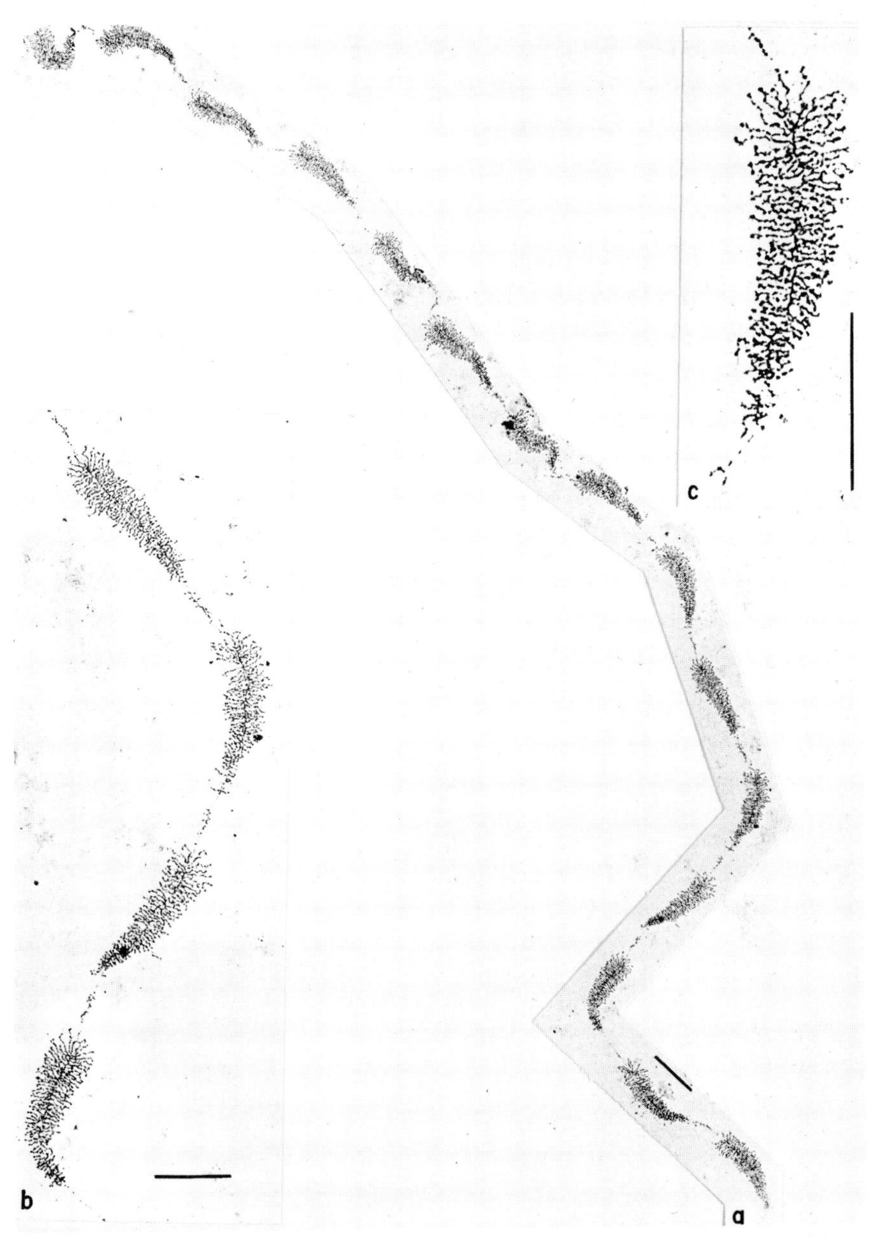

Fig. 9.11. (a) Well spread nucleolar material showing the regular pattern of 'matrix' units and 'spacer' segments. Note the length gradient of lateral fibrils within each matrix unit. (b) Note occasional groups of small fibrils associated with the spacer units. (c) At higher magnification, the dense packing of lateral fibrils within the matrix unit is clearly seen. Scales indicate 1 μm, (From Trendelenburg *et al.*, 1974.)

dispersed contents of a subunit from the nucleolar aggregate of *Acetabularia*. This shows a regular arrangement of fibril-covered 'matrix' units and fibril-free 'spacer' regions arranged along a central axis. A higher magnification of a 'matrix' unit is shown in Fig. 9.11. The central axis is the double-stranded ribosomal DNA itself. The 'matrix' units represent ribosomal genes which are being transcribed into RNA at a number of points simultaneously. The fibrils of the 'matrix' unit are growing chains of rRNA precursor molecules (see the section on RNA synthesis). RNA polymerase enzymes can be seen clustered on the DNA. About 120 enzymes are engaged in the simultaneous transcription of the DNA in each 'matrix' unit. The marked polarity reflects the fact the growing chains of attached RNA molecules increase in length as the polymerase enzymes responsible for their synthesis proceed along the DNA axis. In addition there is a tendency for the free ends of the RNA chains to form terminal knobs as they become folded, probably in association with proteins.

Multiple rRNA genes therefore alternate in a precise arrangement with non-transcribed regions of the DNA. Hybridization evidence shows that separate 'cistrons' for the 25s and 18s rRNA are present in the DNA, but there is no evidence for the existence of separate transcription units. In contrast, in *Acetabularia* each transcribed region of double-stranded DNA is approximately 2 μm long. This indicates that it codes for an RNA molecule with a molecular weight of approximately 2×10^6 daltons. This is the combined mass of the 25s and 18s rRNA molecules ($1{\cdot}3 + 0{\cdot}7 \times 10^6$ daltons). There is clear evidence (discussed in the section on RNA synthesis) that ribosomal RNA is first transcribed as a polycistronic precursor molecule. In many plants this precursor has a molecular weight of approximately $2{\cdot}4 \times 10^6$ daltons, and contains one sequence of 25s RNA plus one sequence of 18s RNA together with some extra sequences which are subsequently removed. In order to reconcile such a mode of synthesis with studies on the organization of the rRNA genes it is necessary to propose that each 'matrix' unit consists of an 18s and 25s sequence and some transcribed 'spacer' DNA. Fortunately, in cases where the organization of ribosomal DNA has been studied in detail, for example in *Xenopus*, evidence for precisely this type of arrangement has been obtained (Birnstiel *et al.*, 1968). Furthermore, recent studies with the electron microscope have provided evidence for a common organization of the rRNA genes in all animals examined (Schibler *et al.*, 1975). A model for the repeating unit, which probably applies also to plants, for it is compatible with all the available evidence, is shown in Fig. 9.12.

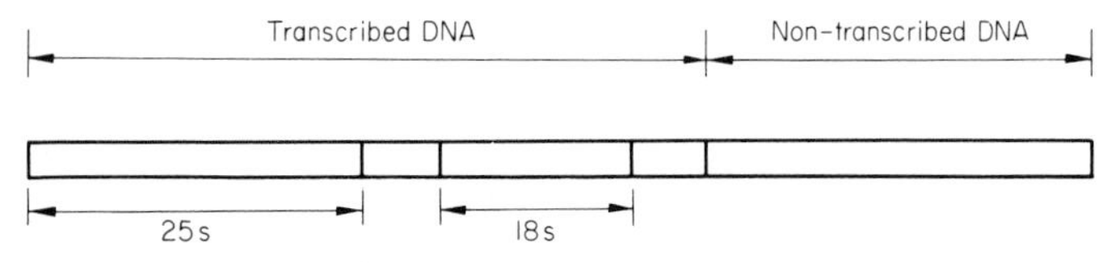

Fig. 9.12. Suggested structure of the repeat unit of ribosomal DNA in plants. (Modified from Schibler *et al.*, 1975.)

By carrying out partial denaturation studies it has proved possible to map certain areas rich in GC or AT base sequences. In addition, it is known that both the non-transcribed 'spacer' regions and the transcribed spacer can vary significantly in length and composition between species. Possible functions for such regions are discussed below. In contrast, however, the regions that code for rRNA are remarkably constant in composition. This probably reflects the importance of their role in providing structural RNA for ribosome formation.

As shown in Fig. 9.9, chloroplast rRNA also hybridizes to nuclear DNA (Ingle *et al.*, 1970; Tewari & Wildman, 1968). This is a surprising result in view of the fact that chloroplasts contain their own DNA and synthesize their own rRNA (see chapter 11). The location of these genes within the nucleus is not known, and at present there is no evidence that they are ever transcribed in the nucleus.

9.6 ORGANIZATION AND FUNCTION OF THE GENOME

9.6.1 REPETITIVE DNA

Reassociation studies indicate that between 40% and 90% of the genome of higher plants is composed of repetitive DNA. This fraction is not homogeneous but consists of a number of types of sequence with different lengths and repetition frequencies. In hexaploid wheat, for example, 70%–80% of the genome is made up of families of sequences present in from 100 to 100,000 copies (Smith & Flavell, 1975). The question of the composition, organization and function of these sequences is still under investigation but at least three different classes can be distinguished. These are (a) clustered sequences of repetitive DNA which are transcribed into RNA, (b) clustered sequences which are apparently not transcribed, and (c) reverse-repeats distributed at intervals throughout a substantial part of the genome.

9.6.1.1 *Transcription of repeated sequences*

DNA-RNA hybridization has shown that a small part of the repeated sequences represents genes for rRNA. These consist of regularly arranged alternating regions of transcribed and non-transcribed DNA. The transcribed sequences can be subdivided into those where the products are stable, the 18s and 25s rRNA; and those that have metabolically unstable products, the transcribed spacer DNA in the rRNA precursor. The role of the transcribed and non-transcribed spacer sequences has not yet been unequivocally established although obvious possibilities are that they may be involved in genetic organization or regulation of transcription. Different groups of rRNA genes may sometimes be present on a number of chromosomes but as far as is known they are all practically identical.

Other families of repeated genes which have been shown to be transcribed include those for 5s RNA, tRNA and, at least in animals, the genes for histone proteins. What proportion of the genome do these account for? The actual value is higher than at first appears because hybridization only detects one strand of the DNA and normally takes no account of either the transcribed or non-transcribed spacer. This means, for example, that a 0·5% hybridization level for rRNA probably accounts for about 2·5% of the total DNA. Similar calculations for the 5s, tRNA and histone genes show that the total number of known sequences of this type only account for something like 5% of the genome.

9.6.1.2 *Palindrome sequences*

A small fraction of rapidly-renaturing DNA is known to be interspersed throughout a large part of the genome. The evidence for this comes from the 'zero-time binding' of DNA to hydroxyapatite. When DNA is denatured and immediately passed through hydroxyapatite a fraction of the DNA binds, suggesting that it is double-stranded. This happens even at very low Cot values where reassociation probably does not occur. The most convincing explanation of this behaviour is that it is due not to strand reassociation but to intra-strand 'snap-back'. This would occur if double-stranded DNA contained reverse-repeat sequences of the following type:

– – A – T – A – G – G – C – G – C – C – T – A – T – –

– – T – A – T – C – C – G – C – G – G – A – T – A – –

With this arrangement, termed a 'palindrome' sequence, the potential complementarity between adjoining regions of the same single-stranded DNA molecule results in rapid hairpin formation. In consequence any single-stranded DNA molecule will bind to hydroxyapatite providing it contains a double-stranded hairpin region. This provides a means of deducing the organization of palindrome sequences in the genome. The amount of DNA that binds to hydroxyapatite at zero time is markedly dependent upon the fragment size of the DNA used. This suggests that the palindrome sequences are interspersed with other regions of DNA. It has been calculated that in wheat the average distance between these palindrome sequences is 8,500 nucleotide pairs. Between 4% and 10% of the DNA is palindromic and such sequences are spread throughout at least 40% of the genome (Smith & Flavell, 1975).

There are several possible functions for such reverse-repeats. Any RNA molecule transcribed from DNA interspersed with sequences of this type would be expected to have double-stranded hairpin regions. These are known to occur in RNA but the proportion of palindromes in DNA is probably too high to be accounted for solely on this basis. One obvious property of palindromes that may have functional significance is that of symmetry. This may be important for the recognition of particular regions of double-stranded DNA by enzymes or regulatory elements.

9.6.1.3 *Other types of repetitive DNA*

Satellite DNAs also consist of repeated sequences. It is clear from Table 9.3 that satellites vary substantially in buoyant density, and therefore base composition, in different plants. This variation is also reflected in sequence content and repetition frequency. Although apparently homogeneous in caesium chloride the melon satellite has been shown to consist of at least two components, judged by melting experiments, caesium sulphate centrifugation and renaturation behaviour. Preliminary estimates are that the repeat lengths of the two sequences are respectively several hundred and a million or more nucleotide pairs (Bendich & Anderson, 1974; Sinclair *et al.*, 1975). It is of interest to ask to what extent the copies within one family of repeated sequences diverge in sequence because this knowledge may provide clues about the origin and function of repetitive DNA. Indirect information on this point comes from melting experiments carried out with native and renatured satellite DNA. When similar, but not identical, sequences renature the possibility exists for the formation of duplexes with a certain proportion of mis-matched bases. For every 1% mismatching in renatured DNA the Tm is decreased by $1{\cdot}0 \pm 0{\cdot}3$°C below that of native DNA (Britten *et al.*, 1974). Judged by this criterion the sequences within the two types of melon satellite DNA are remarkably similar. Many animal satellite DNAs appear to have evolved by duplication of a rather short sequence of DNA followed by divergence from the basic sequence. They are believed not to be transcribed. In contrast, the melon satellite DNAs appear to show very little sequence divergence and furthermore one component at least has a suprisingly large potential coding capacity. However, this may prove to be exceptional and further experimental work is necessary before it can be established whether or not plant satellites are more complex than those found in animals and whether they are ever transcribed into RNA.

The division of repetitive DNA into satellite and non-satellite components is probably an arbitrary distinction. All plants contain repeated sequence DNA irrespective of whether or not they possess satellites. This emphasises the considerable variation in properties of repeated DNA sequences. However, very little is known about their function. The fact that they sometimes form satellites suggests that they are clustered to some extent, otherwise they would remain attached to, and therefore band in association with, main-band DNA in caesium chloride. In the mouse, hybridization to cytological preparations has shown that satellite sequences are concentrated around the centromeres of many of the chromosomes, but this is not always true of satellite DNA.

Where satellites are present the amounts may vary in different cell types of the same plant (Ingle & Timmis, 1975). Furthermore there is some evidence that at least part of the repeated DNA can be dispensed with altogether. For example, Beridze (1972) has investigated six species of *Phaseolus* and shown that three contain substantial amounts of satellite DNA whereas three others possess little if any satellite. Clearly, in these cases, the satellite is not essential and its presence or absence appears to have little effect on the plant phenotype. It is quite possible that at least some of the repetitive sequences in DNA are manifestations of evolutionary 'doodling'.

9.6.2 UNIQUE DNA SEQUENCES

One might conclude that the non-repetitive portion of the genome represents that fraction of the DNA that codes for proteins. However, taking into account the repetitive DNA fraction, the *amount* of unique sequence DNA in different species is found to vary at least ten-fold (Table 9.4). One way to estimate the proportion of DNA transcribed is to carry out hybridization experiments between mRNA and unique-sequence DNA. Due to the very high Cot values required the rate and extent of hybridization using the conventional filtered technique are unsatisfactory. An alterative approach is to carry out DNA renaturation at high concentrations, using only a trace of highly radioactive RNA for hybridization (Bishop *et al.*, 1974). However, another technical difficulty is encountered in obtaining RNA preparations sufficiently enriched with mRNA and of sufficiently high specific radioactivity for this purpose and no estimates are yet available concerning the extent of transcription of the unique fraction of the plant genome.

9.7 RNA SYNTHESIS

Most types of RNA are single-stranded molecules transcribed by RNA polymerase enzymes from one of the two strands of a DNA template. RNA molecules are, therefore, the direct products of genes. The following section deals with the transcription of nuclear DNA and the processing and transport of RNA to the cytoplasm. The transcription of chloroplast and mitochondrial DNA and the special case of the synthesis of RNA on a RNA template by plant viruses are excluded.

9.7.1 GENERAL PROPERTIES OF RNA

Different types of RNA are characterized by their subcellular distribution and properties such as molecular weight and nucleotide composition. Approximately 70% of cellular RNA is contained in the ribosomes where it plays a structural role. Each 80s cytoplasmic ribosome contains three different types of ribosomal RNA (rRNA). These molecules are most commonly referred to by 's values' which are related to their rates of sedimentation during ultracentrifugation.

The larger ribosomal subunit contains two rRNA molecules of 25s and 5s and the smaller subunit contains a single rRNA of 18s. The approximate molecular weights are $1{\cdot}3 \times 10^6$ daltons for 25s rRNA (about 3,900 nucleotides), $0{\cdot}7 \times 10^6$ daltons for 18s rRNA (about 2,100 nucleotides) and 4×10^4 daltons for 5s RNA (about 120 nucleotides). The nucleotide compositions of a range of rRNAs are shown in Table 9.6. Transfer RNA (tRNA or 4s RNA) accounts for a

Table 9.6. Base composition of ribosomal-RNA and poly (A)-containing RNA from different organisms.

		Moles per cent				
Organism	rRNA	C	A	G	U	G+C
Yeast	26s	19·2	26·4	28·4	26·0	47·6
(*Saccharomyces cerevisiae*)	18s	19·1	26·6	26·1	28·1	45·2
Mushroom	26s	20·1	27·7	27·8	24·3	47·9
(*Psalliota campestris*)	18s	18·0	24·9	30·1	27·1	48·1
Artichoke	26s	21·6	28·1	31·5	18·9	53·1
(*Helianthus tuberosus*)	18s	20·5	27·6	28·1	23·9	48·6
Pea	26s	21·8	24·9	32·2	21·1	54·0
(*Pisum satisum*)	18s	21·1	25·9	28·5	24·4	49·6
Mung bean	26s	23·0	24·5	32·4	20·1	55·4
(*Phaseolus aureus*)	18s	20·2	25·4	30·2	25·2	50·4
Corn	26s	28·0	21·3	33·9	16·9	61·9
(*Zea mays*)	18s	25·07	22·0	33·1	20·2	58·17
		Poly (A)-containing RNA				
French bean (*Phaseolus vulgaris*)		16·6	39·0	21·6	22·7	38·2
Sycamore (*Acer pseudoplatanus*)		14·6	41·0	19·3	25·1	33·9

further 25% of the total. In contrast to rRNA this fraction is heterogeneous. It consists of a range of similar but not identical molecules approximately 80 nucleotides long, each capable of being charged with a specific amino acid. By interacting with messenger-RNA appropriately charged tRNAs are responsible for aligning amino acids in the correct order during protein synthesis. In addition to the four major bases tRNA contains a very high proportion of modified or unusual bases. The most interesting type of RNA, mRNA, is present in small amounts and constitutes less than 5% of the total. It is this fraction that directs the synthesis of specific proteins by the ribosomes. The length and nucleotide sequence of mRNAs are related to the properties of the proteins for which they code. As most plant cells synthesize a variety of different proteins the mRNA fraction generally consists of a range of molecules of different size and nucleotide composition. The fractionation, by polyacrylamide gel electrophoresis, of total RNA from suspension cultures of sycamore cells is shown in Fig. 9.13a. The rRNA and tRNA are clearly distinguished but the small amount of mRNA present is not detected. In Fig. 9.13b purified radioactive mRNA from polyribosomes of sycamore cells is compared with non-radioactive rRNA. The distance moved by each RNA through the gel is, with a few exceptions, inversely related to the log of the molecular weight (Loening, 1968, 1969).

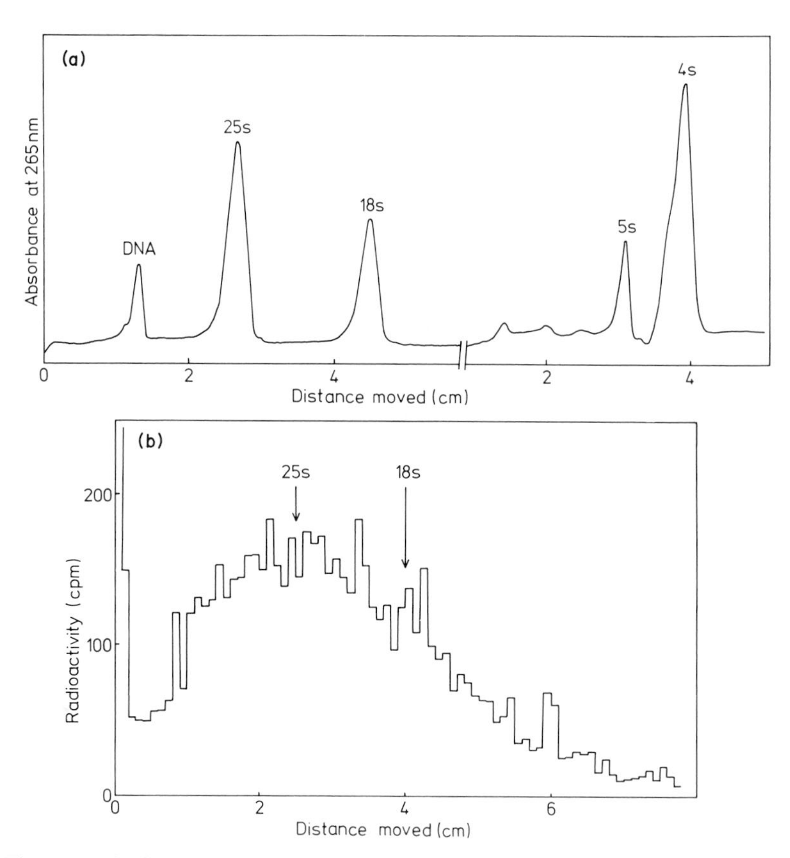

Fig. 9.13. The fractionation of sycamore (*Acer pseudoplatanus*) RNA by gel electrophoresis.

(a) Composite of two gel profiles. On the left the fractionation of DNA and 25s and 18s rRNA in a 2·4% acrylamide gel. On the right the separation of 5s from tRNA in a 7·5% gel. (b) Gel electrophoresis of poly (A)-containing mRNA purified from polyribosomes of sycamore cells labelled with [^{3}H]-uridine for 30 minutes. The positions of unlabelled 25s and 18s rRNA used as molecular weight markers in the same gel are indicated. (Unpublished results of D. Grierson.)

9.7.2 RIBOSOMAL RNA SYNTHESIS

9.7.2.1 *The ribosomal RNA precursor*

The DNA that codes for rRNA is known to be associated with nucleoli and the transcription of nucleolar DNA has been studied with the electron microscope (Fig. 9.11). More detailed information has been obtained by the labelling of intact seedlings, plant segments, or cultured cells, with [^{32}P]-phosphate or [^{3}H]-uridine for various times followed by the extraction and characterization of the rapidly-labelled RNA. The mechanism of rRNA synthesis is fairly well understood and a generalized scheme outlining the main features of the process is given in Fig. 9.14. The first stable gene product which can be detected is an rRNA precursor molecule. The molecular weight of this RNA varies in different plants but is generally within the range $2{\cdot}2$–$2{\cdot}6 \times 10^6$ daltons. The rRNA precursor from mung bean roots, detected by polyacrylamide gel electrophoresis,

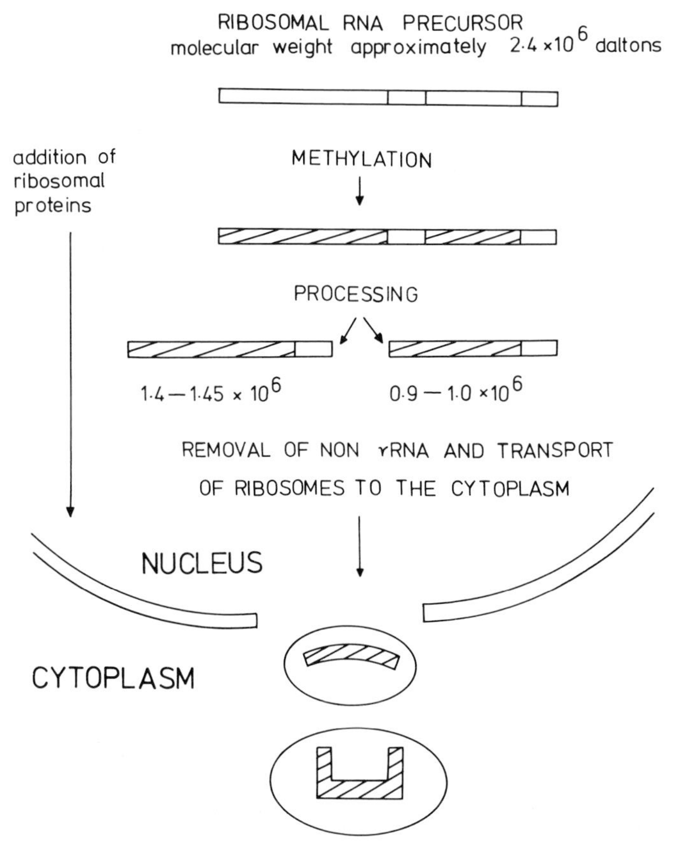

Fig. 9.14. The synthesis and processing of the precursor to rRNA. (Modified from Grierson *et al.*, 1975.)

is shown in Fig. 9.15. The precursor undergoes considerable post-transcriptional modification in the nucleolus before it is converted into 'mature' rRNA and enters the cytoplasm. Shortly after synthesis the molecule becomes methylated at selected sites. This can be demonstrated by incubating plants in methyl-labelled methionine for a short time. The methyl groups become incorporated into the rRNA precursor *via* s-adenosyl methionine (Cox & Thurnock, 1973). Ribosomal-RNA is known to contain 2′-methyl ribose at certain positions. Only a small number of the residues are modified in this way and it is probable that methylation is completed during the rRNA precursor stage.

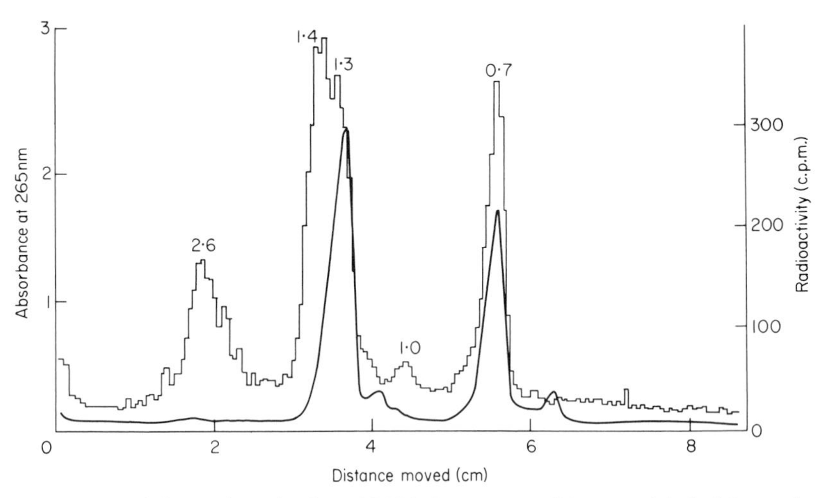

Fig. 9.15. Gel electrophoresis of total RNA from roots of *P. aureus* labelled for 1·5 h with [^{32}P]-phosphate. The molecular weights of the rRNA precursor and processing intermediates are shown in millions. (Unpublished result of D. Grierson.)

The precursor molecules correspond to the lateral fibrils attached to the matrix units of active nucleolar DNA (Fig. 9.11). Each molecule contains one sequence of 25s rRNA, one sequence of 18s (total molecular weight $=2{\cdot}0\times10^6$ daltons) together with some non-rRNA. The evidence for this is as follows. Firstly, although it is rapidly labelled, the precursor does not accumulate in large amounts. This suggests, therefore, that it must either be rapidly degraded or converted to some other molecular form (Rogers *et al.*, 1970; Leaver & Key, 1970; Grierson *et al.*, 1970; Cox & Turnock, 1973; Grierson & Loening, 1974). This latter explanation would account for the fact that radioactivity appears in mature rRNA slightly later than in the precursor. Secondly, the nucleotide composition of the rRNA-precursor is similar to that of rRNA (Rogers *et al.*, 1970; Leaver & Key, 1970; Grierson & Loening, 1974). Thirdly, competition hybridization experiments have shown that both unlabelled 25s and 18s rRNA compete with radioactive precursor-RNA for sites in the DNA, suggesting that they share common sequences (Fig. 9.16). Finally, partial digestion of radioactive 25s, 18s and precursor RNA by T1 ribonuclease, followed by two-dimensional fractionation of the fragments shows that the 'fingerprint' pattern of the precursor is similar to that expected of a mixture of 25s and 18s rRNA (Sen *et al.*, 1975).

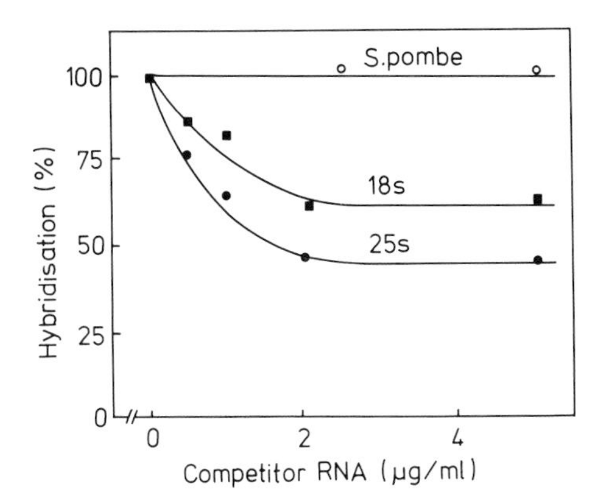

Fig. 9.16. Competition hybridization between unlabelled rRNA and radioactive rRNA precursor for similar sites in *P. aureus* DNA. (From Grierson & Loening, 1974.)

In some plants more than one precursor is detected (Leaver & Key, 1970; Cox & Turnock, 1973). The two precursors from cultured sycamore cells are shown in Fig. 9.17. In some instances the second precursor may originate in chloroplasts or mitochondria (Grierson & Loening, 1974; Kuriyama & Luck, 1973), but in carrot and sycamore both precursors are thought to be in the nucleus (Leaver & Key, 1970; Cox & Turnock, 1974). It is not clear in these cases whether the two RNAs contain similar sequences or whether they are transcribed from different genes. It is possible that the larger RNA is converted to the smaller one, but careful analysis of the rate of labelling of both molecules does not support this suggestion (Cox & Turnock, 1973; Leaver & Key, 1970).

9.7.2.2 *Processing of the precursor*

In addition to becoming methylated, and while still in the nucleolus, the precursor becomes associated with protein molecules and is processed by nucleolytic enzymes which remove the excess non-rRNA in stages to produce mature rRNA in ribosomal subunits ready for transport to the cytoplasm. Some RNA processing intermediates are present in the sample shown in Fig. 9.15. The scheme for processing outlined in Fig. 9.14 is consistent with some, but not all, the studies on rRNA synthesis in plants. For example, no immediate precursor to 25s rRNA is detectable in rye embryos (Sen *et al.*, 1975). Furthermore the $1{\cdot}0 \times 10^6$ daltons RNA is not always detected wheras certain additional RNAs

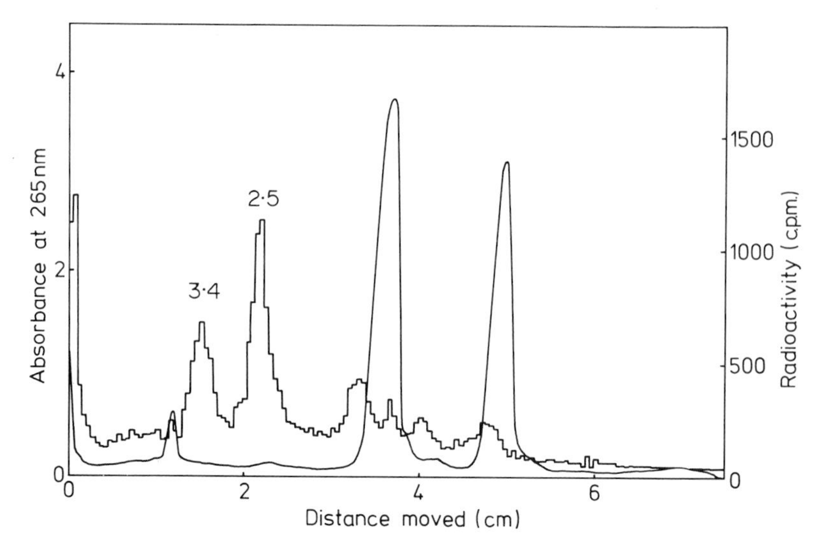

Fig. 9.17. Pulse-labelled RNA from sycamore cells showing two rRNA precursor molecules. The cells were labelled for 40 minutes with [^{3}H]-uridine and total nucleic acid fractionated by gel electrophoresis. The DNA was not removed but an identical distribution of radioactivity is observed when this is carried out. The approximate molecular weights of the precursor RNAs are indicated in millions. (Unpublished result of D. Grierson.)

are sometimes observed. For example, after cycloheximide treatment of cultured parsley cells, which slows down rRNA processing, previously undetected RNAs with molecular weights of 2·0 and 0·9 × 10^6 daltons were found (Gebauer *et al.*, 1975). This suggests that if certain processing events occur very rapidly they may normally go undetected. It is very probable that the complete sequence of processing steps will only be worked out when it becomes possible to study accurately RNA metabolism in isolated nucleoli. Apparent differences between closely related species may be explained by assuming that certain critical stages of processing occur at different rates. In addition, it should be realised that the processing enzymes are functioning at different, possibly distant, sites within the same molecule. There seems no compelling reason to expect these modifications always to occur in the same chronological order and, if they do not, variations in the pattern of processing intermediates might be expected under altered conditions or in different species.

The function of the excess RNA in the precursor is not known. It is too variable in size in different organisms to be a mRNA for the ribosomal proteins (Loening, 1970) and in any case it appears to be degraded very rapidly after synthesis. It may play some role in the regulation of transcription but more probably it is necessary for ensuring that rRNA adopts the correct secondary and tertiary structure required for methylation and for the addition of proteins during ribosome assembly.

It is pertinent to ask whether the multi-stage mechanism that operates for rRNA production is unique or whether it represents a general principle common to the synthesis of all types of RNA. Precursors to tRNA and 5s RNA have been identified and characterized in bacteria and animals (Hecht *et al.*, 1968; Bernhardt & Darnell, 1969) and precursors of mRNA have also been detected, although there is uncertainty about the actual details of processing (Brawerman, 1974). Although RNA metabolism has not been so intensively studied in plants the indications are that precursors to 5s RNA, tRNA and mRNA do occur. When considering the question of the regulation of RNA metabolism, therefore, it is necessary to take account of post-transcriptional modifications as possible control points.

9.7.3 MESSENGER-RNA SYNTHESIS AND METABOLISM

The question of mRNA synthesis has proved extremely difficult to investigate due largely to problems of identification and purification of mRNA. In the past, rather imprecise criteria for mRNA identification, such as size heterogeneity, similarity in nucleotide composition to DNA, rapidity of labelling and association with polyribosomes, were adopted. More recently, two additional approaches have become available. Firstly, the development of an *in vitro* protein synthesizing system has made it possible to test the capacity of an RNA fraction to direct the synthesis of specific proteins. Secondly, the discovery that many mRNAs contain sequences of poly (adenylic) acid (poly (A)) has provided a simple approach to the purification of mRNA.

9.7.3.1 *Poly (A)-containing RNA*

It has been known for some considerable time that a small fraction of total plant RNA has a very high AMP content. This fraction, often referred to as DNA-like RNA or D-RNA, is considered to consist, at least in part, of mRNA and is partially purified by chromatography on columns of methylated albumin adsorbed to kieselguhr (Ingle *et al.*, 1965; Lin *et al.*, 1966; Jackson & Ingle, 1973). More recent work with animal mRNA has shown that the high AMP content is due to the presence of regions of poly (A) in the mRNA molecules. These sequences of poly (A) are generally 50 to 250 nucleotides in length. In animals they are attached to the 3′ end of the mRNA molecules *after* transcription (Brawerman, 1974) although there is some evidence that shorter regions of poly (A) may also be introduced into RNA as part of the transcription process (Jacobson *et al.*, 1974). Poly (A)-containing RNA has been identified in a variety of animals (Brawerman, 1974; Matthews, 1973) and plants (Sagher *et al.*, 1974; Verma *et al.*, 1974; Higgins *et al.*, 1973) although it has not been found in higher plant chloroplasts (Wheeler & Hartley, 1975). It is apparently present in mitochondria from some organisms (Perlman *et al.*, 1973; Avadhani *et al.*,

1973; Ojala & Attardi, 1974) but not others (Groot *et al.*, 1974). This may reflect true differences between organisms or alternatively may be a question of the experimental technique. For example, contrary to earlier reports that poly (A) is absent from mRNA from bacteria, it has recently been shown to be present at a low level in *E. coli.* (Nakazato *et al.*, 1975).

The function of the poly (A) sequences is not known but their presence can be used to purify mRNA. There are a number of methods for doing this. Poly (A) will bind to nitrocellulose or cellulose at high salt concentrations. RNA without poly (A) can be washed off and the poly (A)-containing RNA is then eluted at low ionic strength. However, cellulose and nitrocellulose do not necessarily retain all the poly (A)-containing RNA. A more specific type of binding is achieved using columns of oligo (deoxythymidylic) acid attached to cellulose or poly (uridylic) acid attached to sepharose. Both these polymers rapidly form stable nucleic acid hybrids with the poly (A) sequence attached to the mRNA. Other types of RNA pass straight through the column, although it is sometimes necessary to wash through extensively with a buffer. The poly (A)-containing RNA is only released after lowering the salt concentration or raising the temperature to cause melting of the hybrid between poly (A) and either oligo (dT) or poly (U).

There is considerable evidence that poly (A)-containing RNA from plants is mRNA. It is associated with polyribosomes (Higgins *et al.*, 1973; Grierson *et al.*, 1976) and its synthesis is inhibited by the antibiotic α-amanitin which prevents transcription of non-rRNA in the nucleus (Higgins *et al.*, 1973). When purified from cells synthesizing a large number of proteins of different sizes poly (A)-containing RNA is polydisperse, but when extracted from soybean root nodules, which synthesize substantial amounts of leghaemoglobin, two peaks of RNA are detected, each presumably consisting of similar or identical RNA molecules. These RNA fractions stimulate protein synthesis by a cell-free system from wheat germ. Furthermore, leghaemoglobin has been positively identified among the products of *in vitro* synthesis (Verma *et al.*, 1974). Supporting evidence that poly (A)-containing RNA stimulates protein synthesis comes from a large number of studies and the small number of individual proteins identified as *in vitro* products is growing steadily. However, it should be stressed that not all mRNA molecules contain poly (A). In animals, for example, about 70% of mRNA is estimated to contain poly (A) (Milcarek *et al.*, 1974) whereas only about 40% of polydisperse RNA from polyribosomes of cultured sycamore cells can be purified by oligo (dT)-cellulose (Covey & Grierson, 1976). The functional significance of the two mRNA populations is not clear.

9.7.3.2 *Postranscriptional modification of mRNA*

Poly (A)-containing RNA is synthesized in the nuclei of higher plant cells. Total rapidly-labelled RNA from nuclei consists of the rRNA precursors and processing intermediates together with polydisperse RNA (Fig. 9.18). Approximately 15% of this RNA is retained by oligo (dT)-cellulose and by this criterion is judged to contain poly (A) sequences (Grierson & Covey, 1976). It is polydisperse in size and has a high AMP content (Table 9.6). There is little direct evidence concerning the mechanism of poly (A) addition to nuclear polydisperse RNA in plants. It is assumed to be added to the 3′ end of the molecules by the enzyme ATP-poly-nucleotidylexotransferase after transcription by RNA polymerase (Maale *et al.*, 1975). The poly (A) sequences in higher plant RNA have been estimated to range from 50–250 nucleotides in length (Sagher *et al.*, 1974; Covey & Grierson, 1976). However, plant DNA appears to contain short sequences of AT base pairs which could give rise to short poly (A) regions by transcription (Grierson, 1975) and there is evidence that similar regions are transcribed in slime moulds (Jacobsen *et al.*, 1974).

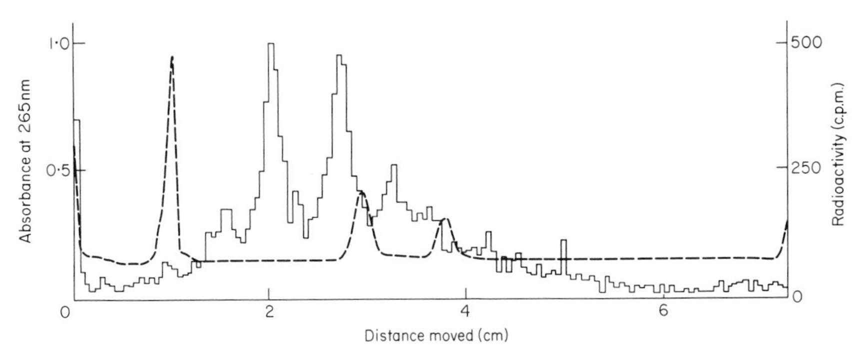

Fig. 9.18. Gel electrophoresis of total nucleic acid from nuclei of sycamore cells. The cells were labelled for 1 h with [^{3}H]-uridine, a nuclear fraction was prepared and total nucleic acid extracted and fractionated by gel electrophoresis. The radioactivity is shown as a histogram. Note that the nuclear fraction contains very little rRNA by comparison with Fig. 9.17. (Grierson & Covey, 1976.)

Comparison of the nuclear and cytoplasmic poly (A)-containing RNA by gel electrophoresis suggests that nuclear RNA is reduced in size before entering the cytoplasm (Fig. 9.19). This suggests that there may be a processing mechanism analogous to that which operates for rRNA synthesis. In slime moulds the nuclear precursor appears to be approximately 20% larger than cytoplasmic mRNA. (Firtel & Lodish, 1973). In higher plants the average molecular weight of the nuclear poly (A)-containing RNA is 1×10^6 daltons (Grierson & Covey, 1976). This is in contrast to the giant precursors to mRNA earlier claimed to occur in animal cells. Very little is known about the processing mechanism, but Firtel and Lodish (1973) have shown that the nuclear precursor contains a sequence of about 300 nucleotides near the 5′ end of the molecule which is lost before the RNA enters the cytoplasm. This is of considerable interest

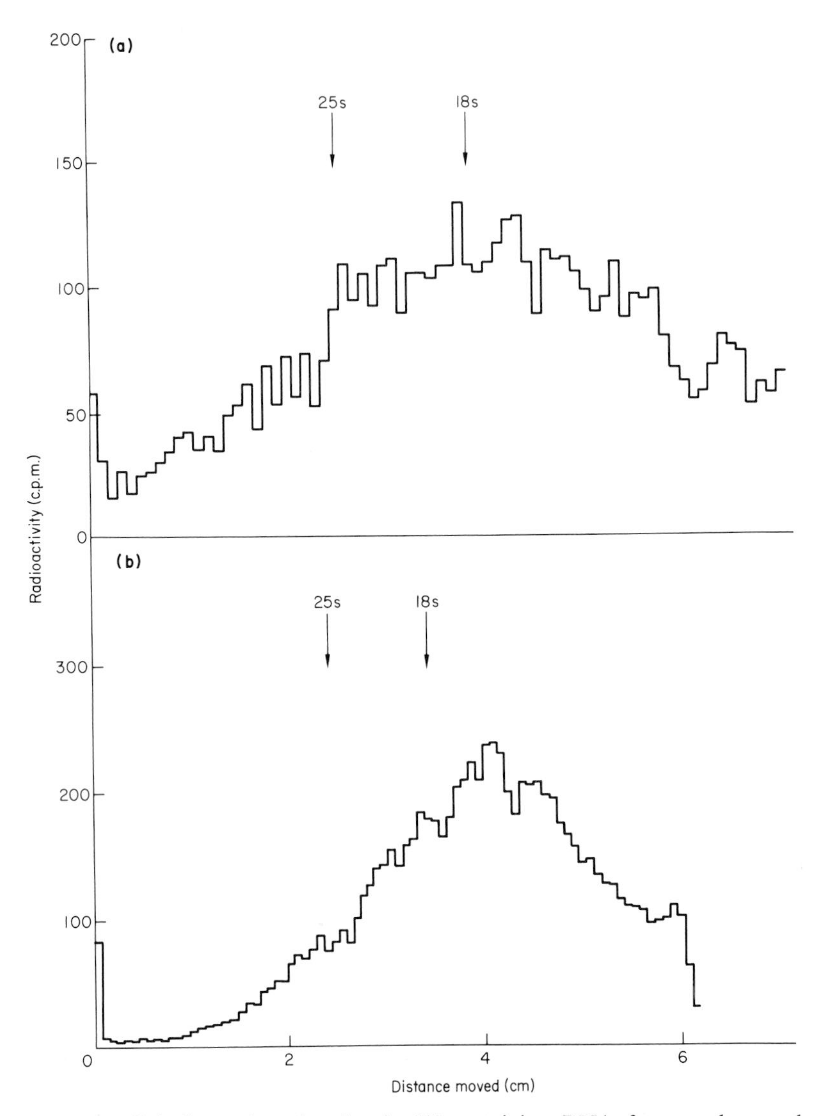

Fig. 9.19. Gel electrophoresis of poly (A)-containing RNA from nucleus and cytoplasm.

Pulse labelled sycamore cells were separated into (a) nuclear, and (b) cytoplasmic fractions and poly (A)-containing RNA purified by oligo (dT)-cellulose chromatograph. The two RNA preparations were denatured in 8 B urea at 60°C to disrupt aggregates and fractionated by gel electrophoresis in the presence of marker rRNA. The cytoplasmic RNA is distinctly smaller, on average, than the nuclear RNA. (Grierson & Covey, 1976.)

because this region appears to be transcribed from *repetitive* DNA. It therefore seems that some of the repeated DNA sequences interspersed between non-repetitive DNA are transcribed and may play a regulatory role in processing or transport of mRNA.

9.8 NUCLEIC ACID-PROTEIN INTERACTIONS

9.8.1 HISTONES

Chromosomal DNA is invariably associated with several different types of proteins (Table 9.7). These include histones which occur in most eukaryotes.

Table 9.7. Chemical composition of varied chromatins

	Content, relative to DNA, of			
Source of chromatin	DNA	Histone	Non-histone protein	RNA
Pea embryonic axis	1·00	1·03	0·29	0·26
Pea vegetative bud	1·00	1·30	0·10	0·11
Pea growing cotyledon	1·00	0·76	0·36	0·13
Rat liver	1·00	1·00	0·67	0·043
Rat ascites tumour	1·00	1·16	1·00	0·13
Human HeLa cells	1·00	1·02	0·71	0·09
Cow thymus	1·00	1·14	0·33	0·07
Sea urchin blastula	1·00	1·04	0·48	0·039
Sea urchin pluteus	1·00	0·86	1·04	0·078

From Bonner *et al.* (1968).

There are five main histone proteins designated F_1, F_{2b}, F_{2a1}, F_{2a2} and F_3, with molecular weights from 11,000 to 22,000 daltons. They are strongly basic and generally have a high content of arginine and lysine. The basic amino acids are often clustered towards either the N or C terminus of the proteins (Macgillivray & Rickwood, 1974) suggesting that these regions interact directly with the DNA. Stedman and Stedman (1950) proposed that histones are responsible in some way for the regulation of gene expression but it now seems that they lack the specificity for recognition of different gene sequences and it is more likely that they fulfil a more general role in the organization of chromosomes. Comparison of the amino acid sequences of histones from different species suggests that they have been highly conserved during evolution. For example, the sequences of F_{2a1} from calf, pig and rat are identical and differ only in two substitutions from F_{2a1} from pea (Delange *et al.*, 1966). This suggests these proteins evolved to fulfil an essential role with strict requirements for a precise sequence, although other histones, notably F_{2b} and F_{2a2} show greater variability. Histones frequently undergo modifications such as methylation, acetylation or phosphorylation. These are normally reversible and would be expected to alter the association of histones with DNA. In general these modifications occur at about the time of DNA synthesis although they may also herald an increase in RNA synthesis. During the cell cycle the synthesis of histones is closely linked to that of DNA and phosphorylation of F_1 has been shown to precede cell division (Bradbury *et al.*, 1974). Basic non-histone proteins and acidic proteins also occur in association with DNA. These form a more heterogeneous group than the histones and include components of different molecular weight, charge and tissue distribution. They are less stable metabolically than histones and do not behave as a single class of proteins. Several lines of evidence suggest that this group may include regulatory elements (Macgillivray & Rickwood, 1974).

9.8.2 SUBUNIT STRUCTURE OF CHROMATIN AND CHROMOSOME ORGANIZATION

The double helical structure of DNA is widely known and it is equally widely recognized that a further degree of organization is responsible for the folding of 1 m or more of DNA into each set of chromosomes. Early observations on the structure of chromosomes indicated that DNA is organized in fibres of the order of 10 nm thick and it was suggested that the double helix, in association with proteins, is supercoiled. In addition, X-ray studies suggested that there is a repeating structure in chromatin at intervals of approximately 10 nm and more recently a linear arrangement of beads of the order of 7 nm diameter has been observed in chromatin with the aid of the electron microscope (Olins & Olins, 1974). Further evidence for a regular arrangement of DNA in chromatin comes from the observation that following nuclease digestion reproducible fragments of DNA approximately 205 base pairs long are produced (Noll, 1975). Kornberg (1974) has suggested a repeating subunit structure of approximately 200 DNA base pairs associated with 2 molecules of each of the four major histones (F_{2b}, F_{2a1}, F_{2a2} and F_3) and one molecule of F_1. In this model the DNA double helix is envisaged to be folded to approximately one seventh its length and Crick and Klug (1975) have suggested that this could occur by 'kinking' of the DNA helix at intervals of perhaps 20 base pairs. Bradbury and his colleagues (Baldwin *et al.*, 1975) have proposed that repeating units of DNA approximately 200 base pairs long are coiled on the outside of aggregates of the four major histones which associate with each other by apolar interactions, leaving their basic regions free to associate with the DNA. Histone F_1, which is believed not to be directly involved in the subunit, may perform a cross-linking function. Only about 87% of the DNA in chromatin is protected against deoxyribonuclease digestion (Noll, 1975) and it is possible that not all the DNA participates in formation of this type of subunit.

Further insight into the functional organization in chromatin comes from studies on the giant chromosomes from certain insects. These chromosomes consist of a series of densely coiled bands connected by less dense regions. These are much larger than the repeating unit of 200 DNA base pairs. Each band plus interband region is considered to correspond to a gene because there is good numerical agreement between their number and the number of complementation groups determined by breeding experiments. However, calculations show that only a fraction of the DNA in each band plus interband region is necessary in order to code for a protein (Beerman, 1972) and the question of the function and organization of the DNA remains unresolved. Regions of the DNA being actively transcribed appear as chromosome puffs composed of DNA, RNA and protein. Crick (1971) has suggested that the interband regions contain DNA that is transcribed and that each band consists of regulatory DNA sequences. In contrast, Paul (1972) has proposed a different type of arrangement.

Different regions of chromosomes often contain chromatin condensed to different extents (Macgillivray & Rickwood, 1974). Euchromatin appears to be loosely packed. It is known to contain both unique and repeated sequences and is actively transcribed. Heterochromatin is more tightly packed and consists of two types. Constitutive heterochromatin contains sequences that are never transcribed and includes repeated sequences. In some species satellite DNAs are known to be clustered in heterochromatic regions near the centromeres. Finally, facultative heterochromatin constitutes those regions of the chromosome that sometimes appear as euchromatin.

9.8.3 REGULATION OF TRANSCRIPTION

Thus there is evidence for the coiling of DNA and formation of regular DNA-protein complexes which may be further folded into larger units. Different regions of the chromosome are folded to different extents and this presumably affects the availability of DNA for transcription. Comparison of the extent of transcription of pure DNA and isolated chromatin by added bacterial RNA polymerase suggests that in animals only about 10% of the genome is available for RNA synthesis. Hybridization experiments suggest that the RNA synthesized from isolated animal chromatin is tissue specific and resembles in sequence the type of RNA synthesized *in vivo* (Macgillivray & Rickwood, 1974). It is probable that the non-histone proteins are more important in determining which segments of the genome are repressed but further characterization of the properties and mechanism of action of these proteins is necessary before their function becomes clear (see Mcgillivray & Rickwood, 1974). It has also been suggested that chromosomal RNA may be important in regulating gene expression, possibly in conjunction with repeated sequences in DNA (Georgiev, 1969; Ryskov *et al.*, 1971; Britten & Davidson, 1969). There is undoubtedly RNA associated with chromatin, including growing chains of polydisperse-

RNA and rRNA-precursor, but the existence of an additional type of regulator RNA has been proposed. However, at the present time there is a certain amount of controversy concerning the characterization and properties of chromosomal RNA (Jacobson & Bonner, 1971; Huang, 1967; Heyden & Zachau, 1971; Holmes *et al.*, 1972). As yet there is no evidence that this RNA regulates transcription.

In addition to repression of the genome by interaction with proteins, template selection may also be controlled by RNA polymerase enzymes and by other protein factors that interact with the polymerases to regulate transcription. As far as studies on plant chromatin are concerned *in vitro* RNA synthesis has been achieved using the endogenous RNA polymerases that remain attached to the isolated chromatin (Guilfoyle & Hanson, 1974). However, there have been few attempts to characterize the RNA products in detail. There are several distinct RNA polymerases with different properties. These enzymes are probably responsible for transcribing different regions of the genome. For example, RNA polymerase I, which is insensitive to the antibiotic α-amanitin, is located within the nucleolus (Brandle & Zetsche, 1973). This enzyme appears to be responsible for transcription of the rRNA genes. The activity of this enzyme and the rate of RNA production are significantly enhanced in tissue treated with auxin (Guilfoyle *et al.*, 1975). Enzymes sensitive to α-amanitin are present in the nucleoplasm and presumably are responsible for transcribing other parts of the genome. These polymerases may also be affected by hormones *via* the mediation of hormone-binding factors (see chapter 13). In addition, evidence has been obtained for the existence of other protein factors which govern the activity of RNA polymerases by stimulating initiation of RNA synthesis (Mondal *et al.*, 1972a, 1972b).

9.9 CONCLUSIONS

Now that we have a general idea of the organization and properties of DNA it is possible to pose more precise questions about the function of the different parts of the genome. For example it would be interesting to establish what proportion of the DNA is transcribed into RNA in plants with different nuclear DNA contents. There is also the question of the function of the repeated DNA sequences: are they involved in regulation of transcription; to what extent are they transcribed; and what is the function of the repetitive sequences in the nuclear precursors to mRNA? Further characterization of the various RNA polymerase enzymes and the protein factors which influence RNA synthesis will doubtless increase our knowledge of the control of gene expression. However, it does seem that ultimately such studies should be carried out using chromatin as template in order to understand the precise molecular interactions involved. Finally, the possibility of post-transcriptional regulation should not be overlooked. It is important to establish the role of post-transcriptional modification of RNA molecules in relation to mRNA selection and transport to the cytoplasm.

FURTHER READING

BISWAS B.B., GANGULY A. & DAS A. (1975) Eukaryotic RNA polymerases and the factors that control them. In *Progress in Nucleic Acid Research and Molecular Biology* (ed. W.E. Cohen), Vol. 15, pp. 145–84. Academic Press, New York.

BRYANT J.A. (1976) Nuclear DNA. In *Molecular Aspects of Gene Expression in Plants* (ed. J.A. Bryant), pp. 1–51. Academic Press, London.

DUDA C.T. (1976) Plant RNA polymerases. *Ann. Rev. Pl. Physiol.* **27**, 119–32.

GRIERSON D (1976) RNA structure and metabolism. In *Molecular Aspects of Gene Expression in Plants* (ed. J.A. Bryant). pp. 53–108. Academic Press, London.

KORNBERG A. (1974) *DNA Synthesis*. W.H. Freeman & Co., San Francisco.

LEWIN B. (1974) *Gene Expression, Volume 2, Eukaryotic Chromosomes*. J. Wiley & Sons Ltd., London.

NAGL W. (1976) Nuclear organisation. *Ann. Rev. Pl. Physiol.* **27**, 39–69.

PARISH J.H. (1972) *Principles and Practice of Experiments with Nucleic Acids*. Longman, London.

RAE P.M.M. (1972) The distribution of repetitive DNA sequences in chromosomes. In *Advances in Cell Biology and Molecular Biology* (ed. E.J. Dupraw). Vol. 2, pp. 109–49. Academic Press, New York.

STEWART P.R. & LETHAM D.S. (eds.) (1973) *The Ribonucleic Acids*. Springer-Verlag, Berlin.

CHAPTER 10

PROTEIN SYNTHESIS IN THE CYTOPLASM

10.1 INTRODUCTION

Since gene expression occurs almost exclusively *via* the synthesis of proteins, an understanding of the mechanism of this process is important. Although living organisms make several thousand different proteins with a variety of functions, present evidence suggests that they are all made, by the same basic mechanism, on polysomes. These structures consist of a template RNA molecule, messenger-RNA (mRNA), with several associated ribosomes, and protein synthesis is a complex, energy requiring, multi-enzymic process. A template mechanism of synthesis is dictated by the fact that the protein product has a specificity which depends upon the precise sequence of the constituent amino acid residues in the polypeptide chain. This automatically rules out the normal enzymic method by which most other cell constituents are made, since each enzyme of a set specific for the synthesis of one protein would, in turn, require a further set for its synthesis, *ad infinitum*. However, since some small polypeptides are synthesized by a non-template mechanism, it is possible that some proteins with highly repetitive sequences of amino acid residues, may be synthesized similarly.

Because all organisms probably synthesize proteins by a similar mechanism, the tendency has been to use an organism which is particularly amenable for basic studies. This organism is the prokaryote, *Escherichia coli*, and higher plant material has not been much used because of associated technical difficulties. However, recognition of the importance of the deposition of proteins in plants as our major source of protein food, is leading to an increased spate of work in the desire to understand, and possibly manipulate, this important process in the major food plants.

Proteins are synthesized on free and membrane-bound polysomes in the cytosol or in various sub-cellular organelles, e.g. mitochondria, chloroplasts. Only the biochemistry of protein synthesis by cytoplasmic ribosomes will be reviewed in this chapter; organelle protein synthesis is dealt with in chapter 11. Different aspects of the control of protein synthesis are covered in chapters 13 and 15, but it should be re-iterated here that the level of protein in the cell depends on a dynamic balance between synthesis and degradation.

10.2 POLYSOMES

10.2.1 STRUCTURE OF RIBOSOMES

Ribosomes are classified as eukaryotic or prokaryotic in type, on the basis of: (a) sensitivity to various antibiotics; (b) functional interchangeability of soluble factors and ribosomes from different sources; and (c) structure and sedimentation characteristics, e.g. 70s (prokaryotic) or 80s (eukaryotic).

Bacterial ribosomes, plant chloroplast and mitochondrial ribosomes are classified as prokaryotic, even though they may not always exhibit all of the above characteristics, e.g. plant mitochondrial ribosomes have sedimentation values of about 80s.

Plant and animal cytoplasmic ribosomes are of the eukaryotic type and have sedimentation values of about 80s. This class of ribosomes are heterogeneous in size however, with most of the differences accounted for by differences in the large subunit, the small subunit having been conserved during evolution (Cammarano *et al.*, 1972a,b & c).

Measurements of the size of plant cytoplasmic ribosomes in the electron microscope vary, but they are approximately 25×20 nm. Miller *et al.* (1966) have described them as being acorn-shaped, and several workers have described a cleft in the small subunit. In the rat liver ribosome model of Nonomura *et al.* (1971) there is a tunnel between the two subunits, directly under the cleft in the small subunit, which is thought to accommodate the mRNA. Ribosomes require Mg^{2+} for structural integrity and dissociate at low Mg^{2+} concentrations into a large and a small subunit with sedimentation values of about 60s and 40s respectively (Ajtkhozhin *et al.*, 1972). There is good evidence that mRNA attaches to the small subunit before the addition of the large subunit completes the ribosome structure, and the different roles of the two subunits are a feature of all the models of ribosome structure and function (see Noll *et al.*, 1973). The 80s ribosome of plants corresponds to a molecular weight of $3{\cdot}9 \times 10^6$ daltons, with molecular weights for the large subunit of $2{\cdot}4 \times 10^6$ daltons and the small subunits, $1{\cdot}5 \times 10^6$ daltons; the small subunit is of the same molecular weight as its mammalian counterpart, the large subunit is smaller. The large subunit of plant ribosomes contains one molecule of 25s RNA, hydrogen-bonded to 1 molecule of 5·8s RNA (Payne & Dyer, 1972), and 1 molecule of 5s RNA; the small subunit contains 1 molecule of 18s RNA. The molecular weight of the 25s RNA is $1{\cdot}3 \times 10^6$ daltons, and that of the 18s is $0{\cdot}7 \times 10^6$ daltons (Loening, 1968). The large subunit contains 46% protein, and the small 54%.

The large subunit of pea seedling ribosomes contains 44–45 proteins and the small subunit 32–40 proteins, of which most are basic and of molecular weights between 20×10^3 and 30×10^3 daltons. Some proteins may be represented by more than one copy per ribosome. Proteins extracted from cytoplasmic and chloroplast ribosomes of the same species, show little similarity in two-dimensional electrophoresis or by immunological comparison and are significantly less similar than are the cytoplasmic ribosomes of different species, e.g. beans and wheat (Gualerzi *et al.*, 1974).

Before a complete understanding of the mechanism of protein synthesis is elucidated, it will be necessary to know the spatial relationships and three-dimensional structures of the proteins and the RNAs of the ribosome, as well as those of the associated molecules which also play a part in protein synthesis. Information on the structure of the proteins and the RNA molecules of the *E. coli* ribosome, together with reconstitution experiments, is now well advanced (Traub & Nomura, 1968; Nashimoto *et al.*, 1971; Wittmann, 1973; Anderson *et al.*, 1974; Nierhaus & Dohme, 1974).

10.2.2 FREE AND MEMBRANE-BOUND POLYSOMES

Ribosomes associate with mRNA to form polysomes, as can be seen in the electron micrograph (Fig. 10.1); the size of the polysome varies according to the length of the mRNA and the number of attached ribosomes. Polysomes are found either free in the cytoplasm or attached to the surface of membranes of the endoplasmic reticulum (ER) and the nucleus. In animal embryonic cells, most of the polysomes are free and engaged in protein synthesis for internal use, whereas in those differentiated animal cells from which large amounts of protein are exported, most of the polysomes are found attached to the ER. Thus, the generalization arose that membrane-bound ribosomes synthesize protein for export and free ribosomes for intracellular use. It soon became clear however, that membrane-bound ribosomes occur in some tissues which do not export protein, and that in cells where all the protein synthesized is for internal use, different classes of protein are synthesized on the two types of ribosomes. Less information is available for plants. In the developing broadbean seed the highly vacuolate, relatively membrane-free cells of the cotyledon are transformed just prior to the onset of storage protein synthesis, into cells whose cytoplasm is vesicular and in which ER with attached ribosomes is very prominent. This new protein synthesis machinery is assembled at a precise time in the course of seed development for the production, in large amounts, of the storage proteins of the seed (Bailey *et al.*, 1970; see Fig. 10.2). Later, during dehydration of the seed, ribosomes become detached and the population of free ribosomes thereby increased. A similar series of events has been recorded for the developing seeds in many other plants.

The way in which ribosomes attach to membranes is not clear; the ribosomes themselves, the membranes and the protein being synthesized, have all been suggested as being involved in the binding. There is evidence from animals that the large subunit is in close contact with the membranes, and Baglioni *et al.* (1971) have postulated that the large subunit binds directly to the membrane, probably at a specific binding site (Sunshine *et al.*, 1971), and that a protein on

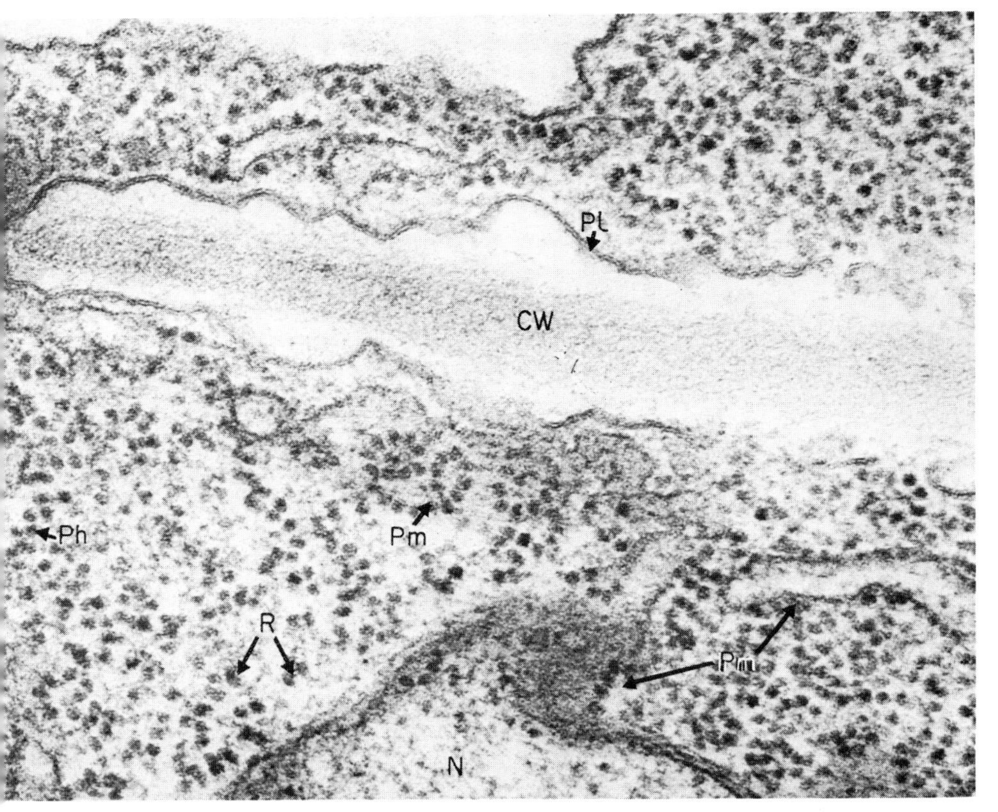

Fig. 10.1. Electron micrograph of a thin section of parts of two adjacent cells in a shoot apex of pea showing ribosomes and polyribosomes; CW = Cell Wall, Pl = Plasmalemma, N = Nucleus, Pm = Membrane bound polysomes, R = Ribosome Ph = Polysome helix.

By courtesy of A. D. Greenwood, Department of Botany and Plant Technology, Imperial College of Science and Technology, London.

the membranes is responsible for the attachment (James *et al.*, 1969). Alternatively, it has been suggested that one of the proteins of the large subunit is responsible, whereas other observations suggest that the binding may be dependent on the nascent polypeptide chain. It now seems likely that, *in vivo*, membrane-bound ribosomes synthesize a different class of protein to free ribosomes, whereas *in vitro*, proteins of both classes are synthesized by both types of ribosomes, suggesting the control is not a function of the ribosome itself.

10.2.3 ISOLATION AND PURIFICATION

Ribosomes and polysomes are normally isolated from plants and purified by sedimentation through sucrose cushions as originally described by Wettstein

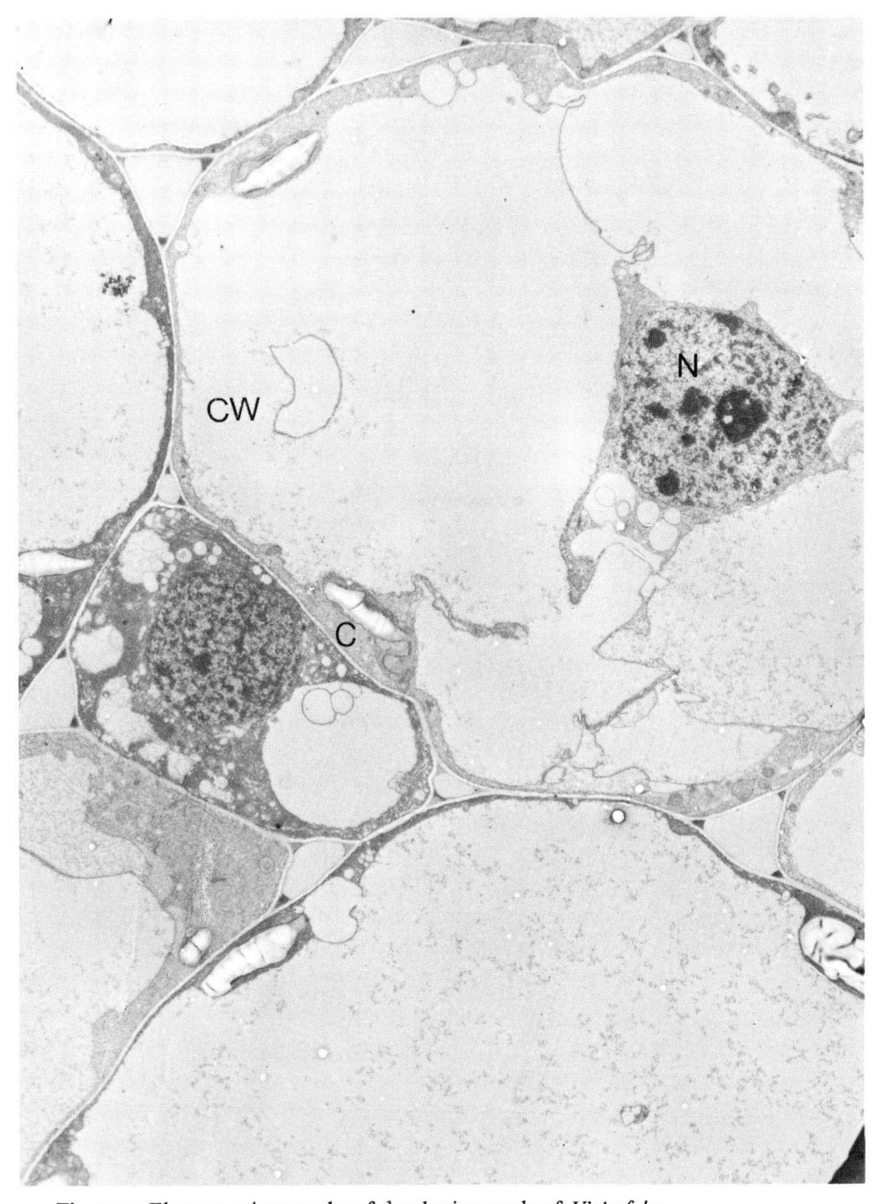

Fig. 10.2. Electron micrographs of developing seeds of *Vicia faba*.
(a) 25 days after fertilization. Ribosomes free in the cytoplasm. Very little ER present.

et al. (1963). In a typical procedure, the material is homogenized in 0·25 M sucrose containing 200 mM-Tris-HCl, pH 8·5 at 2°C, 500 mM-KCl and 15 mM-$MgCl_2$, with a Willems Polytron for 3 seconds at a speed setting of 8. The homo-

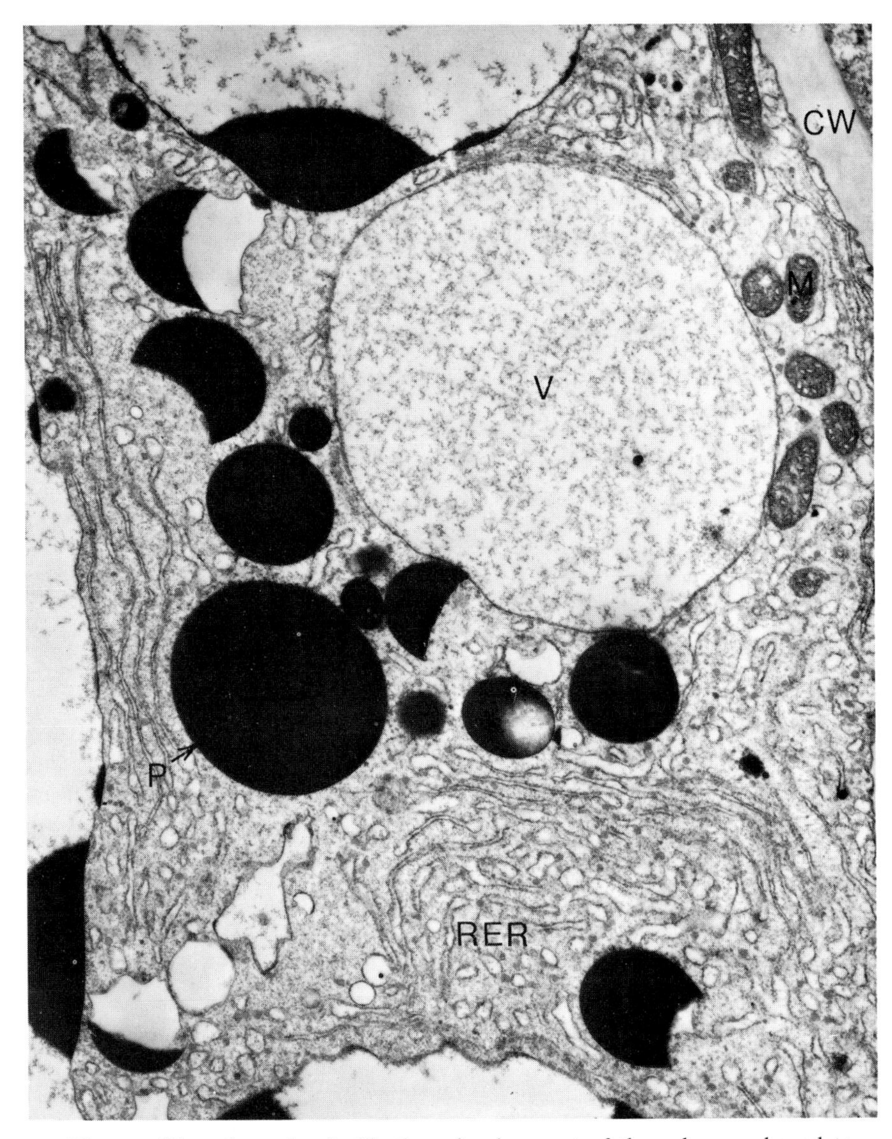

Fig.10.2. (b) 55 days after fertilization, showing most of the polysomes bound to ER and storage protein being laid down.

C=Cytoplasm, P=Protein body, CW=Cell Wall, RER=Endoplasm reticulum with polysomes attached, N=Nucleus, V=Vacuole.

From Boulter *et al.* (*Qual. Plant.* XXIII, 239–50, 1973).

genate is filtered through Miracloth and Triton X-100 is added to a final concentration of 2% (v/v). The filtrate is centrifuged at $10^4 \times g$ for 10 min at 2°C and the ribosomes and polysomes are recovered from the supernatant by layering over a 3 ml cushion of 0·1 M sucrose in 50 mM-Tris-HCl, pH 8·5, 50 mM-KCl and 10 mM-$MgCl_2$, followed by centrifugation for at least $4{\cdot}4 \times 10^7$ *g*-min, as described by Leaver & Dyer (1974) (Fig. 10.3). The inclusion of the detergent

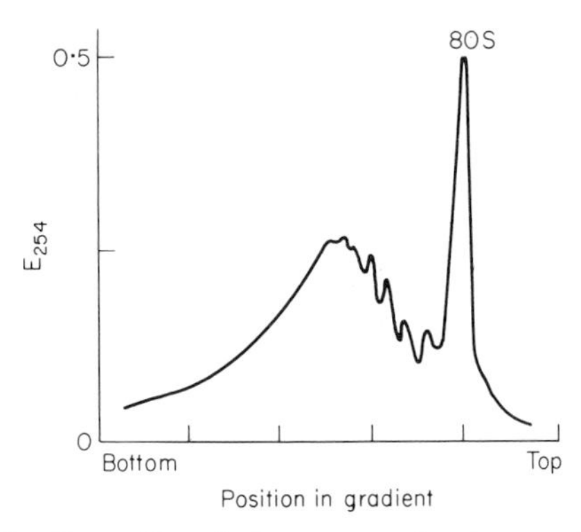

Fig. 10.3. Sucrose density profiles of ribosomes and polysomes from *Pisum sativum*. Ribosomes recovered after 6 h. centrifugation through 1 M-sucrose cushion. E_{254}= absorbance at 254 nm (From Leaver & Dyer *Biochem. J.* **144**, 165–7, 1974.)

Triton X-100 is necessary since plant materials contain a proportion of polysomes which are membrane-bound and the addition of Triton X-100 solubilizes the endoplasmic reticulum, so releasing them. The method gives a preparation of both polysomes and ribosomes and the latter can be removed by centrifugation techniques or, alternatively, the whole preparation can be converted to ribosomes by exposure of the plants to nitrogen gas for at least 1 hour prior to extraction. Since different plants and even different tissues from the same plant contain different amounts of membrane-bound to free polysomes and different free polysomes to free monoribosome ratios, isolation conditions may need to be varied for optimum results with different experimental materials. The situation is further complicated by the fact that plant mitochondrial ribosomes, although not functionally of the 80s type, sediment at 80s (Leaver & Harmy, 1973), and can contaminate cytoplasmic preparations to varying extents, depending on the experimental material. Damage to ribosomes, both structural and by the removal of associated proteins, can occur during isolation and preparation. The extent of this damage can be assessed, to some extent, by extracting the RNA and fractionating it by polyacrylamide gel electrophoresis (see Leaver & Key, 1970).

10.3 THE BIOCHEMICAL MECHANISM OF PROTEIN SYNTHESIS

Protein synthesis, i.e. the assembly of polypeptide chains from their amino acid constituents, occurs on the ribosomes. The information specifying the amino acid sequence of the polypeptide chain is contained in mRNA molecules, which in the case of eukaryotes are complexed with specific proteins *in vivo*. Messenger-like RNA-protein particles called 'informosomes' have been isolated from the cytoplasm and from polysomes of germinating wheat embryos (Ajtkhozhin *et al.*, 1973; Ajtkhozhin & Akhanov, 1974). So far, however, the RNA of these particles has not been definitely established as mRNA. When an mRNA molecule associates with ribosomes, the information it contains is translated; the number of ribosomes involved depends largely upon the size of the polypeptide chain being synthesized, since generally, ribosomes are evenly spaced along mRNA molecules.

The translation process can be arbitrarily split into initiation, elongation, and termination and release. However, prior to the initiation of the process on the ribosome, the twenty different protein amino acids must be activated and attached to specific tRNA molecules. Also, after release of the polypeptide chain from the ribosome, a series of post-translational changes may occur.

10.3.1 AMINO ACID ACTIVATION AND AMINOACYL-tRNA SYNTHESIS

This energy-requiring process is accomplished in the cytosol by enzymes called aminoacyl-tRNA synthetases or ligases. It occurs as a two-stage reaction, both stages being catalysed by the same aminoacyl-tRNA synthetase enzyme (Fig. 10.4). The product of the reaction, aminoacyl-tRNA molecules, contain in the

(a) Amino acid + ATP $\xrightleftharpoons{\text{aminoacyl-tRNA synthetase}}$ Enzyme-aminoacyl-AMP + Pyrophosphate

(b) Enzyme-aminoacyl-AMP + tRNA $\rightleftharpoons$ Enzyme + aminoacyl-tRNA + AMP

Fig. 10.4. The reactions of aminoacyl-tRNA synthesis.
(In the cell a pyrophosphatase acts on the pyrophosphate formed to make reaction (a) irreversible in practice).

aminoacyl-tRNA bond, sufficient energy for the subsequent formation of peptide bonds to occur spontaneously and, in addition, the anticodon triplet of bases of the tRNA ensures that the amino acid will be located at a particular residue position in the amino acid sequence of the protein product. This reaction, therefore, controls the proportion of free to aminoacylated-tRNAs in the cytoplasm, and theoretically therefore, could affect the rate and/or type of protein being synthesized.

10.3.1.1 *tRNA*

The code words (codons) of the genetic code were first established for *E. coil.* This work showed that there were 61 codons shared between twenty protein amino acids, i.e. there is more than one codon for some amino acids and these can attach to more than one tRNA species. Chemically different tRNA species which can be acylated by the same amino acid are called isoacceptor tRNAs.

However, it was soon realized that not every codon has a corresponding tRNA with a specific anticodon and that some tRNAs recognize more than one codon. Crick's 'wobble hypothesis' (1966) to account for this, proposed alternative base-pairing to occur between the base in the third position of the mRNA codon and the corresponding base in the tRNA anticodon. The proposed rules are set out in Fig. 10.5. Thus, inosine in the 5′ position in the anticodon of an

5′-Anticodon base	mRNA codon
I	XXU, XXC, XXA
G	XXU, XXC
U	XXA, XXG
A	XXU
C	XXG

Fig. 10.5. The basis of the 'wobble' hypothesis. (X=Any nucleotide; I=Inosine; G=Guanine; U=Uridine; A=Adenine; C=Cytosine;) (From Boulter *et al.* (*Biol. Rev.* **47**, 113–75, 1972.))

aminoacyl-tRNA molecule allows this base to pair with either U, C or A in the third 3′ position of a messenger codon. Similarly, with G or U in the 5′ position in the anticodon, two codons may be recognized by a single aminoacyl-tRNA, whereas with C or A only a single codon is recognized. Since alternative base-pairing is only possible with the base in the third position of the codon, codons for the same amino acid which differ in either of the first two base positions, base-pair with different tRNAs. The only example of 'wobble' in the first base of a codon occurs with $tRNA_i^{Met}$, which recognizes both the initiator AUG or GUG codons (see later). Even so, separation of tRNAs by counter current-distribution, MAK columns, BD-cellulose columns and reversed phase chromatography has shown that there are more isoacceptor tRNAs than can be accounted for by the degeneracy of the genetic code. Similarly, isoacceptor tRNAs for amino acids exist in plants; for example, soyabean seedlings have at least six different $tRNA^{Leu}$ species (Anderson & Cherry, 1969). Although few experiments have been carried out, it has been found in every case investigated that plants use the same codon in amino acid assignments as those of *E. coli*. However, plants may not use all of the possible codons for a particular amino acid, since some isoacceptor tRNA species may be absent, or because of the 'wobble' effect; e.g. Caskey *et al.* (1968) have shown that the isoleucyl tRNA from guinea pig liver recognizes AUU, AUC and AUA codons, whereas the corresponding isoleucyl-tRNA from *E. coli* only recognizes two of these, AUU and AUC. This response to different codons in the two organisms is explained if the liver tRNA had 5′ inosine in the anticodon, while *E. coli* tRNA had guanine.

The whole question of the number and the activity of tRNAs in different tissues of plants is complicated by the fact that, (a) it is not certain whether the methods for separating tRNAs are completely effective; (b) it is known that during isolation and purification, partial modification of tRNA molecules can occur, which may affect their charging ability with amino acids or their ability to transfer amino acids to proteins, or both (see Sueoka & Kano-Sueoka, 1970); (c) the relative importance of the three protein synthesizing systems, cytoplasmic, chloroplast and mitochondrial, each with its associated tRNAs and synthetases may differ in different tissues. Nevertheless, different complements of isoacceptor tRNAs may occur in different tissues and changes in pattern have been demonstrated during development (Bick *et al.*, 1970; Littauer & Inouye, 1973), and it has been suggested that isoacceptor tRNAs are involved in the control of protein synthesis. Garel *et al.* (1973) have shown that the complement of tRNAs synthesized in the silk gland is related to the mRNAs being translated and they have suggested defining this phenomenon as 'modulated tRNA biosynthesis'.

10.3.1.2 *Aminoacyl-tRNA synthetases*

Aminoacyl-tRNA synthetases have been isolated and purified from a variety of plants (see Boulter 1970; Boulter *et al.*, 1972; Zalik & Jones, 1973), and although a considerable amount of work has been done on their properties, several problems remain. For example, the question of the detailed substrate specificity of different aminoacyl-tRNA synthetases is still under investigation. Since activation is a two-step process, both amino acid and tRNA specificity have to be considered. The synthetase is normally absolutely specific towards the tRNA; few mismatches have been reported in homologous systems (Arca *et al.*, 1967, 1968). However, a synthetase specific for one amino acid can, in some instances, activate to a lesser extent another amino acid. Even so, synthetases are remarkably amino acid specific. A second amino acid attached to a tRNA molecule normally specific for another amino acid, will be located in the polypeptide chain according to the tRNA anticodon, i.e. where the first amino acid would have been placed. With regard to amino acids not normally found in the cell, specificity is not always so pronounced, e.g. azetidine-2-carboxylic acid is activated and transferred to $tRNA^{Pro}$ by the proline enzyme of mung bean (Peterson & Fowden, 1965), whereas *Polygonatum multiflorum*, which contains a high level of this amino acid in nature, contains a prolyl-tRNA synthetase which will not activate it. In *Mimosa* and *Leucaena*, where mimosine, an analogue of phenylalanine, occurs naturally and where it is non-toxic, the phenylalanyl-tRNA synthetase discriminates against mimosine. In several other species where mimosine is toxic, it is activated by the phenylalanine-tRNA synthetase but the enzyme does not attach it to $tRNA^{Phe}$ (Smith & Fowden, 1968).

It is not clear if there is one or more than one synthetase whenever there are a number of isoacceptor tRNAs. In soybean seedlings where there are at least three different leucyl aminoacyl-tRNA synthetases (Kanabus & Cherry, 1971) which have different specificities towards the six leucine isoacceptor tRNAs, it would appear that there is more than one cytoplasmic leucyl-tRNA synthetase. However, in some instances, a single aminoacyl-tRNA synthetase can recognize different isoacceptor tRNAs. When both cytoplasmic and chloroplast tRNA species are present, there are probably at least two sets of aminoacyl-tRNA synthetases. In light-grown *Euglena* for example (Reger *et al.*, 1970; Krauspe & Parthier, 1973), there are two aminoacyl-tRNA synthetases for each of phenylalanine and leucine, whereas dark-grown *Euglena*, which is without chloroplasts, contains only one; there are various reports of chloroplast aminoacyl-tRNA synthetases which only acylate chloroplast tRNAs and not their cytoplasmic counterparts and *vice versa*. However, in heterologous systems such as chloroplast tRNA plus cytoplasmic activating enzymes, specificity may range from none to complete (Burkard *et al.*, 1970; Boulter *et al.*, 1972), although the biological significance of these findings is not clear.

There is evidence that, as well as the tRNAs, the complement of aminoacyl-tRNA synthetases may change during the life cycle of plants (Bick & Strehler, 1971).

10.3.2 TRANSLATION OF mRNA

10.3.2.1 *Initiation of the polypeptide chain*

Initiation of protein synthesis on plant cytoplasmic ribosomes starts with the assembly of a mRNA•40s subunit complex which requires ATP for its formation and this step is followed by the addition of methionyl-$tRNA_i$ to form a mRNA•40s subunit•methionyl-$tRNA_i$ complex. The process requires GTP and at least two initiation factors (Weeks *et al.*, 1972; Marcus *et al.*, 1973). Initiation is completed by the addition of the 60s subunit and the 80s ribosome so formed is then able to accept the next aminoacyl-tRNA, thus starting the elongation process. The $tRNA_i^{Met}$ is a special tRNA charged with unformylated methionine, which can enter the so-called 'P' site of the ribosome complex and base-pair with the initiator codon, AUG or GUG in the mRNA. Although mRNAs are translated from the $5' \rightarrow 3'$ end, evidence from sequence studies of bacteriophage messages shows that the initiator triplet is not at the 5′ terminus, but is set a considerable number of nucleotides in. For example, the first initiator codon of bacteriophage R17 RNA is preceded by 91 nucleotides which are not translated (Adams & Cory, 1970). In the case of polycistronic messengers, one or more initiator triplets will occur intramolecularly.

Evidence that plants have a special type of initiator tRNA, $tRNA_i^{Met}$, comes from the wheatgerm system (Leis & Keller, 1970; Marcus *et al.*, 1970a; Tarrago *et al.*, 1970; Ghosh *et al.*, 1971), and from *Vicia faba* (Yarwood *et al.*, 1971a), where it has been shown that two major and one minor $tRNA^{Met}$ species are present. The minor and one of the major $tRNA^{Met}$ species function as chain initiators as shown by AUG-dependent binding (Tarrago *et al.*, 1970; Yarwood *et al.*, 1971a), and by N-terminal analyses of either the product of tobacco mosaic viral RNA-directed (Marcus *et al.*, 1970a) or poly-AUG, poly-GU and endogenous messenger-directed incorporation (Yarwood *et al.*, 1971a). Neither of the major $tRNA^{Met}$ species from either wheat or beans is formylated, in contrast to the initiator $tRNA_i^{Met}$ of prokaryotes, and it is presumed that these are the cytoplasmic $tRNA^{Met}$s for initiator and internal methionine residues. The minor, formylatable $tRNA^{Met}$, is presumed to be the initiator of chloroplast protein synthesis (see chapter 11).

The hydrolysis of ATP precedes the formation of the first peptide bond and may be required for recycling of the methionyl-$tRNA_i$ binding factor. The requirement for ATP has also been demonstrated in the rabbit globin synthesizing system (Schreier & Staehelin, 1973), but not in those of prokaryotes. The order in which the initiator tRNA and mRNA bind to the small subunit in mammalian and prokaryote systems is in dispute; the order suggested above for wheatgerm by Marcus and his coworkers, therefore, whilst agreeing with some workers on those systems disagrees in particular with the suggestion of Legon *et al.* (1973) and Schreier and Staehelin (1973) that the initiator binds prior to the mRNA in rabbit globin synthesis.

Knowledge of the two plant initiator factors involved in binding methionyl-$tRNA_i$ and mRNA is much less complete than for the similar factors in bacterial and mammalian systems; details of the latter are, therefore, relevant. In bacteria there are three factors, IF-1, IF-2 and IF-3 involved in the initiation process. Factor IF-1 has a molecular weight of 9,400 daltons and is the most basic of the three proteins. Factor IF-2 has been shown to consist of two sub-components, both active, with molecular weights of $91 \times 10^3 - 100 \times 10^3$ daltons and 80×10^3 daltons although the latter is probably derived from the former by proteolysis during purification. Factor IF-3 is a protein or group of proteins with molecular weight(s) of $21{\cdot}5 \times 10^3 - 23{\cdot}5 \times 10^3$ daltons (Grunberg-Manago *et al.*, 1973). The three factors attach to the small subunit; IF-2 binds GTP and formyl-methionyl-RNA, while IF-3 binds mRNA. Factor IF-1 increases the affinity of factors IF-2 and IF-3 for the 30s subunit, and is also necessary for the release of IF-2 from the ribosome.

The order of the attachment of the different factors to the small subunit and the order of the binding of mRNA and acylated initiator tRNA, is still a matter of debate. Some results suggest that IF-1 is attached before IF-2 and IF-3, whilst others indicate that it can only be attached after. Similarly, it is possible that an unstable intermediary complex consisting of formyl-methionyl-$tRNA_i$•30s subunit•initiation factors is formed, i.e. that mRNA is bound after initiator-tRNA, or that a 30s subunit•mRNA•IF-3 complex is formed, which is then stabilized by the addition of formyl-methionyl-$tRNA_i$ and IF-1 and IF-2. The next step in the process, the addition of the large subunit, does not require GTP, but

GTP hydrolysis is required to release IF-2 from the 70s ribosome complex. The role of GTP hydrolysis in initiation is still not fully understood, but it is not only required for the release of IF-2. Factor IF-3 is not only required for the binding of natural mRNAs, but also functions later either to dissociate the subunits after chain termination, or to keep dissociated subunits apart until another cycle of initiation is instituted. This activity of the IF-3 factor is referred to as its DF (ribosome dissociation factor) activity. None of the factors occurs on the polysomes and they recycle in protein synthesis, as shown in Fig. 10.6.

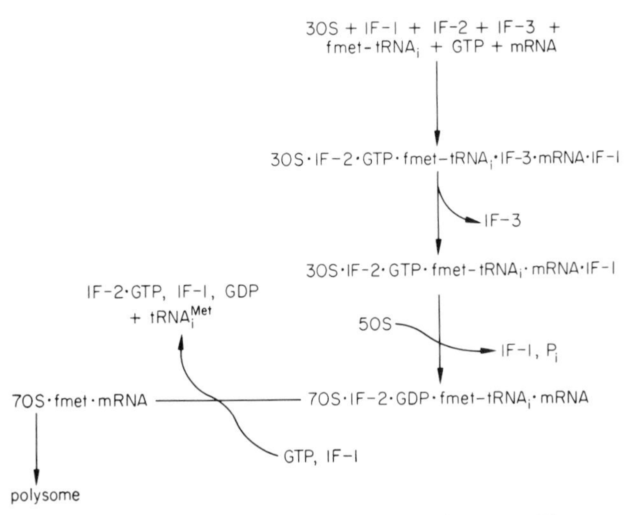

Fig. 10.6. Recycling scheme for initiation factors in prokaryotes. The exact order in which the factors interact with the 30s subunit is still not clear, nor is whether fmet-tRNA$_i$ attaches before or after mRNA. IF-2·GTP·fmet-tRNA$_i$ may exist independent of the 30s subunit. (fmet-tRNA$_i$=acylated formylmethionyl-tRNA$_i$; tRNA$_i^{Met}$=deacylated formylmethionyl-tRNA$_i$). (Modified from Haselkorn & Rothman-Denes (1973).)

Several proteins which form an integral part of the small subunit have also been shown to be involved in initiation. Small subunit proteins, S1, S11, S13, S19, participate in IF-2 binding and S1, S11, S12, S13, S14, in IF-3 binding; S12 is responsible for recognition of mRNA (Haselkorn & Rothman-Denes, 1973; Anderson *et al.*, 1974). Shine and Dalgarno (1974) have proposed a model of how the large subunit·initiation factor complex may recognize and attach mRNA. It would appear that the sequence (5′) GGAGGU (3′) is present in the same relative position with respect to the first translatable AUG triplet in all prokaryotic messengers so far analyzed. Furthermore, the 3′ proximal end of 16s RNA has the sequence GAUC<u>ACCUCCU</u>UA (OH), so that the underlined nucleotides could potentially base-pair with the GGAGGU sequence in the mRNA.

Further complexity of the initiation process is indicated by the work of

Schreier and Staehelin (1973), who have characterized five initiation factors ($IF\text{-}E_1$ to $IF\text{-}E_5$) for mammalian protein synthesis. Their results with rabbit globin mRNA are consistent with the formation initially of a methionyl-$tRNA_i$•$IF\text{-}E_2$•GTP complex, which is bound independently of mRNA to the 40s ribosome subunit by $IF\text{-}E_3$. Subsequent to this, mRNA and the 60s subunit are joined by the cooperative action of $IF\text{-}E_4$, ATP, $IF\text{-}E_1$ and $IF\text{-}E_5$. The binding of natural mRNA requires $IF\text{-}E_4$ and ATP. $IF\text{-}E_5$ promotes the joining of the 40s complex with the 60s subunit and $IF\text{-}E_1$ inhibits complex formation in the absence of mRNA binding. However, it has been suggested that the apparent need for additional factors results from deproteinization of ribosomal subunits and mRNAs, i.e. the requirement may be for structural proteins rather than initiation factors. The probable relationship between the different factors is given in Table 10.1. The little information available suggests that eukaryotic factors are not exchangeable with their prokaryotic counterparts.

Table 10.1. Probable relationship between factors.

Eukaryotes			Prokaryotes
Schreier & Staehelin (1973)	Anderson	Anderson *et al.* (1974)	Haselkorn & Rothman-Denes (1973)
$IF\text{-}E_2$	M_1	EIF-1	IF-2
$IF\text{-}E_3$	M_3	EIF-3	IF-3
$IF\text{-}E_4$	M_{2B}		
$IF\text{-}E_5$	M_{2A}		

Columns 1–3 are the various nomenclature designations given to the equivalent factors of mammalian eukaryotic systems, the last column to prokaryotes.
For original literature see references in Schreier & Staehelin (1973).

Factor EIF-3, contrary to its prokaryotic counterpart, consists of a number of different sized polypeptide chains; the question as to whether there are different EIF-3s or whether selectivity is controlled by additional mRNA specific factors, is still unresolved. A similar uncertainty to that described for the prokaryotic system surrounds the question as to whether or not methionyl-$tRNA_i$ binds to the 40s subunit prior to mRNA.

The initiation process is extremely complex and still not fully understood. A variety of proteins are involved, some being ribosomal structural proteins and others proteins which recycle during the process. These proteins interact, changing the conformation of the ribosome and thereby allowing the different steps of initiation to proceed in a sequential and orderly manner. Different mRNAs probably differ in their rate of attachment to the ribosome and/or in their efficiency in other stages of the initiation process, so affecting the frequency and rate of translation. With proteins which contain prosthetic groups their absence may inhibit initiation, since it has been shown that lack of haemin in globin synthesis, results in the formation of an inhibitor of the binding of methionyl-tRNA$_i$ to 40s subunits (Adamson *et al.*, 1972; Gross & Rabinovitz, 1973; Legon *et al.*, 1973). Furthermore, a variety of mRNA specific factors have been proposed called 'interference' or 'i-factors', which affect the specificity of the initiation factor IF-3 (Groner *et al.*, 1972). Thus, Strycharz *et al.* (1973) have identified a supernatant factor in Krebs II ascites cells, which is specifically required for the initiation of the translation of encephalomyocorditis viral RNA. However, Lodish (1974), from a kinetic analysis of protein synthesis, has proposed a model which precludes the necessity for mRNA specific factors. He points out that these have often been observed in systems which are less active than the corresponding *in vivo* system, and that the three eukaryotic systems available which translate exogenous mRNAs efficiently *in vitro*, do so with a variety of mRNAs.

10.3.2.2 *Elongation of the polypeptide chain*

The elongation process starts once the 80s plant ribosome containing mRNA and methionyl-tRNA$_i$, has been assembled and takes place by the following repeated cycle of events, each cycle being separated into:
(a) codon-directed binding of aminoacyl-tRNA;
(b) peptide bond formation; and
(c) translocation.
The ribosome has two binding sites; a 'P' site for the binding of methionyl-tRNA$_i$ and an 'A' site where all other aminoacyl-tRNAs bind. After the initiation complex has been formed, the next aminoacyl-tRNA, i.e. the one carrying the anticodon to the next codon of the mRNA, binds at the 'A' site. Once the aminoacyl-tRNA has been bound in the 'A' site, a peptide bond is formed by a peptidyl-transferase enzyme, which is part of the large subunit, such that the 'A' site now carries the aminoacyl-tRNA joined to methionine by a peptide bond. The 'P' site now carries the deacylated tRNA$_i^{Met}$, which is subsequently removed leaving an open 'P' site, and the peptidyl-tRNA moves from the 'A' to the 'P' site, so moving the message relative to the ribosome. The 'A' site is now empty and is ready for the next aminoacyl-tRNA to enter according to the next codon in the message. This cycle of reactions requires K^+, Mg^{2+}, GTP and various elongation factors, proteins, which are found in the soluble fraction of cell homogenates.

Our knowledge of polypeptide chain elongation is most complete with *E. coli*, where three elongation factors are known, EF-Tu, EF-Ts and EF-G (Fig. 10.7). Factor EF-Tu binds stepwise with GTP and aminoacyl-tRNA and this ternary complex transfers the aminoacyl-tRNA to the 'A' site of the ribosome, releasing an EF-Tu•GDP complex and inorganic phosphate. Factor EF-Ts displaces GDP to form the complex EF-Ts•EF-Tu, from which GTP displaces EF-Ts to regenerate EF-Tu•GTP. The third elongation factor, EF-G, and free GTP, are responsible for the translocation of the peptidyl tRNA from the A' to the 'P' site with the prior removal of the deacylated tRNA from the 'P' site; GTP is hydrolyzed during the process. The elongation factors, like the initiation factors, recycle during protein synthesis (Fig. 10.7). The protein chain grows from the N-terminal end, and probably starts to fold into its three-dimensional conformation whilst the process of elongation is proceeding.

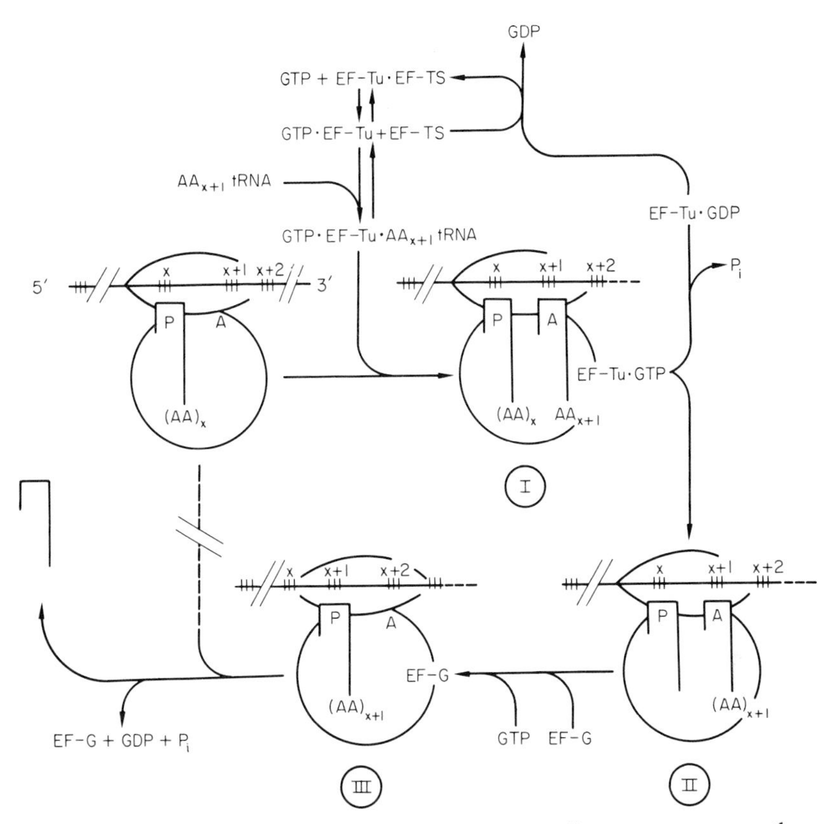

Fig. 10.7. Polypeptide chain elongation in prokaryotes. Ribosomes cover more than two codons on mRNA. Reaction I → II is catalysed by peptidyl transferase. (⅂ = tRNA, $(AA)_x$ = peptide with N-terminal amino acid x. AA_{x+1} = amino acid x + 1.) (Modified from Haselkorn & Rothman-Denes (1973).)

Fewer details of the elongation factors are available in eukaryotic systems. However, two elongation factors, designated as the binding enzyme EF-1 and the translocase EF-2 (Hardesty *et al.*, 1963; Gasior & Moldave, 1965) have been purified from ungerminated wheat embryos (Golinska & Legocki 1973; Twardowski & Legocki, 1973; Legocki, 1973; Allende *et al.*, 1973). The mechanism of binding of aminoacyl-tRNA to wheat ribosomes is on the whole similar to that in *E. coli*, the main difference being the absence in the eukaryotic system of a factor with the properties of EF-Ts, i.e. EF-1 = EF-Tu and EF-2 = EF-G.

Elongation factor EF-1, which requires the presence of Mg^{2+} at low concentration, K^+ and GTP, is responsible for the binding of aminoacyl-tRNA to wheat ribosomes. It contains three active forms differing in molecular weight. The larger EF-1 forms dissociate to a single species of molecular weight 6×10^4 daltons during gel electrophoresis in the presence of SDS. The three forms observed may represent a monomer, trimer and tetramer of a single protein unit, of which the trimer seems to be the most stable form. Some experimental evidence supports the view that the light species is the active form of the EF-1 enzyme. For example, in calf brain and wheat embryo, it has been shown that the interaction of EF-1 with GTP and aminoacyl-tRNA yields a ternary complex which contains the light form of the enzyme, i.e. molecular weight 5×10^4 to 6×10^4 daltons (Moon *et al.*, 1972; Legocki, 1973). It is suggested that the amino group of the esterified amino acid plays an important role in aminoacyl-tRNA binding (Jerez *et al.*, 1969) and this differs from the bacterial system where EF-Tu can react with the deaminated product of phenylalanyl-tRNA, phenyl-lactyl-tRNA (Fahnestock *et al.*, 1972).

The second elongation factor, EF-2, is involved in translocation of the peptidyl-tRNA from the 'A' site to the 'P' site during elongation. It is a protein of molecular weight 7×10^4 daltons and requires thiol compounds for activity (Twardowski & Legocki, 1973).

Ribosomes and elongation factors from the cytoplasm of eukaryotic organisms are not functionally interchangeable with their *E. coli* counterparts (Krisko *et al.*, 1969). Furthermore, Perani *et al.* (1971) have shown absolute ribosome specificity, either for 70s or 80s ribosomes, of two sets of elongation factors, T and G (EF-1 and EF-2), isolated from yeast, and for factor EF-2 isolated from the alga, *Prototheca zopfii*. It is generally accepted that elongation factors within each ribosomal type (70s or 80s) are exchangeable between eukaryotes (Ciferri, 1972).

10.3.2.3 *Termination and release of the polypeptide chain*

No reliable information is available on chain termination in plants, most of the work having been done with *E. coli*. However, as there are some differences between the prokaryote and mammalian eukaryote systems, these will be mentioned briefly. In *E. coli* the elongation cycle is repeated until certain termination codons on the mRNA come into the 'A' site; these codons are UAG, UAA and UGA. At least three release factors, proteins, which recognize the terminator codons and bring about the release of the completed polypeptide chain from the tRNA, have been identified (Caskey *et al.*, 1969; Capecchi & Klein, 1969; Milman *et al.*, 1969). Originally designated R-1, R-2 and S these are now called RF-1, RF-2 and RF-3. RF-1 recognizes UGG or UAG, RF-2 recognizes UAA or UGA, and RF-3 affects the rate of release of the polypeptide chain. In mammalian systems, only a single release factor which responds to all three codons has been identified, and furthermore the prokaryotic termination requirement for GTP appears to be absent (Haselkorn & Rothman-Denes, 1973).

In addition, several other termination factors have been proposed from different organisms (see Haselkorn & Rothman-Denes, 1973). Although little is known about chain termination in plants, the presumed universality of the code suggests that similar terminator triplets and proteins are involved in plants also.

Originally, it was thought that on chain termination ribosomes dissociated into subunits and that the free ribosomes found in the cell were inactive in further protein synthesis. Noll *et al.* (1973) have presented evidence that at least some of the free ribosomes found in the cell interact with a factor and partially dissociate, so that the small subunit with bound IF-1, IF-2 and IF-3, can initiate protein synthesis.

In conclusion, it may be said that small differences exist between the mechanism of protein synthesis in the pro- and eukaryotic systems, e.g. the initiator tRNA, the involvement of GTP in different steps, as well as in the possible number of elongation factors. In view of the technical difficulties involved, work on the detailed mechanisms of eukaryotic systems could be considered hardly worthwhile. However, it is possible that translation level controls are much more important in eukaryotes than they are in prokaryotes. This possibility, together with the importance of plant proteins as a source of food, are justification enough for an attempt to characterize the process in detail in plants and, thereby develop suitable assay systems.

10.3.2.4 *Cell-free systems*

In order to gain detailed knowledge of the mechanism of protein synthesis in plants, it is necessary to develop *in vitro* assay systems. Such systems must satisfy the following criteria:

(a) be dependent on the exogenous component(s) to be tested;
(b) be efficient, i.e. of comparable activity to *in vivo* rates, in order to allow quantitative analysis;
(c) the product, whose synthesis is monitored, must be clearly defined.

Two basic types of cell-free systems are used: fractionated systems, in which enzymes and components are isolated, purified and then reconstituted into an assay system, and unfractionated systems, in which the cells are broken open and cell-debris, nuclei and mitochondria removed by centrifugation leaving the ribosomes and the various components for protein synthesis in the supernatant, which is then used as the assay system. If all of the components of the assay are isolated from one organism, the system is said to be homologous as opposed to heterologous when they are not.

In addition to the fractionated *E. coli* prokaryotic system, there are four fractionated eukaryotic systems which satisfy the above criteria. These are the rabbit reticulocyte (Lockard & Lingrel, 1969; Gilbert & Anderson, 1970), the mammalian liver (Prichard *et al.*, 1971), the Krebs II ascites cell (Mathews & Korner, 1970), and the only plant system, that of wheatgerm (Allende & Bravo, 1966; Allende, 1970; Leis & Keller, 1970; Marcus *et al.*, 1970a,b; Tarrago *et al.*, 1970; Ghosh *et al.*, 1971; Legocki & Marcus, 1970; Klein *et al.*, 1972; Lundquist *et al.*, 1972). Other fractionated cell-free systems from plants, such as those from developing legume seeds (Gumilevskaya *et al.*, 1971; Payne *et al.*, 1971a,b; Yarwood *et al.*, 1971a,b; Beevers & Poulson, 1972; Wells & Beevers, 1973), are not well characterized and assays often lack quantitative accuracy. This is because in isolating the constituent enzymes and components, damage is caused in breaking the cell wall by hydrodynamic sheer, by cell vacuoles breaking and releasing acids, phenolics, tannins and other substances, and by activation of proteases and nucleases (Payne & Boulter, 1974). If the preparations used contain several components and structural damage to ribosomes has occurred, interpretation of the results may be qualitatively ambiguous, since enzyme preparations may contain proteins needed for the reconstitution of protein-leached ribosomes.

Unfractionated cell-free systems cannot be made from tissues which have a high nuclease or protease activity, and until better methods for the inhibition of degradative enzymes become available, most unfractionated systems from plants will be of limited value. Recently, Davies *et al.* (1972) have used media of high ionic strength and high pH with some success, and Gray and Kekwick (1973) have developed a system from pea seedlings using 0·2 mM vanadyl sulphate as an inhibitor. This system has been shown to synthesize the small subunit of ribulose bisphosphate carboxylase (Fraction I protein), since the tryptic peptides of the *in vitro* product were similar to those of the naturally occurring protein. The identity of the product was also proved by immunoprecipitation. The requirement of product identification is essential in complete cell-free assays (see also chapter 11, where the synthesis of the large subunit of Fraction I protein by the chloroplast cell-free system is described).

The wheatgerm system is by far the most successful unfractionated cell-free system from plants. Its preparation is as follows (Marcus *et al.*, 1974):

> Dry wheat embryos are ground thoroughly with a small amount of sand in a precooled mortar in a total volume of 3·3 ml of 90 mM KCl, 2 mM $CaCl_2$, 1 mM Mg $(Ac)_2$, 6 mM $KHCO_3$. The embryos are initially ground in 1·0 ml with 0·5 and 1·8 ml increments added subsequently. The slurry is then centrifuged for 10 minutes at 23,500 × *g* and the supernatant is removed with a Pasteur pipette, taking care to leave behind as much as possible of the upper lipid layer. Just prior to use, 0·5–3·0 ml are dialyzed against 500 ml of 1 mM Tris-acetate, pH 7·6, 50 mM KCl, 2 mM $Mg(Ac)_2$, 4 mM 2-mercapto-ethanol for 1·75 hours; this preparation is termed S23.

The wheatgerm system is attractive in spite of needing some additions, (e.g. tRNA), since it shows great promise as a 'translation' system for various mRNAs. This is an important development as the isolation of mRNAs from plants is now feasible using binding to oligo-dT columns (since a proportion of eukaryote messengers contain a poly-A sequence), by gradient centrifugation and by immunoprecipitation of polysomes (Haselkorn & Rothman-Denes, 1973; see also chapter 9).

Four RNAs of brome mosaic virus (BMW) induce amino acid incorporation into proteins when used as messengers in the wheatgerm system. RNA4 is translated with an efficiency comparable to that of bacteriophage RNA in *E. coli* extracts. The product is homogeneous and indistinguishable from the coat protein of BMV (Shih & Kaesberg, 1973). Satellite tobacco necrosis virus RNA has also been translated in this system and the product shown to be coat protein (Klein *et al.*, 1972). Several natural eukaryotic messengers have also been translated in the wheatgerm system, e.g. rabbit globin mRNA (Efron & Marcus, 1973) and leghaemoglobin RNA (Verma *et al.*, 1974). The latter is the only naturally occurring plant messenger to be isolated so far. It is a poly A-containing 9s–12s mRNA which was isolated from soybean root nodule polysomes. When used to programme the wheatgerm system by Verma *et al.* (1974), the product was identified serologically as mainly leghaemoglobin S (Lb_S) with a little leghaemoglobin F (Lb_F); there are two types of leghaemoglobin found in soybean nodules and the reason for the preferential synthesis of the Lb_S *in vitro*, is not understood. Factors other than those already present in the wheatgerm system were not required, and there was no need for the addition of heme.

There are several unfractionated cell-free systems from animals (Haselkorn & Rothman-Denes, 1973). Of particular interest is that of the *Xenopus* oocyte since intact eggs are used to translate exogenous messages (Gurdon *et al.*, 1971). Natural plant messages have not been used but there is evidence that tobacco mosaic virus RNA can be translated (Knowland, 1973). Coat protein is not synthesized, however, and the main product is a polypeptide with molecular weight 14×10^4 daltons. Although no function has been assigned to this protein, it is known that plant cells infected with TMV also make a protein of the same molecular weight (J. Gurdon, personal communication).

The use of heterologous systems assumes the interchangeability of the various enzymes and components between different systems and as mentioned previously, it is generally accepted that this is not usually possible between prokaryotic and eukaryotic organisms. Generally, attempts to demonstrate interchangeability between different eukaryotic systems have been positive, but several workers have suggested that this is because discrimination can only be seen under optimum conditions. The general conclusion is that tissue and organism specificity may occur but only in exceptional cases (Mathews, 1973).

10.3.2.5 *Inhibitors*

Several antibiotics and other inhibitors block various steps in protein synthesis, and are, therefore, extremely useful in establishing the mechanism of the process (Pestka, 1974). Table 10.2 lists the more common inhibitors and their sites of action; many of these have been used with plants. It is important to realize

Table 10.2. Inhibitors of Protein Synthesis on 80s Eukaryotic Ribosomes

Supernatant	40s	60s
Aminoalkyl adenylates Guanylyl-5′-methylene diphosphonate Fusidic acid (elongation) Diphtheria toxic (elongation)	Pactamycin (initiation; methionyl-tRNA$_i$ binding) Aurintricarboxylic acid (initiation; mRNA binding) 2-(4-methyl-2,6 dinitroanilino)-N-methylpropionamide (initiation; 40s•mRNA complexing with 60s)	Puromycin (transpeptidation) 4-Aminohexose pyrimidine nucleosides (transpeptidation) Sparsomycin (transpeptidation; translocation, ribosome subunit exchange) Tetracyclines (inhibit various steps including codon recognition, transpeptidation and termination) Cycloheximide Actidione (inhibits various steps, including initiation, transpeptidation and translocation) Ipecac alkaloids (translocation)

Modified from Pestka (1974). Brackets indicate steps involved.

that the concentration of the inhibitor is often critical for the production of a specific effect; it is essential to establish the conditions for the correct use of an inhibitor with each tissue and to maintain effective controls. It has been suggested by Glazer and Sartorelli (1972) that in rat liver, membrane-bound 80s ribosomes are more susceptible to a range of inhibitors than are the 80s ribosomes free in the cytosol.

10.3.3 POST-TRANSLATIONAL CHANGES

Changes which take place after the synthesis of the polypeptide chain are called post-translational. The most common of these are concerned with the many proteins which consist of more than one polypeptide chain. Since there is no evidence that unpartnered subunits occur in any quantity, machinery must exist in the cell to ensure that approximately correct numbers of each constituent polypeptide chain are synthesized and correctly assembled, although little is known about the mechanisms. Some mRNAs may be polycistronic but in other cases mRNAs for each polypeptide chain are translated independently. Self-assembly by interaction at specific sites on different polypeptide chains may occur.

Several examples are known where some subunits of a protein are made in the cytoplasm and others on the chloroplast or mitochondria. For example, ribulose bisphosphate carboxylase is a large molecule (also known as Fraction I protein) consisting of two types of subunit; the small subunits are synthesized on 80s cytoplasmic ribosomes, whilst the large subunits are synthesized on 70s chloroplast ribosomes and assembly of the two types to form the complete molecule occurs in the chloroplast (see chapter 11).

Post-translational changes also occur by the addition of prosthetic groups or by modification of the polypeptide chain. Such modifications may be the formation of disulphide linkages, the removal of N-terminal portions, the removal of intramolecular segments, the derivation of amino acids and the addition of carbohydrate or lipid units to form glycoproteins and lipoproteins respectively. These changes are all brought about by specific enzymes and probably after the polypeptide chain has left the ribosome.

The formation of disulphide linkages is often part of the assembly of multi-chain proteins, e.g. legumin in *Vicia faba* (Wright & Boulter, 1974), although the enzymes involved have not been demonstrated from plants. Few plant proteins have methionine at their N-terminus when isolated, although this amino acid must be presumed to be the N-terminus when synthesized, and, therefore, removal of one or more residues at the N-terminal region must have occurred subsequently. Enzymes for this purpose have been demonstrated in bacteria but not in plants, and it is possible that the process in plants is non-specific and mediated by leucine aminopeptidase, a ubiquitous plant enzyme which removes amino acids from the N-terminus of proteins and may continue to do so until the conformation of the protein inhibits its activity. The removal of intramolecular segments of plant proteins similar to the situation with pro-insulin in animals, has not been demonstrated, but there is no reason to suppose that it does not occur. Modification of individual amino acids is quite common in plant proteins; examples are, Σ-N-trimethyllysine in some cytochromes *c* (Ramshaw *et al.*, 1974), and hydroxyproline in extensin, a protein of cell walls (Lamport, 1965). Where investigated, the protein when synthesized contained the parent amino acid, the substituted amino acid arising secondarily, probably after the protein had left the ribosome. Some residue positions of a particular amino acid may be substituted and others not, though the specificity of the changes is not understood.

Synthesis of the protein moiety of glycoproteins takes place on the membrane-bound ribosomes of the ER, but the sugar groups are added sequentially at different sites in the cell. Those closely linked to the polypeptide chain such as glucosamine and mannose, are added immediately after release from the ribosome or in the ER itself, other more terminal groups may be added later, some in the Golgi apparatus (Whaley *et al.*, 1972).

10.4 PROTEIN SYNTHESIS *IN VIVO*

The rate of protein synthesis *in vivo* can be measured by determining the amount of protein in a plant, organ or tissue at two different times; this will give the net rate since proteins 'turnover' and the amount of protein found would represent the balance between synthesis and degradation (Huffaker & Petersen, 1974).

Various equations describing the relationship between the synthesis and degradation of protein have been proposed but there are many technical and interpretive difficulties in this type of experiment. The turnover rate for an individual protein is not the same as the value obtained for the total protein and although the latter may be of interest agriculturally, most biochemical and physiological studies require information on individual proteins. This information has only been obtained so far with a few plant proteins, although most are thought to turnover (Huffaker & Petersen, 1974; see also chapter 12). Possible exceptions are storage proteins which are sequestered from the metabolically active cytoplasm and stored in membrane-bound protein bodies until subsequently required at a more active phase of the life-cycle.

A variety of studies using acrylamide gel electrophoresis has shown that the protein complement of organs and tissues changes during the life-cycle of a plant, and also in response to environmental conditions. Organ-specific proteins have been demonstrated serologically and often one or a few proteins may predominate in an organ, e.g. ribulose bisphosphate carboxylase, which often accounts for 50% of the protein of leaves. The changing complement of proteins during development is due to changes in the type or rate of protein synthesis and/or degradation. For example, Beevers and Poulson (1972) followed changes in the protein content of pea cotyledons during the period 9–33 days after flowering. Initially, protein content increased gradually, then rapidly between 21–27 days to decline once more as the seed dehydrated and matured. Incorporation experiments with ^{14}C-leucine indicated that albumin proteins were synthesized early in cotyledon development whilst globulin synthesis predominated with increasing maturity. During the period of rapid protein synthesis, greater amounts and a higher percentage of polysomes to monosomes were extracted than at other stages in the development of the seed. However, unlike in bacteria, where induced enzyme synthesis is widespread and the product is not 'turned over' but diluted out by a rapid rate of cell division, plants have a much slower rate of cell division and protein degradation appears to be a common method for regulating the amount, and hence the activity of proteins.

The relative amounts of different seed proteins synthesized may vary considerably in different varieties and breeding lines of a crop; for example, in maize where the major storage proteins of the seeds are prolamins and glutelins normal maize contains a relatively high proportion of prolamin, whereas in the mutant Opaque 2, the proportion of prolamin is greatly reduced. This is of particular significance since prolamin is very deficient in the essential amino acid lysine, so that seed meals obtained from Opaque 2 have a higher content of lysine and are consequently more nutritious than those from normal maize.

Protein synthesis is a concerted process involving the biochemical machinery of the cell as a whole, since it consists of a series of co-ordinated reactions, the constituents of which may be synthesized in the nucleus (mRNA, tRNA), the mitochondria (ATP), the cytosol (amino acids and aminoacyl tRNAs), the ribosomes (enzymes). Although changes in the rate of protein synthesis occur, very little is known about the mechanisms involved. Theoretically, control could be exerted at any one of the many constituent reactions, this being the pacemaker reaction for protein synthesis in that particular cell/tissue. However, the control mechanisms must allow the overall process to adjust in response to genetic, physiological and environmental stimuli, so that the rate and/or the type of proteins being synthesized can change.

FURTHER READING

Allende J.E. (1970) Protein biosynthesis in plant systems. *Techniques in Protein Biosynthesis*, vol. 2 (eds. P.N. Campbell & J.R. Sargeant), pp. 55–100. Academic Press, London and New York.

Anderson W.F., Bosch L., Gros F., Grunberg-Manago M., Ochoa S., Rich A. & Staehelin Th. (1974) Initiation of protein synthesis in prokaryotic and eukaryotic systems. *FEBS Letters* **48**, 1–6.

Boulter D. (1970) Protein synthesis in plants *Ann. Rev. Pl. Physiol.* **21**, 91–114.

Boulter D., Ellis R.J. & Yarwood A. (1972) Biochemistry of protein synthesis in plants. *Biol. Rev.* **47**, 113–75.

Haselkorn R. & Rothman-Denes L.B. (1973) Protein synthesis. *Ann. Rev. Biochem.* **42**, 397–438.

Kurland C.G. (1972) Structure and function of the bacterial ribosomes. *Ann. Rev. Biochem.* **41**, 377–408.

Mans R.J. (1967) Protein synthesis in higher plants. *Ann. Rev. Pl. Physiol.* **18**, 127–46.

Stewart P.R. & Letham D.S. (Eds.) (1973) *The Ribonucleic Acids*. Springer-Verlag.

Wittman H.G. (1973) Structure and function of bacterial ribosomes. In: *Regulation of Transcription and Translation in Eukaryotes*, (eds. E.K.F. Bautz, P. Karlson & H. Kersten), pp. 211–212. 24 Colloquium der Gesellschaft für Biologische Chemie. Springer-Verlag Berlin, Heidelberg, New York.

Zalik S. & Jones B.L. (1973) Protein biosynthesis. *Ann. Rev. Pl. Physiol.* **24**, 47–68.

CHAPTER 11

THE GENETIC INFORMATION OF ORGANELLES AND ITS EXPRESSION

11.1 INTRODUCTION: THE CONCEPT OF ORGANELLE AUTONOMY

The most notable event in biology at the dawn of the present century was the rediscovery of the Mendelian laws of inheritance by de Vries, Correns, and Tschermak. The realization by Sutton that the results of Mendel's experiments with garden peas could be explained in terms of the visible behaviour of chromosomes during meiosis led to the nucleus being regarded as the sole carrier of the hereditary material. As often happens in science, this splendidly simple view did not survive for long. The results of experiments by Baur (1909) and Correns (1909) on the inheritance of plastid defects in variegated plants were difficult to explain on the basis that the genes concerned were located in the nucleus. In certain cases, the inheritance of such defects occurred only through the maternal line, i.e. it was uniparental. In other cases the inheritance was biparental, but did not obey the rules of Mendel. For a time it appeared that, far from being 'nuclear', some aspects of inheritance were 'unclear'. It was Baur who pointed out that these results were explicable on the assumption of the genetic continuity of an extrachromosomal entity located in the plastid. Much more extensive evidence for this concept was later provided by the work of Renner (1929) using the genus *Oenothera*; he suggested the term 'plastome' to describe the genetic system in the plastid. Maternal inheritance is thus explicable in terms of the lack of plastids in the pollen tube of some species; the plastome of the new generation is consequently derived entirely from the mother.

These studies of variegated plants provided the first firm evidence for the existence of extrachromosomal inheritance. Since then, further evidence has accumulated with examples known in representatives of most major groups of organisms. Today we realize that organelle genetic systems are a fundamental feature of the organization of eukaryotic cells. The most detailed work has concentrated on plastids and mitochondria, but there are indications that genetic determinants may occur in other organelles found in eukaryotic cells, especially centrioles, basal granules and kinetoplasts. There is no evidence that either peroxisomes or glyoxysomes contain any genetic material. In this chapter, the biochemical evidence relating to the concept of the genetic autonomy of plastids and mitochondria only will be considered.

The concept of chloroplast autonomy was founded on the observation that, in the algae, chloroplasts can be seen to divide and to be passed to the new cells in cell division (Strasburger, 1882; Green, 1964). This cytological evidence led to the view, first proposed by Schimper (1885) and Meyer (1893), that plastids do not arise *de novo*, but are formed by the division of pre-existing plastids; this view was supported by the genetic experiments of Baur and Correns carried out in the following two decades. The discovery in 1962 that chloroplasts contain both DNA and ribosomes brought the idea of chloroplast autonomy back into vogue, and this notion has so dominated our concepts of chloroplast development in recent years that several attempts have been made to grow isolated chloroplasts in culture. It is now clear that such attempts are ill-founded.

Our current dogmas maintain that for any biological system to be autonomous it must contain four components; (a) DNA to code for its *entire* structure; (b) DNA polymerase to replicate the DNA; (c) RNA polymerase to transcribe the DNA; (d) protein-synthesizing machinery to translate the messenger RNAs into *all* the necessary proteins. Intermediary metabolism is not necessary in principle since a supply of small molecules could be taken up from the environment. An extensive literature establishes that both chloroplasts and mitochondria do in fact contain these four components; the properties and functions of these are discussed in the rest of this chapter. It is equally clear that the DNA does not code for all the organelle proteins nor does the protein-synthesizing system make all the organelle proteins. For example, many genes concerned with chloroplast and mitochondrial structure and function are inherited in a Medelian fashion, and are therefore presumed to be located in the nucleus, while there is increasing evidence that the majority of chloroplast and mitochondrial proteins are synthesized on cytoplasmic ribosomes. It is now realized that the demonstration of the cytological continuity of plastids from generation to generation is not sufficient to establish that they replicate independently of nuclear control. How, then, are we to regard the concept of organelle autonomy?

It is my contention that chloroplasts and mitochondria are not autonomous in any rigorous sense; the term is useful only as a quick way to describe the fact that these organelles contain *some* genes and make *some* proteins. We must regard the formation of these organelles as resulting from a complex interplay between the genomes of the organelles and the genome of the nucleus, and the fascination of this subject lies in unravelling the details of this interplay at the molecular level. Much more information is available about the autonomy of chloroplasts than about the autonomy of plant mitochondria; research on mitochondrial autonomy has concentrated almost entirely on animal and fungal cells. This imbalance is reflected in this chapter, which is concerned largely with chloroplast autonomy.

The current state of knowledge can be summarized by saying that, while some information has been accumulated about the properties of chloroplast nucleic acids and protein synthesis, only recently has any hard evidence emerged about their biological function. Developments at the genetic level in higher plants, and the establishment of isolated chloroplast systems which synthesize specific protein and RNA molecules, promise to provide increasing understanding of the precise functions of chloroplast nucleic acids and protein synthesis.

11.2 CHLOROPLAST AUTONOMY

11.2.1 CHLOROPLAST DNA

11.2.1.1 *Discovery*

The existence of DNA in chloroplasts has been established by both cytological and biochemical criteria. Ris and Plaut (1962) provided the first convincing evidence that chloroplasts contain DNA. Electron microscopy of chloroplasts in sections of both algal and higher plant cells revealed areas of low electron density which contained fibrils 2·5–3·0 nm in diameter. These fibrils were not seen if the sections were treated with deoxyribonuclease. There are earlier reports that chloroplasts contain regions which can be stained with the Feulgen reagent for DNA, but in general these reports have not proved reproducible. It is now clear that in most species the concentration of DNA inside the chloroplast is below the limit of detection by the Feulgen method. It can however be detected by autoradiography after exposure of cells to -H-thymidine; the report of Rose *et al.* (1974) on the distribution of DNA in dividing spinach chloroplasts contains an excellent example of this method. It is important when evaluating such studies to consider the controls that are used. If the incorporated label is all present in DNA, it should be removed by treatment of sections with deoxyribonuclease, but not by treatment with ribonuclease. This control works well for tobacco and spinach leaves and several algae, but not or maize leaves, where deoxyribonuclease fails to remove the label from chloroplasts.

Most studies on chloroplast DNA since 1962 have been carried out using isolated chloroplasts. This approach suffers from the difficulty of distinguishing chloroplast DNA from DNA originating from nuclei, mitochondria, and microorganisms which may contaminate the chloroplast pellet. The resolution of these problems is still not complete, and depends mainly on the differing properties of the DNA from the various sources. There are four characteristic properties of chloroplast DNA that have been used to identify it: namely buoyant density, ease of renaturation, lack of histones, and the absence of 5-methylcytosine. These properties are discussed below. It must be remembered that there is a possibility some chloroplast DNA may not meet these criteria. If, for example, chloroplasts in some species contain a minor DNA component which is a nuclear transcript, contains 5-methylcytosine and renatures poorly, it is doubtful whether present techniques could identify it as chloroplast in location. In one species the problem of contamination by nuclei and microorganisms can be entirely avoided; DNA has been identified in chloroplasts isolated from sterile enucleated plants of *Acetabularia* (Gibor & Izawa, 1963; Baltus & Brachet, 1963).

11.2.1.2 *Buoyant Density*

It is now clear that chloroplast DNA from the algae *Euglena*, *Chlamydomonas*, and *Chlorella* has a buoyant density sufficiently different from that of nuclear DNA to allow resolution by analytical ultracentrifugation, but that in many higher plants the densities of the two DNA types are often, but not always, too close to permit this. This information has been hard-won; the story of the way in which the interpretation of band patterns in neutral caesium chloride density gradients has changed since 1963 has been told by Kirk (1971) and Tewari (1971). A summary is given here since it illustrates the very real problem of establishing the identity of chloroplast DNA from higher plants by biochemical methods.

The first report of the chemical characterization of higher plant chloroplast DNA was provided by Kirk (1963). He found that chloroplast DNA from *Vicia faba* has a GC content (37·4%) which is slightly but significantly different from that of nuclear DNA (39·4%). In the same year a contrasting report was published by Chun *et al.* (1963), who found that chloroplast preparations from *Spinacia oleracea* and *Beta vulgaris* contained two components of much higher densities than the nuclear DNA as judged by caesium chloride equilibrium density gradient centrifugation. However the bulk of the DNA in these chloroplast preparations had a density similar to that of nuclear DNA; this band was attributed to contamination of the chloroplast pellet by nuclear fragments, and thus the chloroplast DNA was regarded as the two high density components. There followed a spate of papers which supported the claim that, in higher plants, chloroplast DNA has an appreciably higher density than nuclear DNA. This picture changed when re-examination by more rigorous methods supported the earlier view of Kirk, and the higher density components are now attributed to contamination by mitochondria and bacteria.

The present position can be summarized by saying that, in all the higher plants examined, chloroplast DNA had a base composition of $37{\cdot}5 \pm 1\%$ GC and a buoyant density of $1{\cdot}697 \pm 0{\cdot}001$ g cm^{-3}. Table 11.1 lists some values for the buoyant densities of chloroplast and nuclear DNA which have been agreed by Kirk and Tewari. Nuclear DNA is seen to have a smaller, larger, or similar density to chloroplast DNA from the same plant, depending on the species. The GC contents in Table 11.1 have been calculated from the buoyant densities, but a more accurate method has been devised by Kirk (1967). This method releases the purine bases by gentle acid hydrolysis and separates the adenine and guanine by ion-exchange chromatography.

The moral to be drawn from this story is that the purity of subcellular fractions from higher plant tissues must be established by positive methods if meaningful interpretations are to be made. Microbial contamination can be reduced by surface sterilization of fresh tissues and by the use of sterile solutions, while mitochondrial contamination can be assayed by succinic dehydrogenase or cytochrome oxidase activities. Nuclear contamination can be reduced by incubation of intact chloroplasts with deoxyribonuclease and phosphodiesterase. The extent of nuclear contamination is difficult to measure, but the ease of renaturation and the absence of 5-methylcytosine serve to distinguish chloroplast from nuclear DNA in all cases so far studied.

Table 11.1. Buoyant densities and base composition of higher-plant and algal chloroplast and nuclear DNA*.

Species	Nuclear DNA Density in CsCl (g cm^{-3})	Nuclear DNA % GC** content	Chloroplast DNA Density in CsCl (g cm^{-3})	Chloroplast DNA % GC content
Chlorella ellipsoidea	1·716	60	1·692	33·0
Euglena gracilis	1·707	52	1·685	25.0
Chlamydomonas reinhardi	1·723	67	1·695	39·0
Vicia faba	1·695	38·8	1·697	37·8
Vicia faba	1·696	39·8	1·696	36·8
Spinacia oleracea	1·694	37·8	1·697	37·8
Lactuca sativa	1·694	37·8	1·697	37·8
Pisum sativum	1·695	38·8	1·697	37·8
Nicotiana tabacum	1·697	40·8	1·697	37·8
Nicotiana tabacum	1·697	40·8	1·700	40·8
Nicotiana tabacum	1·695	38·8	1·697	38·0
Phaseolus vulgaris	1·697	40·8	1·696	36·7
Phaseolus vulgaris	1·696–1·697	40·0	1·696–1·697	37–38
Allium cepa	1·691	34·7	1·696	37·2
Triticum vulgare	1·702	47·0	1·697	38·4
Oenothera hookeri	1·703	47·0	1·697	38·0
Antirrhinum majus	1·691	34·7	1·697	38·0

* This table is taken partly from a compilation by Kirk (1971).

** In order to correct for the lowering of the buoyant density of the nuclear DNAs caused by methylation of cytosine residues, a value of 0·003 was added to all the observed densities of nuclear DNA before calculating the GC content.

(Reproduced with permission from MTP International Review of Science, Biochemistry Series One, Vol. 6, '*Biochemistry of Nucleic Acids*', 1974, ed. by K. Burton, and published by Butterworths, London.)

11.2.1.3 *Ease of Renaturation*

A more useful criterion than the buoyant density for detecting chloroplast DNA from higher plants is the ease with which it will renature after heat or alkali denaturation. Nuclear DNA renatures only to a slight extent in a few hours,

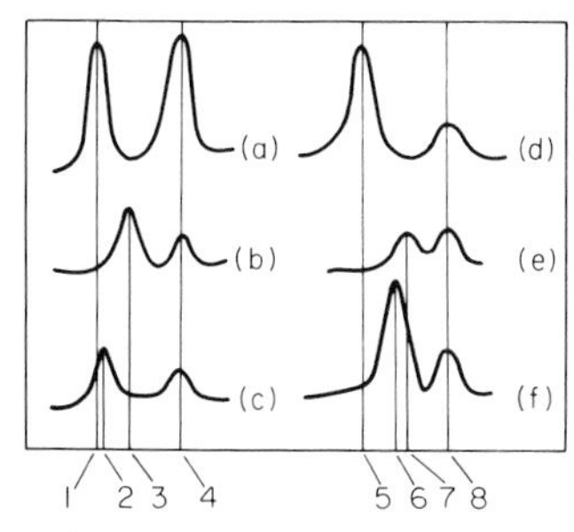

Fig. 11.1. Densitometer tracings of ultraviolet absorption photographs of *Vicia faba* DNA banded in caesium chloride gradients. (a) Native chloroplast DNA; (b) heat-denatured chloroplast DNA; (c) renatured chloroplast DNA; (d) native nuclear DNA; (e) heat-denatured nuclear DNA; (f) renatured nuclear DNA. The numbers refer to densities in g cm^{-3} as follows: (1) 1·696; (2) 1·699; (3) 1·712; (4) 1·728; (5) 1·696; (6) 1·709; (7) 1·712; (8) 1·728. The peaks at 1·728 g cm^{-3} represent marker DNA from *Micrococcus radiodurans*. (From Kung & Williams, 1968, by courtesy of Elsevier.)

depending on its content of reiterated sequences. Figure 11.1 illustrates this difference in the case of *Vicia faba*. The rapid renaturation of chloroplast DNA has encouraged attempts to estimate its genome size from measurements of kinetic complexity. The kinetic complexity of a DNA sample is a measure of the size of the unique set of nucleotide sequences it contains, as judged from the rate at which the DNA renatures. A rapid rate of renaturation implies that like sequences are present in high concentration, and therefore the number of unique sequences, or kinetic complexity, is small. Table 11.2 lists the kinetic complexity, in terms of molecular weight, for the chloroplast DNA from some algae and higher plant species. Some of these values have been corrected from the published figures since the estimate of the molecular weight of the bacteriophage T4 DNA, used as standard, has been revised from $1{\cdot}3 \times 10^8$ to $1{\cdot}06 \times 10^8$ (Dubin *et al.*, 1970). It is striking that the corrected values for the kinetic complexity of chloroplast DNA from the few algae and higher plants examined so far are all in the range $0{\cdot}9$–$1{\cdot}0 \times 10^8$. This may be a coincidence, or it could mean the information content of chloroplast DNA is basically similar throughout the plant kingdom. All the species so far examined belong to the Chlorophyta, with the exception of *Euglena*. It would be interesting to make such measurements of chloroplast DNA from other plant groups, especially from algae with unusually shaped chloroplasts. Table 11.2 also lists the analytical complexities of chloroplast DNA, i.e. the amount of DNA per chloroplast. Since the kinetic complexities are always much less than the analytical complexities, there must be between 20 and 60 copies of the DNA sequences in each chloroplast. These renaturation studies cannot rule out the possibility of microheterogeneity in nucleotide sequence between the copies but, if it does exist, such heterogeneity is beyond detection by current techniques.

Table 11.2. Corrected kinetic and analytical complexities of chloroplast DNA.

Species	Kinetic Complexity (Daltons $\times 10^8$)	Analytical Complexity (Daltons per chloroplast $\times 10^9$)	Reference
Euglena gracilis	0·9	5·8	Stutz (1970)
Chlamydomonas reinhardi (*gamete*)	0·99	4–5	Wells & Sager (1971); Bastia *et al.* (1971)
Pisum sativum	0·95	—	Kolodner & Tewari (1972)
Lactuca sativa	0·98	2	Wells & Birnstiel (1969)
Nicotiana tabacum	0·93	3	Tewari & Wildman (1969)

11.2.1.4 *Absence of Histones and 5-Methylcytosine*

When released from intact isolated chloroplasts by gentle lysis, chloroplast DNA is not combined with basic proteins. This finding confirms the microscopic observations of Ris and Plaut (1962) that chloroplasts contain DNA fibrils in the same form as they appear in the nucleoplasms of prokaryotic cells. It is also characteristic of chloroplast DNA from both higher plants and algae that it contains no detectable 5-methylcytosine, whereas nuclear DNA invariably does. Whitfeld and Spencer (1968) regard the absence of this base as the most reliable criterion for establishing the purity of chloroplast DNA.

11.2.1.5 *Circularity*

A recent discovery is that, when precautions are taken to minimize shearing, a proportion of chloroplast DNA can be isolated as circles. Circular DNA has been reported from chloroplasts of *Euglena* (Manning *et al.*, 1971), *Pisum sativum* (Kolodner & Tewari, 1972), *Spinacia oleracea* (Manning *et al.*, 1972), *Antirrhinum majus*, *Oenothera hookeri*, and *Beta vulgaris* (Hermann *et al.*, 1975). Figure 11.2 shows a molecule of chloroplast DNA from *Spinacia oleracea.* The significance of circularity in DNA is not known, but it is a useful property since it implies the molecule has not been degraded on isolation. The most interesting aspect of this finding is that in all the species listed the contour length of the majority of the circles is in the range 37–45 μm; this length of double-stranded DNA has a calculated molecular weight of $0{\cdot}85$–$1{\cdot}0 \times 10^8$ daltons, which is in the same range as the kinetic complexity of chloroplast DNA (Table 11.2). This correspondence of length and kinetic complexity suggests that the genetic information carried by the chloroplast DNA is accommodated by the length of the circular molecule.

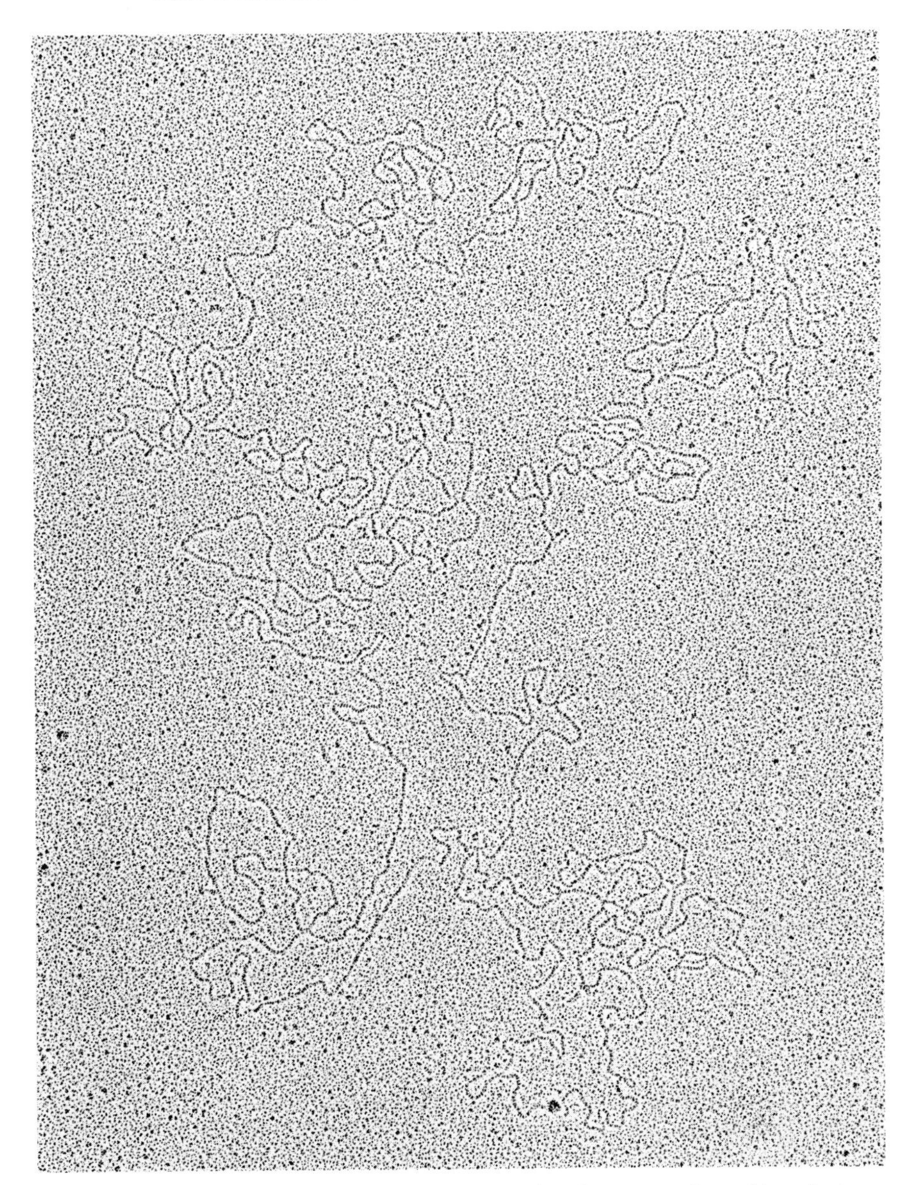

Fig. 11.2. Open circular DNA molecule, contour length 44·7 μm, from chloroplasts of *Spinacia oleracea*. (Reprinted from Hermann *et al.* (1975) by courtesy of Elsevier.)

11.2.1.6 *Ploidy*

It is clear from the data in Table 11.2, and from the evidence for circular DNA, that each chloroplast may contain many copies of a circular genome, and thus may be genetically polyploid. It has been pointed out by Kirk (1972) that this multiplicity increases the probability that a mutation will appear in a chloroplast in a given time, but there is likely to be a long delay before the mutation can be expressed, since a mutation in one copy will be swamped by all the other wild-type copies. It is therefore not surprising that known non-Mendelian mutations affecting chloroplasts are small in number, and difficult to induce with mutagens. In *Chlamydomonas reinhardi* a number of mutations are known which alter the sensitivity of chloroplast ribosomes to inhibition by antibiotics such as erythromycin and carbomycin; some of these mutations show uniparental inheritance (Sager, 1972; Mets & Bogorad, 1972). Those genes which are inherited in a uniparental fashion have been shown by recombinational analysis to form one linkage group which behaves as if it were diploid in vegetative cells. It is difficult to reconcile this diploid behaviour with the evidence which suggests that there are many copies of the genome in each chloroplast. It may be that the uniparental linkage group does not reside in chloroplast DNA, or if it does, that a reduction in the number of genomes in each chloroplast occurs at some stage in the cell cycle, so that only a 'master copy' is passed during sexual reproduction. There is as yet no evidence to resolve these questions.

11.2.1.7 *Functions*

A 40 μm circle of double-stranded DNA of unique base sequence is sufficient in principle to code for about 125 proteins each of molecular weight 50,000. Since our knowledge depends ultimately on the validity of the techniques which can be used, the evidence for the functions of chloroplast DNA will be considered in terms of the four methods tried so far:

The selective inhibition of chloroplast DNA transcription

The antibiotic rifampicin is a potent inhibitor of the initiation of RNA synthesis in bacteria. This compound has been reported to inhibit the incorporation of labelled precursors into chloroplast ribosomal-RNA, but not into cytoplasmic ribosomal-RNA, in several unicellular algae, namely *Chlamydomonas*, *Chlorella*, and *Acetabularia*. Rifampicin has also been reported to inhibit chloroplast RNA polymerase activity in extracts of *Chlamydomonas*. These results suggest that the functional genes for chloroplast ribosomal-RNA are located in chloroplast DNA and are transcribed by the chloroplast RNA polymerase.

The effect of rifampicin in higher plant systems is controversial. There have been some reports that it specifically inhibits chloroplast ribosomal-RNA synthesis, but other workers could not reproduce this result (Bottomley *et al.*, 1971). There now seems general agreement that chloroplast RNA polymerase from higher plants, as normally prepared and assayed, is insensitive to rifampicin, but this may mean only that the preparations are not initiating RNA synthesis.

Genetic analysis of mutants

Many mutations are known which affect chloroplast components, but the vast majority are inherited in a Mendelian fashion and are therefore presumed to be located in nuclear genes. For example, seven different genes are known which control steps in the chlorophyll biosynthetic pathway, and nuclear genes have been shown to be involved in the synthesis of phosphoribulokinase and at least five components of the photosynthetic electron transport chain (Kirk, 1972; Kirk & Tilney-Bassett, 1967; Levine & Goodenough, 1970).

In *Chlamydomonas* many mutations affecting chloroplast ribosomes and photosynthetic capacity are inherited as a single linkage group in a uniparental fashion (Sager, 1972; Schlanger & Sager, 1974). In this alga, like gametes fuse completely, and the physical basis for uniparental inheritance has been suggested to be the destruction of the chloroplast DNA of one of the gametes by a modification-restriction system of the type known in bacteria. As pointed out earlier, there are difficulties in concluding that this uniparental class of mutations resides in chloroplast DNA, but it is tempting to believe that it does. A simple interpretation suggests that some of the proteins of the chloroplast ribosomes are encoded in chloroplast DNA; there is also evidence that at least one other protein of the chloroplast ribosome is encoded in a nuclear gene (Mets & Bogorad, 1972).

Rigorous proof that a mutation inherited in a maternal or uniparental fashion actually alters the amino acid sequence of a protein has been obtained in only one case. Chan and Wildman (1972) studied the inheritance of a mutation in the large subunit of Fraction I protein in *Nicotiana tabacum* at the tryptic peptide level; this mutation was inherited *via* the maternal line only. By contrast, mutations in the small subunit of Fraction I protein are inherited in a Mendelian fashion (Kawashima & Wildman, 1972). This type of combined biochemical-genetic approach is very promising and deserves further use, especially with cultivated plants where many varieties are available. It must be emphasized that this approach must be conducted at the level of the tryptic peptide analysis of a purified protein. It is not sufficient to show that particular proteins are absent in a mutant variety, since absence of a protein might result from a mutation in a chloroplast gene which controls the formation of a protein encoded in the nucleus.

DNA-RNA hybridization studies

If RNA isolated from chloroplasts can be shown to hybridize to chloroplast DNA but not to nuclear DNA, this is good grounds for believing that such RNA is both encoded in and transcribed from chloroplast DNA. A number of hybridization studies have been carried out with chloroplast ribosomal-RNA, which hybridizes to 0·5–6·0% of chloroplast DNA (Ellis & Hartley, 1974). The most recent study is that of Thomas and Tewari (1974); they found that in a number of higher plants each circle of chloroplast DNA contains two genes for chloroplast ribosomal-RNA. There are several reports that chloroplast ribosomal-RNA will also hybridize to nuclear DNA. The significance of these reports awaits further study, but the possibility is raised that there are two types of chloroplast ribosomal-RNA, one encoded in chloroplast DNA and the other in nuclear DNA. This arrangement could be a means whereby the nucleus exerts control over events in the chloroplasts, especially if it is further postulated that chloroplast ribosomes containing nuclear RNA translate only messengers originating in the nucleus.

There is evidence that some of the transfer RNA species found in chloroplasts will also hybridize to chloroplast DNA. Tewari and Wildman (1970) found that tRNA from tobacco chloroplasts would hybridize to 0·4–0·7% of chloroplast DNA. This amount of DNA would be enough to code for 20 to 30 transfer RNA molecules, each of molecular weight 25,000. This work used unfractionated tRNA however, and more studies need to be done to establish the site of encoding of individual tRNA species.

Identification of RNA and protein molecules synthesized by isolated chloroplasts

This is the most direct method; if transcription coupled to translation can be obtained in isolated chloroplasts, identification of the products would simultaneously determine both the structural genes present in chloroplast DNA and the function of chloroplast ribosomes. Work in the author's laboratory has shown that isolated intact chloroplasts from *Pisum sativum* and *Spinacia oleracea* will synthesize discrete protein and RNA molecules, but transcription and translation are not coupled (Hartley & Ellis, 1973; Ellis, 1974; Ellis & Hartley, 1974). It is therefore not possible to infer that the proteins synthesized by isolated chloroplasts are encoded in chloroplast DNA. The RNA synthesized by isolated chloroplasts has been analyzed by polyacrylamide gel electrophoresis (Fig. 11.3). The chief product is a species of molecular weight about $2{\cdot}7 \times 10^6$; this has been shown by competitive hybridization to chloroplast DNA to be a precursor to chloroplast ribosomal-RNA. Isolated chloroplasts have not been shown convincingly to synthesize ribosomal-RNA; presumably the processing system which trims the precursor molecules does not survive the trauma of isolation (see also chapter 9).

The known genes in chloroplast DNA can be summarized as follows: chloroplast transfer-RNA, chloroplast ribosomal-RNA, several chloroplast ribosomal proteins, and the large subunit of Fraction I protein. These genes account for less than 10% of the total potential coding capacity of a 40 μm circle of DNA. The elucidation of the function of the remaining 90% of chloroplast DNA is the most important problem in this field.

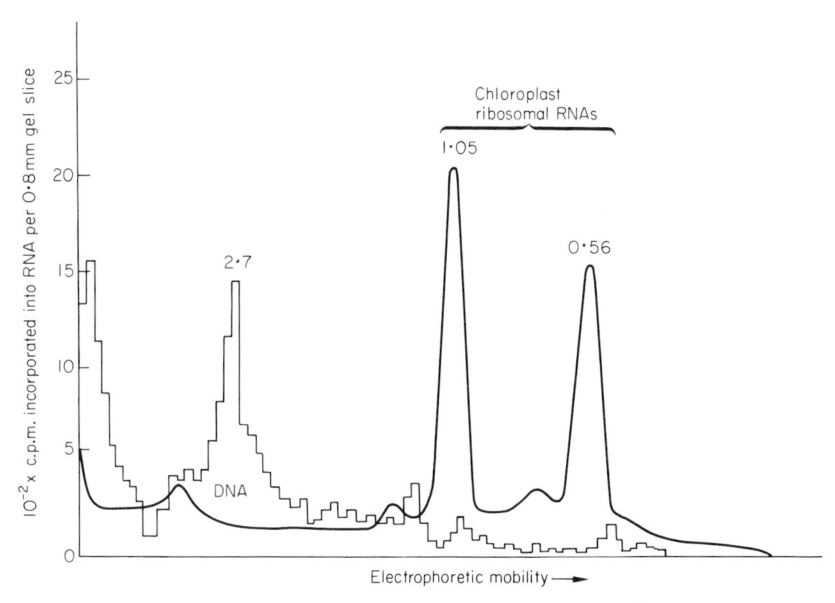

Fig. 11.3. Analysis by polyacrylamide gel electrophoresis of RNA synthesized by chloroplasts isolated from *Spinacia oleracea.* Isolated chloroplasts were incubated with ^{3}H-uridine and 15,000 lux of red light for 20 min at 20°C. Nucleic acid was extracted and run on polyacrylamide gels. The solid line represents the A_{260}, while the histogram shows the radioactivity. The figures are the molecular weights $\times 10^{-6}$ of the RNA components. (Reprinted from Hartley & Ellis, 1973, by courtesy of the Biochemical Journal).

11.2.2 CHLOROPLAST DNA POLYMERASE

DNA polymerase activity has been detected in chloroplast suspensions prepared from *Spinacia oleracea*, *Nicotiana tabacum*, and *Euglena gracilis*. The product renatures readily after heat denaturation, and hybridizes to a much larger extent with chloroplast DNA than with nuclear DNA. It has not been established whether the chloroplast polymerase is identical with any nuclear polymerase or whether it carries out either a replicase or a repair function. The enzyme is bound to the chloroplast membranes, but in the case of *Euglena*, it has been solubilized by treatment with high concentrations of salt and highly purified; the purified enzyme is inhibited by ethidium bromide (Keller *et al.*, 1973). Flechtner and Sager (1973) have made the interesting observation that treatment of *Chlamydomonas* cells with ethidium bromide induces a selective and reversible inhibition of chloroplast DNA replication. Thus nuclear DNA synthesis proceeds normally while replication of chloroplast DNA is impaired; the pre-existing chloroplast DNA decreases by at least 80% in one cell generation, but this loss is reversible if the drug is removed within 12 hours. This result suggests that one or a few of the chloroplast DNA copies may be protected in some way, perhaps by close attachment to chloroplast membranes. Such sequestered DNA could act as a 'master' copy, and might account for the diploid behaviour of the uniparental linkage group in this organism. The chloroplast DNA of *Euglena* and *Chlamydomonas* has been shown to replicate in a semi-conservative fashion at a different time in the growth cycle from nuclear DNA, but the factors controlling the time and rate of synthesis of chloroplast DNA are unknown.

11.2.3 CHLOROPLAST RNA POLYMERASE

DNA-dependent RNA polymerase activity has been demonstrated in chloroplast preparations from several species of algae and higher plants. Most workers study lysed chloroplast preparations because the nucleotide triphosphates used as substrates penetrate the chloroplast envelope at slow rates. Unlike intact

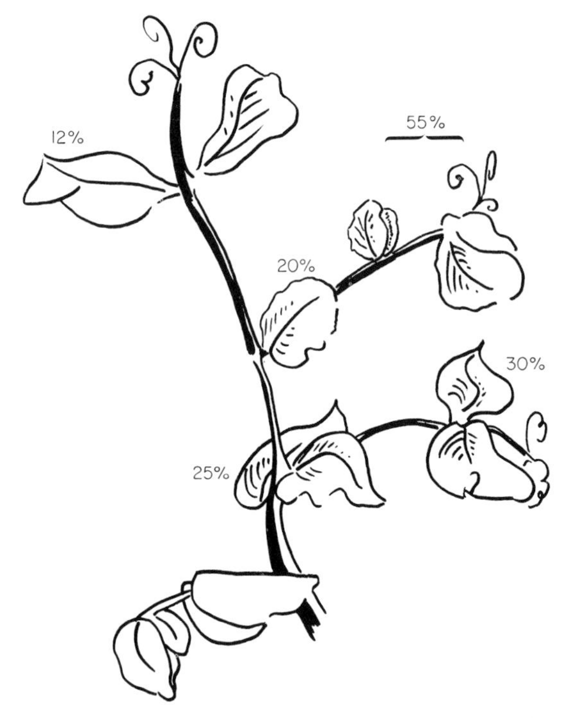

Fig. 11.4. Distribution of chloroplast RNA polymerase activity in leaves of young shoots of *Pisum sativum*. Pea seedlings were grown for 14 days and the leaves homogenized. Total RNA polymerase activity was measured in the low-speed pellet containing nuclei and chloroplasts, and chloroplast RNA polymerase activity in purified chloroplast membranes. The figures are the activities of the chloroplast RNA polymerase as percentages of the total activity. (Unpublished work of J. Bennett.)

chloroplasts, which use light to phosphorylate added ^{3}H-uridine (Fig. 11.3), lysed preparations do not synthesize discrete species of RNA; instead a polydisperse pattern of products ranging in size from 5s to 23s is obtained (Spencer & Whitfeld, 1967). This difference between the products from intact and lysed chloroplasts may result from the dilution of controlling factors on lysis.

The RNA polymerase activity of chloroplast preparations can represent a high proportion of the total RNA polymerase activity measurable in leaf extracts. Figure 11.4 illustrates this point in the case of the growing shoot of *Pisum sativum*; the chloroplast polymerase activity can account for half of the total activity in the youngest leaves at the stem apex, but accounts for progressively less in the older leaves.

There are two key questions about chloroplast RNA polymerase which need to be answered: firstly, how similar is it to any nuclear polymerase, and secondly, how is its activity regulated? The enzyme is bound to the chloroplast lamellae, but a technique for its quantitative removal has been devised (Bennett & Ellis, 1973). The solubilized chloroplast polymerase has been purified, and appears very similar to a nuclear polymerase (Bogorad *et al.*, 1973), but a detailed characterization has yet to be reported. It has been shown that the synthesis of chloroplast ribosomal-RNA in dark-grown plants is phytochrome-mediated (Scott *et al.*, 1971), but the nature of the mechanism which links phytochrome to RNA polymerase is unknown.

11.2.4 CHLOROPLAST PROTEIN SYNTHESIS

Chloroplast preparations capable of incorporating labelled amino acids into protein have been isolated from higher plants and algae. Table 11.3 illustrates some results obtained with preparations from young bean and tobacco leaves; preparations from mature leaves have little activity. These results are typical of those reported in many studies with many species (Ellis *et al.*, 1973). In such work it is vital to establish that the incorporation is due to chloroplasts and not to microorganisms, intact leaf cells, or cytoplasmic ribosomes. The best criterion is the dependence of incorporation on an added energy source; this is ATP in the case of lysed chloroplasts, or light in the case of intact chloroplasts which can carry out photophosphorylation. Dependence on an added energy source eliminates intact leaf cells and microorganisms as the agents of incorporation. Activity by contaminating cytoplasmic ribosomes can be ruled out by the different sensitivity of the two types of ribosome to antibiotic inhibitors (see 11.2.4.5). Other criteria are the sensitivity of incorporation to added ribonuclease, and to variation in the concentration of added Mg^{2+} ions; these criteria are useful when lysed chloroplasts or isolated ribosomes are being tested, but not when the incorporation is due to intact chloroplasts.

Some of the characteristics and functions of protein synthesis by chloroplasts will now be considered.

Table 11.3. Characteristics of [^{14}C]-leucine incorporation into protein by isolated lysed chloroplasts. A pellet containing chloroplasts was resuspended in sterile 50 mM Tris-HCl, 10 mM $MgCl_2$, 100 mM KCl, 5 mM 2-mercaptoethanol (pH 7·8), and incubated for 40 min at 25°C with 2 mM ATP, 0·2 mM GTP, and 1 μC [^{14}C]-leucine. Total protein was extracted and counted. The incorporation obtained in the complete system is called 100. (Unpublished results of R.J. Ellis.)

Assay Conditions	*Phaseolus vulgaris*	*Nicotiana tabacum*
Complete	100	100
Zero time	0·5	1·5
GTP and ATP omitted	3	5
Ribonuclease added	5	6
Puromycin added	21	22
Chloramphenicol added	23	25
Actinomycin D added	100	100
Cycloheximide added	90	95

11.2.4.1 *Ribosomes*

Lyttleton (1962) was the first to show that green plant cells contain two classes of ribosome which differ in their sedimentation coefficients. Chloroplasts contain 70s ribosomes while the cytoplasm contains 80s ribosomes. Figure 11.5 shows a density gradient analysis of an homogenate of spinach leaves, from which it can be seen that chloroplast ribosomes constitute a high proportion of the total cellular complement of ribosomes; values as high as 60% have been reported.

Chloroplast ribosomes resemble those from prokaryote cells in their s value. Their RNA components are also of the same size as those found in prokaryote ribosomes, i.e. 16s and 23s. The 80s cytoplasmic ribosomes by contrast, contain 18s and 25s RNA molecules. These similarities between chloroplast and prokaryote ribosomes have been much stressed in the past, but it is now clear that these similarities are of size only, and not in the primary structure of the RNA and protein components. For example, ribosomal-RNA from *Escherichia coli* does not compete with chloroplast ribosomal-RNA from *Euglena* for hybridization to chloroplast DNA. The protein complement of prokaryote ribosomes is also quite different from that of chloroplast ribosomes as judged by gel electrophoresis and immunological tests. Similarly, the proteins of chloroplast and cytoplasmic ribosomes from the same plant differ distinctly in the patterns they give on gel electrophoresis.

Some of the chloroplast ribosomes are bound to the internal membranes. Electron microscopy has revealed whorl-like polyribosomes attached to the granal and intergranal thylakoid membranes, while analysis of isolated chloroplasts shows that up to 50% of the ribosomes cannot be removed from the membranes by washing with hypotonic buffer. Tao and Jagendorf (1973) have

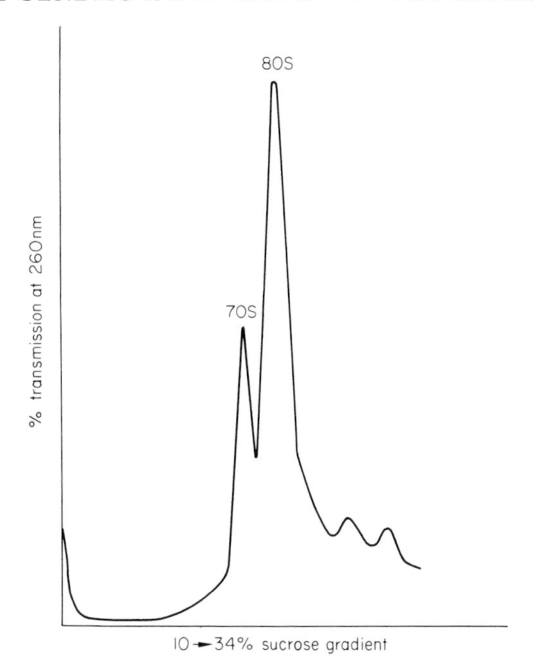

Fig. 11.5. Proportion of chloroplast ribosomes in leaves of *Spinacia oleracea*. Extracts of spinach leaves were treated with detergent to release membrane-bound ribosomes, and all the ribosomes collected by centrifugation. The ribosomes were analysed on a linear sucrose density gradient. (Unpublished work of M. R. Hartley.)

shown that some free ribosomes are lost during chloroplast isolation without the chloroplasts lysing irreversibly; when this loss is taken into account, they estimate that in chloroplasts from *Pisum sativum*, about 20% of the ribosomes are membrane-bound while the remainder are free in the stroma. There is evidence that the two classes of chloroplast ribosome synthesize different types of protein (see 11.2.4.6).

11.2.4.2 *Amino Acid Activation*

Chloroplasts isolated from several algae and higher plants have been found to contain aminoacyl-transfer-RNAs and the corresponding synthetases (Ellis *et al.*, 1973). These transfer RNA and synthetase molecules can often, but not always, be distinguished from those involved in the activation of the same amino acids in the cytoplasm. Studies of the compatibility between transfer-RNA and synthetases from the two cellular compartments show that the specificity of interaction ranges from none to absolute. For example, the cytoplasm of leaves of *Phaseolus vulgaris* contains a leucyl-transfer-RNA synthetase which can add leucine only to the two leucyl-transfer-RNA species common to the cytoplasm and chloroplasts, but is not able to recognize the three leucyl-transfer-RNA species found only in the chloroplasts. The chloroplast enzyme, on the other hand, can aminoacylate all five leucyl-transfer-RNA species. It is difficult to attach any biological significance to these variations in specificity. The simplest presumption is that in the intact cell the chloroplast synthetases aminoacylate only the transfer-RNA species located in the chloroplast and these are sufficient for chloroplast protein synthesis to proceed, while the cytoplasmic synthetases aminoacylate only the transfer-RNA species located in the cytoplasm. A more interesting possibility is that some of the transfer RNA species required to translate chloroplast messenger-RNA are acylated in the cytoplasm; such a requirement would provide one way in which protein synthesis in the cytoplasm and chloroplast could be integrated, but there is no direct evidence to support this suggestion. Experiments with *Euglena* suggest that at least some of the chloroplast aminoacyl-transfer-RNA synthetases are encoded in nuclear DNA and are synthesized on cytoplasmic ribosomes (Hecker *et al.*, 1974).

11.2.4.3 *Initiation of Chloroplast Protein Synthesis*

Protein synthesis in bacteria is initiated by N-formyl-methionyl transfer RNA; there is evidence that this is also true in chloroplasts, but not in the cytoplasm of eukaryotic cells. Schwarz *et al.* (1967) found that chloroplast ribosomes from *Euglena* would translate RNA from bacteriophage f2 into viral coat protein with N-formylmethionine at the amino terminus. Detailed studies of initiating methionyl-transfer-RNA species have since been reported for several higher plants. For example, there are two methionyl-transfer-RNA species in wheat chloroplasts. One of these can be formylated by a chloroplast transformylase, whereas the other cannot, and may serve to direct methionine into internal positions in the polypeptide chains (Leis & Keller, 1971). Two other methionyl-transfer-RNA species are found in the cytoplasmic fraction, neither of which can be formylated. It is probable from this type of evidence that the initiation of protein synthesis by chloroplasts is similar to that in prokaryotes in that it uses a formylated methionyl-transfer-RNA, but distinct from that in the cytoplasm which uses an unformylated methionyl-transfer-RNA. The significance of this distinction is not clear; it may reflect the evolutionary origin of the compartments in eukaryotic cells, and have no contemporary functional value.

11.2.4.4 *Energy Source*

In most studies of protein synthesis by isolated chloroplasts, added ATP has been used as an energy source (see Table 11.3). Since isolated chloroplasts can generate ATP by photophosphorylation, it should be possible to drive protein synthesis in chloroplasts with light. Spencer (1965) found that spinach chloroplasts would use light to incorporate amino acids into protein provided that ADP, inorganic phosphate and pyocyanine were added. The necessity for pyocyanine as catalyst indicates that the chloroplasts were broken, and had lost their natural catalyst, ferredoxin, by dilution. If precautions are taken to isolate chloroplasts which have their outer envelopes intact, protein synthesis is stimulated twenty-fold by light in the absence of either cofactors or catalysts of photophosphorylation (Fig. 11.6). The rates of light-driven protein synthesis

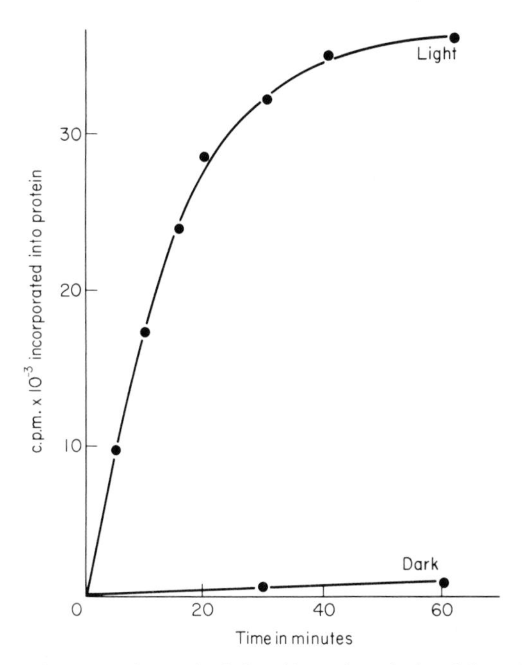

Fig. 11.6. Light-driven protein synthesis by chloroplasts isolated from *Pisum sativum*. Intact isolated chloroplasts were incubated with ^{14}C-leucine and 4,000 lux red light at 20°C. Total protein was then extracted and its radioactivity measured. (Reprinted from Blair & Ellis, 1973, by courtesy of Elsevier.)

by such chloroplasts are the highest yet recorded for any isolated chloroplast system. The use of intact chloroplasts for studies of protein synthesis has a major advantage over the use of broken chloroplasts in that conditions around the polysomes are more normal with respect to controlling factors. It is thus more likely that correct termination and release of polypeptide chains occurs in intact chloroplasts than in lysed preparations; such intact chloroplasts make discrete protein molecules rather than incomplete polypeptide chains (see 11.2.4.6). A further advantage of using light as the energy source is that the broken chloroplasts, which inevitably contaminate the preparation of intact chloroplasts, cannot contribute to the incorporation in the absence of added cofactors.

It is possible that protein synthesis by chloroplasts *in vivo* uses ATP provided by the chloroplast itself. However, this is not the case in developing chloroplasts, since organisms such as *Chlorella*, *Chlamydomonas*, and *Pinus* do not require light to form chloroplasts. The formation of chloroplasts must therefore depend on ATP supplied by the rest of the cell. This conclusion is confirmed by the ability of intact isolated etioplasts from *Pisum sativum* to use ATP, but not light, as their source of energy for protein synthesis (Siddell & Ellis, 1975).

11.2.4.5 *Inhibitors*

It is well established that protein synthesis by isolated chloroplast ribosomes is inhibited by the same antibiotics which inhibit protein synthesis by prokaryote ribosomes. The best studied example is chloramphenicol, which inhibits protein synthesis by isolated chloroplasts from all the species so far tested; inhibition is shown only by the D-threo isomer of chloramphenicol, and not by the other stereo isomers. Three other unrelated antibiotics, spectinomycin, lincomycin, and erythromycin also inhibit protein synthesis by chloroplast ribosomes. Light-driven protein synthesis by chloroplasts from *Pisum sativum* is especially sensitive to lincomycin, 50% inhibition being given by 0·2 μg ml^{-1} (Ellis & Hartley, 1971). Protein synthesis by cytoplasmic ribosomes from plants is not inhibited by any of these antibiotics. On the other hand, cytoplasmic ribosomes are inhibited by cycloheximide, which does not affect the activity of chloroplast ribosomes. It must be emphasized that all these findings relate only to the activity of isolated sub-cellular systems; it cannot be assumed from these data alone that it is valid to use these compounds on intact cells with the expectation of inhibiting protein synthesis in one compartment but not in the other.

The protein complements of chloroplast and bacterial ribosomes are different, and thus the above similarity between them must reside in their antibiotic-binding sites, and not in the properties of the proteins as revealed by gel electrophoresis or immunological tests. There is no evidence that the mechanism of action of bacterial antibiotics on chloroplast ribosomes is similar to that on bacterial ribosomes; nor is there any understanding of the selective pressure which maintains the sensitivity of chloroplast ribosomes to these antibiotics in the face of mutations to resistance, which can, for example, be found in *Chlamydomonas* (see 11.2.1.6).

11.2.4.6 *Functions*

The high proportion of plant ribosomes located inside chloroplasts raises the question as to their function. Are they required in such quantities because they make a large number of different chloroplast proteins, or because they make a small number of different chloroplast proteins in large amounts? The available evidence suggests that the latter is the case (Ellis *et al.*, 1973; Ellis, 1974).

The problem of identifying which proteins are synthesized by chloroplast ribosomes has been tackled in two ways. The first is to supply antibiotic inhibitors, especially chloramphenicol and cycloheximide, to intact cells making chloroplasts, and determining which proteins are no longer synthesized. The validity of results from experiments with any inhibitor depends absolutely on the specificity of its action in intact cells, and there is evidence that both chloramphenicol and cycloheximide have effects on activities other than protein synthesis in some higher plants (Ellis & MacDonald, 1970). Processes such as ion uptake, oxidative phosphorylation and photophosphorylation are inhibited by all four stereoisomers of chloramphenicol, whereas the inhibition of protein synthesis by isolated chloroplast ribosomes is specific for the D-threo isomer (Ellis, 1969). This stereospecificity provides a means of establishing for any particular tissue whether chloramphenicol is inhibiting protein synthesis directly at the ribosomal level, or in addition, is affecting some other process such as energy supply. It is strongly recommended that only if an inhibition is produced specifically by the D-threo isomer should an interpretation directly involving protein synthesis be invoked. Another problem is that in most of the inhibitor experiments on chloroplast ribosomal function, increases in specific proteins have been measured as enzymic activities rather than as amounts of protein. Failure to observe an effect by a particular inhibitor might therefore mean that the increase in enzymic activity is due to activation of a precursor protein, rather than to *de novo* synthesis by either chloroplast or cytoplasmic ribosomes.

Bearing these difficulties in mind, this author interprets the bulk of the published inhibitor experiments as suggesting that most of the chloroplast proteins are synthesized on cytoplasmic ribosomes (Ellis *et al.*, 1973; Ellis, 1975). In all the studies reported on several algae and higher plants, the synthesis of Fraction I protein was found to be inhibited by 70s ribosomal inhibitors. In most cases, the synthesis of the other soluble enzymes of the photosynthetic carbon dioxide reduction cycle appears to occur on cytoplasmic ribosomes; the same is true for ferredoxin and the chloroplast RNA polymerase. Besides Fraction I protein, the only other proteins which appear to be synthesized by chloroplast ribosomes are some of the chloroplast ribosomal and lamellar proteins, including the photosynthetic cytochromes. It must be emphasized, however, that these inhibitor experiments are never more than suggestive. Strictly interpreted, they never say more than that the activity of a particular group of ribosomes is required for a particular protein to accumulate in the chloroplast. This is not the same as saying that these ribosomes actually synthesize that protein, because it is possible that the apoenzyme is synthesized by one class of ribosomes but requires for its appearance in the chloroplast in an active state, additional protein(s) which are synthesized by another class of ribosomes.

That this is true for Fraction I protein has been shown by the second approach to the problem of chloroplast ribosomal function, namely, the synthesis of specific proteins by isolated sub-cellular systems. This is the most direct approach to this problem but it has been successful only recently because of the difficulty of isolating sub-cellular systems which will carry out complete protein synthesis with fidelity; algal systems are especially difficult in this respect. Isolated intact chloroplasts from *Pisum sativum*, however, synthesize at least six discrete proteins (Fig. 11.7); similar results have been obtained with

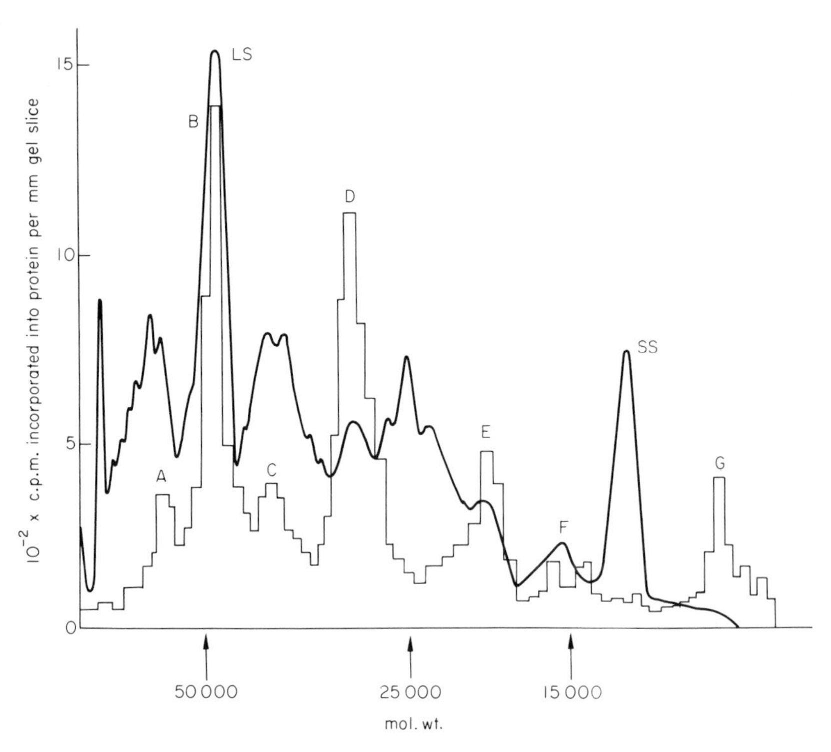

Fig. 11.7. Products of protein synthesis by chloroplasts isolated from *Pisum sativum*. Chloroplasts were incubated as in Fig. 11.6, and then dissolved in sodium dodecylsulphate. The solubilized extract was electrophoresed on gels containing 15% polyacrylamide. The smooth line represents the absorbance of stained protein bands, and the histogram is radioactivity. The letters A–G represent the discrete radioactive peaks. LS and SS mark the large and small subunit of Fraction I protein respectively. (From Eaglesham & Ellis, 1974, by courtesy of Elsevier.)

chloroplasts from *Hordeum vulgare* and *Spinacia oleracea*. Only one of these proteins (peak B in Fig. 11.7) is soluble; it has been identified as the large subunit of Fraction I protein by tryptic peptide analysis (Blair & Ellis, 1973). Recent work has shown this large subunit to be made by the free ribosomes of the chloroplast, but not by the membrane-bound ribosomes. The high proportion of chloroplast ribosomes found in leaves may thus be required because Fraction I protein is one of the most abundant proteins (Ellis, 1973). The small subunit of Fraction I protein is not synthesized by isolated chloroplasts, but it has been detected as the product of protein synthesis by isolated cytoplasmic ribosomes from *Phaseolus vulgaris* leaves (Kekwick & Gray, 1973). The other proteins made by isolated chloroplasts are all membrane-bound, and have so far resisted identification; it is clear they are minor components of the thylakoids. Two proteins associated with the chloroplast envelope are also labelled in isolated pea chloroplasts (Joy & Ellis, 1975). The same pattern of proteins synthesized by intact pea chloroplasts has been found in studies of protein synthesis by isolated pea etioplasts (Siddell & Ellis, 1975).

This *in vitro* approach to the study of chloroplast ribosomal function gives direct and unambiguous results, and it should be extended to species lower on the evolutionary scale to see how far the pattern found in angiosperms extends. In view of the uniformity of the size of the chloroplast genome in algae and higher plants (Table 11.2), it seems likely that the pattern seen in isolated pea chloroplasts will be universal. However, this assumes that proteins synthesized by chloroplast ribosomes are also encoded by chloroplast DNA; this may not be true if messenger-RNA can cross the chloroplast envelope, but so far this is only a theoretical possibility.

Another area which may repay attention is the study of protein synthesis by plastids other than chloroplasts, especially proplastids and amyloplasts. It seems probable that these plastids synthesize a different range of proteins from chloroplasts. If these proteins are encoded in plastid DNA, they would account for some of the functions of the 90% of this genome which have not been identified (see 11.2.1.7).

11.2.5 CO-OPERATION BETWEEN CHLOROPLAST AND NUCLEAR GENOMES

It is clear from the available evidence that the chloroplast genome requires for its expression the co-operation of the nuclear genome. The best information about the details of this co-operation concerns the synthesis of Fraction I protein. By combining the results of Wildman's genetic studies with those of the *in vitro* studies of protein synthesis, a model for the synthesis of Fraction I protein can be constructed (Fig. 11.8). In this model, the large subunit is both encoded and synthesized within the chloroplast, while the small subunit is both encoded and synthesized outside the chloroplast. This model is tidy, and requires protein, but not nucleic acid, to cross the chloroplast envelope. This transport of protein must be on a large scale because it involves not only the small subunit of Fraction I protein but all the other proteins which, from the inhibitor evidence, are made on cytoplasmic ribosomes. The mechanism of this transport is unknown, but it must be able to distinguish between different proteins. One suggestion is that a protein exists in the outer envelope of the chloroplast which recognizes a site common to all those proteins made on cytoplasmic ribosomes, but destined to function in the chloroplast.

Besides structural and enzymic proteins, regulatory proteins probably also cross the chloroplast envelope. How light regulates chloroplast development is not known, but it would be reasonable to speculate in terms of proteins entering the organelle to trigger nucleic acid and protein synthesis. There is the possibility of proteins also passing out of the organelle, since recent work in *Chlamydomonas* has implicated chloroplast protein synthesis in the regulation of nuclear DNA replication (Blamire *et al.*, 1974). This question of protein transport into and out of the chloroplast is a crucial one to study if we are to understand how the chloroplast and nuclear genomes interact; almost no research has been carried out in this area. The movement of nucleic acids across the envelope remains a possibility, but no compelling evidence to support it has been published.

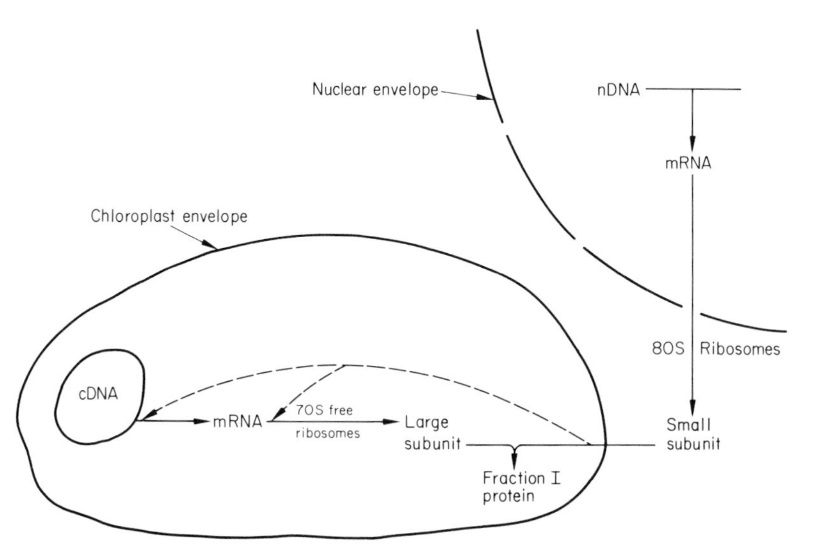

Fig. 11.8. Model for the co-operation of nuclear and chloroplast genomes in the synthesis of Fraction I protein. cDNA and nDNA stand for chloroplast and nuclear DNA respectively. The dashed lines indicate possible sites at which small subunits may control the synthesis of large subunits. (Reprinted from Ellis, 1975, by courtesy of Pergamon Press.)

Another problem needing study is the nature of the mechanism which integrates the synthesis of the two subunits of Fraction I protein in different cellular compartments. The dashed lines in Fig. 11.8 illustrate one possibility; the small subunit is postulated to be required as a positive factor for the initiation of either transcription or translation of the large subunit messenger (Ellis, 1975). The recent detection of the large subunit messenger should allow this possibility to be tested (Hartley *et al.*, 1975).

11.3 MITOCHONDRIAL AUTONOMY

Research on the biogenesis of plant mitochondria has lagged far behind that on mitochondria from animal and fungal cells. A summary of the position is given here in the hope of stimulating readers to undertake work in this field; recent reviews have appeared on the biogenesis of both animal and fungal mitochondria (Borst, 1971, 1972), and of those in plants (Boulter *et al.*, 1972; Leaver & Harmey, 1973; Leaver, 1975).

The first point to emphasize is that mitochondria constitute a much smaller compartment in plant cells compared to chloroplasts. The amount of mitochondrial DNA and ribosomes is correspondingly much smaller as a percentage of the total cellular complement. The technical difficulties of extracting small quantities of intact, uncontaminated mitochondria from plant tissues make studies on their genetic system more difficult than in the case of chloroplasts. Another difference from chloroplasts is that the DNA and ribosomes of mitochondria are more variable in their properties between species, so the extrapolation of results must be done with caution.

In a range of higher plants the density of mitochondrial DNA is constant at 1·706–1·707 g cm^{-3}, whereas the nuclear DNA from the same species varies in density between 1·691–1·702 g cm^{-3}. In many organisms mitochondrial DNA has been isolated in circular form; in contrast to the situation for chloroplasts (see 11.2.1.5), the contour length of the circles varies with the species. Thus, animals have mitochondrial DNA circles of about 5 μm contour length; this corresponds to a molecular weight of about 10^7, or 15,000 base pairs. In Ascomycetes the contour length of circular mitochondrial DNA is in the range 20–26 μm, while in higher plants lengths of 30 μm have been reported. The small size of the mitochondrial DNA in animals means that it can code for only a few components. Hybridization studies in *Xenopus* indicate that mitochondrial ribosomal RNA and at least some mitochondrial transfer RNA are encoded in mitochondrial DNA. The remaining mitochondrial DNA is sufficient to code for only about ten proteins each of molecular weight 50,000. However, kinetic complexity measurements on mitochondrial DNA from higher plants indicate that it has between six and ten times the coding potential of animal mitochondrial DNA. This observation raises the possibility that mitochondria from plants possess a higher degree of autonomy than those from animals, and this reinforces the plea that more research be carried out with plant cells.

The sedimentation coefficients of mitochondrial ribosomes and their component ribosomal-RNAs vary greatly with the species. Animal mitochondria contain so-called 'miniribosomes' which sediment between 55S and 60S, while fungal mitochondria contain 70–74S ribosomes. Leaver and Harmey (1972) have shown that mitochondrial ribosomes from several higher plants sediment at 77S–78S, while their major RNA components sediment at 25S and 18S. Although mitochondrial ribosomes differ in size in different species, they are all similar in that protein synthesis by them is sensitive to the same set of antibiotics which inhibit prokaryote and chloroplast ribosomes.

Protein synthesis by isolated mitochondria has been studied extensively in animals and fungi, but hardly at all in higher plants. Several proteins of the inner, but not the outer, mitochondrial membrane are synthesized by animal and fungal mitochondria. These proteins have been identified as subunits of the ATPase complex, cytochrome *b*, and cytochrome *c* oxidase; the other subunits of these insoluble complexes are synthesized by cytoplasmic ribosomes. This evidence suggests the synthesis of mitochondrial membranes requires the co-operative activity of both mitochondrial and nuclear genomes. This idea of the co-operation between the several genetic systems in eukaryotic cells is already familiar to the reader from section 11.2.5. The same problem of the transport of proteins across the bounding membranes applies to mitochondria as well as to chloroplasts. In contrast to chloroplasts, mitochondria from animal and fungal cells synthesize no soluble proteins; whether higher plant mitochondria synthesize any soluble proteins is unknown.

No protein has so far been rigorously identified as being encoded in mitochondrial DNA in any organism. A simple presumption is that those proteins synthesized by isolated mitochondria are encoded in mitochondrial DNA. It is likely that the integrative controls between nuclear and mitochondrial genomes in higher plants differ from those in animals which contain less than one sixth as much DNA. Why do plant mitochondria apparently contain so much more genetic information to perform the same functions as animal mitochondria? Answering this question is a challenge which must be taken up if our understanding of the origin of this organelle in plants is to improve.

FURTHER READING

Borst P. (1971) Size, structure and information content of mitochondrial DNA. In *Autonomy and Biogenesis of Mitochondria and Chloroplasts*, (eds N.K. Boardman, A.W. Linnane & R.M. Smillie), pp. 260–6. North Holland, Amsterdam.

Borst P. (1972) Mitochondrial nucleic acids. *Ann. Rev. Biochem.* **41**, 333–76.

Boulter D., Ellis R.J. & Yarwood A. (1972) Biochemistry of protein synthesis in plants. *Biol. Rev.* **47**, 113–75.

Ellis R.J. (1975a) The synthesis of chloroplast membranes in *Pisum sativum*. In *Membrane Biogenesis: Mitochondria, Chloroplasts & Bacteria*, (ed. A. Tzagoloff). Plenum Publishing Co., New York.

Ellis R.J., Blair G.E. & Hartley M.R. (1973) The nature and function of chloroplast protein synthesis. *Biochem. Soc. Symp.* **38**, 137–62.

Ellis R.J. & Hartley M.R. (1974) Nucleic acids of chloroplasts. In *Biochemistry of Nucleic Acids*, (ed. K. Burton). MTP International Review of Science Biochemistry Series One, Vol. 6, 323–48. Butterworths, London & University Park Press, Baltimore.

Kirk J.T.O. (1972) The genetic control of plastid formation: recent advances and strategies for the future. *Sub-Cell. Biochem.* **1**, 333–61.

Leaver C.J. (1975) The biogenesis of plant mitochondria. In *The Chemistry and Biochemistry of Plant Proteins*, (ed. J. Harborne). pp. 137–65. Academic Press, New York & London.

LEAVER C.J. & HARMEY M.A. (1973) Plant mitochondrial nucleic acids. *Biochem. Soc. Symp.* **38**, 175–93.

LEVINE R.F. & GOODENOUGH U.W. (1970) The genetics of photosynthesis and of the chloroplast in *Chlamydomonas reinhardi*. *Ann. Rev. Genet*, **4**, 397–408.

SAGER R. (1972) *Cytoplasmic Genes and Organelles*. Academic Press, New York & London.

TEWARI K.K. & Wildman S.G. (1970) Information content in the chloroplast DNA. In *Control of Organelle Development*, (ed. P.L. Miller), pp. 147–79. Symposium 24 of the Society for Experimental Biology, Cambridge University Press, Cambridge.

CHAPTER 12

REGULATION OF ENZYME LEVELS AND ACTIVITY

12.1 INTRODUCTION

The first theory of metabolic control was introduced into plant physiology in 1905 by F. F. Blackman and has become known as the 'Law of Limiting Factors' which states that 'when a process is conditioned as to its rapidity by a number of separate factors the rate of the process is limited by the pace of the slowest reaction'.

It is not possible to derive a rate equation from this statement and discussions of the 'Law' tend to be by analogy, e.g. the strength of a chain is the strength of its weakest link. Experimentally, the statement has led to the fruitless search for metabolic master reactions. Mathematically, we can derive rate equations for multistep reactions and establish that *every* step contributes to the overall rate. Theoretically Blackman's statement is a denial of the steady state; if a metabolic sequence is in a steady state then the concentration of each intermediate is constant and the individual reactions proceed at the same rate—hence no reaction can be described as the slowest.

12.1.1 PACEMAKERS

If in a metabolic sequence all reactions are proceeding at the same pace then a single reaction may be almost entirely responsible for determining that pace; such a reaction has been called a 'bottleneck' or pacemaker and much work in metabolic control is directed towards the identification of such control points. Krebs and Kornberg (1957) in a consideration of pacemakers stated 'there is a principle which may guide the search for pacemakers. As pacemakers are reactions of variable rate, the level of substrate concentration of the pacemaker must vary inversely with the rate: it must increase when the reaction rate decreases'. It should be possible to take published data for a metabolic pathway, apply the Krebs-Kornberg principle and so determine which reaction is the pacemaker. It turns out that in some cases the Krebs-Kornberg principle cannot be applied. For example in some tissues when the rate of glycolysis is increased not a single compound shows the expected reduction in concentration.

12.1.2 OCCAM'S RAZOR

William of Occam, the mediaeval controversialist, formulated a procedural rule, that when a number of possible solutions can be proposed for a problem, one should accept the simplest solution until it is shown to be untenable. The Krebs-Kornberg principle is a simple solution to a complex problem and in certain cases it yields a valid solution. However, when it is not applicable we must examine more complex solutions.

12.1.3 SYSTEMS PROPERTIES

The rate equation for a metabolic pathway includes parameters for all the enzymes and variables for all the metabolites (Waley, 1964). The flux through the pathway is a systemic property in which all the parameters interact as a system. If a single parameter is altered, say the activity of a single enzyme, then the whole system responds and adjusts to that change. The extent to which an enzyme can be considered a control step is the extent to which a fractional change in its activity produces a fractional change in the flux through the system. The ratio

$$\frac{\text{fractional change in flux}}{\text{fractional change in enzyme activity}}$$

has been termed the sensitivity coefficient for that step (Kacser & Burns, 1973). If a 1% change in activity of an enzyme produces a 1% change in flux through the system, then the sensitivity coefficient is 1 and the enzyme must be fully controlling the flux and is a pacemaker. However, the sum of all the sensitivity coefficients for a metabolic sequence is equal to unity and only in special circumstances can a single pacemaker be identified. That the sensitivity coefficient of a particular enzyme is a system property, only in part determined by its own parameters, can be intuitively understood by an example. In a given situation an enzyme may have a very small sensitivity coefficient and contribute little to the overall control of the system. If however the activity of this enzyme is drastically reduced its sensitivity coefficient may be increased and it may contribute significantly to the overall flux. Since the sum of sensitivity coefficients must always equal unit the sensitivity coefficients of all the other enzymes must have changed despite the fact that their parameters are unchanged.

Much work on metabolic control is concerned with the identification of control points. Since each reaction contributes to the overall control a sense of judgement is necessary to identify major control points and it must be remembered that control may pass from one reaction to another as the flux through the system changes.

12.2 THE IDENTIFICATION OF CONTROL POINTS

12.2.1 EQUILIBRIUM CONSIDERATIONS

Bücher and Russmann (1964) have proposed a model for metabolic control based on a controlled waterway (Fig. 12.1).

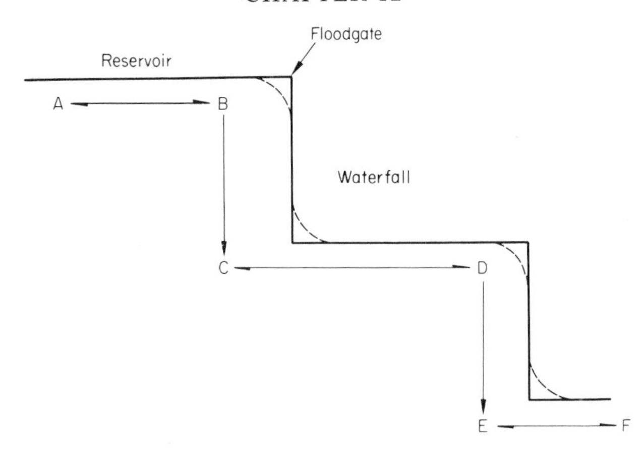

Fig. 12.1. Schematic representation of a regulated waterway. The solid line represents the situation when the flood gates are raised and there is little flow through the system. The dashed line represents the increased flow when the flood gates are lowered.

The flow of water through the system is controlled by the flood gates. With a flood gate up, there is a negligible gradient across the reservoir and a slight flow of water over the dam. In metabolic terms the reaction $A \leftrightharpoons B$ represents the reservoir and can be considered as a fast reaction close to equilibrium ($\Delta G \simeq 0$). The reaction $B \rightarrow C$ represents the waterfall and is a slow reaction far from equilibrium (ΔG is large).

When the flood gate is lowered a slight gradient is produced across the reservoir and there is a rapid flow of water over the dam, which could be utilized to perform work. In metabolic terms, as the flux through the system increases, the reaction $A \leftrightharpoons B$ moves away from equilibrium and the controlled reaction $B \rightarrow C$ increases in rate and simultaneously moves towards equilibrium. The model thus identifies control reactions as reactions far from equilibrium and which move towards equilibrium as the flux increases. Since ΔG for such reactions is large, control points will tend to be associated with reactions involved in the production of useful work by a coupled reaction. However, the useful work function is not an essential feature of the model and in theabsence of a coupled reaction to utilize the large ΔG, the loss of energy can be considered as the price paid for control.

12.2.2 THE CROSSOVER THEOREM

The crossover theorem states that: the variations of the concentrations of the metabolites before and after an enzyme which is a control point have different signs. It will be noted that this is only a formal statement of the situation at the waterfall in the Bücher and Russmann analogy, i.e. when the flux increases the concentration of the substrate falls and the concentration of the product increases. However, the theorem is widely used, possibly because it presents data in a graphical form. Metabolic intermediates are plotted on the **x** coordinate in the sequence in which they are formed in the metabolic pathway and the concentration of each substrate, expressed as a percentage of the initial value is plotted on the **y** coordinate. An example is shown in Fig. 12.2 for the situation usually referred to as the 'Pasteur Effect'. Pasteur observed that 'oxygen inhibits fermentation' so that when carrot discs are transferred from air to nitrogen glycolysis increases with associated changes in the levels of intermediates. These changes are plotted for a period some 6 minutes after transfer from air to nitrogen (Fig. 12.2).

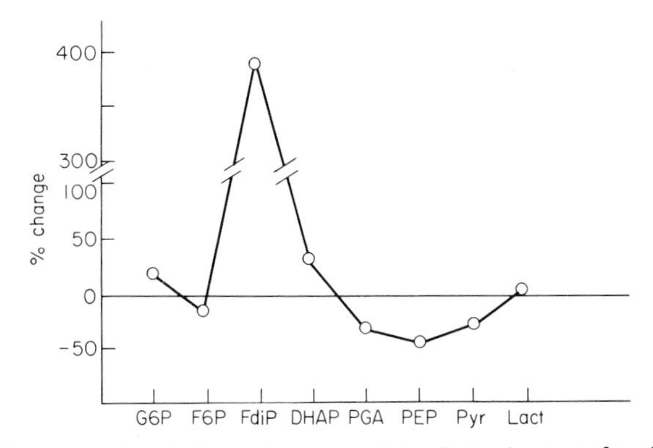

Fig. 12.2. Crossover plot of glycolytic intermediates following transfer of aged carrot discs from air to nitrogen (Data of Faiz-ur-Rahman, Trewavas & Davies, 1974). Intermediates were measured before transfer and 6 minutes after transfer from air to nitrogen.

The positive crossover observed between fructose-6-phosphate and fructose diphosphate suggests that phosphofructokinase is a control point. There is some confusion concerning the interpretation of negative crossovers. Some authors argue that a negative crossover does not indicate a control point. However, it must be remembered that when the flux through a system is changed, control can pass from one point to another. Thus in the case illustrated in Fig. 12.2 the increased flux through the glycolytic system involves an increase in activity of phosphofructokinase but with the increased rate of glycolysis, phosphoglycerate is removed more rapidly than it is formed, suggesting that a bottleneck exists between dihydroxyacetone phosphate and phosphoglycerate when phosphofructokinase is highly active.

The crossover theorem should be used with caution; the absence of a crossover should not be taken to mean the absence of a control point and there are many conditions which can produce crossovers indicating spurious control points. Before applying the crossover theorem, the reader is advised to consult the critique offered by Heinrich and Rapoport (1974) and to remember that control is a property of the system; the identification of a control point is to focus attention on one component knowing that to varying degrees all other enzymes in the sequence contribute to the overall control.

12.3 CONTROL MECHANISMS

Having identified possible control points it becomes necessary to determine the mechanism of control. Over a relatively long time scale control may be achieved by changing the amount of an enzyme but over a short time scale control is likely to be achieved by changing the activity of the enzyme.

12.3.1 CONTROL BY PRODUCT INHIBITION

Every enzyme has a certain affinity for its products and if the enzyme has a high affinity for its product then product inhibition will be pronounced. It appears likely that this form of control is restricted to some synthetic routes and to minor metabolic sequences.

12.3.2 CONTROL BY NEGATIVE FEEDBACK

The discovery of negative feedback in metabolic systems appears to have been made by Dische in 1940 who noted that 3-phosphoglycerate inhibits the phosphorylation of glucose in red cells and he proposed that this inhibition might play a regulatory function in glucose metabolism. His paper did not receive the attention it deserved and the idea of feedback had to be rediscovered.

In 1955 Roberts and his coworkers published the result of their extensive studies of the metabolism of *E. coli* grown on a medium containing ^{14}C-glucose. They found that the addition of one of a number of amino acids (known to be products of specific synthetic pathways) to the growth medium resulted in the virtually complete inhibition of incorporation of ^{14}C into that amino acid.

In 1956 Umbarger reported that isoleucine inhibited threonine deaminase—an enzyme which initiates the set of reactions leading to isoleucine biosynthesis. In the same year Yates and Pardee observed that cytidine monophosphate inhibits aspartate transcarbamylase—the first reaction of the pathway leading to the biosynthesis of the pyrimidines. These results offered a biochemical explanation for the end-product inhibition observed by Roberts with intact bacteria.

In 1959 Sir Hans Krebs introduced the first symposium to be held on metabolic control. Ideas on the regulation of carbohydrate and amino acid metabolism were discussed and the general ideas of feedback formulated. Control by negative feedback is an engineering concept and familiar examples include governers of steam engines, thermostats and automatic volume control in amplifiers. The limitations of the analogy between electrical feedback amplifiers and metabolic reactions have been stressed by Britton-Chance. However, as with other terminology developed for use with metabolic control systems (see discussion of allosteric enzymes) the tendency has been to dispense with rigorous definitions and the general definition of negative feedback is 'increased output decreases the input'.

12.3.3 PATTERNS OF CONTROL

The most frequently observed pattern of control is when the end product of a metabolic sequence inhibits the first reaction belonging to that sequence. For example serine can be formed by the reaction sequence shown in Fig. 12.3.

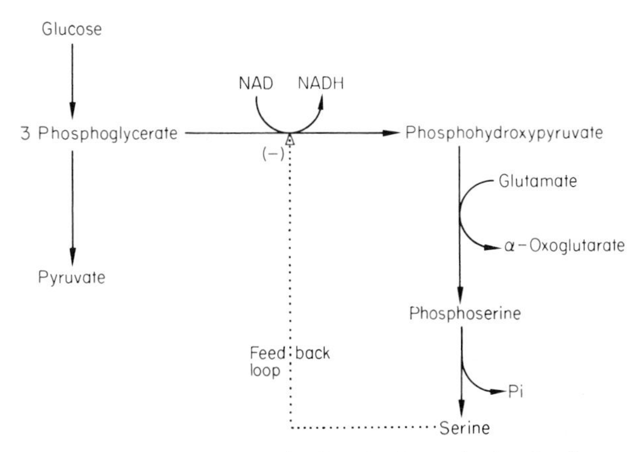

Fig. 12.3. Reactions involved in the biosynthesis of serine.

Serine inhibits 3-phosphoglycerate dehydrogenase and so controls its own biosynthesis. This simple mechanism is effective, presumably because a relatively small amount of the flux through the glycolytic pathway is directed towards serine and thus variations in the rate of serine production will not seriously perturb the glycolytic flux.

12.3.4 CONTROL OF BRANCHED PATHWAYS

Inhibition of the first step in a metabolic pathway by an end-product leads to special problems in the case of branched biosynthetic pathways. Consider the sequence

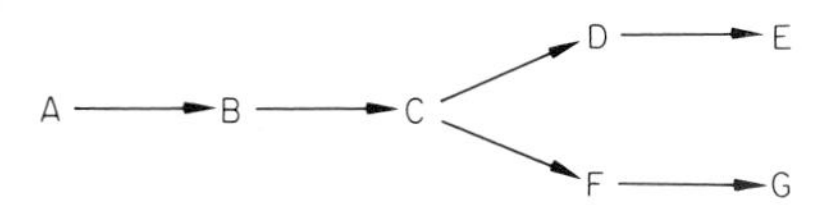

It the first common step (A→B) is inhibited by either or both end-products, then an excess of one could inhibit the step A→B and lead to a deficiency of the

other end-product. Nature has evolved several mechanisms which avoid, to varying degrees, these difficulties. Stadtman (1970) has classified these mechanisms; here we discuss a few examples which have been studied in plants.

12.3.4.1 *The Aspartate Family*

The formation of some amino acids from aspartate is outlined in Fig. 12.4 together with feedback controls which have been demonstrated.

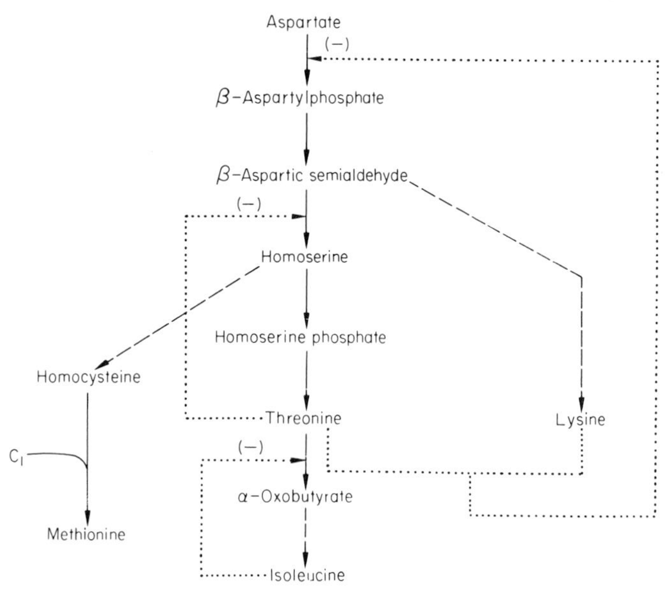

Fig. 12.4 Reactions involved in the biosynthesis of the aspartate family of amino acids. →Single step reaction; --->multi-step reaction; ···>feed back loop.

Aspartokinase which catalyses the first reaction of the pathway is inhibited by lysine and threonine in a concerted or synergistic manner. The term concerted feedback inhibition has been used by Stadtman to describe cases of inhibition by two end products when the single end products produce no inhibition. The term synergistic inhibition is used when the total inhibition is much greater than the sum of their independent effects. The aspartokinase from plants appears to be inhibited by low concentrations of lysine but not by low concentrations of threonine—when both are present a much greater inhibition is observed. Although this effect would seem best to fit Stadtman's definition of synergistic inhibition it has been called concerted inhibition. Whatever the terminology, this mechanism does not involve methionine and is thus only a partial solution to the control of the branched pathway. The non-involvement of methionine means that in conditions where threonine and lysine accumulate they will tend to inhibit the synthesis of methionine—even in situations where methionine may be in short supply. This imperfection is compounded by the feedback inhibition of homoserine dehydrogenase by threonine which further reduces the flow of intermediates available for the biosynthesis of methionine.

These imperfections in the control mechanism can be invoked to explain the effect of amino acids on the growth of *Lemna minor* (Table 12.1).

Table 12.1. Effect of amino acids on the growth of *Lemna minor* in 10 days (after Wong & Dennis 1973.)

Lemna was grown in a growth medium containing the indicated amino acid (0·25 mM).

Addition	Frond number
None	373
Aspartic acid	400
Lysine	230
Threonine	84
Homoserine	98
Isoleucine	73
Methionine	After 5 days signs of death
Lysine and threonine	Death before five days
Lysine and threonine and aspartic acid	,,
Lysine and threonine and isoleucine	,,
Lysine and threonine and homoserine	203
Lysine and threonine and methionine	48

The relatively small inhibition observed with lysine can be interpreted in terms of inhibition of aspartokinase (but see section on *metabolic interlock*). The inhibition produced by threonine is consistent with the inhibition of homoserine dehydrogenase which would prevent the flow of carbon to methionine. The inhibition produced by homoserine is difficult to explain since as far as known it does not inhibit any of the enzymes involved. Furthermore in some plants e.g. peas, homoserine is produced in large quantities and transported in the phoem. The inhibition produced by isoleucine is consistent with a control pattern known as *sequential inhibition.* Isoleucine has been shown to inhibit threonine deaminase which would be expected to lead to an accumulation of threonine which in turn inhibits homoserine dehydrogenase and so reduces the flow of carbon to methionine. The inhibition produced by methionine is difficult to explain in terms of the reactions shown in Fig. 12.4. The pronounced inhibition observed with lysine and threonine is consistent with their synergistic effects on aspartokinase producing a reduction in the flow of carbon to methionine, which is reinforced by the inhibition of homoserine dehydrogenase by threonine. Aspartic acid and isoleucine are unable to reduce the inhibition but methionine and homoserine are able to do so, presumably by supplying methionine directly or indirectly.

12.3.4.2 *Aromatic Biosynthesis*

The control of aromatic amino acid biosynthesis has been extensively studied in fungi and bacteria and it is clear that nature has evolved many solutions to the problem of control in branched pathways. Some of the solutions which have been demonstrated in higher plants are shown in Fig. 12.5.

12.3.4.3 *Enzyme Multiplicity*

In this control mechanism the first common step is catalysed by two or more enzymes which are under feedback control by compounds formed after the branching point. Four examples are given in Fig. 12.5, the control of chorismate mutase being particularly complicated. Three isozymes have been demonstrated in plants (Woodin & Nishioka, 1973); CM_1 and CM_3 are inhibited by phenylalanine and *activated* by tryptophane. CM_1 and CM_2 are inhibited by caffeic acid and chlorogenic acid whilst CM_3 is inhibited by ferulic acid.

12.3.4.4 *Enzyme Aggregation*

A number of fungi possess an aggregate of the five enzymes necessary for the conversion of 3-deoxy-D-heptulosonate-7-phosphate to enolpyruvyl-shikimate-5-phosphate. In *Neurospora* the five enzymes form a complex (M.W. 230,000) coded by the *arom* gene cluster. This large complex has not been demonstrated in higher plants, but a bifunctional enzyme consisting of dehydroquinase and shikimate dehydrogenase has been isolated (Boudet, 1971). *Neurospora* contains two dehydroquinases, the one in the aggregate is involved in an anabolic sequence, the other maps outside the *arom* cluster and is thought to function as a component of a catabolic sequence. The physiological significance of the multienzyme cluster could be to separate the degradative and synthetic routes. This could be achieved if the dehydroquinic acid which is formed in the aggregate is not released as a free product. This idea has become known as *metabolic channelling* and in a number of cases it has been shown that metabolites involved in channelling do not leave the enzyme surface. For example fungi possess a bifunctional enzyme consisting of carbamyl phosphate synthetase and aspartate transcarbamylase.

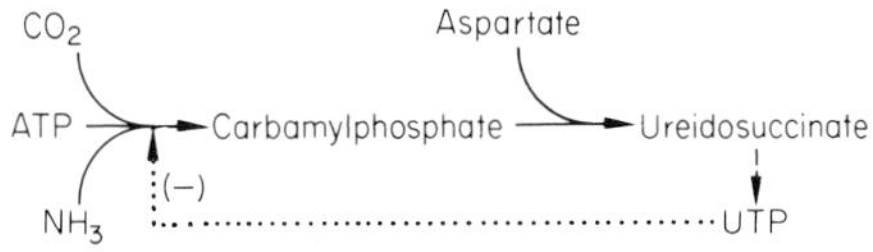

CO_2 Aspartate
ATP → Carbamylphosphate → Ureidosuccinate
(−)
NH_3 .... UTP

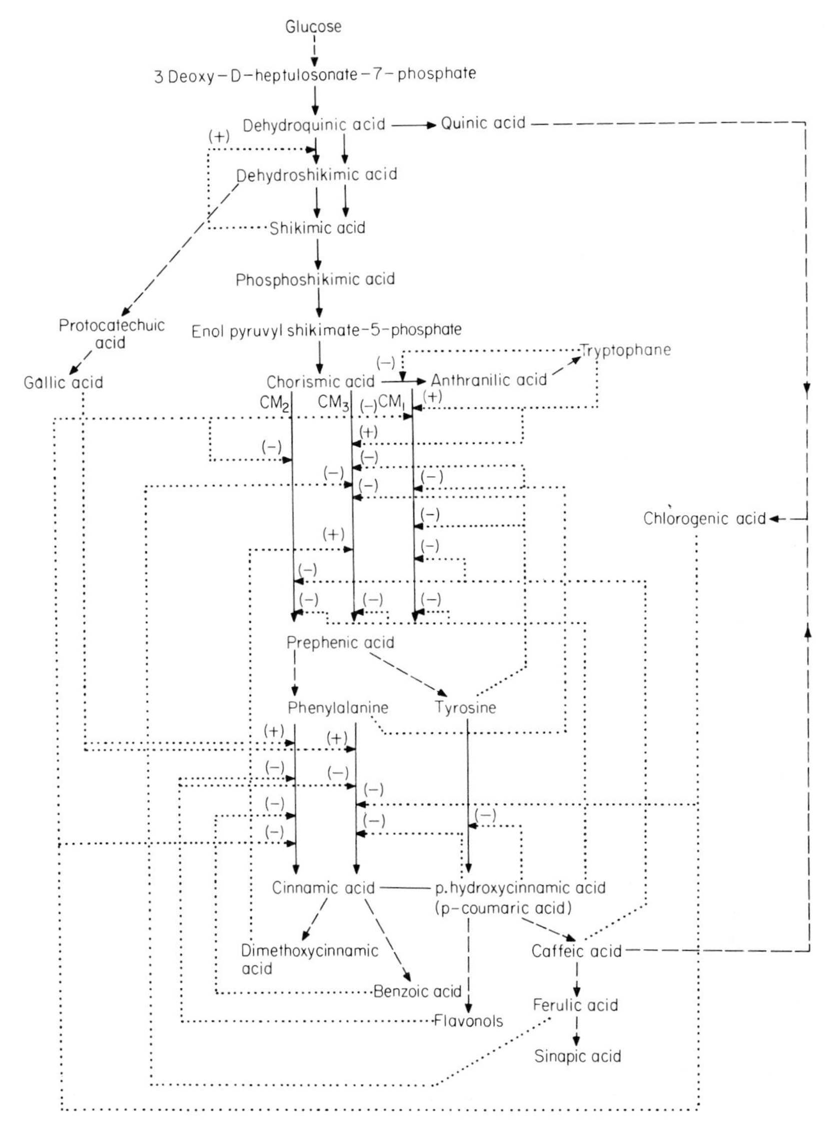

Fig. 12.5. Reactions involved in the biosynthesis of aromatic compounds. →Single step reactions; --->multi-step reactions; ···>feed back loop.

When the bifunctional enzyme is synthesizing ureidosuccinate from $^{14}CO_2$, the addition of carbamyl phosphate does not dilute the incorporation of ^{14}C into ureidosuccinate, showing that carbamyl phosphate formed by the double enzyme does not equilibrate with free carbamyl phosphate.

Higher plants contain two dehydroquinases one forming a bifunctional association with shikimate dehydrogenase, the other being activated by shikimic acid. Such a system involving a specific regulation of isoenzymes and molecular compartmentalization may function to control the partitioning of dehydroquinic acid between the pathways to phenolcarboxylic acids and aromatic amino acids.

12.3.5 METABOLIC INTERLOCK

The reactions and feedback loops shown in Fig. 12.5 give some idea of the complexity of control within a metabolic pathway. When the feedback loops of other metabolic pathways are taken into consideration it seems highly likely that the control must be interlocked. As an example, consider the reactions involved in the biosynthesis of methionine. In Fig. 12.4, this is shown as the addition of a C_1 group to homocysteine. The synthesis of this C_1 group is shown in Fig. 12.6.

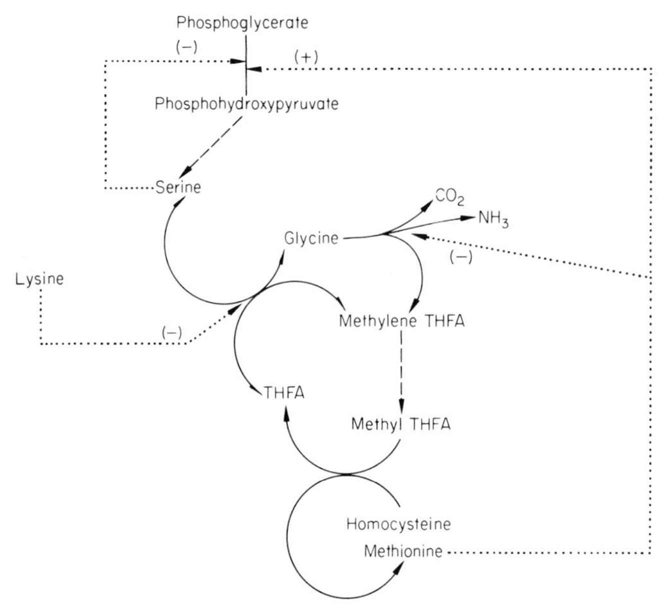

Fig. 12.6. Reactions involved in the biosynthesis of methionine. →Single step reaction; ---→multi-step reaction; ...→feed back loop.

The active C_1 group is generated in the serine hydroxymethyl transferase reaction as methylene tetrahydrofolate. After reduction to methyl THFA the methyl group is transferred to homocysteine to form methionine. Homocysteine is derived from the aspartate family (cf. Fig. 12.4) and so is lysine. The metabolism of the aspartate family and the serine glycine family are thus interlocked *via* lysine, which inhibits both aspartokinase and serine hydroxymethyl transferase.

12.3.6 ENZYMES AS CONTROL ELEMENTS

Until the discovery of feedback inhibition in 1956 enzymes were considered as biological catalysts and most enzymologists were concerned with the properties of the active site. When it became apparent that enzyme activity was under the control of specific metabolites, which were structurally unlike the active site, it became necessary to postulate a control site to which metabolites could bind and modulate the activity of the enzyme. Recognizing the lack of structural similarity between the substrate and the effector the term allosteric site was introduced (Monod, Changeux & Jacob, 1963). Subsequently Monod, Wyman and Changeux (1965) noted that many regulatory enzymes with specific allosteric sites exhibit cooperative-type kinetics and they developed a concerted transition hypothesis to explain the observed kinetics. Many workers assume that any enzyme showing sigmoidal kinetics is an allosteric enzyme and some workers imply that the mechanism of activation of all allosteric enzymes follows the mechanism postulated by Monod *et al.* (1965). The purist may object but the original precise definition has been replaced by a broader but less precise usage.

12.3.6.1 *Allosteric enzymes*

Before 1956, the kinetics of numerous enzymes were shown to follow the rate law of Michaelis and Menten giving a hyperbolic plot of **v** against **S**. An examination of enzymes which were known to be subject to feedback regulation showed that in some cases the relationship between **v** and **S** was sigmoid. In other cases the addition of a feedback inhibitor led to a change from a hyperbolic to a sigmoid relationship.

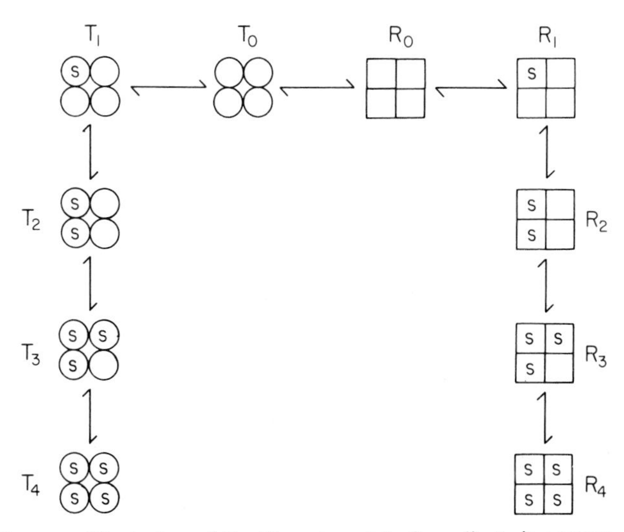

Fig. 12.7. Illustration of the Monod model of an allosteric enzyme.

Monod *et al.* (1965) proposed a simple model to account for the observed kinetics in which they took account of the observation that allosteric enzymes are composed of subunits. They assumed that the protein subunits were arranged in such a way that they occupy equivalent positions so that the enzyme must have at least one axis of symmetry. They further assumed two interconvertible states—*R* and *T*—each maintaining the symmetry principle. A simple illustration of the model is given in Fig. 12.7 based on the symbols used by Koshland (1970). The reader is referred to the original article by Monod *et al.* (1965) for the derivation of the rate law. However a consideration of the illustration leads to a basic understanding of the underlying causes of sigmoid kinetics. Two types of allosteric behaviour may be distinguished—one the variable *K* system in which the two states of the enzyme have different affinities for the substrate, the other, the variable *V* system in which the *R* state is catalytically more active (Fig. 12.8).

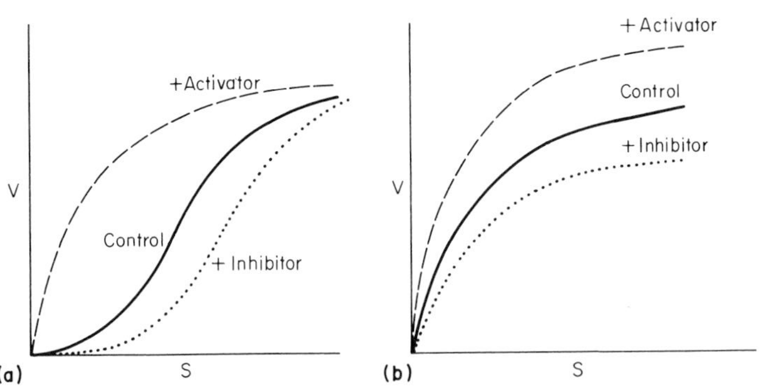

Fig. 12.8. Relationship between **v** and **s.** (A) a variable *K* system; (B) a variable *V* system.

Variable K Systems

In the variable *K* system, the *R* form is assumed to have a higher affinity for **S** than the *T* form of the enzyme. If in the absence of ligands, the equilibrium favours the *T* form then increasing concentrations of S will shift the equilibrium towards the *R* state giving sigmoid **v** against **S** plots. Another way of looking at this situation is to imagine that the interconversion of the *R* and *T* forms can be blocked and the enzyme isolated as the *R* or *T* form. The *T* form would give a normal Michaelis-Menten curve with a high $Km^{(T)}$ and the *R* form would also give a Michaelis-Menten curve but with a low $Km^{(R)}$. Allowing *R* and *T* to be interconvertible allows *Km* to vary.

If an activator binds preferentially to the *R* state it will tend to put the enzyme entirely in the *R* state and so remove the sigmoidicity of the **v** against **S** plot. Similarly if an inhibitor binds preferentially to the *T* form it will increase the sigmoidicity.

Variable *V* Systems

In the variable *V* system the *R* form is assumed to be catalytically more active than the *T* form though both forms have the same affinity for the substrate so that the **v**/**S** plot is hyperbolic in the presence or absence of activators and inhibitors. An activator is assumed to have a greater affinity for the *R* form than for the *T* form whilst the reverse holds for an inhibitor.

12.3.6.2 *Alternative models for allosteric enzymes*

This superficial consideration of the Monod model concentrates on only one aspect of the model—the two configurational states. This limited aspect gives a qualitative explanation of sigmoid kinetics described by the equation:

$$\mathbf{v} = \frac{V_{\mathrm{Max}}\,(\mathbf{S})^{\mathrm{N}}}{Km \pm (\mathbf{S})^{\mathrm{N}}}$$

where $N = 2$.

To explain higher order reactions the Monod model includes some highly restrictive assumptions which produce mathematical simplicity but are nevertheless based on the fundamental properties of protein structure.

An alternative model proposed by Koshland (1970) assumes an induced fit so that the protein subunit undergoes a conformational change when it binds a ligand. A simple sequential model may be illustrated:

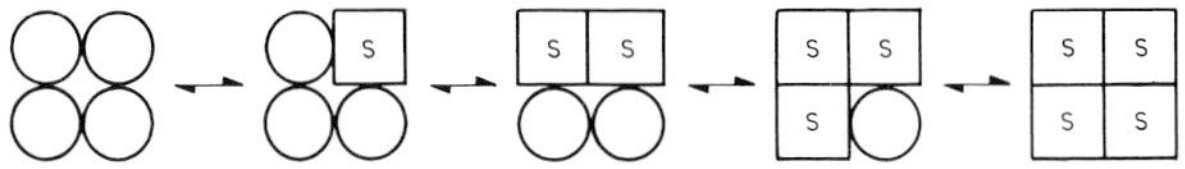

S S S S S S S S S S

This simple model is not a necessary part of the induced fit model (which in fact suggests a diversity of forms of association between enzyme and ligand—25 for a tetrameric protein). Nevertheless it simplifies the mathematical treatment and in many cases provides a close fit to experimental values.

12.3.6.3 *Kinetic constants for allosteric enzymes*

When an electronic engineer constructs a control circuit he needs to know the characteristics of the components—the amplification factor, grid bias, etc. Similarly if we wish to understand the control of a system we need to know the characteristics of its components, that is, the kinetic constants of the enzymes. Clearly a variable *K* system cannot be defined by a *Km* value nevertheless the same operation value can be used, that is the concentration of substrate producing half maximum velocity which is in this case designated $\mathbf{S}_{0\cdot5}$. More generally the term $(X_{0\cdot5})$ is used to designate the ligand concentration at half saturation.

The constant N is operationally the order of the reaction, its physical meaning may relate to the number of substrate binding sites on the enzyme or to the degree of cooperativity of binding. Whatever the meaning of N it can be estimated by means of a Hill plot (Fig. 12.9). Another kinetic term, R_s is the ratio between the 90% and 10% saturation values. R_s will be 81 for all curves following Michaelis-Menten kinetics and less than 81 for cases of positive cooperativity—a few cases of negative cooperativity are known and these give R_s values greater than 81.

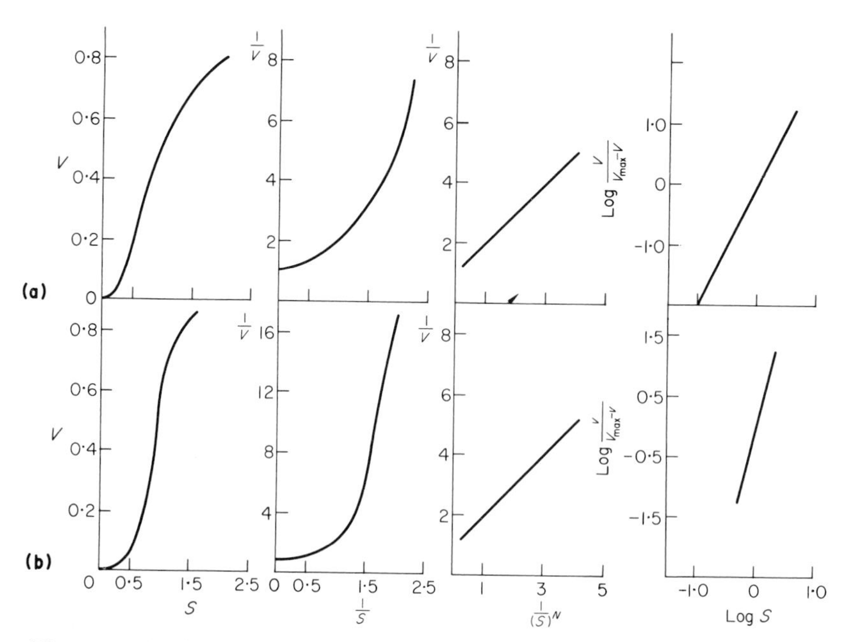

Fig. 12.9. Kinetic data for an enzyme with variable 'K' properties. The curves are drawn assuming

$$v = \frac{V_{Max}\,(S)^N}{Km + (S)^N}.$$

(A) $N = 2$;
(B) $N = 4$.

Valve characteristics define the properties of the valve but do not determine the function of the valve—the function of the valve is determined by the circuit in which it operates, in one circuit it may function as a cathode follower, in another as an amplifier. Similarly, the kinetic constants of a control enzyme do not determine its role in metabolic organization. The properties of an enzyme in a metabolic pathway are modified by its association with other enzymes. The study of control in metabolic pathways relates enzymology to physiology and illustrates the somewhat unfashionable view that the whole is more than the sum of the parts.

12.4 CONTROL MECHANISMS INVOLVING CHANGES IN THE AMOUNT OF ENZYME

The brilliant work of Jacob and Monod (1961) on the regulatory mechanisms governing β-galactosidase activity in *E. coli* led to many attempts to demonstrate similar control systems in plants. A long series of negative experiments has made it unlikely that operons exist in higher plants. There is, however, evidence for the interdependent regulation of some groups of enzymes. For example in parsley and soybean cell cultures, the three enzymes catalysing the synthesis of *p*-coumaroyl CoA from phenylalanine show strictly concomitant changes in activity. Large changes in activity are initiated by light but there is no satisfactory explanation for the strict coordination in activity. Mohr (1972) has suggested that genes can be divided into four classes—active genes which function in the same way in light and dark, inactive genes which function only at special times in the life history of the plant (e.g. flowering genes), potentially active genes whose activation is under the control of the phytochrome system and repressible genes which can be repressed by the phytochrome system. Positive photo-responses are identified with gene activation and negative photoresponses with gene repression. This model is so basic that it must contain some truth, but it offers no explanation of the mechanism of regulation other than to identify regulation as occurring at the gene level. Much of the evidence for Mohr's model is concerned with the increase in phenylalanine ammonia lyase which occurs when mustard cotyledons are exposed to far-red light. Recently however, it has been suggested (Attridge *et al.*, 1974) that this increase in activity is due to an activation of pre-existing enzyme rather than synthesis.

Following the success in understanding the regulatory system in prokaryotes it was clearly sensible to see if the system was applicable to plants. It now appears that the regulatory systems are very different, but the technical vocabulary used in the study of prokaryotes is still used in studies with plants. It should be noted that terms such as repression and derepression have precise meanings in relation to molecular models of regulation in prokaryotes, but when applied to plants, these terms should not be taken to imply a mechanism similar to that found in bacteria.

Many cases are known where a metabolite reduces the level of an enzyme involved in its metabolism (Table 12.2). Most of these cases are reported as

Table 12.2. End-product control of enzyme activity.

End-product	Tissue	Enzyme showing reduction in activity
Inorganic phosphate	Wheat scutellum	Phytase
Glucose	Sugar cane	Invertase
Glucose	Wheat embryo	Glucose-6-phosphatase and fructose-1, 6-diphosphatase
Arginine	Soybean culture	Arginosuccinate synthetase
Ammonia	*Lemna* plants	Nitrate reductase

examples of repression but the mechanism involved varies from case to case—for example arginine probably exerts its effect by stimulating proteolysis, the effect of ammonia may be due to proteolysis, though the formation of a specific protein inactivator has been proposed, and the effect of glucose on invertase has been reported as due to the destruction of m-RNA required for invertase synthesis.

The regulation of enzyme levels may involve one or more of the following:

(a) Degradation
(b) Inactivation
(c) Activation
(d) Synthesis

When examining a particular system it is necessary to establish which control system is involved and to measure its rate—this is frequently surprisingly difficult.

12.4.1 PROTEIN TURNOVER AND DEGRADATION

The view that plant proteins may undergo a cycle of synthesis and degradation was first proposed by Gregory and Sen in 1937. The terminology is somewhat confused; protein turnover implies synthesis *and* degradation but usually only degradation is measured.

Three methods are available for determining the overall rate of protein degradation.

(a) Incorporation of ^{14}C-amino acids.
The tissue is pulse-labelled with a ^{14}C-amino acid and the incorporation into protein measured. During the chase period the rate of protein degradation is estimated from the fall of radioactivity in the protein. The method has two serious disadvantages: (1) The recycling of protein amino acids—that is the reincorporation of a ^{14}C-amino acid released by degradation tends to underestimate the rate of degradation. This disadvantage can be minimized by using a ^{14}C-amino acid which is rapidly metabolized; (2) The existence of pools of amino acids means that the specific activity of the amino acid at the site of protein synthesis is unknown. Furthermore the incorporation of a labelled amino acid into a metabolic pool during the pulse period may be followed by a slow release of ^{14}C-amino acid during the chase. This disadvantage can be minimized by using an amino acid whose pool size is small, e.g. leucine.

(b) Measurement of the specific activity of aminoacyl t-RNA.
Trewavas (1969) has developed a method in which the specific activity of the amino acid entering protein during a labelling experiment is determined by isolating the appropriate aminoacyl t-RNA.

(c) Measurement of the incorporation of ^{3}H.
Humphrey and Davies (1975) established that when *Lemna* fronds are exposed to $^{3}H_2O$, the tritium rapidly enters and equilibrates with H on the α-carbon atoms of amino acids. Protein synthesized during the exposure to $^{3}H_2O$ is thus

labelled in a stable position. When the plants are transferred to H_2O the amino acids rapidly lose their ^3H, the ^3H in protein is stable, but as soon as the protein is degraded to amino acids the ^3H exchanges with H_2O. Thus the rate of protein degradation can be determined simply by measuring the loss of ^3H from protein. The method depends upon the speed with which transaminases catalyse the exchange reaction between water and the hydrogen on the α-carbon atom of amino acids. If it can be established for a particular tissue that this reaction is fast relative to the rate of protein synthesis the method overcomes problems of pools and recycling. The method can in principle be applied to a specific enzyme, but this requires the isolation of the pure enzyme.

12.4.1.1 *Degradation of specific enzymes*

If the degradation of proteins were a random process then a low rate of protein hydrolysis would almost completely destroy overall enzyme activity. This is because a 5% per day rate of degradation means that *each* enzyme molecule would be subject to an average degradation of 5% and for many enzymes this would mean a complete loss of catalytic activity. Consequently we must look for specificity in degradation.

A NADH-nitrate reductase-inactivating enzyme has been isolated from maize roots which appears to have a high degree of specificity (Wallace, 1974). The enzyme has been purified 460-fold, the purified preparation has proteolytic activity but no peptidase activity and the capacity to inactivate nitrate reductase is blocked by phenylmethylsulphonyl fluoride suggesting the involvement of a serine residue at the active site.

12.4.1.2 *The measurement of enzyme half-lives*

If protein synthesis is blocked by an appropriate inhibitor and at various time intervals the level of enzyme is measured, we can determine the fall in activity of the enzyme. If we assume that the inhibitor completely blocks protein synthesis without interfering with the reactions involved in inactivating or degrading the enzyme we can estimate its half-life (Table 12.3).

Table 12.3. Half-lives of some plant enzymes.

Enzyme	Tissue	Half-life
Vacuolar invertase	Sugar cane	2 hrs.
RNase	Sugar cane	2–4 hrs.
Nitrate reductase	Maize seedlings	4·2 hrs.
Phenylalanine ammonia-lyase	Mustard seedlings (dark)	6 hrs.
Cellulase	Pea epicotyls	24–30 hrs.
Isocitrate lyase	Water melon seedlings	2–3 days
Malate synthetase	Water melon seedlings	2–3 days
Peroxidase	Sugar cane	Stable

This method is open to criticism since in a number of cases inhibitors of protein synthesis have been shown to interfere with enzyme degradation. Thus cycloheximide stops the decline in phenylalanine ammonia lyase which follows the light induced elevation of the enzyme in potato discs and also the decline in phosphoglucomutase which occurs on slicing potato tubers. In these cases, cycloheximide is assumed to block the synthesis of the specific proteolytic enzymes necessary for the removal of the lyase and the mutase.

12.4.2 ENZYME INACTIVATION

Experimentally it is difficult to distinguish between enzyme degradation and inactivation. This is reflected in the terminology, thus the enzyme described in the previous section which is thought to degrade nitrate reductase is referred to as a nitrate reductase inactivating enzyme by Wallace. A simple kinetic test enables an investigator to determine if enzyme activity is changing during the period of an assay. The rate equation for an enzyme catalysed reaction can be written:

$$\frac{d(P)}{dt} = (E).f.(S)(A)(I)(P)$$

Provided that (S) substrate, (A) activator and (I) inhibitor are constant in a series of assays, the integral form of the equation is

$$f(P) = (E)t.$$

Thus for a series of assays when the amount of product formed is plotted not against time, but against time multiplied by enzyme concentration, all the values should fall on a single curve. If the points do not fall on a single curve then it is likely that the enzyme is being degraded or inactivated in some way (Selwyn, 1964).

12.4.2.1 *Inactivation by protein-protein interaction*

The modulation of enzyme activity by the interaction of protein subunits is widely recognized. A number of reports suggest that interaction between different proteins may result in enzyme inactivation. For example the naturally occurring inhibitor of potato tuber invertase appears to be a protein which forms an essentially undissociable complex with invertase. The inactivation of *Lemna* nitrate reductase has been attributed to protein-protein interaction and ammonium adapted plants are said to contain a protein which inhibits nitrate reductase.

However, the most clearly established case is the interaction between ornithine transcarbamylase and arginase in *Saccharomyces* (Wiame *et al.*, 1973). The activity of ornithine transcarbamylase was shown to decline following the addition of arginase. It was established that arginase binds to the ornithine transcarbamylase and inhibits its activity. The binding is on a one to one basis and does not reduce the arginase activity. This interaction has not been observed in other organisms.

12.4.2.2 *Chemical modification*

A number of enzymes are inactivated by low concentrations of pyridoxal phosphate at physiological pH. The mechanism of inactivation is the formation of a Schiff base between pyridoxal phosphate and an amino group of the protein, usually the ε-amino group of lysine.

Glutamate dehydrogenase and aldolase have been shown to be inactivated by pyridoxal phosphate and to be protected by their substrates, suggesting that lysine residues are involved in the active centre of these enzymes. It is difficult to assess the physiological significance of these reactions; the concentration of free pyridoxal phosphate in plants is low but the ratio pyridoxal phosphate: pyridoxamine phosphate changes with the nitrogen status of the plant and the possibility of this ratio monitoring the nitrogen status of the plant and controlling the activities of key enzyme has been discussed.

Carbamyl phosphate also reacts with ε-amino groups of lysine, the rate of carbamylation varying from protein to protein. Glutamate dehydrogenase is readily inactivated by carbamyl phosphate but it is doubtful that carbamylation of enzymes has physiological significance in plants.

Enzyme catalysed chemical modification is well documented in animals and bacteria and among the important examples we note glycogen phosphorylase and glycogen synthetase of *E. coli* which is adenylated by ATP. Adenylation has not been observed in plants but phosphorylation, methylation and acetylation of proteins has been demonstrated. The extent to which these modifications are important regulatory mechanisms remains to be determined.

The classification of chemically modified enzymes in terms of activation or inactivation is somewhat artificial. For example phosphorylation leads to an activation of glycogen phosphorylase but an inactivation of glycogen synthetase. These chemical modifications could equally well be considered as dephosphorylation in which case glycogen phosphorylase is inactivated and glycogen synthetase activated. However, from an operational point of view, whether an enzyme is activated or inactivated is of paramount importance.

12.4.3 *Enzyme activation*

Four enzymes of the Calvin cycle and two enzymes of the Hatch-Slack pathway are light activated whilst glucose-6-phosphate dehydrogenase, an enzyme of the pentose phosphate pathway is light inactivated. It is highly probable that a number of these responses reflect a reductive process, since dithiothreitol treatment of crude extracts gives results similar to light treatment of intact seedlings. However, in some cases the reaction is certainly more complicated than a direct response to a change in the SH/SS redox potential. A scheme for the activation/inactivation of pyruvate phosphate dikinase is shown in Fig. 12.10.

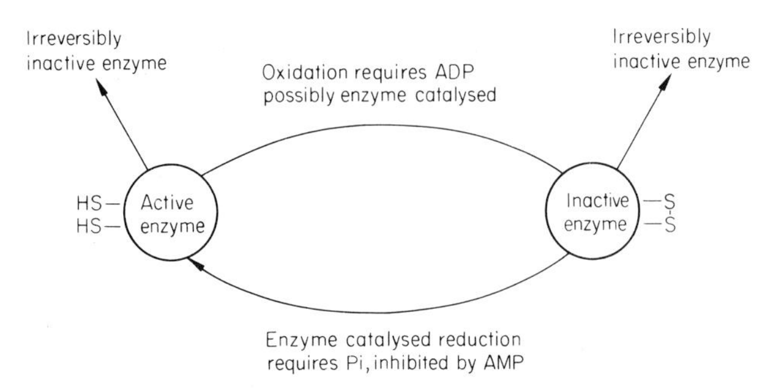

Fig. 12. 10. Schematic representation of the processes of activation and inactivation of pyruvate, P: dikinase. The process leading to the irreversible inactivation of both active and inactive enzyme *in vitro* apparently include an O_2-dependent oxidative reaction involving thiol groups. (After Hatch & Slack, 1969.)

The mechanism of activation may vary from species to species. Thus the light-activation of pea NADP glyceraldehyde phosphate dehydrogenase appears to involve a conformational change produced by reduction and leads to an increase in V_{max}. The spinach enzyme however, appears to need NADPH or NADP or ATP in addition to a reducing agent for activation. The activation involves dissociation of a tetramer leading to an increased affinity for NADP.

An interesting suggestion for the blue light activation of phenylalanine ammonia lyase has been proposed by Engelsma (1974). He suggests that the enzyme is present in gherkin hypocotyls in an inactive form due to inhibition by trans-hydroxycinnamic acid. Blue light (in the presence of a photoreceptor such as riboflavin) converts the trans- to the cis-hydroxycinnamic acid which is much less inhibitory. A phytochrome-mediated activation of phenylalanine ammonia lyase has also been reported in mustard cotyledons by Attridge *et al.* (1974). It is not clear if the cis-trans isomerization is applicable to the phytochrome-mediated activation which is inhibited by cycloheximide.

12.4.4 ENZYME SYNTHESIS

In the absence of activation or inactivation, the net synthesis of an enzyme is given by gross synthesis minus degradation. Net synthesis is measured as the change of activity so that gross synthesis can be measured if the rate of degradation is known. Because of the technical difficulties involved in obtaining quantitative data most workers have been concerned to develop methods which establish that the increase in activity of an enzyme is due to synthesis rather than degradation.

12.4.4.1 *Density labelling*

Gibberellic acid stimulates the synthesis of a number of enzymes in the aleurone layer of barley seeds. Filner and Varner (1967) introduced a method of demonstrating that the bulk of α-amylase which appears after treatment with gibberellic acid is due to protein synthesis. They incubated aleurone cells with $H_2{}^{18}O$ so that when storage proteins were hydrolysed ^{18}O-labelled amino acids were formed.

$$\text{Protein} \xrightarrow{H_2{}^{18}O} RCHNH_2C^{16}O^{18}OH \longleftrightarrow RCHNH_2C^{18}O^{16}OH$$

When new protein is synthesized from the ^{18}O-labelled amino acids it will have a greater density due to the incorporation of ^{18}O into the peptide links. The density difference between ^{18}O and ^{16}O containing proteins allows separation by equilibrium density gradient centrifugation. The observed increase in density was consistent with the bulk of the new α-amylase being synthesized from amino acids. A major limitation is that proteolysis is required to introduce ^{18}O into amino acids and this restricts the method to studies of seed germination.

An alternative method is to introduce ^2H into the protein. This can be achieved by incubating a tissue in ^2H_2O so that ^2H labels the amino acids by entering at the α-carbon atom in a transaminase catlysed exchange reaction. With prolonged incubation in ^2H_2O, deuterium enters the amino acids at other points producing large changes in the density of newly formed protein.

The major advantage of the density labelling methods is that they enable the synthesis of a particular enzyme to be demonstrated without purifying the enzyme. However, the method requires expensive equipment and is time consuming. Consequently many investigators rely on data obtained by the use of inhibitors of protein synthesis. Such data is rapidly obtained and at low cost but its interpretation must always be equivocal.

12.4.4.2 *Enzyme induction*

Filner *et al.* (1969) have listed the numerous examples of increases in enzyme activity in response to age, hormonal or environmental changes. In many cases these increases may be due to enzyme synthesis, though the evidence is often little more than that gained by the use of inhibitors of protein synthesis. The evidence for substrate-induced enzyme synthesis is extensive in the case of nitrate reductase but in other cases, for example the induction of a specific isoenzyme of glutamate dehydrogenase, the case is not yet proven.

12.5 FINAL COMMENTS

The basic similarity of living organisms is reflected in their biochemistry, the major metabolic pathways being common to all living organisms. The diversity of living organisms is reflected in their specialized control systems. The control mechanisms of one species may be very different from those of a closely related species. The investigator has a daunting task when he considers the amount of work necessary to establish control mechanisms over a wide range of species.

The difficulties of comprehending control mechanisms even in a single organism are no less daunting. The difficulties stem from the number of assays necessary to establish the complete rate law for a regulatory enzyme. For example the enzyme glutamine synthetase from *E. coli* is known to be affected by eight reactants and modifiers. Hence if 6 points per curve are necessary for a kinetic analysis, then to establish the rate law for glutamine synthetase will require in excess of $1{\cdot}6 \times 10^6$ assays. The task of analysing the many enzymes involved in metabolism is frightening and when we recognize that all the enzymes interact and affect one another the system defies analysis even with modern computers.

The way ahead seems to require inspired guesses at the nature of the control mechanisms involved, followed by model building and a systems analysis—probably making linear approximations—to see if the behaviour of the biological system can be predicted.

The difficulty in analysing control systems may be compared with the difficulties in understanding economics. The main control step of glycolysis appears to be phosphofructokinase, a reaction which involves a loss of free energy of 4 Kcals per mole. This loss may be considered the cost of controlling glycolysis. Protein degradation involves a large loss of free energy and this may be the price the cell has to pay for its complicated but effective control mechanisms. The situation can be likened to the economy of surplus production in society—in the Brave New World of Aldous Huxley products were destroyed to maintain the economy—perhaps nature is a capitalist!

FURTHER READING

BAYER P.D. (1970) (Ed.) *The Enzymes*. See articles by KOSHLAND D.E., STADTMAN E.R., & ATKINSON D.E. Academic Press.

COHEN G.N. (1968) *The Regulation of Cell Metabolism*. Holt, Rhinehart & Winston.

DAVIES D.D. (1973) (Ed.) Rate control of biological processes. *Symposia of the Society for Experimental Biology*. Vol 27.

KUN E. & GRISOLIA S. (Eds.) (1972) *Biochemical Regulatory Mechanisms in Eukaryotic Cells*. Wiley & Sons.

MILBORROW B.V. (Ed.) (1973) *Biosynthesis and its Control in Plants*. Phytochem. Soc. Sym. Series No. 9.

NEWSHOLME E.A. & START C. (1973) *Regulation in Metabolism*. Wiley & Sons.

CHAPTER 13
HORMONE ACTION

13.1 INTRODUCTION

The genesis of hormone research in plants occurred with the discoveries of Charles Darwin and his son Francis, who published, almost a century ago, their book entitled 'The Power of Movement in Plants'. Their experiments dealing with phototropism revealed that an 'influence' produced in the tip of a canary grass coleoptile was translocated to the basal portion where it caused the organ to grow towards the light. Almost fifty years later Went isolated a substance with just these properties which was produced in the tip of coleoptiles and transported to the basal portion of the organ. This substance was named auxin. Ten years after Went's (1934) discovery, the structure of this material was identified as indoleacetic acid (IAA).

The nature of auxin effects on plant growth and development was studied in many laboratories in the late 1930's and 1940's. However, it was not until World War II that the selective action of indoleacetic acid served as a basis for developing new chemicals for the control of growth in specific plants. At that time phenoxyacetic acids were discovered and their derivatives were carefully studied under the cover of military secrecy, both in the United States and England. The results of those studies were first published in scientific journals in 1945 and 1947. Today the most widespread uses of growth regulators are as herbicides.

Plant hormones may be divided into five general groups. Auxins include the native indoleacetic acid, as well as the synthetic phenoxyacetic acids, notably 2,4-D (2,4-dichlorophenoxyacetic acid). The second group is the gibberellins which are steroid-like compounds comprising over forty different structures. The third group, the cytokinins, are all N^6 substituted adenosine compounds. The fourth class is represented by a single substance, abscisic acid, which is also derived from isoprenoid units as are the gibberellins. Ethylene, the simplest of all the plant hormones, is a gas and therefore is easily spread from organs of production to organs of sensitivity.

It is interesting that of these five classes of hormones, two are generated from mevalonic acid. Ethylene is produced by the metabolism of methionine and indoleacetic acid is produced by the removal of carbon and nitrogen from tryptophane. Cytokinins, in terms of synthesis, are probably the most complex hormones. They are possibly produced from the breakdown of transfer RNA (tRNA). In the intact tRNA, the adenosine moiety of the RNA is modified by the addition of an isoprenoid group; subsequently, the tRNA is probably degraded, yielding isopentenyl adenosine, (IPA) the endogenous cytokinin.

One might reasonably imagine that an understanding of the physiological role and biochemical mechanisms of action of the plant hormones would have been elucidated quite some time ago. This of course, is in relation to what is known about the many and complex hormones in animals. Unfortunately, however, the plant hormones are not as well understood. Probably the main obstacle to the elucidation of the action of plant hormones has been the fact that they have overlapping and complementary effects on the actions of each other.

During the years between 1960 and 1970 substantial progress appeared to be made on the biochemical mechanisms of action of plant hormones. Much of this research dealt with the effect of the hormones on nucleic acid metabolism, protein synthesis and the synthesis of specific enzymes, since it was felt that the mechanism of hormone action would involve the control of the production of messenger-RNA (mRNA) and the *de novo* synthesis of enzymes. In the late 1960's the general nature of the research was radically changed because many investigators had found that the plant hormones mediated an effect on cell growth within a few minutes. This information implied a very rapid primary response mechanism of hormone action; consequently, these effects could not be mediated by changing the rate or type of synthesis of nucleic acids or of specific proteins. Many years of research have since been devoted to studies dealing with short-time growth responses to plant hormones. It now seems most likely that plant hormones regulate plant growth through both short-term and long-term growth controls.

It is fairly well understood that animal hormones first interact with the animal cell by binding to, or reacting with, a receptor site of some type within the target cell. Plant physiologists and biochemists are now beginning to consider that the action of many of the plant hormones similarly involves their binding to receptor agents which in turn amplify the action of the hormone and thereby bring about specific changes in nucleic acid synthesis, protein synthesis, enzyme activity and possibly other physiological responses, such as changes in permeability of membranes.

This chapter will summarize the historical evidence which has led to the current level of understanding of the action of plant hormones. Emphasis is placed on the reactive sites within target cells and how the interaction of the hormones with their specific receptors may regulate the growth and development of plant cells.

When a hormone is applied to a responsive plant system it brings about a specific change which results eventually in a measurable biochemical or physiological effect. Two distinct aspects of the measured effect are involved: the specific change in metabolism and the series of steps which lead to the physiological effect. Usually, the molecular interaction of the hormone at its site of action is referred to as the *mechanism* of action. The subsequent sequence of reactions leading to the physiological effect is referred to as the *mode* of action. Therefore, by definition each hormone has its own distinctive mechanism of action even though the manifestation of the hormone mechanism may depend upon prior action by other factors. Thus, it is possible for the hormonal *mechanisms* in one plant system to lead to a series of physiological responses which may be completely different from those in a second system. This difference in *mode* of action can be brought about because the second system has more or less of another hormone, has other biochemically rate-limiting components or possesses structural and cytological differences.

13.2 AUXIN ACTIONS

13.2.1 INTRODUCTION

Amongst plant growth regulators, the auxins have been studied the longest and have involved the greatest amount of research. This class of hormones is known to promote cell enlargement or cell elongation, a process which requires extension of the cell wall. Figure 13.1 shows the chemical structures of the major natural and synthetic auxins.

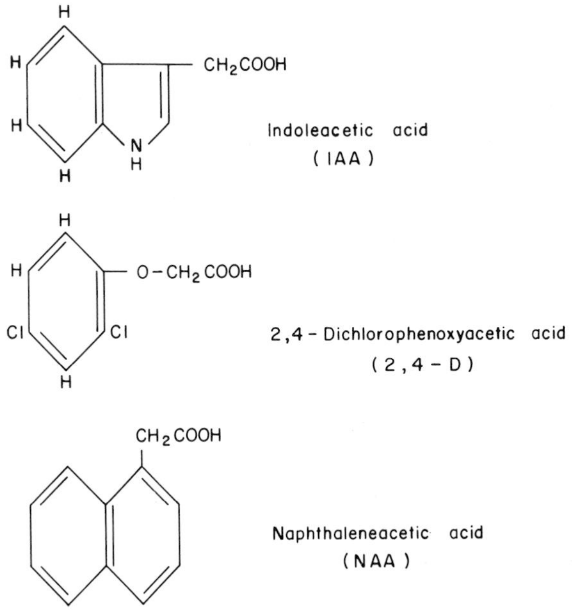

Fig. 13.1. Structure of various auxins.

Commencing in 1940, the majority of studies involving auxins were begun in an attempt to determine the molecular interaction with the plant cell. Most of this research was directed toward the cell wall because it was believed that auxin action required a change in the cell wall to allow for cellular expansion. Thus, the effects of the hormone were thought either to change the cell wall deposition or to hydrolyse certain cross-linkages in cell walls making them more elastic.

This type of research was carried out for several years without achieving a clear understanding of how auxin might control such processes within the cell wall. In the 1950's a new area of research was begun. Under the direction of F. K. Skoog in Madison, Wisconsin, a report (Silberger & Skoog, 1953) was published which showed that the auxin, indole-3-acetic acid, remarkedly affected the RNA and DNA contents of plants. Auxin increased the content of nucleic acids in tobacco tissue cultured on a sucrose-agar medium. This increase occurred prior to the auxin-induced growth of the tissue at concentrations of IAA which were optimal for cell enlargement. For many years following this discovery much research was devoted to the general mechanism by which auxin increased the synthesis of nucleic acids.

A great share of this research was begun in J. Hanson's laboratory in Urbana, Illinois by West and Key (West *et al.*, 1960) who showed that the synthetic auxin, 2,4-D, (see (Fig. 13.1), produced a wide range of morphological and physiological changes in the hypocotyl of soybean and the mesocotyl of corn.

Within 15–24 hours after 2,4-D treatment, cellular enlargement was noted concomitant with an increase in the size of the nucleus. Accompanying these changes were dramatic increases in RNA content, most of which were due to increased ribosomal-RNA. Chrispeels and Hanson (1962) suggested that auxin acts on the nucleus causing it to revert to a meristematic type of metabolism. The role of the nucleus in such a sequence of events is of obvious importance since the nucleus has been shown to accumulate RNA in response to auxin. Beginning in the 1960's, Key began experiments on auxin regulation of nucleic acid synthesis. He and Ingle (1964) demonstrated that auxin controls the synthesis of nucleic acids other than that of ribosomal-RNA. Their data revealed that auxin causes production of nucleic acid which appears to be of the messenger-RNA type.

Subsequent experiments by O'Brien *et al.* (1968) showed that treatment of soybean with auxin caused a large increase in chromatin-directed RNA synthesis. It was of interest to note that auxin-induced RNA synthesis produced a type of RNA which was different from the control RNA as judged by molecular size and nearest neighbour analysis. Following this area of research other studies (Hardin & Cherry, 1972; Hardin *et al.*, 1970; Hardin *et al.*, 1972) showed that 2,4-D increased the activity of RNA polymerase. Current research suggests that particular cytoplasmic or membrane bound factors may enhance the activity of RNA polymerase. It is hypothesized (Hardin *et al.*, 1972) that the mechanism of action of auxin is to bind to a receptor molecule located on the plasma membrane. The receptor is then released and moves into the nucleus where it modulates the activity of the RNA polymerase. The increase in RNA polymerase activity leads to increased synthesis of messenger-RNAs, which in turn regulate, or control, the synthesis of specific proteins within the target cells.

A primary challenge is to isolate the factor which binds to, or reacts with, auxin, and determine whether this is the primary action of the hormone. Secondarily, research is needed to determine what series of events this interaction puts into metabolic play. It is likely, once the auxin has bound to the receptor molecule within the target cells, that the changes in nucleic acid synthesis cause an increased activity of the various enzymes associated with the cell wall. Many other activities which have been measured over the many years are probably secondary effects resulting from the primary action of the auxin reacting with its receptor molecule.

13.2.2 REGULATION OF CELL WALL EXTENSION ABILITY

When plant cell walls elongate, an accompanying increase in weight and volume also takes place. This increase in size requires that the cell wall be increased in mass as well as area. Even though the dry weight of wall material greatly increases due to cellular enlargement, the thickness and density of the wall remains constant. Thus, wall synthesis appears to be a fundamental requirement during cell elongation. In addition, an increase in wall mass must come from a deposition of polysaccharides. It was believed for many years that auxin, in this particular case indoleacetic acid, increased cell elongation by promoting cell wall synthesis. Auxin, in order to allow for the increased wall extensibility, may also regulate the activity of enzymes involved in cell wall loosening which catalyse the breaking of various cross-linkages between the wall microfibrils.

Plant tissues which respond to auxin by an increased rate of cellular elongation also exhibit an increase in cell wall loosening. However, the kinetics of auxin-induced wall loosening vary considerably from tissue to tissue. For example, in dark grown maize coleoptiles wall loosening is induced by concentrations of auxin around 10^{-4} M, and the increase in loosening proceeds very slowly with time. However, the same tissue responds to auxin very dramatically as maximal increases in total length occur between two and six hours after exposure to auxin. It is of interest, therefore, that the rate of cell wall elongation, as measured by increased section length, is much greater than the increase in cell wall extensibility.

Thus, considerable changes in cell wall extensibility in a given tissue as caused by auxin, may affect growth in a different manner. In certain tissues the rate of growth is constant whilst total wall extensibility changes. In other tissues wall extensibility remains constant during a period of changing growth rate. At the present time it seems that auxin causes a biochemical change in the wall probably by breaking or modifying the cross-links between the wall polysaccharide chains. These changes in the cell wall are then translated under cell turgor pressure into wall elongation. It is now relevant to discuss how auxin affects the cross-links in a cell wall and what the biochemical changes are which take place in a cell wall during cell elongation. In this regard, it is necessary to look at the effect of auxin on various enzymes associated with the cell wall in the context of how these enzymes affect cell wall loosening properties.

13.2.3 ACTION ON CELL WALL ASSOCIATED ENZYMES

The most dramatic auxin effect on enzyme formation is the induction of cellulase, polygalacturonase and other hydrolases in pea cotyledons treated with a high concentration of IAA (Datko & Maclachlan, 1968). Detection of *in vitro* formation of cellulase by a ribosomal preparation from peas has been claimed. This was reported to be enhanced using a ribosomal system prepared from IAA-treated peas (Davis & Maclachlan, 1969). The large IAA effect on cellulase formation *in vivo* is a slow response occurring over several days and cannot be involved in a rapid auxin action on elongation. However, it could be a cause of the lateral swelling response seen in pea cells following IAA treatment.

Auxin stimulates not only the synthesis of cellulase but also the level of glucan synthetase. Masuda (1968) and Masuda and Yamamoto (1970) found that the fungal β-1,3-glucanase, isolated from cultures of *Sclerotinia libertans* induced rapid elongation of excised oat coleoptile segments. This enzyme was shown to increase cell wall extensibility as measured by a stretching method. In other studies, Masuda *et al.* (1970), compared the activity of the endo-β-1,3-glucanase and its activity on *Avena* coleoptile cells with those of the exoenzyme. They found that exoglucanase enhanced elongation and extensibility of the cell wall. But the effect was not additive to the effect of IAA. Furthermore, at least three hours of incubation with exoglucanase was required for enhancement of elongation. The endoglucanase showed no effect on cell wall elongation. Cell wall turnover and auxin effects thereon in pea stem tissue have been studied using pulse-chase wall labelling experiments. Considerable turnover of galactan occurs but this is not influenced by IAA (Labavitch & Ray, 1974).

The action of auxin on soluble xyloglucan can be assayed with relative ease and precision and has been positively demonstrated to be in progress within thirty minutes after exposure to IAA and probably within fifteen minutes, placing it amongst the most rapid metabolic effects of auxin. This auxin effect is blocked by metabolic inhibitors that are known to block elongation, but persists under complete osmotic inhibition of elongation by mannitol. It seems likely that it is involved in the action of the cell wall that leads to elongation in pea cells. Involvement of xyloglucan is understandable in terms of a recent model for cell wall structure, according to which xyloglucan serves to bind matrix polysaccharides to cellulose microfibrils (Bauer *et al.*, 1973; see also chapter 1).

13.2.4 HYDROGEN-ION PUMP

Since the 1930's it had been known that acidic media can stimulate elongation in auxin-sensitive tissue. Recent work showed that acidic buffers or CO_2 solutions in the pH range from 3 to 4 induce a rate of coleoptile cell elongation as great as or greater than that obtainable with auxin (Rayle & Cleland, 1970; Hager *et al.*, 1971; Evans *et al.*, 1971). Acid-induced elongation starts almost immediately upon treatment rather than after the 10–15 minute latent period characteristic of auxin action. Acid-induced elongation is not suppressible by metabolic inhibitors, such as cyanide, mercurials, and cycloheximide, or by lack of oxygen, all of which block auxin-induced growth, and appears to be a passive process independent of metabolism. In later work it was found that when the epidermis is removed to improve H^+ entry into the tissue, the full acid-pH stimulation of elongation develops over the pH range from about 6·0 to about 5·0 (Rayle, 1973; Cleland, 1973).

Rayle *et al.*, (1970b) found that cell wall skeletons of frozen-thawed, dead coleoptile tissue would elongate dramatically if treated with acidic media while being held under tension by an applied load. Rayle and Cleland (1972) concluded that acid pH- and IAA-induced elongation must occur by the same mechanism, because the two kinds of elongation had similar rate, similar temperature dependences, and a similar yield threshold. While suggestive, these similarities do not establish identity of biochemical mechanism.

Hager *et al.* (1971) also felt, on grounds that IAA-induced elongation could be suppressed by alkaline media, that IAA-induced growth involves the same biochemical mechanism as acid-induced elongation. In support of their proposed auxin-stimulated H^+ pump they offered experiments showing rapid stimulation of coleoptile elongation by ATP, ITP, and GTP under anaerobic conditions, although these results may have represented merely the acid pH effect itself, since these compounds are strong acids and were applied in unbuffered, pH 5 media.

In recent work an auxin-induced release of H^+ ions from coleoptiles stripped of their epidermis (Cleland, 1973; Rayle, 1973) and from other auxin-sensitive tissues (Marrè *et al.*, 1973; Ilan, 1973) has been detected. This is measured simply as a gradual fall in pH of the medium bathing the tissue, upon treatment with IAA, from near-neutral pH to pH 5 or below, i.e., into the range that by itself induces elongation. Detectable H^+ secretion by coleoptiles begins within 20 to 30 minutes after exposure to IAA and may be regarded as a 'rapid response'. To the extent so far studied, H^+ secretion by coleoptiles has a specificity for auxin analogues and antagonists similar to the specificity seen in growth, and a similar dependence on metabolism. Secretion is sensitive to inhibitors and uncouplers of energy metabolism and, perhaps unexpectedly for a transport process but just like cell enlargement, H^+ secretion is quickly inhibited by low concentrations of cycloheximide (Rayle, 1973; Cleland, 1973). Comparable results with pea stem segments have been reported by Marrè *et al.*, (1973).

These findings, coupled with the observations that induction of elongation by auxin is prevented by sufficiently well-buffered media of pH 6 to 8 provided the cuticle is removed or rendered permeable by gentle abrasion (Rayle, 1973), constitute presumptive evidence that the auxin effect on cell enlargement is mediated by externally secreted H^+ ions.

The simplicity and directness of the acid secretion theory of auxin action is appealing and the theory has quickly become popular. Various auxin effects on cation and anion transport noted previously were considered by their authors to be consistent with the H^+ secretion theory. H^+ secretion could involve either a parallel flow of anions, or a counterflow of, or exchange with, cations as has been inferred for H^+ pumping by beet cells (Poole, 1973). However, as yet no immediate dependence on external ions of either auxin-induced H^+ secretion or growth has been found.

Cleland (1973) and Rayle (1973) observed that auxin-treated coleoptiles cease to release H^+ when the medium reaches pH 5·0 or slightly below, suggesting that auxin-stimulated H^+ secretion is inhibited by acid pH, like other H^+ pumps (Poole, 1973), or alternatively that, at pH 5, back-diffusion of H^+ into the cell offsets the action of the pump. This feature should allow extension-promoting pH values in the cell wall to be reached rapidly under auxin treatment, while shutting down further acidification which would probably be injurious to the cell (Cleland, 1973).

Some objections to the H^+ secretion theory of auxin action and weaknesses in the evidence for it should be mentioned. For example, it has not yet been shown that suppression of auxin elongation by neutral buffers is not due to inhibited uptake of auxin, or to failure of primary action (the auxin binding found by Hertel *et al.*, (1972) had a pH optimum below pH 6·5).

The most important piece of evidence still needed for the acid secretion theory is that the pH in the tissue's free space falls to an extension-stimulating value at the time that rapid auxin-induced elongation commences, i.e., within 10–15 minutes after exposure to auxin, as against the hour or more that is required for an external bathing medium to reach pH 5·0 to 5·5 (Cleland, 1973; Rayle, 1973). At the very least it must be demonstrated that IAA induces H^+ secretion as quickly as it induces rapid elongation. Existing data fail this test and show an increase in H^+ efflux from coleoptiles beginning, at the earliest, 20 or 30 minutes after exposure to IAA (Cleland, 1973; Rayle, 1973). These authors feel that the discrepancy may be attributed to time lags imposed by the free space volume and the diffusion path length for H^+ to reach the external medium. A worse timing discrepancy was seen with sunflower hypocotyl (Ilan, 1973; cf. Uhrström, 1969). A more sensitive method for measuring H^+ efflux needs to be employed to resolve these discrepancies. Better still, the pH within the free space of the tissue should be measured directly during response to auxin treatment.

13.2.5 REGULATION OF GENETIC MATERIAL

After the discovery of messenger-RNA it was suggested that the regulation of gene action by auxin required the specific control of messenger-RNA synthesis. Using a series of inhibitors, Key (1964) and Key and Ingle (1964) noted that treatment of soybean hypocotyl with 5-fluorouracil (5-FU) inhibited total RNA synthesis by 80% whilst growth proceeded normally. Further studies on the fractionation of nucleic acid from soybean hypocotyls on a methylated albumin Kieselguhr column showed that 5-FU inhibited the incorporation of labelled adenosine diphosphate into ribosomal-RNA but had little effect on the labelling of messenger-RNA. Additional experiments with soybeans showed that 5-FU did not affect the tenaciously bound RNA whose synthesis is promoted by 2,4-D.

Sen and his colleagues (Roychoudury & Sen, 1964; Roychoudury *et al.*, 1965) showed that nuclei isolated from coconut milk responded to auxin by making more RNA *in vitro*. On the basis of these observations, it was proposed that auxin directly affected the nucleus and thereby regulated gene expression. It was thought that in the presence of the hormone a greater amount of DNA template was exposed and allowed to be transcribed into RNA. In an attempt to test this hypothesis, Cherry (1967) showed that nuclei isolated from peanut cotyledons did not respond *in vitro* to 2,4-D at physiological concentrations (10^{-8} to 10^{-6} M). From many other experiments, Cherry demonstrated that only in approximately 1 out of 10 attempts could nuclei be shown to respond to the hormone *in vitro* by the synthesis of more RNA. In general, absolutely no effect of the auxin on the capacity of the isolated nuclei to produce RNA was found. On the other hand, it was found routinely that isolated nuclei from soybean seedlings pretreated with 2,4-D produced twice as much nucleic acid as did control nuclei. Furthermore, experiments of O'Brien *et al.* (1968) showed that chromatin isolated from soybean hypocotyls did not respond *in vitro* to 2,4-D. In all cases the tissues needed to be treated with the growth regulator for at least two hours before any effect on chromatin-directed RNA synthesis could be noted. It is to be noted, however, that in a few experiments an effect of the hormone could be observed within thirty minutes. Subsequently, a progressive increase in chromatin-directed RNA synthesis was noted as a function of time. From those experiments, it was suggested that the auxin-enhancement was a result either of a more active RNA polymerase or of gene derepression leading to an increased availability of DNA template. A third possibility was that both of these effects were involved. Subsequent experiments demonstrated that in the presence of *E. coli* RNA polymerase, chromatin from both control and 2,4-D treated plants allowed the synthesis of similar amounts of RNA at saturation with the polymerase. Even though the total *E. coli* RNA polymerase activity with chromatin from 2,4-D treated plants was slightly higher than that with control chromatin, it was concluded that the auxin primarily promoted the endogenous RNA polymerase, rather than the chromatin template availability.

When it became possible to solubilize RNA polymerase from chromatin (Hardin & Cherry, 1972) two important aspects were noted. First of all, it has not been possible to show that the addition of auxin *in vitro* to solubilized RNA polymerase, or to chromatin, increases the rate of RNA synthesis (Hardin *et al.*, 1970). Secondly, of the many experiments that have been performed, it appears that treatment of sensitive plants with auxin leads to a greater production of RNA polymerase I (Hardin & Cherry, 1972; Hardin *et al.*, 1972). This is the enzyme which is thought to be present in the nucleoli, as judged from comparisons with animal RNA polymerase, and is the enzyme thought to transcribe ribosomal into ribosomal-RNA. These data agree with the fact that large increases in ribosomal-RNA and ribosomes are observed after auxin treatment. However, they are inconsistent with the idea that auxin increases the activity of the RNA polymerase enzyme which transcribes unique DNA sequences into messenger-RNA.

13.2.6 ACTION ON MEMBRANES

A current, interesting view of auxin action is that the hormone rapidly effects its action at, or on, cellular membranes, possibly involving the regulation of the export of growth active materials across the plasma membrane into the cell wall space. Auxin action at the plasma membrane was hypothesized by Hertel and Flory (1968) and by Rayle *et al.* (1970) partly in the belief that auxin transport and auxin action on growth involve closely related, if not identical, interactions, presumably with some carrier site located in the plasma membrane. Evidence for auxin action at the plasma membrane comes from observed IAA effects on membrane potentials (Tanada, 1970). This effect, however, has not yet been adequately studied nor shown to be related to the elongation process. Another indication of auxin action at the plasma membrane is the induction of cell elongation by certain fungal toxins which are known to increase membrane permeability (Evans, 1973).

It is assumed that specific auxin receptor sites must be located in or on the plasma membrane if the hormone is to change the permeability or to change any physical property within the membrane. With this general assumption in mind, Lembi *et al.* (1971) began to look at the binding of labelled NPA (naphthylphthallamic acid) a powerful competitive inhibitor of polar auxin transport which was considered to be likely to bind to specific auxin binding sites. Using maize particle preparations, they tentatively showed that there was a greater amount of NPA binding to fractions rich in plasma membranes in comparison to all other fractions obtained. IAA did not compete with NPA for the sites which thus do not seem to be the actual carrier sites for polar transport of IAA. It is of course conceivable that the sites binding NPA are not the same sites which would normally bind the auxin, IAA.

Hertel *et al.* (1972) detected a specific binding of labelled IAA and naphthaleneacetic acid (NAA) to certain particles in maize homogenates. Specific binding was assayed as radioactivity retained by particles exposed to 2 to 10×10^{-7} M labelled auxin minus radioactivity retained when, in addition, 10^{-4} M unlabelled IAA or NAA was added to compete with and displace labelled auxin from saturateable binding sites. The radioactivity which remained in the latter case then was termed unspecific binding. This difference-assay gave specific binding with binding constants in the rage of 10^{-}; to $10^{-}{}_{5}$ M. Specificity of the binding as an auxin phenomenon was indicated by the evidence that analogues of IAA or NAA which are active on growth, and are transportable, can displace labelled IAA or NAA from the binding sites, whereas chemically similar, but biologically inactive, molecules do not compete in the binding assay.

One defect of these data was that 2,4-D, a very active auxin, competed relatively weakly with IAA or NAA in the assay, nor was specific binding of labelled 2,4-D itself detected. Based on the evidence that auxin-specific binding sites are localized in plasma membranes, Hardin *et al.* (1972) proposed that the reaction of auxin, either the native IAA or the synthetic 2,4-D, with the plasma membrane would probably lead to conformational changes within the membrane, causing the release of a receptor molecule into the cytoplasm. This receptor molecule may or may not be the same binding agent with which the auxin interacts. Nevertheless, the receptor released from the plasma membrane could then travel through the cytoplasm and into the nucleus where we know auxin ultimately increases the activity of RNA polymerase.

From previously published information from animal systems, (Shyamala & Gorski, 1969; Jensen *et al.*, 1968) a good analogy can be drawn with hormone interaction with receptor molecules which then migrate into the nucleus where nucleic acid synthesis is controlled. Following these lines of investigation, Hardin *et al.* (1972) isolated a plasma membrane-rich fraction in which the plasma membrane was thought to be 70% homogeneous. The addition of plasma membranes to soybean RNA polymerase caused an increase in RNA synthesis *in vitro*. Subsequently, it was shown that pretreatment of the membranes with 10^{-7} M 2,4-D greatly increased RNA polymerase activity. Still further, when the plasma membranes were pretreated with 2,4-D followed by removal by centrifugation, the supernatant fraction contained the RNA polymerase stimulus. The chemical release of this factor appears to be specific for auxin, i.e., IAA and 2,4-D bring about this release, but a non-auxin such as 2,5-D is totally inactive.

In other studies an attempt was made to determine which of the multiple RNA polymerases were stimulated by the auxin-released factor. It was found that the addition of plasma membrane factor to a semi-purified RNA polymerase preparation enhanced the α-amanitin sensitive polymerase. By analogy with animal RNA-polymerase studies, these data imply that auxin increases the nucleoplasmic RNA polymerase, the enzyme which is thought to transcribe DNA into messenger-RNA. If this is true then it is sensible to believe that auxin, through the release of a receptor molecule, may regulate the transcription of unique DNA sequences through the control of a specific enzyme.

13.2.7 SUMMARY

Because of rapid auxin responses on plant cells, over the last few years a whole new field of research has begun which relates not only to the gene concept but also to rapid responses. Therefore, the new area of research has dealt with actions of auxins on receptors, membranes and other binding surfaces which could lead to a rapid growth response, as well as a continued and long response involving nucleic acid synthesis and protein synthesis.

The apparent binding of auxin to plasma membrane sites has stimulated interest in the concept that the hormone might bring about conformational changes within the membrane structure. This in turn might lead to changes in permeability, causing an increased ion flux across the membrane. If a receptor molecule is located on the plasma membrane, a change in conformation of the membrane might bring about a release of the receptor. It has been proposed that this receptor moves into the nucleus of the target cell and once there, increases the activity of RNA polymerase. This, in turn, leads to the synthesis of messenger-RNA which codes for proteins and this brings about the net result of auxin increased growth.

In this particular section the effects of cyclic-AMP on auxin controlled growth have not been covered. When cyclic-AMP was first found to stimulate plant growth, considerable excitement was generated (Solomon & Mascarenhas, 1971). However, that excitement was short-lived, and it now appears that cyclic-AMP has little or no effect on plant cells (Ownby *et al.*, 1975).

A number of enzymes have been shown to be affected by auxin. These include the hydrolases, particularly cellulase, and other enzymes associated with cell wall degradation as well as cell wall synthesis. Whilst these enzymes are affected by the auxins, it is very likely that this result is a secondary effect of the auxin and is not related to the primary mechanism of action. In essencc then, it appears that auxins control a multitude of physiological and biochemical responses in plants. It is possible that the primary site of action is localized within the plasma membrane or some surface binding site of the cell. A change at this particular level, either in structure or function, could bring about the release of a factor which moves into the nucleus. As a secondary amplification, the hormone thus modulates the activity of RNA polymerase. This in turn leads to the synthesis

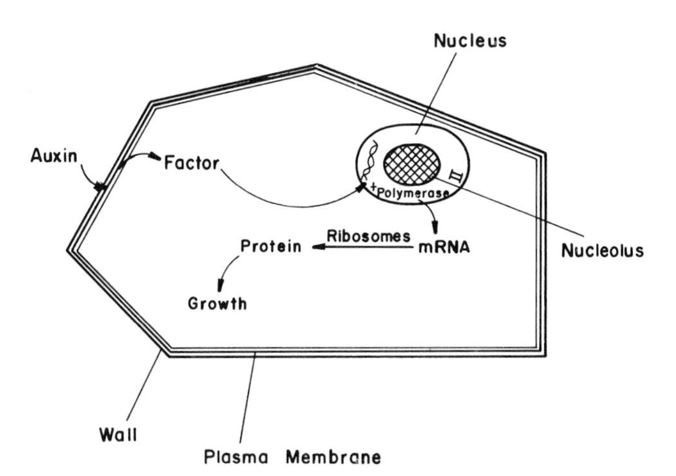

Fig. 13.2. A hypothetical model of auxin action. The interaction of auxin with the plasma membrane results in the release of a factor which moves through the cytoplasm and into the nucleus. The factor controls the activity of RNA polymerase II in the nuclei which leads to synthesis of mRNA. The new species of mRNA translated in the cytoplasm into new proteins leads to enhanced cellular growth.

of RNA from parts of the genome which were either not transcribed at all, or just to a low level in the absence of the receptor. A change in DNA transcription in this manner would lead to the synthesis of new proteins which would bring about the change in the growth potential (Fig. 13.2).

13.3 GIBBERELLIN ACTION

13.3.1 INTRODUCTION

Gibberellins may be defined as compounds having a gibbane skeleton and biological activity in stimulating cell division or cell elongation or both (various structures are given in Fig. 13.3). Gibberelins also may possess other biological

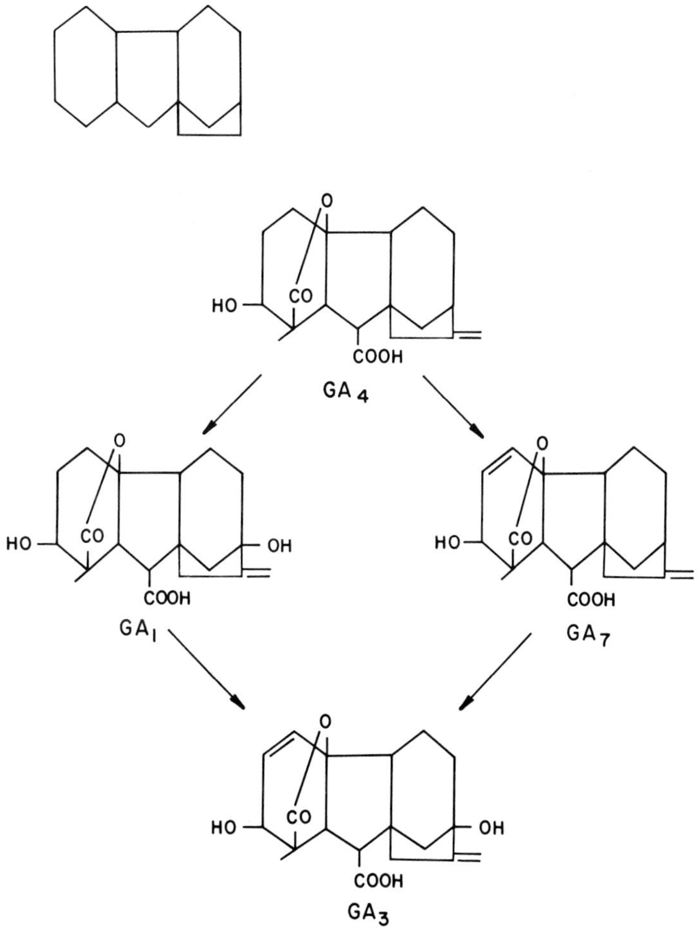

Fig. 13.3. Structure of gibbane and the conversion of GA_4 to common gibberellins.

activities such as induction of nucleic acid and enzyme synthesis. Presently there are more than forty known gibberellins which have been isolated from green plants or the fungus *Gibberella fujikuroi* (Cherry, 1973). Since the late 1960's three to four new gibberellins have been identified each year. Thus, it seems that great progress is being made in the identification of naturally-occurring gibberellins; on the other hand the mechanism of action of gibberellins is still unknown. Plant physiologists are therefore confronted with a very wide array of gibberellins, several of which often occur in the same tissue, and the question arises as to their function within the cells.

Work beginning in the 1960's on the possible regulation of gene transcription by gibberellins was initially successful and achieved wide-spread acceptance. Later work raised many doubts, however, and at the present time, many workers in the field reject the concept that gibberellins regulate physiological processes through the control of gene transcription. Alternative mechanisms involving an action of gibberellins on lipids and membranes are now being proposed.

13.3.2 GIBBERELLIN CONTROL OF ENZYME SYNTHESIS

The aleurone layer of cereal grain is composed of a relatively homogeneous population of non-dividing cells which respond as a target tissue to gibberellins in the concentration range of 10^{-7} to 10^{-10} M (Varner & Johri, 1968). It is, therefore, an ideal tissue for a study of the hormonal control of enzyme activity. Haberlandt in 1890 first discovered the secretion of starch liquifying enzymes by the aleurone layer in rye grain in response to some substance obtained from the germinating embryo. Since that early discovery, over the many years following, many workers have found embryo substances or factors and gibberellin-like materials which bring about an increase in activity of several hydrolase enzymes (Varner & Johri, 1968). Some of the enzymes which have been shown to increase in activity in the aleurone layers of barley grains in response to treatment with gibberellic acid (GA_3) include the following: α-amylase, proteinase ribonuclease, β-glucanase, and pentosanases. A similar control of amylase activity by gibberellins has been observed in wheat and in wild oats. In barley aleurone layers GA-induced increase in α-amylase is prevented by the lack of oxygen, by cytochrome *c* inhibitors and inhibitors of RNA and protein synthesis. Varner and Johri (1968) have shown that gibberellin control of amylase appears to involve the control of nucleic acid and protein synthesis.

GA_3 enhances the development of α-amylase activity in cereal grains. The enzyme is synthesized *de novo* and its synthesis blocked by cycloheximide and other inhibitors of protein synthesis. It is of interest to note that actinomycin D (an inhibitor of RNA synthesis) only inhibited the production of α-amylase when the tissue was exposed to the inhibitor early in the incubation process. If, however, the tissue was exposed to actinomycin D at some point after enzyme synthesis had begun, little or no inhibition occurred. Thus, it was not certain whether the hormone actually controlled the synthesis of RNA at the level of transcription or controlled the translation of pre-existing messenger-RNA. Many attempts were initiated to answer this question; unfortunately, to date, a definitive decision has not been possible. The most likely conclusion is that the hormone does not control the synthesis of α-amylase at transcription (Chrispeels & Varner, 1967).

13.3.3 THE EFFECTS ON RNA SYNTHESIS

Considerable indirect evidence has shown that GA_3 influences the synthesis of RNA in many plant systems (Varner & Johri, 1968). If this RNA synthesis is a primary effect of the reaction of GA_3 within a cell, it is then most likely that the primary action occurs in the nucleus. In many other plant tissues also, either an increase in protein and RNA synthesis or in the incorporation of labelled precursors into these macromolecules in the target tissues is evoked by hormones (Key, 1969).

It seems obvious that if GA interacts directly with the nucleus it would be advantageous to work with isolated nuclei or chromatin obtained from the nuclei. Therefore, attempts were made to work with nuclei isolated from light-grown dwarf peas (Varner & Johri, 1968). To determine if the RNA synthesized in the presence of GA_3 was qualitatively different, the nuclei were incubated in the presence of one of the four nucleoside triphosphates labelled in the α position with ^{32}P and the radioactivity then determined in the 2′,3′ nucleotides. It was found that a significant difference existed in the relative distribution of radioactivity among the four nucleotides of the RNA synthesized in the presence and absence of gibberellic acid. This nearest-neighbour analysis indicated that RNA synthesized by GA-treated nuclei is in part qualitatively different from the RNA synthesized by untreated nuclei.

Fractionation of the labelled RNA on MAK columns showed differences also in the types of nucleic acids produced in the presence of GA_3. The results indicated that GA preferentially enhances the synthesis of a small fraction of RNA. However, fractionation of the labelled nucleic acids on sucrose gradients indicated that the *in vitro* synthesized RNA sediments in a peak corresponding approximately to 6–10s. These results indicate that the RNA synthesized by GA-treated nuclei has a higher average molecular weight than that synthesized by the control nuclei. Varner and Johri (1968) concluded that the enhanced RNA synthesis could be due (a) to an increase in template DNA availability, (b) to an increase in the RNA polymerase activity, or (c) to a combination of the above two possibilities. An enhanced activity of RNA polymerase could be due to activation or to protection against degradation of the enzyme. Another explanation might be that the hormone changes the permeability of the nuclear membrane and thereby makes the RNA precursors more available for use in transcription.

13.3.4 GIBBERELLIN CONTROL OF ENZYME SECRETION

The gibberellin-enhanced synthesis of hydrolases is preceded by a proliferation of rough endoplasmic reticulum (Yomo & Varner, 1971). The control of rough endoplasmic reticulum seems more likely than the control of hydrolase synthesis and secretion as a candidate for the primary site of gibberellin action since it could more readily be a general response of plant tissues to gibberellins. Varner and his associates (Evins & Varner, 1971; Koehler & Varner, 1973) have therefore examined rough endoplasmic reticulum in detail. There is a gibberellin-enhanced incorporation of labelled choline into the lipid-soluble material of the endoplasmic reticulum and of other cell fractions. An enhanced incorporation of ^{32}P into the phosphoryl-lipids of several cell fractions is also seen. This enhancement is first measurable about four hours after the addition of gibberellin and is proportional to the logarithm of the concentration of the hormone.

GA_3 also enhances the activity of phosphorylcholine cytidylate-transferase and phosphorylcholine-glyceride transferase (Johnson & Kende, 1971). These responses are measurable as early as two hours after treatment. Gibberellin-evoked increases in these membrane-bound enzyme activities are prevented by the inhibitors, abscisic acid, actinomycin D and cycloheximide. These early effects on phosphoryl-lipid metabolism suggest that hydrolase synthesis occurs only on polysomes bound to newly synthesized endoplasmic reticulum. Thus the synthesis of hydrolase before the addition of GA_3 may be limited not by messenger-RNA but by the availability of appropriate membranes for the attachment of polysomes which carry hydrolase-specific messenger-RNAs. An examination of the secretion of hydrolases is of interest because it appears that gibberellin is required for the secretion of β-glucanase which had been synthesized before hormone addition. Gibberellin-enhanced release of β-glucanase into the medium surrounding the aleurone cells is prevented by inhibitors and uncouplers of oxidative phosphorylation, by actinomycin D, 6-methylpurine, and by cycloheximide. This is in contrast to the secretion of α-amylase which requires phosphorylative energy transformation but is not affected by 6-methylpurine nor by cycloheximide. These differences in the secretion of β-glucanase and α-amylase may represent different modes of secretion for the two enzymes. Or, it may be that the cellular apparatus required for the secretion of both enzymes is made during the first few hours of gibberellin treatment.

13.3.5 EFFECTS OF GIBBERELLINS ON MEMBRANES

For the past few years Paleg and his associates have studied the effects of GA_3 on various membranes and membrane properties (Wood & Paleg, 1974; Wood *et al.*, 1974). Their studies were initially encouraged by the work of Bangham *et al.* (1965) who provided evidence that steroids have potent effects on model membranes composed only of lipids. Correlation between steroid activity *in vivo* and in the model membrane systems was good. A suggestion was made that steroids produced their biological effect through a direct interaction with lipid, independent of polysaccharides, proteins or even active cell metabolism. The influence of GA_3 on the permeability of lipid model systems was first described by Wood and Paleg (1972) who suggested that gibberellins might similarly exert their physiological action through an effect on lipids. Recently, Wood *et al.* (1974) have demonstrated complex formation between GA_3 and phosphatidyl choline thus providing definitive evidence of a molecular association between the hormone and a common membrane component. In this study, complex formation was dependent on hormone concentration. Wood and Paleg (1974) also showed that the process of glucose diffusion from liposomes (i.e. microvesicles composed of a bimolecular lipid membrane surrounding aqueous contents) is not affected by the presence of GA_3. The same energy is required for a mole of glucose to permeate the lipid at each temperature in the presence or absence of the hormone. However the barrier to diffusion is perturbed in such a way that more glucose can pass per unit of time in the presence of GA_3. Furthermore, the temperature at which the lipid membrane undergoes a change in conformation (the transition point) is lowered by several degrees by hormone treatment.

By comparison with living systems, GA_3 could, by its influence on lipid polarity, affect two main processes *in vivo*; (a) the permeability of the membranes to regulatory ions which act as limited substrates; and (b) the activity of membrane-bound enzymes such as ATPase which are activated by sodium, potassium, etc. Both of these effects could be brought about by the same mechanism—greater membrane fluidity. Since GA_3 influences not only the transition point at which the liposome changes in structure but also the diffusion rate of glucose above this transition point, one or both of two mechanisms may explain its action. Either its presence directly reduces the viscosity of the central core of the membrane or, more likely, it perturbs the still-ordered polar regions of the liquid crystal phase leading to greater fluidity.

However, the temperature zone of greatest physiological interest is that where the untreated system is below the transition point while in the presence of hormone the system has switched to high temperature sensitivity. The relationship between biological activity of different gibberellins and their ability to influence the thermal transition and membrane permeability has not been explored extensively. However, the methyl ester of gibberellic acid, a derivative with only about 10^{-3} of the biological activity of unsubstituted GA, was found to be without effect on the temperature at which the transition of the membrane occurs. In this regard it is of interest that it also induces very little alteration of the nuclear magnetic resonance spectrum of lecithin, which indicates little complex formation between the hormone and the fatty acid. This new area of research may suggest that the mechanism of GA action is through the direct alteration of membrane structure and/or function.

13.3.6 SUMMARY

The gibberellins comprise a complex and large group of related compounds which control cell elongation and enzyme secretion. The mechanism by which the hormone turns on the synthesis of enzymes has been studied in fairly close detail. Initially, it was felt that the hormone regulated the synthesis of RNA at the level of transcription. Through the careful and dedicated work of Varner and his associates it appears reasonably clear that the hormone does not control the secretion of enzymes by controlling the synthesis of messenger-RNA. However, it is clear primarily from the work on amylase, that inhibitors of nucleic acid synthesis and protein synthesis disrupt the production of the enzyme. The recent experiments of Paleg and his associates bring a new dimension to the question of gibberellin action by their demonstration that GA_3 affects the membrane permeability of fluidity. Thus, a new range of possible hormonal mechanisms can now be envisaged and explored.

13.4.1 ACTION OF CYTOKININS

13.4.1 INTRODUCTION

The term cytokinin has been accepted practically universally as a generic name for substances which promote cell division and exert other growth regulatory functions in the same manner as kinetin. The synthesis and testing of compounds for cytokinin activity began with the discovery of kinetin (6-furfurylaminopurine). Today there are probably at least one hundred known synthetic and native cytokinins (Fig. 13.4).

Structural requirements for a high order of cytokinin activity generally include an adenine molecule with the purine ring intact and with an N^6 substituent of moderate size. One exception to the requirement for a modified purine exists, notably diphenylurea and its derivatives. Certain other substances which lack a true purine ring, such as 8-azakinetin, 6-benzylamino-8-azapurine and 6-(3-methyl-2-butenylamino)-8-azapurine are active, but each of them is less than 10% as active as its corresponding purine derivative. Substitution of either O or S for N in the N^6 position of adenine in each case results in a more than 90% loss in growth promoting activity in the tobacco bio-assay.

Cytokinins play a role in practically all phases of plant development from cell division to the formation of flowers and fruits. They affect metabolism including the activities of enzymes and the biosynthesis of growth factors. They also influence the biogenesis of organelles and the flow of assimilates and nutrients within the plant. Cytokinins defer senescence and may protect the plant against adverse environments, such as water stress. These many diverse effects presumably stem from some primary anabolic function of the cytokinins that remains to be elucidated.

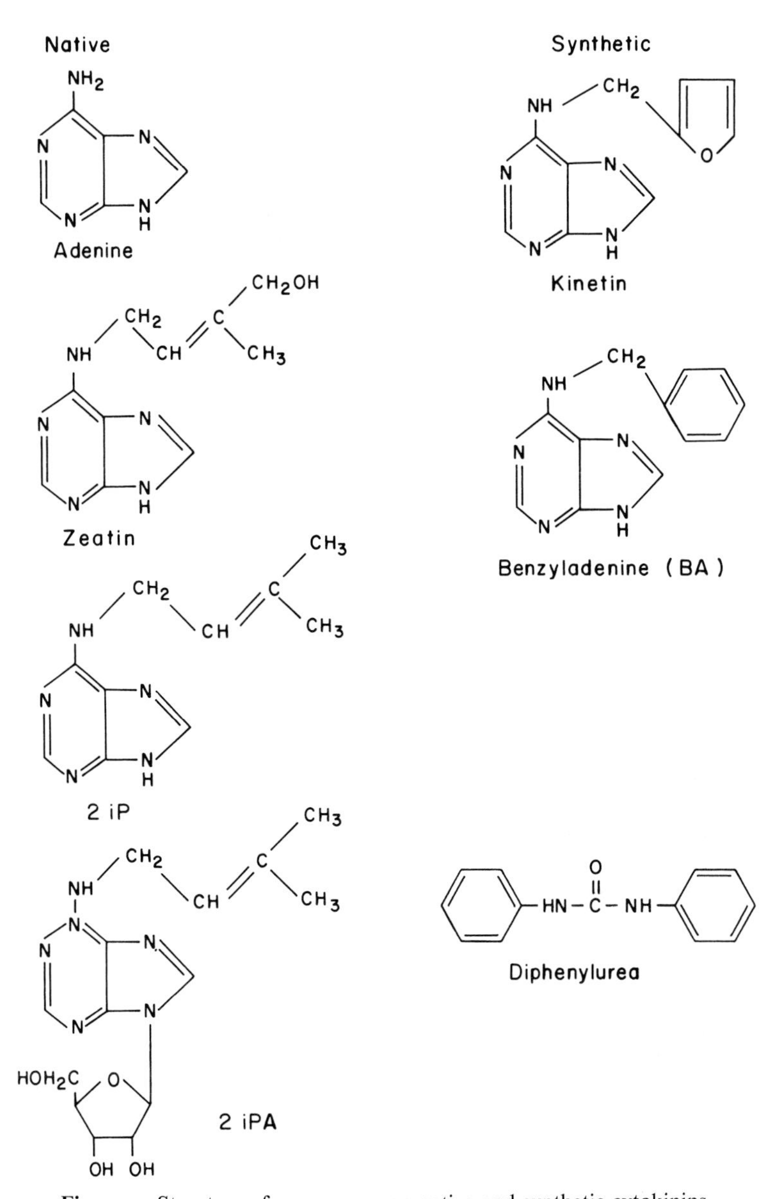

Fig. 13.4. Structure of some common native and synthetic cytokinins.

Cytokinins have been found in certain transfer-RNA (t-RNA) molecules from a large number of organisms including bacteria, yeast, plants and animals (see Cherry & Anderson, 1971). In sequence analysis the native cytokinin, isopentenyl adenine, always occurs adjacent (3′ side) to the anticodon (Fig. 13.5). Furthermore, the location of the isopentenyl group is required for the

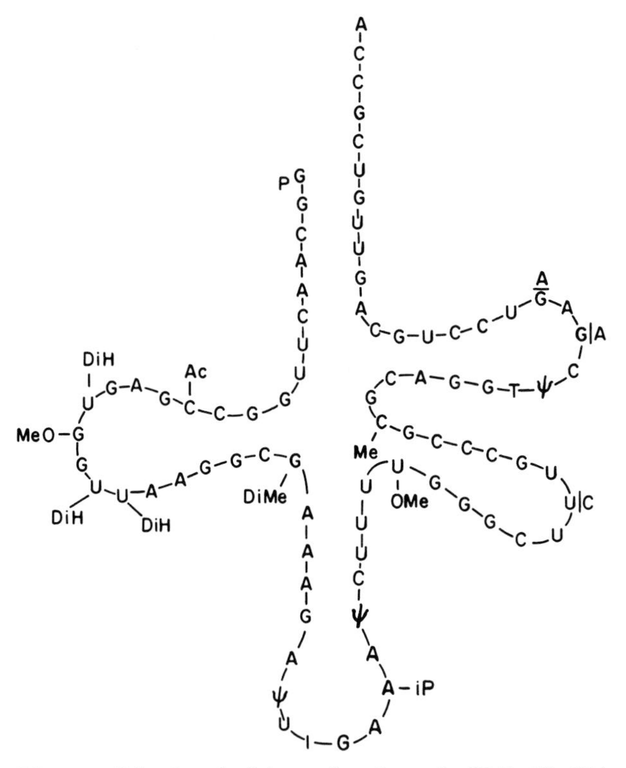

Fig. 13.5. Diagram of the cloverleaf shape of seryl-transfer RNA. The IPA residue is located adjacent to the anticodon.

specific tRNA to function efficiently in protein synthesis. Results involving bacteriophage infection of bacteria indicate that cytokinin in tRNA may have an important regulatory function in protein synthesis. In soybean seedlings (Anderson & Cherry, 1969) the application of a cytokinin (6-benzyladenine) results in increased levels of two species of leucyl-tRNAs. Although the relationship of cytokinins to tRNA may not be directly associated with the primary action of cytokinins on cellular activity, the alteration of specific tRNAs could control protein synthesis. Such a regulatory mechanism would be manifested in the control of growth and morphogenesis as the ultimate expression of the hormone.

13.4.2 ACTIONS OF CYTOKININS ON ENZYMES

Cytokinins have been shown to influence the activities of a number of specific enzymes. A most thoroughly studied case is the promotion by cytokinins of tyramine methylpherase activity in roots of germinating barley embryos. This appears to involve *de novo* synthesis although this conclusion is based solely on the results of inhibitor experiments (Mann *et al.*, 1963; Steinhart *et al.*, 1964). Activities of four other enzymes tested were unaffected by kinetin treatment. The selective action on methylpherase is most likely related to the developmental stage of the seedling material and not to a specific affinity of cytokinin for the enzyme. Substances other than cytokinins (auxins and lysine) which also stimulated the appearance of tyramine methylpherase activity either were less effective or acted only during the early stages of germination. An analogous effect of kinetin in promoting synthesis of alkaloids in the roots of lupin has also been reported (see Skoog & Armstrong, 1970).

The induction of α-amylase activity in half-seeds without embryos, a well known effect of gibberellins in barley, is achieved in wheat also by treatment with cytokinins (Boothby & Wright, 1962). However, the effect of cytokinins on wheat is much less marked than that of gibberellin in the barley. In barley seeds, cytokinins cannot replace gibberellins, but do counteract an inhibition of gibberellin action by abscisic acid (Khan, 1969). The formation of amylase activity in excised hypocotyls of pea is stimulated by cytokinins but not by gibberellins or auxins (Chum, 1967). In germinating squash seeds cytokinins, but not gibberellins, at least partly substitute for the embryo in promoting the formation of isocitrate lyase and proteinase activity in the cotyledons.

The synthesis of deoxyisoflavones in soybean callus tissue is promoted by both cytokinins and auxins (Miller, 1969). The ease with which the compounds can be detected and the rapidity of the response suggests that this system might be useful as a rapid bioassay for cytokinins.

The rates of synthesis of ribulose bisphosphate carboxylase and NADP-dependent glyceraldehyde-phosphate dehydrogenase, studied in rice seedlings, appear to be under cytokinin control (Feierabend, 1969). Kinetin did not affect these enzymes until 96 hours after the beginning of germination. By contrast the activity of the cytoplasmic enzyme 6-phosphogluconic acid dehydrogenase was promoted by kinetin only during early stages of germination. It was concluded that cytokinins probably are not directly involved in gene derepression although the level of cytokinin may determine the extent to which the appropriate genes are expressed.

13.4.3 NUCLEIC ACID SYNTHESIS

Early studies with tobacco pith tissue showed that a proportion of the tetraploid cells present would be stimulated by kinetin treatments to enter mitosis and undergo cytokinesis without incorporation of thymidine. Thus the effect of kinetin on cell division was not always associated with DNA synthesis. In *Lemna* cultures 6-benzyl-aminopurine reverses the effect of abscisic acid in inhibiting growth and nucleic acid synthesis (van Overbeek *et al.*, 1967). DNA synthesis in response to these growth substances appears to be more rapid than RNA synthesis. In axillary buds of tobacco which are normally inhibited by the apex, treatment with 6-benzylaminopurine initiates DNA synthesis and growth (see Skoog & Armstrong, 1970).

Some inhibitors such as chloromycetin and 6-azauracil uncouple the two processes, preventing growth but not the cytokinin-induced DNA synthesis. Nevertheless, in view of the early work, it seems likely that cytokinin effects on DNA synthesis are indirect. In a detailed study of tuberization, Palmer and Smith (1969) have shown that, in excised stolons of potato, cytokinins inhibit stolon elongation and induce tuber formation. Other factors which inhibit elongation may also increase the inductive response in response to cytokinins, but they are not themselves inductive factors. Inhibitor studies suggest that cytokinin action is associated with protein synthesis or activation but not with DNA or messenger-RNA synthesis. Effects of cytokinin on RNA and protein synthesis and isolated cell organelles have been reported, but have not yet been analyzed in depth (Skoog & Armstrong, 1970). Nuclei preparations separated from coconut milk and incubated in the presence of kinetin showed an increased incorporation of labelled precursors into RNA and protein (Datta & Sen, 1965).

The effect of kinetin on RNA synthesis in an *in vitro* system containing purified chromatin and *E. coli* RNA polymerase has been studied by Matthysse and Phillips (1969). In the presence of a particular protein fraction, kinetin stimulated RNA synthesis. A similar effect of auxin was also mediated by a protein fraction. The effects of cytokinin in this system have not yet been studied in great detail. Fellenberg (1969) has observed that *in vitro* bindnig of kinetin to chromatin of pea epicotyls caused a slight decrease in the melting point of the nucleoprotein complex.

3.4.4 ROLE OF CYTOKININS ON TRANSFER RNA

One suggestion has been that the endogenous cytokinins somehow exert their activity by being incorporated into transfer-RNA (Skoog & Armstrong, 1969). For example, the synthetic cytokinin, benzyladenine, was reported to be incorporated into the tRNA of callus tissue (Fox & Chen, 1967). Richmond *et al.* (1970) specifically looked for incorporation of labelled benzyladenine into the RNA of senescing leaves, but found no evidence for incorporation. On the other hand, under similar conditions kinetin was incorporated into the RNA. Recently Walker *et al.* (1974) have shown that double-labelled benzyladenine is incorporated into tRNA at a frequency of one per 10^4 molecules.

Chen and Hall (1969) have shown that cytokinin-dependent tobacco callus tissue contains the enzyme system which attaches the Δ^2-isopentenyl group to the appropriate adenine residue in tRNA. In the case of tobacco callus tissue grown in the presence of radioactive benzyladenine, the tRNA still contained its complement of isopentenyl adenine derivatives even though the benzyladenine was incorporated into the tRNA. In general, various purine analogues can be incorporated into RNA in a rather non-specific way.

It is also quite possible that incorporation occurs into RNA fractions other than tRNA, and unless the tRNA is rigorously characterized one cannot exclude the possibility of error arising by contamination with other RNA fractions. To establish whether the incorporation of N^6-substituted adenosine derivatives into tRNA has a biological significance it would be necessary to fractionate the tRNA and identify the incorporated material at a specific location in the tRNA sequence.

13.4.5 MODEL FOR CYTOKININ ACTION

Cytokinins are found in transfer-RNAs of animals, microorganisms and plants (see Cherry, 1973). Furthermore, in those cases where the nucleotide sequence is known the native cytokinin, IPA (isopentenyladenosine), or a modified form, is found adjacent to the anticodon. At the moment it appears that all IPA containing species of tRNA recognize codon triplets which begin with U (Skoog & Armstrong 1970). Thus, the anticodon loop of a cytokinin-containing RNA could be one of at least sixteen codon sequences of the following type: A— —IPA. By eliminating the nonsense terminator and the phenylalanine codons, it seems that IPA-containing tRNAs would be confined to less than ten codons (see Table 13.1).

Table 13.1. Codons beginning with U.

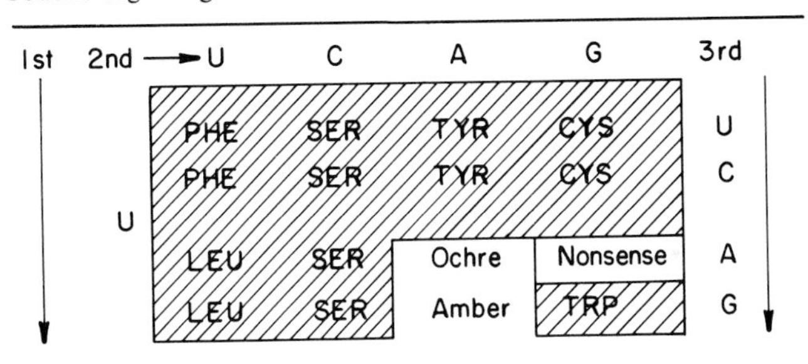

It may be that IPA adjacent to the anticodon of these tRNA species plays some role in messenger-RNA-tRNA-ribosome recognition, and that these tRNA species are required for the translation of some messenger-RNAs into specific proteins. Based on these assumptions it is possible to speculate on the mechanism of action of cytokinins in plants. From the observation that cytokinins appear not to be incorporated into tRNA, as precursor units, but rather the isopentenyl group comes from mevalonic acid (Chen & Hall, 1969), it seems likely that cytokinins have no effect on the synthesis of tRNA containing IPA. Furthermore, a number of chemicals, including various isomers of IPA, kinetin, 6-benzyladenine and even diphenylurea, have cytokinin biological activities, and it appears that these cytokinins do not participate in tRNA synthesis. If this is true, then what is the mechanism of cytokinin action?

A model presented in Fig. 13.6 is based on the speculation that specific

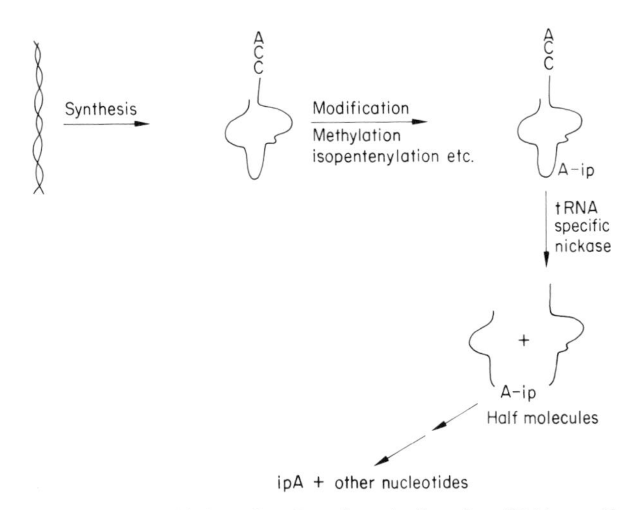

Fig. 13.6. A model of cytokinin action. Based on the fact that tRNA specific nucleases are known, it is suggested that free cytokinins protect cytokinin-containing tRNAs by inhibiting nuclease activity, specific as well as general.

nucleases break the primary structure of the IPA containing tRNA (Cherry & Anderson, 1971). The IPA would provide a unique attachment for the enzyme to bind to and subsequently to break the phosphodiester bonds. The action of this enzyme would destroy the function of the tRNA and yield free cytokinin. Cytokinins including 6-benzyladenine and diphenylurea added exogenously to plants or tissue cultures would defer senescence by essentially protecting the cytokinin-containing tRNA species. This protection would be mediated by the exogenously added cytokinin binding to the nuclease and competitively inhibiting its action. Alternatively, the cytokinin could bind to other agents which in turn prevent an increase in nuclease activity which would destroy the cytokinin-containing tRNA. Therefore, tissue treated with cytokinin would retain its IPA-containing tRNA species and continue to synthesize essential proteins as long as sufficient free cytokinin was present within the tissue.

Preliminary unpublished work of Locy (1974) provides suggestive evidence of a tRNA-specific nuclease. We have found that chloroplasts of soybeans contain a nuclease which specifically degrades one of the three tyrosyl-tRNAs. At the moment we do not have any evidence to suggest that free cytokinin controls the activity of this tRNA nuclease. In regards to this model it is to be emphasized that it is hypothetical and is presented as a working base upon which to investigate further the mechanism of action of cytokinins at the molecular level.

13.4.6 SUMMARY

Cytokinins have been shown to increase the activities of a number of enzymes. These increases in activity do not appear to be a direct effect of the cytokinin but rather to be an indirect effect of some kind. Cytokinins increase the amount of DNA produced in a number of tissues. Furthermore, cytokinins increase the rate of RNA synthesis in ageing tissue. However, a number of studies have pointed out that the cytokinin effect on nucleic acid synthesis most likely is not due to a direct effect on the derepression of the DNA within the genome.

Over the past few years there has been a great deal of excitement generated relating to the fact that the native cytokinin is found in some transfer-RNA isoaccepting species. For a few years data was presented which showed that cytokinins, particularly the synthetic cytokinin, 6-benzyladenine, was incorporated directly into transfer-RNA. That excitement, however, appears to be shortlived since other studies have shown that the incorporation of the labelled synthetic cytokinins is not into tRNA species but into heavier forms of RNA. Therefore, it appears quite clear that cytokinins are likely not incorporated into tRNA directly. A model is presented (Fig. 13.6) which suggests that the free cytokinins act as a protective agent within the target cells of plants to protect the transfer-RNA molecules which contain the native cytokinin. It is suggested that there are specific nucleases which degrade only certain species of transfer RNA. The model implies that free cytokinin inhibits the nucleases through product inhibition, or possibly through some other mechanism which prevents the enzyme from attacking the tRNAs which contain the cytokinin. At present there is only a small amount of evidence to substantiate this model.

13.5 ACTIONS OF ETHYLENE

13.5.1 INTRODUCTION

Ethylene is the simplest hormone which regulates plant growth (Fig. 13.7). It is a natural constituent of plants and affects a wide array of physiological processes. For many years, however, investigations of ethylene physiology were concerned primarily with fruit ripening. With the advent of gas chromatography many experimenters have begun research on the biochemical and physiological action of ethylene. The production of ethylene by germinating seeds and seedlings suggests that the hormone may be involved in the normal regulation of growth and development. Evidence for such a proposal includes the fact that low concentrations of applied ethylene blocks photo-induced apical bud expansion and hook opening in etiolated pea seedlings. It was shown that apical tissues of

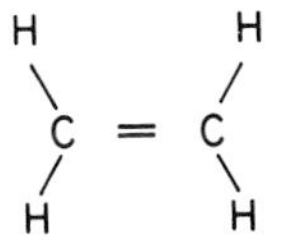

$$\begin{matrix} H & & & & H \\ & \diagdown & & \diagup & \\ & C & = & C & \\ & \diagup & & \diagdown & \\ H & & & & H \end{matrix}$$

Fig. 13. 7. Structure of ethylene.

etiolated seedlings are the major site of ethylene production. Light has been found to decrease the tendency of pea stem segments to produce ethylene in response to high concentration of auxin.

On the other hand, red light appears to stimulate ethylene production in dormancy. Another important role of endogenous ethylene in etiolated seedlings is the regulation of radial expansion of the pea epicotyl in a region below the apical hook. The exposure of the plant to red light or to CO_2 inhibits an ethylene-mediated increase in the diameter of the epicotyl.

Growth, flowering, abscission and fruit ripening all are affected by ethylene. It is currently popular to speculate that the mode of action of ethylene involves a mechanism which regulates some aspect of the transcription of DNA or the translation of RNA. Studies of the effects of ethylene on abscission and growth indicate sizeable changes in RNA and protein contents. There is indeed evidence of changes in the activities of catalase, peroxidase, and other hydrolases. Furthermore a study showed that exposure of soybean plants to ethylene significantly altered RNA polymerase activity associated with chromatin. As judged by nearest neighbour analysis the RNA produced from ethylene treated plants has a different base composition than that from the control. Even though ethylene at very low concentrations affects a wide array of physiological and biochemical processes in plants, much more work is required to know the mechanism of action.

13.5.2 EFFECT OF ETHYLENE ON ENZYMES

A number of investigators have examined the possibility that ethylene has a direct effect on enzyme activity (Abeles, 1973). However, investigations with β-glucosidase, emulsin, α-amylase, invertase, peroxidase, and adenosine triphosphatase have shown that ethylene has no effect on these enzymes. Results reported by Abeles (1973) have similarly shown that ethylene has no effect on cellulase and carbonic anhydrase. Carbonic anhydrase was chosen because it appeared to be a likely candidate to show a positive response. Carbonic anhydrase contains zinc and has the ability to combine with CO_2, both features which suggest potential sensitivities for ethylene action. Nelson (1939) reported that ethylene increased the activity of trypsin. However, this effect was thought to be due to the removal of O_2 since H_2 had the same effect.

13.5.3 ACTIONS ON MEMBRANES

Since ethylene is more soluble in oil than in water and since membranes contain large amounts of lipid material, a number of investigators have tested the idea that ethylene affects some aspect of membrane permeability. However, proponents of this idea have failed to note that CO, an ethylene analogue, does not share the same lipid solubilities as ethylene but nevertheless has similar physiological activities. The idea that ethylene has a disruptive effect on the membrane causing a change in permeability and an alteration in compartmentalization does not appear valid. Ripening fruits exhibit obvious changes in terms of permeability and retention of soluble components and it seems natural to suggest that ethylene leads to changes in membrane permeability which in turn causes softening and increased respiration. However, current evidence suggests that a change in membrane properties is an effect of ripening rather than a cause.

A similar situation exists in flowers. Nichols (1968) pointed out that solute leakage increased from carnations during senescence. Senescence and leakage were promoted by ethylene and reversed by CO_2. Ethylene has no influence on membrane permeability of potato, pea, avocado, banana, and bean. However, on the other hand, von Guttenberg and Beythien (1951) reported that ethylene increased the rate of deplasmolysis of *Rhœo* leaves, but Burg (1968), was unable to confirm their results. Nord and Franke (1928) claimed that ethylene increased the permeability of yeast cells but the effects were small, while Abeles (1973) failed to confirm their results. Even if ethylene regulated permeability of yeast cells it would tell us little about the hormonal mechanism of action since ethylene has no effect yet detected on the physiology of yeast cells. The same criticism is applicable to the work involving red blood cells where ethylene was found to increase permeability. In this case nitrous oxide had the same effect on permeability showing that the ethylene effect was non-specific (Lyons & Pratt, 1964). Similarly, early reports on the response of mitochondria to ethylene suggested an influence of a high concentration of the gas on conformation (Olson & Spencer, 1968). While the effects were real, subsequent work pointed out the effects were not typical of normal ethylene action. First of all, high concentrations of ethylene were required to induce conformational changes and second, saturated gases such as ethane and propane gave the same effects as ethylene (Ku & Leopold, 1970).

13.5.4 ENZYME INDUCTION BY ETHYLENE

Regeimbal and Harvey (1927) were the first to report that ethylene-treated tissues contained higher levels of particular enzymes than the control. They found that ethylene increased the level of protease and invertase extracted from pineapple fruits. Since that time, reports (see Abeles, 1973) have appeared which show effects of ethylene on a number of enzymes. This list includes acid phosphatase, ATPase, α-amylase, cellulase, chitinase, chlorophyllase, cinnamic acid 4-hydroxylase, cytochrome *c* reductase, diaphorase, β-1,3-glucanase, invertase, malic enzyme, pectin esterase, peroxidase, phenylalanine ammonia-lyase, polygalacturonase, polyphenyloxidase, protease and pyruvic carboxylase.

Enzyme induction does not always depend entirely on the action of ethylene. In some cases, cutting the tissue causes an increase in enzyme activity. The function of ethylene is to reduce the lag time or increase the rate of increase in enzyme activity. Examples of enzymes whose increased activity does not strictly depend on ethylene action include β-1,3-glucanase, malic enzyme, phenylalanine ammonia-lyase and peroxidase (Abeles, 1973). Excising tissue can cause wound ethylene production and it is possible that this source of ethylene plays some role in enzyme induction in tissue slices. On the other hand, enzyme induction during abscission was dependent on ethylene and for a reasonable period of time no cellulase synthesis or abscission occurred following excision unless ethylene was added to the gaseous phase.

13.5.5 EFFECTS OF ETHYLENE ON RNA SYNTHESIS

The first report of ethylene effects on RNA synthesis was by Turkova *et al.* (1965) who reported an increase in RNA synthesis during epinasty of tomato leaves. Whether or not RNA synthesis was required for epinasty was not shown, although the idea is intriguing. Studies with actinomycin D indicated that RNA synthesis occurred during abscission and was required for the process to occur (Abeles & Holm, 1966). Support for this interpretation stems from the work of a number of investigators (see Abeles, 1973). It is known that the increase in RNA synthesis precedes that in protein synthesis and is localized at or near the separation layer. The increase in RNA occurred in all fractions including messenger-RNA, ribosomal-RNA and transfer-RNA. Differential extraction of the nucleic acids indicated that the ethylene-stimulated fraction was confined to that portion of RNA extracted by sodium lauryl sulphate with the increase found in ribosomal- and messenger-RNAs. The inhibitor, 5-fluorouracil, which blocks 50% of the ethylene-enhanced incorporation of ^{32}P into RNA did not inhibit abscission. The greatest inhibition occurred in transfer-RNA and ribosomal-RNA which indicates that the synthesis of all fractions of RNA is not required for abscission. Presumably as long as messenger-RNA was being synthesized, enough ribosomal- and transfer-RNAs were already available within the cell to permit abscission to take place. When all RNA synthesis was blocked by actinomycin D, abscission stopped (Holm & Abeles, 1967). Ethylene has also been found to promote RNA synthesis in preclimacteric fruit. Marei and Romani (1971) reported, as in the case of abscission, that the synthesis of all classes of RNA in fig fruit was stimulated. Holm *et al.* (1971) found that ethylene increased RNA synthesis in apples and that the increase in RNA was followed by an increase in protein synthesis. However, Sacher and Salminen (1969) reported that they failed to find an increase in RNA synthesis when preclimacteric bananas or avocados were treated with ethylene.

13.5.6 EFFECTS OF ETHYLENE ON CHROMATIN ACTIVITY

Holm *et al.* (1970) have reported that ethylene inhibited the growth of the apical part of soybean seedlings and caused an increase in the elongating and basal portions of the stem. At the same time, RNA levels in the apical zone were reduced while the levels were enhanced in the elongating and basal regions. Chromatin from the various parts of the seedling were studied to determine if the capacity for RNA synthesis was also modified. They found that the activity in the apex was reduced while the activity in the elongating and basal regions were promoted. The rate of the response was rapid since the increase in chromatin activity was apparent after 3 hours. Nearest-neighbour analysis of the RNA synthesized demonstrated that there was a qualitative difference between RNA synthesized from chromatins of normal and ethylene-treated tissue. It was concluded that ethylene can regulate RNA synthesis as manifested by a change in quantity and kind of RNA.

13.5.7 EFFECTS OF ETHYLENE ON DNA METABOLISM

Plant growth is either promoted, inhibited or unaffected by ethylene depending upon the tissue involved. Examples of growth promotion are swelling, epinasty, hook closure, seedling elongation and seed germination (Abeles, 1973). Bud break is probably a special case. Here no growth takes place as long as ethylene is present. However, after the gas is removed growth of the buds ensues. Growth inhibition is seen as arrested development of buds, leaves or apical meristems. Mature tissue, such as stems and leaves. do not undergo any change of size or weight, although premature senescence usually occurs. Since growth, or the lack of it, may be associated with cytokinesis, it is of interest to learn that DNA synthesis is controlled by ethylene. Holm and Abeles (1967) reported that DNA synthesis or DNA content in bean leaf tissue was not affected by seven hours exposure to ethylene, although abscission was promoted. Later they found that in soybean seedlings treated with ethylene the synthesis of DNA was inhibited in the apex where growth was inhibited but was promoted in the subapical part where swelling took place. Burg *et al.* (1971) found a similar situation in pea seedlings in which inhibition of cell division, measured as a reduction in metaphase figures, occurred within two hours after ethylene was added. However, it is not clear whether the change in DNA synthesis was the cause or the result of inhibited growth. Ethylene slows the growth of pea seedlings very quickly; for example, Warner and Leopold (1971) found growth slowed six minutes after ethylene was introduced into the gas phase and returned to normal sixteen minutes after the ethylene was removed. Kinetic studies on changes in DNA or other postulated sites of action are thus required to establish the relationship between cause and effect. Burg *et al.* (1971) have suggested that ethylene regulates DNA synthesis by some action on microtubule structure essential for spindle fibre formation during mitosis. If the action of ethylene was directed toward microtubules, this might also explain the reorientation of cellulose microfibril deposition that occurs in swelling.

13.5.8 SUMMARY

Even though little is known about the binding site(s) of ethylene, in a few cases such as ripening, abscission, swelling and senescence there is reason to believe that the combination of ethylene with a site results in the regulation of protein or enzyme synthesis which in turn accounts for the observed responses. In these cases RNA synthesis, presumably messenger-RNA, with the accompanying support of soluble- and ribosomal-RNA, is an essential or early step suggesting regulation of gene action. However, in other cases, especially those associated with the physiology of excised tissue, control is probably not exerted at the level of RNA or protein synthesis. In the case of the inhibition of elongation, the action is apparently directed toward blockage of DNA metabolism. The site of action of ethylene in epinasty, root initiation, intumescence formation and floral initiation is even more poorly understood. The only valid conclusion appears to be that a number of essential features of plant growth and development are susceptible to ethylene action. In the final analysis it is concluded that there may be as many mechanisms of ethylene action as there are modes of ethylene operation.

13.6 ACTIONS OF ABSCISIC ACID

13.6.1 INTRODUCTION

Abscisic acid (ABA) is a plant hormone which now ranks in importanance with the auxins, gibberellins, cytokinins, and ethylene. Interest in the physiology and chemistry of ABA has grown greatly since the structure was established in 1965. During the 1950's and early 1960's a number of laboratories were engaged in research on growth-inhibiting substances. ABA was first isolated from cotton plants and was named abscisin II by a team from Addicott's laboratory (Ohkuma *et al.*, 1963). During the same year Wareing's research team (Eagles & Wareing, 1963) isolated an active substance from *Acer* leaves which they named dormin. Abscisin II and dormin are the same substance which is now called abscisic

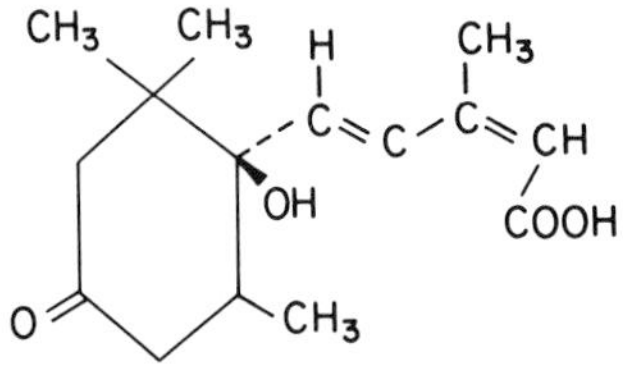

Fig. 13.8. Structure of abscisic acid.

acid (Fig. 13.8). Plant tissues of all ages appear to synthesize and inactivate ABA. The number of various plant responses affected by ABA is very large. Generally, the physiological processes are related to senescence or abscission and growth retardation or inhibition. ABA appears to act as an abscission accelerating hormone in many fruits and leaves. Furthermore, it also tends to induce dormancy in some woody plants. ABA has been shown to move from the leaves to the apical bud to bring about a dormant condition. In potato, the levels of inhibitors, including ABA, decrease during the quiescent period prior to renewed growth. ABA in extremely low concentration, moreover, prolongs the dormancy of excised potato buds.

Recently, ABA has been shown to be involved in the responses of many plants to stress conditions. As the ABA concentration increases in the leaves the stomata close. In this way, it appears that ABA is directly involved in the opening and closing of the stomata, thus regulating the rate of transpiration. Through this mechanism, ABA appears to protect the plant through conditions of water stress or drought.

At present the mechanism of action of ABA is not clearly understood. However, the available evidence indicates that ABA affects transcription as shown by reduced activity of chromatin-associated RNA polymerases. In other cases the mechanism of action of ABA appears to involve regulation of the translation of long-lived messenger-RNA, whilst its effects on stomata probably involve the regulation of membrane permeability.

13.6.2 ROLE OF ABA IN DORMANCY

One of the lines of research which led to the isolation of ABA from sycamore leaves was a change in the growth inhibitory activities in extracts of tree seedlings grown on long- or short-days. An increase in content of the growth inhibitory material in leaves was noted during the late summer and early autumn (Phillips & Wareing, 1958). Furthermore, the idea that bud dormancy in potato tubers may be caused by growth inhibiting substances had been suggested by Hemberg (1949). The original correlation of inhibitory material in leaves grown under short- and long-days and the induction of dormancy were based on measurements obtained by bioassay techniques. Recently, Lenton *et al.* (1973) have attempted to repeat these determinations using gas chromatography to compare the amounts of ABA present. They found that transferring birch, red maple or sycamore plants to short-days had no effect on ABA content. The importance of a balance between growth promoters and growth inhibitors has been stressed frequently. Tinklin and Schwabe (1970) have found considerably more inhibitor in bud scales than in the bud axis of blackcurrant. In these experiments the ABA concentrations were correlated with the degree of dormancy and the levels were reduced by treatments which encouraged bud break.

The most convincing evidence suggesting that ABA induces dormancy is the production of turions in *Lemna polyrrhiza*. When ABA is added to the medium under conditions that allow continuous growth, it not only inhibits growth but causes production of the dense, dormant, fronds known as turions (Stewart, 1969).

ABA is a potent inhibitor of seed germination, and its presence as a major growth inhibitor in dormant seeds of many species has cast it in the role of the maintainer of seed dormancy. ABA has been isolated from seeds of many genera of higher plants, and seeds of an equally large number of species have been prevented from germination by soaking in ABA solutions. When seeds and fruit parts are separated, it is usually found that the concentration of ABA in the fruit is about 5 to 10 times greater than in the seed (Milborrow, 1974).

Lipe and Crane (1966) found that ABA in peach seeds decreased during stratification. Obviously, the part played by endogenous growth promoting compounds and the balance between them and ABA during the breakage of dormancy needs to be explored in more detail. Certainly, the balance between the two kinds of regulators is important. ABA is not an irreversible inhibitor since one of its most striking features is the facility with which it can be leached out of treated seeds thereby allowing the resumption of growth.

13.6.3 EFFECTS ON ABSCISSION

In the last few years several hundred experiments have been carried out in which plants, and parts of plants, of many genera have been treated in a variety of ways with ABA. It is quite unusual for any report to show clearly that the hormone controls abscission of leaves, which leads to the conclusion that exogenously applied ABA has little effect on leaf abscission. Nevertheless, ABA was first isolated by following its abscission-accelerating activity in petiolar stumps of cotton explants. The growth inhibitory action of this factor was believed to be responsible for the premature abscission of immature young lupin fruits (Cornforth *et al.*, 1966). Consequently, the abscission-accelerating effect of ABA is well documented and extensively discussed (Milborrow, 1974).

Many of the experiments reporting leaf abscission have been carried out on tree crops using extremely high concentrations of ABA and often near the end of the growth season in attempts to cause abscission of fruits. The observed stimulation of leaf abscission may be an indirect effect of non-physiological high concentrations which stimulate ethylene production (Edgerton, 1971). For example, Cooper and Henry (1968) treated orange trees with sprays of 500 $\mu g\ ml^{-1}$ of ABA in summer and winter. The summer treatment caused leaves to develop colour and fall, but the winter treatments had no such effects. Olive trees were found to suffer some leaf abscission in one experiment but no effect was observed in another.

It appears, therefore, that ABA is not closely involved in the regulation of leaf abscission. The test system in which it has a stimulatory action consists of isolated petiolar explants containing presumptive abscission zones and maturing leaves nearing the end of their life. Even this tissue requires application of abnormally high concentrations of ABA to manifest an effect.

The role of ABA on fruit abscission is more certain. The early work of Van Steveninck (1959) showed that the inhibitor now known as ABA was intimately involved in the abortion of the young immature fruit of lupin. Application of ABA to mature peach, olive, apple and citrus fruits accelerated abscission and the effect of ABA was also marked on young grape flowers and berries (see Milborrow, 1966).

13.6.4 EFFECTS OF ABA ON WILTING

Wright (1969) found that when cut shoots were wilted there was an increase in the concentration of the so-called β-inhibitor. He went on to identify this inhibitor as ABA and defined the conditions under which the increase occurred. A water loss of about 10% in total fresh weight caused approximately 40-fold increase in ABA content while further water loss had no additional effect. Wright and Hiron (1969) have reported that other stress conditions, such as waterlogging, caused a similar rise in ABA, but they point out that such treatments cause wilting by reducing the efficiency of water uptake. The surprising feature of the increase in ABA is the rapidity with which it occurs. The content in turgid bean leaves has been calculated from the results of Wright and Hiron (1969). They show that ABA content increases from a normal level of 6 $\mu g\ kg^{-1}$ fresh weight to 7 $\mu g\ kg^{-1}$ within 7 minutes after blowing a dry and warm air stream across the plants. This level increases to 33 $\mu g\ kg^{-1}$ within 25 minutes and 68 $\mu g\ kg^{-1}$ within 45 minutes. In other experiments they show that in wheat leaves the ABA content increased from 23 to 171 $\mu g\ kg^{-1}$ fresh weight within 4 hours of wilting. The ABA content of bean leaves remained at 67 $\mu g\ kg^{-1}$ while they were kept waterlogged for 5 days. The implications of these data have yet to be explored in detail, but the observations offers a feasible explanation of the reduced growth of crops suffered during drought.

The extra ABA is probably formed by synthesis rather than by release from a precursor or conjugate because much more labelled mevalonate was incorporated into ABA by a sample of leaves that had been fed and then wilted than similar leaves which were kept moist during the entire experiment. Furthermore, the presence of 40 times the amount of a precursor or conjugate probably would have been detected by extraction and bioassay techniques. Many types of experiments have shown dramatic increases in ABA content on wilting in french beans, brussel sprouts, sugar cane, wheat, avocado spinach, cotton, peas and tomatoes (see Milborrow, 1974).

Perhaps the best information supporting a direct involvement in the direct closure of the stomata is afforded by a wilty tomato mutant produced by X-irradiation (Imber & Tal, 1970). The shoots of this plant contain one-tenth the amount of ABA contained in the normal variety. Applications of 0·1 to 10 μg of ABA caused a rapid and progressive reduction in transpiration rate of leaves and leaf discs. With 10 μg ABA ml^{-1} the stomata closed in darkness.

13.6.5 AFFECTS OF ABA ON ENZYME ACTIVITIES

The ability of plant hormones to affect directly the activity of enzymes has not been extensively investigated. When an effect has been reported it is smaller than would be expected if the hormone specifically regulates at that site. Hemberg (1967) reported that ABA can inhibit α-amylase *in vitro*, but as the enzyme used was extracted from fungi the significance as related to higher plants requires additional investigation. ABA has been shown to complex with fungal α-amylase and thereby change its physical properties. Saunders and Poulson (1968) found a slight stimulation of invertase activity at 10^{-7} M ABA and slight inhibition at 5×10^{-7} M and above. Again the significance is difficult to assess since a fungal enzyme was used.

13.6.6 EFFECTS OF ABA ON NUCLEIC ACID SYNTHESIS

The first observations of the effects of ABA on nucleic acids were made by Van Overbeek *et al.* (1967) using *Lemna*. The incorporation of radioactive phosphate (^{32}P) into nucleic acid fractions was inhibited by ABA but reversed by benzyladenine. The primary site of action of ABA cannot be deduced with certainty because of the time scale and the complexity of the responses. Subsequent work has indicated that DNA synthesis is almost certainly not the primary target of ABA (Villiers, 1968). This has been clearly shown to be the case in dry wheat embryos by Chen and Osborne (1970) who found that protein synthesis commenced from imbibition and was inhibited at 6 hours by ABA, whereas RNA synthesis, as measured by the incorporation of ^{3}H-uridine, was not measurable until 12 hours. In the same experiments the incorporation of ^{3}H-thymidine into DNA was measurable after 24 hours. Experiments by Pearson and Wareing (1969) have shown that chromatin-directed RNA polymerase activity was slightly inhibited by 0.26 $\mu g\ ml^{-1}$ of ABA when added to the grinding medium. However, when ABA was added to the purified chromatin little or no effect was noted. Bex (1972) also reports that ABA has no effect on the binding between nucleohistones and DNA as measured by their melting point measurements. Schwartz (1971) has shown that ABA can alter the balance of alcohol dehydrogenase in maize, but the inhibitor was added during the growth of the cells. Thus, while it appears that ABA has small and random effects on nucleic acid biosynthesis in a number of plants, it is reasonably clear that the hormone has little or no direct effect on the synthesis of nucleic acids at the level of the genome.

13.6.7 THE INVOLVEMENT OF ABSCISIC ACID IN MESSENGER-RNA TRANSLATION

Ihle and Dure (1972) investigated the appearance of various enzyme activities during the germination of cotton seeds and embryos. They showed that protease and iso-citrate lyase appear to be synthesized *de novo* during germination utilizing messenger-RNA which had been transcribed much earlier when only about 60% completion of embryogenesis had been reached. The translation of these messenger-RNAs during the last 40% of embryogenesis was apparently prohibited by the presence of ABA diffusing into the embryo from the ovule wall. The mode of action of ABA in maintaining translation inhibition appears to involve RNA synthesis as judged by the fact that the ABA inhibition was inhibited by the presence of actinomycin D. Translation of the required messenger-RNAs for the germination enzymes may be induced prematurely by simply dissecting the ovules from the plant, suggesting that the breakage of the connection between the ovule and the mother plant may be required for the *in vivo* induction of translation. This hypothesis is further substantiated by the observation that, *in vivo*, the connection between the ovule and the placenta is normally severed at 60% completion of embryo-genesis.

The results coming from Dure's laboratory are the only ones at present which show that ABA may have a direct effect on the translation of messenger-RNA pre-existing in the embryo tissue.

13.6.8 SUMMARY

Even though ABA was first found in cotton plants and dormant sycamore leaves, it seems reasonably clear that this hormone has little or no effect on either the abscission of leaves or the dormancy of buds. It appears that the most likely physiological action of ABA on plants is to control the abscission zone in the fruit pedicel and thereby allow abscission to take place and cause fruit drop. Secondly, ABA very definitely appears to be involved in stomatal closure in response to stress conditions, particularly drought. In terms of a more biochemical approach it appears that ABA has little effect on various enzyme activities. Those enzymes studied show only small enhancement, or in some cases, inhibition, by ABA. Generally it has been shown that ABA inhibits the synthesis of various nucleic acids.

Again, one must be drawn to the conclusion that ABA has little effect directly on nucleic acid synthesis at the genome level. The most excitingp aproach to abscisic acid action appears to be that found in the cotton embryo where it has been shown to inhibit the translation of pre-existing messenger-RNAs found in the developing embryos. Even this response probably involves more complex processes than a direct inhibitory effect, since actinomycin D inhibits the ABA-induced inhibition. These data possibly imply that a suppressor molecule has to be formed to bring about effects.

FURTHER READING

ABELES F.B. (1973). *Ethylene in Plant Biology*. Academic Press, N.Y. 302 pp.

CHERRY J.H. (1973). *Molecular Biology of Plants, A Text-Manual*. Columbia University Press. N.Y. 204 pp.

HALL R.H. (1973). Cytokinins as a probe of developmental processes. *Ann. Rev. Plant Physiol.* **24**, 415–44.

JONES R.L. (1973). Gibberellins: their physiological role. *Ann. Rev. Plant Physiol.* **24**, 571–98.

KEY J.L. (1969). Hormones and nucleic acid metabolism. *Ann. Rev. Plant Physiol.* **20**, 449–74.

MILBORROW B.V. (1974). The chemistry and physiology of abscisic acid. *Ann. Rev. Plant Physiol.* **25**, 259–307.

RUNECKLES V.C., SONDHEIMER E. & WALTON D.C. (eds) (1974). *The Chemistry and Biochemistry of Plant Hormones*. Vol. 7. Recent Advances in Phytochemistry. Academic Press. N.Y. 178 pp.

SKOOG F. & ARMSTRONG D.J. (1970). Cytokinins. *Ann. Rev. Plant Physiol.* **24**, 359–84.

VARNER J.E. & JOHRI M.M. (1968). Hormonal control of enzyme synthesis. In *Biochemistry and Physiology of Plant Growth Substances* (eds. F. Wrightman & G. Setterfield). The Runge Press Ltd. Ottawa. pp. 793–814.

CHAPTER 14
PHYTOCHROME ACTION

14.1 INTRODUCTION

Light, in addition to providing the energy which drives photosynthesis, also acts as a regulatory environmental stimulus. Many light-controlled plant responses are now believed to be mediated by the photoreceptor, phytochrome. This pigment has a regulatory role in all phases of plant growth and development (photomorphogenesis) and is apparently ubiquitous in all taxonomic groups of eukaryotic plants with the exception of fungi.

The central question of the molecular mechanism of phytochrome action is, however, yet to be resolved. Approaches to this problem fall into two broad categories: (a) Examination of the physical and chemical properties, localization and behaviour of the phytochrome molecule itself; (b) Investigations of the diverse array of phytochrome-mediated plant responses demonstrable at almost any level from the molecular to the gross morphogenetic. The principal findings and concepts that have emerged from such studies are summarized next. Some of the evidence that has led to these concepts is examined in subsequent sections.

14.2 PHYTOCHROME DOGMA

Phytochrome is a blue-green biliprotein. Two principal forms of the molecule are readily distinguishable on the basis of their absorption spectra (Fig. 14.1)

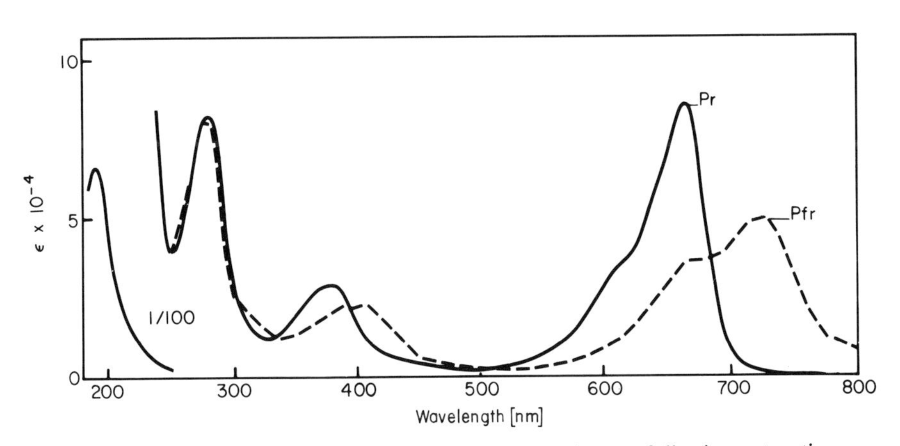

Fig. 14.1. Absorption spectra of purified oat phytochrome following saturating irradiations with red and far-red light (after Anderson *et al.*, 1970).

and biological activity. The *Pr* form (=red-absorbing, λ_{Max}=660 nm) is considered biologically inactive, whereas the *Pfr* form (=far-red-absorbing, λ_{Max}=730 nm) is considered to be the active species i.e. capable of inducing a biological response.

The two forms are reversibly interconvertible by red and far-red light ('phototransformation'). In addition, whereas the *Pr* form is stable in the dark, *Pfr* can revert thermally to *Pr* ('dark reversion') or undergo an irreversible loss of photoactivity ('destruction'). These properties are schematically represented thus:

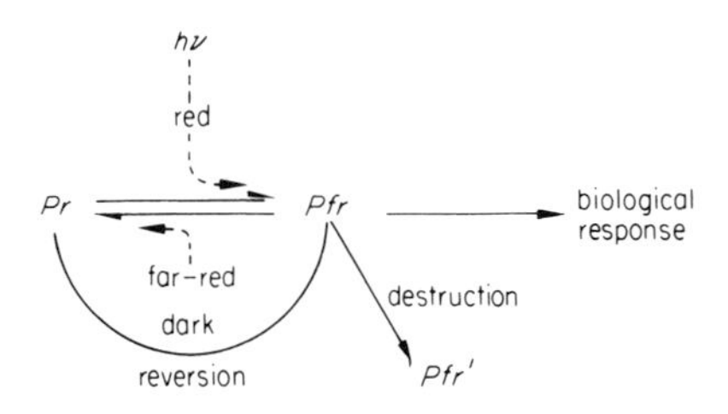

Red light absorbed by *Pr* converts the molecule to the *Pfr* form which in turn induces a measurable biological response. Far-red light, dark reversion and destruction provide alternate pathways for the removal of *Pfr* and thereby the potential for reversing induced responses. These properties endow the molecule with the capacity to function as a reversible biological switch, monitoring the environment for the presence, absence, intensity and spectral quality of photomorphogenically active light.

In molecular terms, phytochrome is considered to mediate the transmission of the light stimulus to the cell via photoconversion to an active effector. Two major theories on how the cellular response system recognizes the effector have been advanced. One proposes phytochrome interaction with the genome; the other, interaction with cellular membranes.

14.3 THE PHYTOCHROME MOLECULE

14.3.1 MOLECULAR PROPERTIES

The purified phytochrome molecule is a water-soluble chromoprotein containing less than 4% carbohydrate (Briggs & Rice, 1972). The native monomer is a polypeptide of 120,000 daltons but can form higher molecular weight aggregates. The chromophore is thought to be a linear tetrapyrrole with some evidence suggesting that there is one chromophore per monomer (Tobin & Briggs, 1973). The proposed chromophore structures, their postulated linkages to the protein and a possible photo-isomerization mechanism are shown in Fig. 14.2 (Rüdiger, 1972).

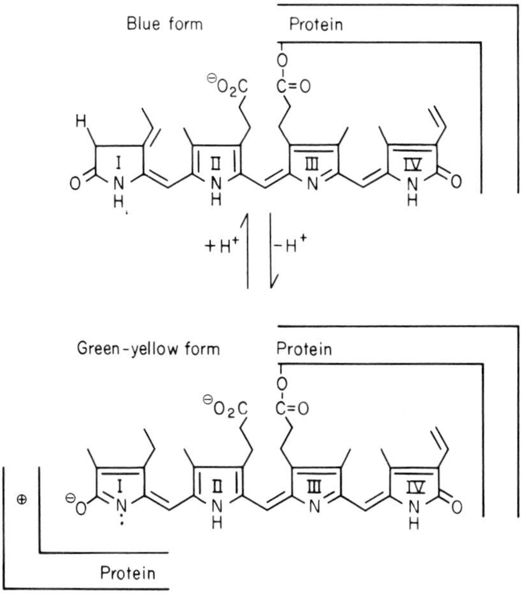

Fig. 14.2. Proposed structure for the phytochrome chromophore, its linkage to the protein and possible phototransformation mechanism (after Rüdiger, 1972). The 'blue' form is thought to correspond to P*r* and the 'green-yellow' form to P*fr*.

Information on differences in the molecular organization of the *Pr* and *Pfr* species has been sought in the hope that this might provide some insight into potential reaction mechanisms. Differences in the chromophore environment are evident from the absorption (Fig. 14.1), circular dichroism and optical rotatory dispersion spectra in the visible region (Kroes, 1970). Changes in protein conformation during photoconversion are also implied from low temperature and freeze-dry studies (Spruit & Kendrick, 1973; Kendrick, 1974). However, the post-conversion differences in the *Pr* and *Pfr* protein configurations are apparently only quite small as revealed by a variety of spectral and chemical methods (Briggs & Rice, 1972). Differences in surface residues are suggested by the ultraviolet difference spectrum (Tobin & Briggs, 1973); the differential reactivity of *Pr* and *Pfr* toward glutaraldehyde (Roux, 1972), and *N*-ethyl maleimide (Gardner *et al.*, 1974); and the differential electrostatic binding of *Pr* and *Pfr* to ribosomal material in plant extracts (Quail, 1975b).

14.3.2 PHOTOCONVERSION REACTIONS

Kinetic analysis following flash excitation of phytochrome indicates that the forward reaction (*Pr*→*Pfr*) requires several seconds to complete, whereas the reverse transformation (*Pfr*→*Pr*) is apparently complete by 20 to 30 msec

(Linschitz *et al.*, 1966; Linschitz & Kasche, 1967). Several intermediates on separate pathways for the forward and reverse reactions have been characterized spectroscopically (Kendrick & Spruit, 1973). The first photochemical intermediate on both pathways appears to result from isomerization of the chromophore only with no change in protein structure. Subsequent dark relaxations apparently involve conformational changes in the protein moeity as well as further chromophore re-arrangements. The actual mechanism involved in chromophore photo-isomerization is uncertain, although tautomerization of the pyrrole group (Fig. 14.2) plus a cross-exhange of protons between chromophore and protein is a currently favoured hypothesis (Lhoste, 1972).

The phototransformation of a static population of phytochrome molecules can be described by the expression (Butler, 1972):

$$\frac{d[Pfr]}{dt} = 2.3\ (I_\lambda\ E_{r\lambda}\ \phi_r[Pr] - I_\lambda\ E_{fr\lambda}\ \phi_{fr}[Pfr]) \qquad \text{Equation 14.1}$$

where λ = wavelength; I_λ = intensity; t = duration of irradiation; $E_{r\lambda}$ and $E_{fr\lambda}$ = the extinction coefficients at λ for *Pr* and *Pfr* respectively; ϕ_r and ϕ_{fr} = the quantum yields for *Pr* and *Pfr* respectively. At $t = \infty$ a photo-equalibrium will be established and the pigment will oscillate ('cyc'e') between the two forms at a rate which is a function of the total absorption of the two species. The ratio of *Pfr* to *Pr* will remain constant. In the absence of any net loss or gain of phytochrome molecules in the population (a 'closed' system), this ratio will be wavelength *dependent* but irradiance *independent* according to the formula:

$$\frac{[Pfr]_\infty}{[Pr]_\infty} = \frac{E_{r\lambda}\,\phi_r}{E_{fr\lambda}\,{}_f\phi_r} \qquad \text{Equation 14.2}$$

In the living cell, however, synthesis and destruction of phytochrome (see 14.3.3.2 below) must be taken into account. This transforms the system into an 'open' one where net loss or gain of pigment molecules can and do occur. For short term irradiations (about 5 minutes), where photo-equilibrium is rapidly established, no significant change in total phytochrome (*Ptot*) occurs and Eq. 14.2 is a good first approximation. Under these conditions the [*Pfr*]:[*Pr*] ratio in the cell will be irradiance independent but wavelength dependent. Under long term continuous irradiations, however, synthesis and destruction become significant parameters with the result that the ratio [*Pfr*]:[*Pr*] becomes *irradiance*, as well as wavelength, dependent (Schäfer & Mohr, 1974; Schäfer, 1975).

Experimentally the kinetics of phototransformation have been shown to be first order both *in vivo* (Schmidt *et al.*, 1973) and *in vitro* (Butler, 1961). Likewise, both the rates of photoconversion (Fig. 14.3) and the short term photosteady state ratio of *Pr* to *Pfr* (Fig. 14.4) have been demonstrated to be wavelength dependent in a manner consistent with the measured absorption spectra (Butler *et al.*, 1964; Hanke *et al.*, 1969).

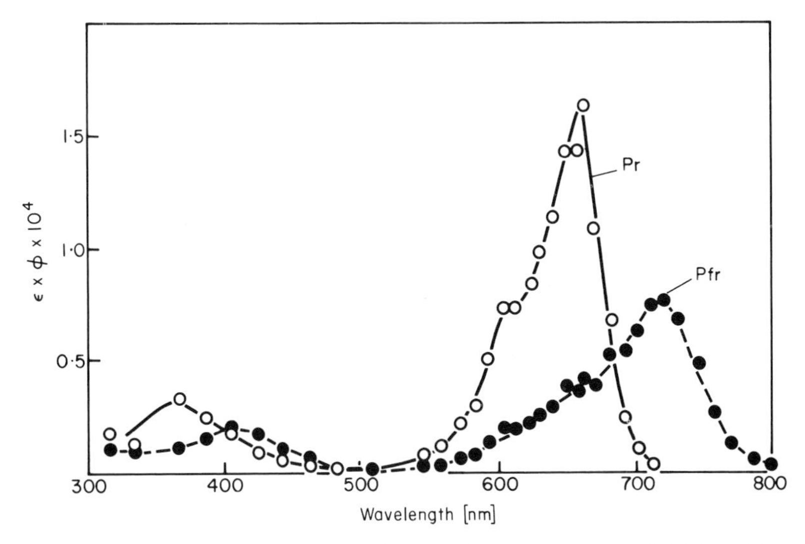

Fig. 14.3. Action spectra of photochemical transformations of P*r* and P*fr* in solution. The extinction coefficient ε is in litre mol^{-1} cm^{-1} and the quantum yield ϕ is in mol $Einstein^{-1}$ (after Butler *et al.*, 1964).

Note that the rate and extent of photoconversion are dependent on the wavelength, irradiance and time of irradiation below photo-equilibrium (Eq. 14.1). The product of time and irradiance determines the total number of quanta or the light 'dose' administered to the system. The wavelength determines how

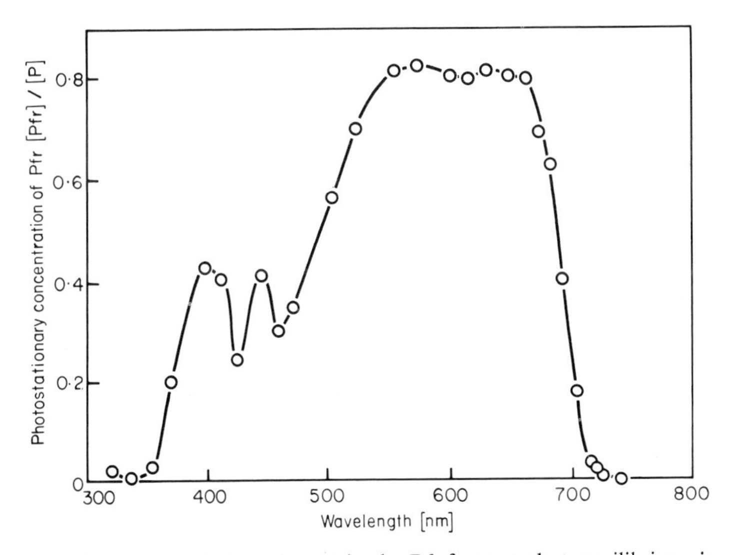

Fig. 14.4. Proportion of phytochrome in the P*fr* form at photoequilibrium *in vivo* (*Sinapis* hooks) as a function of wavelength (after K.M. Hartmann and C.J.P. Spruit in Hanke *et al.*, 1969).

efficiently the incident quanta are absorbed by the pigment (Fig. 14.1). Note also that total photoconversion of *Pr* to *Pfr* is not possible (Fig. 14.4). The maximum attainable is about 80% *Pfr* in the red region of the spectrum. This is because there is no wavelength where *Pr* absorbs and *Pfr* does not (Fig. 14.1). Conversely, more than 97% of the phytochrome can be converted to the *Pr* form by wavelengths longer than 737 nm (Hartmann & Cohnen Unser, 1973) as the *Pfr* absorbance exceeds that of *Pr* in that region of the spectrum. The photosteady state ratio of *Pfr*/*Pr* can be conveniently manipulated with monochromatic light, particularly between 660 and 750 nm (Fig. 14.4). This procedure has been used to advantage in several physiological experiments.

For many years the only quantitative assay for phytochrome has been the spectrophotometric measurement of its photoreversible absorbance changes (Fig. 14.1) (Butler *et al.*, 1959; Spruit, 1972). This procedure can measure phytochrome both *in vivo* and *in vitro*. However, it is unsuitable for use with green tissue and provides no index of the integrity of large regions of the protein moeity. The recent successful immunocytochemical detection of phytochrome now provides a second assay for the pigment (Coleman & Pratt, 1974).

14.3.3 DARK REACTIONS

'Dark reversion' and 'destruction' are the so-called dark reactions of phytochrome. Neither the molecular bases nor the physiological significance of these processes is well understood.

14.3.3.1 *Dark Reversion*

Pr is thermodynamically stable and can only be converted to *Pfr* by light (Lhoste, 1972). *Pfr*, in contrast, is metastable and can therefore revert thermally to *Pr* in the dark.

In vivo, dark reversion occurs in most dicotyledons but not monocotyledons, whereas phytochrome from both sources reverts *in vitro* (Frankland, 1972; Briggs & Rice, 1972). The process *in vivo* appears to be first order and rapid. In several plants reversion is complete within 30 minutes at 20°C although only 15% to 20% of the *Pfr* molecules are involved. The remainder continue to undergo 'destruction' for a considerable period after reversion has ceased. Separate 'reversion' and 'destruction' pools of *Pfr* have been postulated to account for this apparent anomaly (Schäfer & Schmidt, 1974). Discontinuities in Arrhenius plots of the extent of reversion suggest that the process might be membrane associated (Schäfer & Schmidt, 1974).

14.3.3.2 *Synthesis and Destruction*

Dry seeds contain phytochrome (Spruit & Mancinelli, 1969). Rapid, early increases in the photometrically detectable pigment (*Ptot*) during imbibition are apparently due to rehydration of the molecule rather than synthesis (Tobin *et al.*,

1973). The pigment can be stored as either *Pr* or *Pfr* and rehydrated in the stored form (Vidaver & Hsiao, 1972). This suggests an explanation for the appearance of *Pfr* in the dark, sometimes observed in seeds (Rollin, 1972). Further increases in *Ptot* during the growth of etiolated seedlings result from *de novo* synthesis of new molecules in the *Pr* form (Quail *et al.*, 1973b). The pigment accumulates to high levels in the dark ultimately reaching a plateau (Fig. 14.5).

'Destruction' is the disappearance of photometrically detectable *Pfr* without the concomitant appearance of equimolar quantities of *Pr* (Frankland, 1972). This decrease in photoactivity is paralleled by a loss of immunologically detectable phytochrome (Coleman & Pratt, 1974). Since no recycling of the protein moeity occurs, 'destruction' would appear to be a genuine degradative process (Quail *et al.*, 1973b). Some evidence suggests that this process may be enzymatic (Kidd & Pratt, 1973). Destruction is temperature dependent, but the absence of discontinuities in Arrhenius plots suggests that in contrast to dark reversion, it is not membrane associated (Schäfer & Schmidt, 1974). Destruction is observed in both monocotyledons and dicotyledons with half-times ranging from 20 minutes to 4 hours (Frankland, 1972; Schäfer *et al.*, 1973; Kidd & Pratt, 1973).

Destruction of *Pfr* occurs both in the dark following brief irradiations and in continuous light. In the dark, destruction rapidly removes all unreverted *Pfr* leading to a short-term decline in *Ptot*. The reduced pigment levels are replenished, however, as the dark period proceeds by *de novo* synthesis of new *Pr* molecules (Quail *et al.*, 1973b). In continuous light, initially high *Ptot* levels decline at a rate proportional to the photosteady-state *Pfr* concentration (Frankland, 1972) ultimately reaching a new plateau (Fig. 14.5). This new

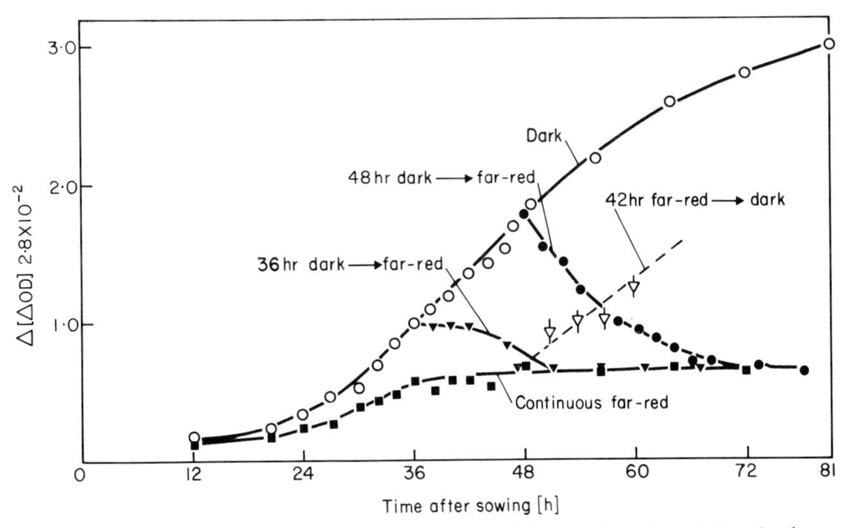

Fig. 14.5. Total phytochrome levels in *Sinapis* cotyledons as function of time in the dark (o); continuous far-red (■); 42 h far-red → dark (△); 36 h dark → far-red (▲); 48 h dark → far-red (●) (after Schäfer *et al.*, 1972).

plateau represents a steady-state equilibrium between synthesis and degradation (Schäfer *et al.*, 1972; Quail *et al.*, 1973b). *Pr* synthesis appears to be a continuous zero order process, itself unaffected by light. The *Ptot* level is then regulated against this background by the disparate first order degradation rate constants for *Pr* and *Pfr*.

The plateau level of *Ptot* established at the steady-state by prolonged irradiations is a function of both wavelength and irradiance (Schäfer & Mohr, 1974; Schäfer, 1975). The *Pfr* level, in direct contrast however, is independent of both irradiance and wavelength under these conditions. This has the extremely important consequence that, whereas the *ratio* of [*Pfr*]:[*Ptot*] will change depending on wavelength and irradiance, the *absolute* level of *Pfr* will be the same under *all* continuous irradiation conditions once the steady-state has been established. This has important implications for the interpretation of the effects of prolonged irradiations.

14.3.3.3 *Physiological Significance of Dark Reactions*

The dark reactions provide a mechanism for the light independent removal of *Pfr* and thereby the potential for a dark period to reverse light-induced responses which require the sustained presence of the effector. Furthermore, as the disappearance of *Pfr* is a time dependent process the system has the potential for measuring time. In principle, therefore, phytochrome should enable the plant to distinguish between light and dark and to time the dark period. Initially it was thought that this so-called 'hour-glass' principle was the basis for phytochrome-controlled photoperiodism (see Vince, 1972). Supporting evidence is scant, however, and this theory has now fallen into disrepute. Some systems do, nevertheless, respond to the timed disappearance of *Pfr* in the dark. An example is accumulation of the enzyme lipoxygenase (Oelse-Karrow & Mohr, 1973).

During prolonged irradiations, synthesis and destruction appear to have the additional role of maintaining a constant *absolute* level of *Pfr* irrespective of wavelength (Schäfer, 1975). This implies that *Pfr per se* is not the effector of high irradiance responses (see 14.4.1.2).

14.3.4 LOCALIZATION

The distribution of phytochrome is highly specific at the tissue and cellular levels as determined photometrically (Briggs & Siegelman, 1965) and immunocytochemically (Pratt & Coleman, 1971; 1974). In dicotyledons, the highest pigment levels are in the apical regions. In etiolated grass shoots, large quantities occur in parenchyma cells near the tip of the coleoptile and in the rapidly differentiating tissue near the shoot apex. High levels of the pigment are also found in root cap cells.

Attempts to establish the subcellular localization of phytochrome fall into two categories: (a) measurement of phytochrome-induced responses having a spatial or vectorial component from which the photoreceptor location can be inferred; (b) direct measurements of the pigment itself, either *in situ* or in subcellular fractions.

The pattern of chloroplast movement in the alga *Mougeotia* in response to polarized red and far-red microbeams has been taken as evidence that phytochrome located and oriented on or near the plasmalemma controls this response (Haupt, 1972b). The directional growth of the germ tubes of *Dryopteris* in polarized light has been interpreted similarly (Etzold, 1965). The change in ion flux associated with phytochrome mediated leaflet movement (Satter & Galston, 1973), root tip adhesion to glass surfaces (Tanada, 1968), and changes in bioelectric potentials (Newman & Briggs, 1972) are also indicative of phytochrome-controlled changes in surface properties but not necessarily that the pigment is a permanent membrane component. A report of phytochrome-controlled development of isolated etioplasts *in vitro* (Wellburn & Wellburn, 1973) implies the presence of functionally active pigment in or on the organelles. More recently a rapid, phytochrome-mediated change in the level of gibberellin extractable from an etioplast-rich fraction in response to *in vitro* irradiations has been demonstrated (Evans, 1975; Evans & Smith, 1976a; Cooke & Saunders, 1975). A red/far-red reversible reduction of NADP *in vitro* in response to irradiation of a mitochondria-rich fraction has also been reported (Manabe & Furuya, 1974). Furthermore, phytochrome has been detected spectrophotometrically in both etioplast- (Evans & Smith, 1975) and mitochondria-rich fractions (Manabe & Furuya, 1974).

Both spectrophotometric and immunological techniques have been used for direct, *in situ* measurements of the intracellular distribution of phytochrome. The presence of the pigment in the nucleus has been claimed on the basis of microspectrophotometric scans of cells (Galston, 1968) but these data have been challenged (Kendrick & Spruit, 1972; Tobin *et al.*, 1973). The cytochemical visualization of phytochrome antibody in non-irradiated tissue sections indicates a general distribution of the photoreceptor throughout the cytoplasm in addition to an association with nuclei and plastids (Pratt & Coleman, 1971). Brief red irradiation prior to fixation causes the pigment in some tissues to concentrate in discrete, as yet unidentified regions of the cytoplasm (Mackenzie *et al.*, 1974). Non-saturating irradiations of maize coleoptile segments with red and far-red light polarized normal to the longitudinal axis were found to photoconvert about 20% more phytochrome than when polarized parallel to this axis (Marmé & Schäfer, 1972). This was interpreted as indicating that phytochrome is located and oriented in the plasmalemma. In considering the locational and orientational rigidity of phytochrome implied from polarized light studies the known rapid and highly fluid lateral and rotational diffusion of other membrane proteins should be borne in mind (Cone, 1972; Singer, 1974).

Cell fractionation procedures have also been used in attempts to localize phytochrome in subcellular components. The well-known precipitation of the pigment protein at low pHs ($\leqslant 6.2$) (Siegelman & Butler, 1965; Hillman, 1967) has been overlooked in some studies leading to claims of associations of phytochrome with mitochondria (Gordon, 1961) and plasmalemma (Marmé *et al.*, 1971). Little of the pigment (< 10%) sediments from homogenates of non-irradiated tissue at neutral pH at forces up to 144,000 x *g* (Rubinstein *et al.*, 1969; Siegelman & Butler, 1965). Red irradiation prior to extraction, however, substantially enhances the level of phytochrome subsequently associated with pelletable material (Quail *et al.*, 1973a). Irradiation of extracts from dark grown material has a similar effect (Marmé *et al.*, 1973). Initially claims were made of the isolation of a phytochrome-containing membrane fraction that could be reversibly 'solubilized' by withdrawal of Mg^{2+} (Marmé *et al.*, 1973; 1974). More recently, however, the pigment in this fraction has been shown to be associated with degraded ribonucleoprotein (RNP) material, probably of ribosomal origin (Quail, 1975b). This association apparently results from the preferential electrostatic adsorption of *Pfr* onto ribosomal material—either free or membrane-bound in the endoplasmic reticulum (Williamson *et al.*, 1975). Whether such an association is artefactual or biologically meaningful is yet to be established. A recent promising variation on this approach is the use of glutaraldehyde in an attempt to immobilize the pigment in the cell prior to extraction (Yu, 1975.)

The existence of meaningful phytochrome-membrane interactions are by no means excluded by the above findings. It has been shown, for example, that phytochrome can mediate photoreversible conductance changes in artificial lipid membranes (Roux & Yguerabide, 1973). The suggestion that phytochrome might function as a stereospecific protein ligand capable of interaction with cellular membranes has been made (Quail & Schäfer, 1974; Boisard *et al.*, 1974).

14.4 PHYTOCHROME PHYSIOLOGY

The biological responses attributed to phytochrome can be usefully characterized in terms of three different but interrelated sets of criteria: (a) the nature of the involvement of light in the inductive process; (b) the temporal expression of the induced response; and (c) the type of cellular or developmental process affected.

14.4.1 INDUCTION-REVERSION AND HIGH IRRADIANCE RESPONSES

Two types of light-controlled phenomena have been attributed to phytochrome —the so-called 'induction-reversion' and 'high irradiance' (HIR) responses (Mohr, 1972). This terminology arises from the irradiation conditions under which the responses are observed and suggests a fundamental difference in the manner in which the phytochrome molecule transmits the light signal to the cell in each case. It does not necessarily reflect an intrinsic property of the actual, biological parameter being monitored. Some parameters display both modes of response, others only one.

14.4.1.1 *Induction-Reversion Responses*

These are the classical phytochrome responses (Borthwick, 1972). A change in the biological parameter being monitored is induced by a brief irradiation of low intensity red light and reversed by a subsequent far-red pulse. The accumulation of anthocyanin in *Sinapis* illustrates this point (Fig. 14.6). Another well known example is lettuce seed germination. This is repeatedly photoreversible for up to 100 alternate red and far-red irradiations (Borthwick, 1972).

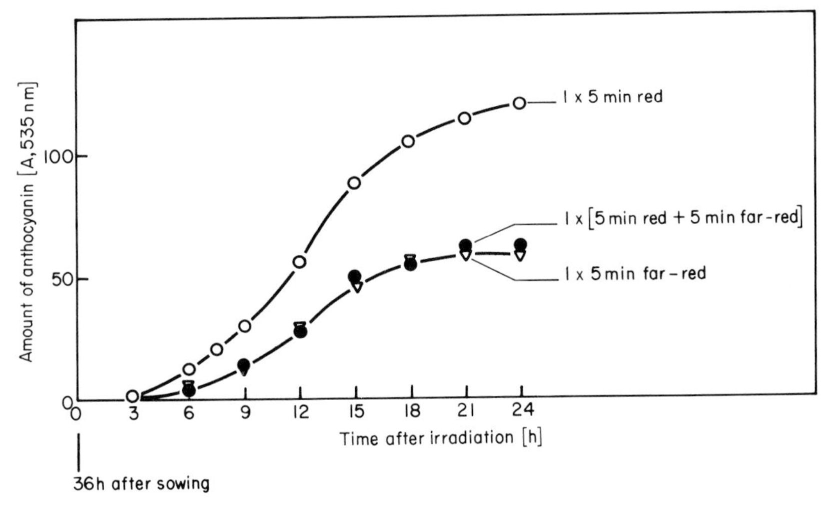

Fig. 14.6. Accumulation of anthocyanin in *Sinapis* in the dark following irradiation treatments at time zero with 5 min red (o), 5 min far-red (▽), or 5 min red followed immediately by 5 min far-red (●) light (after Mohr *et al.*, 1971).

This simple red/far-red photoreversibility forms the basis of the concept that *Pfr* is the biologically active form of the pigment whereas *Pr* is inactive. Attempts to quantify the relationship between the number of *Pfr* molecules formed and the magnitude of the induced response have been both indirect and direct.

Indirect correlations are based on the premise that observed increases in the magnitude of the response with increasing light dose are a function of the degree of photoconversion of *Pr* to *Pfr*, i.e. the more quanta, the more *Pr* is converted to *Pfr* and therefore the greater is the response. The increase in anthocyanin in response to increasing doses of red light (Table 14.1) illustrates this point (Lange *et al.*, 1971). The so-called law of reciprocity (irradiance × time = constant) must hold for the light doses used for this interpretation to be valid (see 14.3.2). This establishes that the magnitude of the response is directly proportional to the total number of incident quanta regardless of the time or irradiance of the irradiation providing those quanta (Table 14.1). The effects of irradiance level during the brief irradiations used in induction-reversion experiments are thus attributed entirely to the degree of photoconversion.

Table 14.1. Reciprocity of authocyanin synthesis in mustard seedlings. (Data within boxes are from seedlings given equal amounts of total energy.)

Red light	Anthocyanin (A, 535 nm)		
Irradiance (I) (μW cm^{-2})	Time (t) of irradiation (sec)		
	300	30	3
675	0·158	0·141	0·107
67·5	0·140	0·110	0·069
6·75	0·107	0·069	

Biological action spectra are an extension of this principle (Fig. 14.7). The magnitude of the response at different wave-lengths is interpreted to be a function of the relative effectiveness of the quanta at those wavelengths in the phytochrome photoconversion process. The close agreement between the action spectra of several biological responses on the one hand (Fig. 14.7) and those of the phototransformation reactions of the isolated pigment on the other (Fig. 14.3) lends strong support to this notion (Borthwick, 1972; Shropshire, 1972). In these cases a seemingly good correlation exists between *Pfr* level and response magnitude.

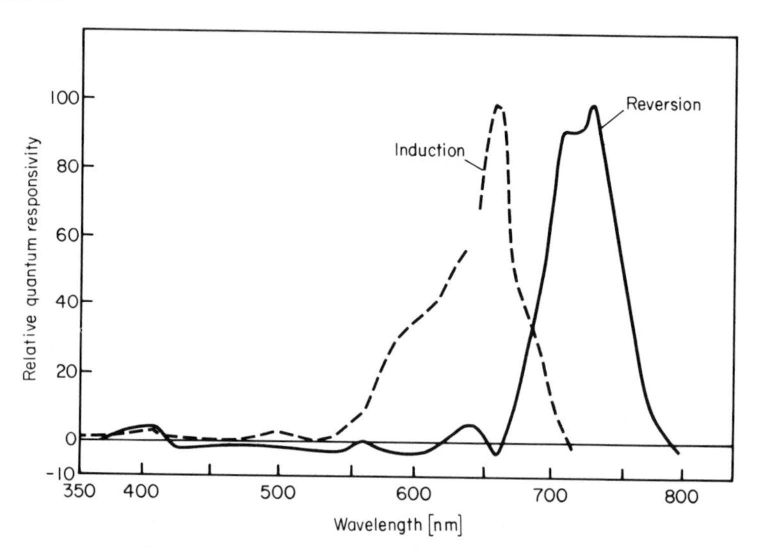

Fig. 14.7. Action spectra for induction and reversion of plumular hook opening in bean seedlings (after Withrow *et al.*, 1957).

In contrast, however, the majority of rigorous attempts to demonstrate a direct quantitative correlation between the *photometrically* detectable *Pfr* level and the relevant biological response in the same system have been unsuccessful (Hillman, 1972). The reasons for this are not understood. An apparent exception is lipoxygenase levels in *Sinapis* (Oelze-Karrow & Mohr, 1973).

Increases in response with increasing doses of quanta must eventually saturate. If the level of *Pfr* is rate-limiting the photoresponse will saturate when the photoconversion process is saturated i.e. when photoequilibrium is reached. If the response system itself is rate-limiting the response may saturate well before photoequilibrium. Examples of both extremes have been observed (Hillman, 1967). Light doses which saturate the photoconversion of *Pr* to *Pfr* do not appear to saturate the inhibition of mesocotyl lengthening in *Avena* (Loecher, 1966), nor the accumulation of anthocyanin in *Sinapis* (Drumm & Mohr, 1974). In contrast, inhibition of lipoxygenase accumulation is saturated by very low ($<3\%$) *Pfr* levels (Mohr, 1972). Other photoresponses fall between these extremes with many being saturated at less than 80% *Pfr* (Hillman, 1967). In addition, whereas many parameters, such as anthocyanin formation, show a graded response, others, such as lipoxygenase accumulation, respond in an all-or-none fashion to changes in *Pfr* level, suggesting some form of cooperative, threshold mechanism (Oelze-Karrow & Mohr, 1973).

Implicit in the far-red reversibility of an induced response is that *Pfr* can act in the dark. Light is strictly a trigger. The magnitude and multiplicity of the responses indicate an extensive amplification mechanism. Unlike photosynthesis where light energy is converted stoichiometrically with quantum yields of less than 1·0, the low irradiances which actuate these phytochrome responses lead to final quantum yields well in excess of unity (Galston, 1974).

14.4.1.2 *High Irradiance Responses* (*HIR*)

If *Pfr* is, as postulated, the active, effector, it is clear that tripartite correlations between light dose, *Pfr* level and response magnitude will only be expected for irradiations terminated prior to photoequilibrium. Once photoequilibrium has been established, *Pfr* levels are no longer irradiance dependent whether for short term (Eq.14.2) or long term (Schäfer, 1975) irradiations. No further increase in the response should result therefore from further increases in irradiance after photoequilibrium.

Some parameters do, however, show a strong irradiance dependence at photoequilibrium. This effect is termed the high irradiance response (HIR) (Mohr, 1969). Such responses are observed with continuous irradiation where a phytochrome photosteady-state is rapidly established and maintained for prolonged periods. The accumulation of anthocyanin in continuous far-red light illustrates this point (Fig. 14.8) (Lange *et al.*, 1971). The rate of accumulation is a function of the irradiance. This effect is only maintained, however, as long as the irradiation continues. The irradiance-enhanced response rate reverts rapidly to that of the dark controls when irradiation ceases, and resumes again upon further irradiation (Fig. 14.9). Reciprocity is not demonstrable for the HIR and, in some cases e.g. lettuce hypocotyl lengthening, the variable itself does not exhibit the classical red/far-red reversible response (Hartmann, 1966). Anthocyanin accumulation, in contrast exhibits *both* modes of response (Figs. 14.6 and 14.8).

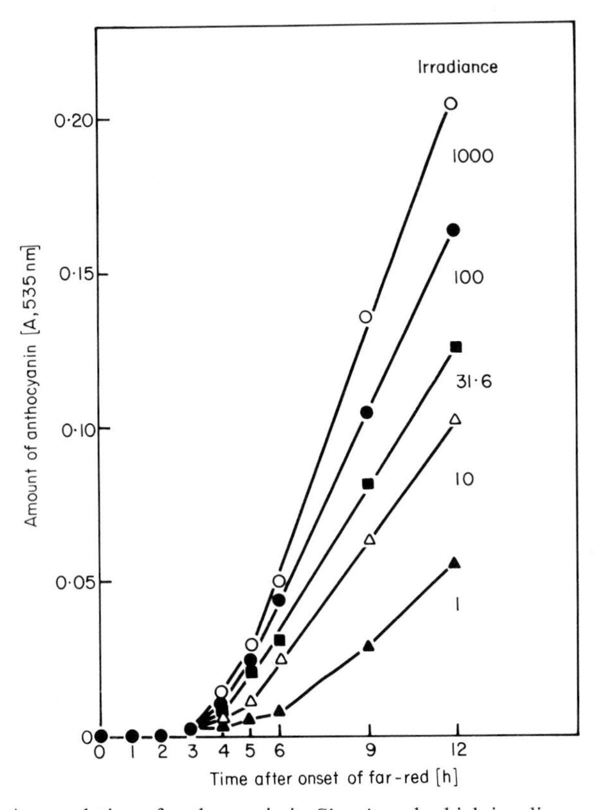

Fig. 14.8. Accumulation of anthocyanin in *Sinapis* under high irradiance conditions. Seedlings were held in continuous far-red light of varying irradiance. All irradiances are expressed relative to the arbitrary value of 1,000 (=350 μW. cm^{-2}) (after Lange *et al.*, 1971).

Action spectra of HIR have peaks in the blue and far-red (Borthwick *et al.*, 1969; Mohr, 1969). The most extensively studied response is that of the inhibition of lettuce hypocotyl lengthening (Hartmann, 1967; see also Fig. 14.10). The seedlings were irradiated continuously for 18 hours with monochromatic light of different wavelengths and irradiances. As with other HIR there is no coincidence of the observed spectrum with the absorption maxima of either *Pr* or *Pfr*.

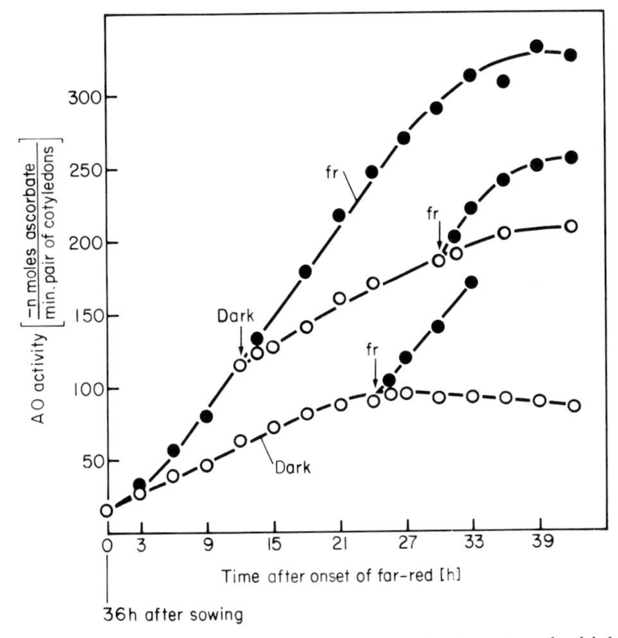

Fig. 14.9. Accumulation of ascorbate oxidase activity in *Sinapis* under high irradiance conditions. Seedlings were either retained in the dark (o) or continuous far-red (●) light. Transfer of seedlings from far-red to the dark or vice versa at various times is indicated by the arrows (↓) (after Drumm *et al.*, 1972).

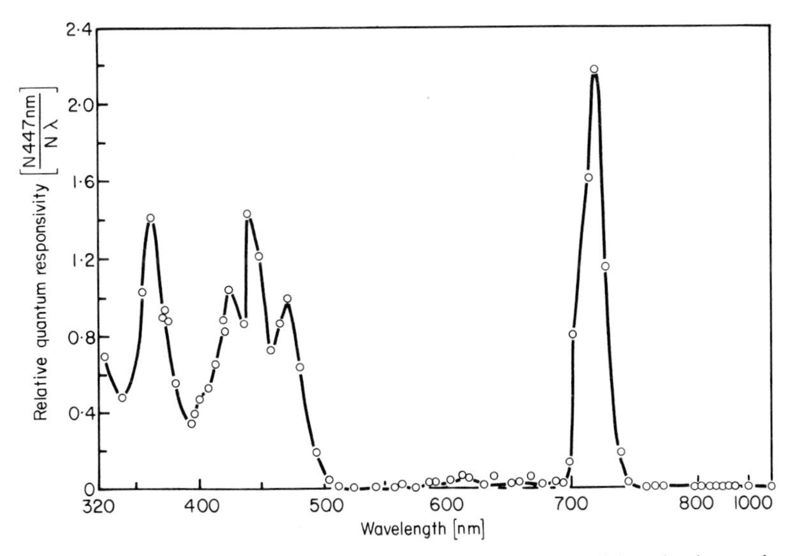

Fig. 14.10. Action spectrum for inhibition of lettuce hypocotyl lengthening under continuous irradiation (after Hartmann, 1967).

The conclusion that phytochrome mediates the high irradiance effects derives from other data using the same plant system (Hartmann, 1966). Prolonged irradiations with two wavelengths which were relatively ineffective when given separately (658 nm and 766 nm) were highly effective when given simultaneously. The maximum effect with these and other wavelength pairs always occurred where the photosteady-state *Pfr* concentration was about 3%. This agrees well with the peak of activity observed at 720 nm with single wavelength monochromatic light (Fig. 14.10) also known to establish about 3% *Pfr* at photoequilibrium (Fig. 14.4). Furthermore, the effectiveness of a single wavelength irradiation at 717 nm could be nullified by simultaneous irradiations of either 658 nm or 759 nm. These wavelengths would shift the photoequilibrium away from 3% *Pfr* towards higher or lower values respectively.

The question remains, however, as to *how* both the irradiance and wavelength dependency of the HIR can be explained in terms of phytochrome. The response is unlikely to be a function of the absolute *Pfr* level as this is irradiance and wavelength *independent* with prolonged irradiations (Schäfer, 1975). The rate at which the pigment molecule oscillates between the two forms at photoequilibrium is, on the other hand, strongly irradiance and wavelength dependent. Accordingly, the HIR has been rationalized to be some function of the cycling rate of phytochrome. The effector molecule has been postulated to be some 'excited form of *Pfr*', denoted *Pfr** (Schopfer & Mohr, 1972). No direct evidence for such a species is available, however, and its purported action has been questioned on the basis of dual wavelength experiments (Hartmann & Cohnen Unser, 1973). High levels of phytochrome intermediates are maintained at high irradiances (Kendrick & Spruit, 1973a) but the possibility that these are HIR effectors is unlikely because of the nature of the action spectra (Hartmann, 1966).

14.4.1.3 *A Unitary Model*

Schäfer (1975) has recently developed a single, formalistic model which can theoretically account for both induction-reversion and HIR responses, whether graded or cooperative, in terms of phytochrome. The pigment is postulated to be a bimodal ligand with the forms *Pr* and *Pfr* which interact with receptor sites also having dual forms, *X* and *X'*:

$$\begin{array}{ccccc} \xrightarrow{^{\circ}ks} & PrX & \underset{k_2}{\overset{k_1}{\rightleftharpoons}} & PfrX & \\ & k_4 \uparrow & & \downarrow k_3 & \\ & PrX' & \underset{k_2}{\overset{k_1}{\rightleftharpoons}} & PfrX' & \xrightarrow{kd} \end{array}$$

where $^{\circ}ks$ and kd are the rate constants for synthesis and destruction of phytochrome respectively; k_1 and k_2 rate constants for the photoconversion reactions; and k_3 and k_4 rate constants for the $X \rightarrow X'$ and $X' \rightarrow X$ transitions.

The irradiance and wavelength dependencies of the HIR response then become explicable in terms of the flux rates through the cycle under continuous irradiation. The basic conclusions reached are that *PfrX′* is the effector element in induction-reversion responses, whereas *PfrX* or the flux k_3.*PfrX* is the effector element in HIR. The molecular nature of *X* and the mechanism of action of the effector elements are, however, unresolved questions.

14.4.1.4 *Modes of Light Signal Transmission*

Given that phytochrome mediates both induction-reversion and HIR, it is clear that the photoreceptor utilizes the incoming light signal differently in each case. Whereas induction-reversion responses are induced by a red pulse and can develop in the subsequent dark period, HIR require a continuous light energy input to sustain the response. Increases in irradiance lead to increases in response magnitude in both cases but the effect is interpreted differently for each. For induction-reversion responses the irradiance effect is only observed prior to photoequilibrium; is interchangeable with time of irradiation (reciprocity holds); and is interpreted as reflecting the effectiveness of the total light dose in determining the degree of photoconversion of *Pr* to *Pfr*. *Pfr* is considered the effector molecule and some form of *Pfr*-response stoichiometry is expected. For HIR, on the other hand, the irradiance effect is observed at photoequilibrium; is not interchangeable with time of irradiation (reciprocity does not hold); and is interpreted as being some function of the phytochrome cycling rate. The precise effector molecule or process is uncertain. For induction-reversion responses light is viewed simply as a trigger and as having no further direct role in the inductive function of *Pfr*. The light signal is ‘stored’ in the *Pfr* form for subsequent utilization. HIR, in contrast, require a sustained, direct interaction of the photoreceptor with the incident excitation energy. This indicates that these effects are light-driven as distinct from being light-triggered. Light appears to have a direct role in the inductive function of the pigment, the energy input being rapidly dissipated.

The suggestion that photosynthesis or cyclic photophosphorylation might in some way be responsible for HIR has been advanced but several pieces of evidence argue against this (Mohr, 1972). Speculative suggestions that phytochrome might function as a photocoupler (Quail, 1975a) or a specific, light-driven permease (Smith, 1970) in the HIR have also been advanced but not substantiated.

14.4.2 RESPONSE KINETICS

Phytochrome is considered to trigger, in some way, a chain of events leading sooner or later to a measurable biological response. The initial triggering of those processes necessary for the development of the response can be termed phytochrome ‘action’; and the appearance of a measurable change in the parameter being monitored can be called response 'expression'. This leads to the recognition of three categories of phenomena: (a) rapid action/rapid expression; (b) rapid action/delayed expression; and (c) delayed action/delayed expression responses. 'Rapid' here arbitrarily means $\leqslant 10$ minutes and 'delayed' $\geqslant 30$ minutes after the initial photoconversion act.

14.4.2.1 *Rapid Action/Rapid Expression Responses*

As phytochrome action must either coincide with or precede expression, rapid action can be implied from the kinetics of the expression alone in these cases. A red/far-red reversible change in electric potential in *Avena* coleoptiles within 15 seconds of the start of irradiation is the most rapid phytochrome-mediated phenomenon thus far reported (Newman & Briggs, 1972). Similar changes have been observed in the biolelectric potential of mung bean root tips exposed to successive red and far-red irradiations (Jaffe, 1968; Racusen & Miller, 1972). Such changes had earlier been inferred from the red/far-red reversible adhesion of root tips to negatively charged glass surfaces (Tanada, 1968). Both responses are detectable within 30 seconds of the start of irradiation (Fig. 14.11). These changes in surface charge are interpreted to indicate changes in the plasmalemma. Fluorescent probe studies support this notion (Racusen, 1973). Red light also induces H^+ efflux from root tips (Yunghans & Jaffe, 1972).

Phytochrome regulates leaflet movement in *Mimosa*, *Albizia* and *Samanea* (Satter & Galston, 1975). This movement is accompanied by an energy-dependent transfer of K^+ ions between the ventral and dorsal motor cells of the pulvinus. Both effects are detectable 10 minutes after red or far-red irradiations. The changes in K^+ are correlated with changes in transmembrane potential (Racusen & Satter, 1974). Furthermore, red/far-red regulated changes in surface charge similar to those of root tips are detectable 30 to 120 seconds after irradiation. These effects are also strongly indicative of phytochrome-mediated changes in plasmamembrane properties.

Phytochrome-regulated changes in the rate of plasmolysis have been detected in *Mougeotia* within 6 minutes of the onset of red irradiations (Wiesenseel & Smeibidl, 1973). A change in plastid orientation is also evident in this alga in less than 10 minutes after a red pulse (Haupt, 1972a). The latter effect can only be fully reversed by far-red during the first minute after red light. Thus potentiation of the response has begun within 1 minute of photoconversion. Available data indicate that the effective phytochrome is located on or near the plasmalemma. Contractile fibrils appear to be responsible for the chloroplast movement (Schönbohm, 1973).

Red light (15 seconds) induces an increase in growth rate in coleoptile tips within 60 seconds of the start of irradiation (Weintraub & Lawson, 1972). The effect is partially reversed if far-red immediately follows the red. Inhibitors of transcription and translation are without effect. As the cell must regulate wall extension through the plasmalemma, this response might also represent a

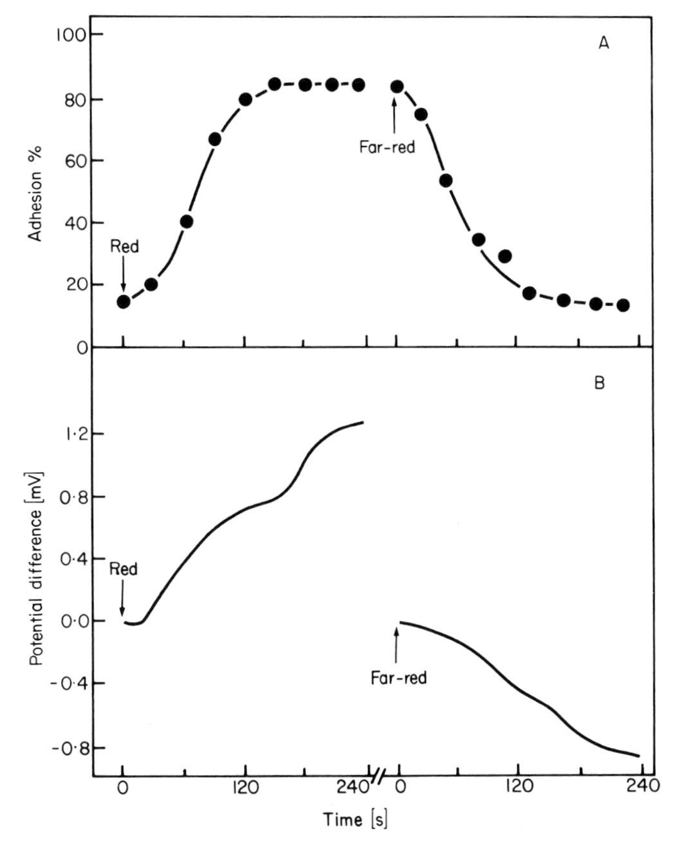

Fig. 14.11. Kinetics of (A) root tip adhesion to a negatively charged glass surface and (B) the development of a biolectric potential across the root tip in response to irradiation with red or far-red light (after Jaffe, 1968).

phytochrome-mediated membrane change. Rapid changes in ATP levels, induced by red and, in some cases, partially reversed by far-red, have been reported but no consistent pattern is obvious (Sandmeier & Ivart, 1972; Yunghans & Jaffe, 1972; White & Pike, 1974).

Rapid transitory increases in hormore levels in response to red light have been known for some time (Reid *et al.*, 1968; Beevers *et al.*, 1970; van Staden & Wareing, 1972). Unequivocal evidence of far-red reversibility has generally been lacking in the past, however, often because of poor experimental design. Recently, the level of gibberellin activity extractable from etioplast-rich preparations from grass leaves has been shown to increase 3-fold within 5 minutes of the termination of a 5 minute red irradiation of the isolated fraction (Evans, 1975; Evans & Smith, 1976a). This effect is reversed by far-red light given immediately after the red. It is postulated that phytochrome causes the movement of gibberellin across the etioplast envelope into the ambient medium.

These observations, and those made with mitochondria-rich fractions (Manabe & Furuya, 1974), are rapid, of potential physiological significance, and suggestive of *in vitro* phytochrome-mediated changes in the functional properties of membranes in the isolated fractions.

Lipoxygenase accumulation in *Sinapis* responds to changes in *Pfr* level in less than 5 minutes under appropriate conditions (Oelze-Karrow & Mohr, 1973). It has been concluded that the level of the enzyme in the cotyledons is controlled by phytochrome in the hypocotyl hook, through a highly cooperative threshold mechanism that responds to, and is saturated by, *Pfr* levels of 1–2% of the total pigment (Oelze-Karrow & Mohr, 1974). Other rapid, inter-organ transfer of phytochrome signals have also been reported (de Greef & Caubergs, 1973). Some form of biophysical transmission system has been postulated, with the membrane continuum of the plasmodesmata as a suggested candidate (Oelze-Karrow & Mohr, 1973).

14.4.2.2 *Rapid Action/Delayed Expression Responses*

Rapid phytochrome action in these cases is deduced from the rate at which the response escapes susceptibility to photoreversal by far-red light following an inductive red pulse. The actual response may not be expressed for days after the irradiation although the inevitability of its appearance has long since been irreversibly established. *Pfr* is said to have 'potentiated' the response (Borthwick, 1972). The escape from reversibility is viewed as the *Pfr*-triggered reaction chain having progressed beyond those steps directly under phytochrome control.

The effect of a red light pulse on flowering in *Pharbitis* is only partially reversed by far-red given 30 seconds after the start of the red irradiation (Fredericq, 1964). After 3 minutes, far-red no longer reverses the effect. Thus although the flowering response itself is not expressed for several days, it is potentiated within seconds by *Pfr*. Flowering in *Chenopodium album* and *Kalanchoe* behaves similarly. A synergism between phytochrome and gibberellin in lettuce seed germination (Bewley *et al.*, 1968) and the de-etiolation response of *Pisum* (Haupt, 1972a) are also in this category.

14.4.2.3 *Delayed Action/Delayed Expression Responses*

These responses show a lag from irradiation to expression but are readily reversed by far-red over relatively long periods in the dark after the red pulse. Gradual escape from reversibility can occur but is relatively slow. This is viewed as indicating that the continued presence of *Pfr* is required over a relatively long period in the dark to maximise expression. An important feature of these responses is that dark reversion and destruction are continually and often rapidly depleting the *Pfr* pool during this period.

The vast majority of recorded red/far-red reversible responses, too numerous to catalogue, are included in this category (see Mitrakos & Shropshire, 1972;

Mohr, 1972; Smith, 1975). Lettuce seed irradiated for 1 minute with red light germinates up to 100% 24 hours later. Far-red reverses this effect up to 12 hours after the red irradiation but with decreasing effectiveness. Anthocyanin formation in *Sinapis* has a lag of 3 hours from irradiation to the onset of accumulation (Fig. 14.6). During this time the effect becomes decreasingly susceptible to reversal by far-red but escape is never complete. The *in vitro* protein synthetic capacity of 80s ribosomes from corn is enhanced by 5 minutes of red light prior to harvest (Travis *et al.*, 1974). This effect is detectable within 30 minutes of red irradiation and escapes far-red reversibility within an hour.

14.4.3 RESPONSE MANIFESTATIONS

Phytochrome has a regulatory role in all major phases of plant growth and development. The changes in cellular biochemistry and physiology which underlie this regulation are detectable at almost any chosen level. Some responses are biophysical in nature (membrane potential), others biochemical (enzyme levels); some require protein synthesis (leaf expansion), others do not (leaf movements); some involve cell division (seed germination), others only cell expansion (plumular hook opening); some are restricted to the irradiated cells (leaf movements), others result from transmissable stimuli (floral induction); some are dependent on phytochrome-endogenous rhythm interactions (flowering), others appear independent (seed germination) (Galston, 1974; Satter & Galston, 1975).

When the nature of the responses is coupled with their kinetic properties a pattern emerges. In general, the most rapid responses are surface or membrane-associated phenomena, often physico-chemical in nature and independent of RNA and protein synthesis. Other cellular processes such as changes in enzyme levels mostly respond more slowly.

14.5 MECHANISM OF ACTION

The molecular mechanism of phytochrome action is the central issue in phytochrome research. The term 'primary reaction of phytochrome' has been used to describe the first reaction in which the physiologically active species becomes involved (Mohr, 1972). Assuming *Pfr* to be the active species this reaction has been formalized:

$$Pfr + X \rightarrow Pfr X$$

where X satisfies the logical necessity of a reaction partner regardless of its nature. The 'primary *action*' of phytochrome then is presumably the induction of a functional *change* in the reaction partner ($X \rightarrow X'$). These are useful formalistic treatments but say nothing of the actual molecular mechanisms involved.

Any molecular hypothesis of phytochrome action must provide an explanation for the specificity, multiplicity and amplification of the induced responses, and attempt to reconcile the seemingly disparate modes of light signal transmission in induction-reversion and high irradiance responses. Specificity here refers to the observed capacity of different plants, organs, adjacent cells in the same tissue and even different processes in the same cell to respond differently to the same effector molecule (Mohr, 1972). All major hypotheses hold the view that this property resides in the fundamental programming of the target cells themselves rather than in the phytochrome molecule.

Multiplicity refers to the vast diversity in form, direction, magnitude and timing of response expression *per se* (see 14.4.3). This can be accounted for by a suitable amplification mechanism. Two major hypotheses of phytochrome action, based on this rationale, have been advanced in recent years: the differential gene activation hypothesis (Mohr, 1966) and the membrane permeability hypothesis (Hendricks & Borthwick, 1967). The possibilities that plant hormones (Black & Vlitos, 1972) or acetylcholine (Jaffe, 1970) might be primary mediators of phytochrome effects have also been investigated but convincing supporting data are sparse (Tanada, 1972).

The gene activation hypothesis in its broadest sense is trivial, as gross morphogenetic changes such as flowering are unlikely to occur without some modification of gene expression. In its strictest sense the hypothesis implies that the primary action of phytochrome is to alter genetic activity. This requires direct *Pfr*-genome interaction. Data relevant to this problem come mainly from measurements of phytochrome-induced changes in enzyme activities and related phenomena (Mohr, 1972).

There are two major criticisms of this hypothesis. First, while the data so far obtained are consistent with the general framework of the concept, they provide no evidence that *Pfr* interacts directly with the genome. Increased incorporation of [-H]-uridine into high molecular weight RNA in continuous far-red light has been reported and interpreted as phytochrome-enhanced transcription of rRNA cistrons (Thien & Schopfer, 1975). Phytochrome-mediated *de novo* synthesis of enzyme proteins has also been demonstrated (Attridge, 1974; Acton & Schopfer, 1974) but this does not establish that control is at the gene level. Indeed, insensitivity of responses such as triose phosphate dehydrogenase (Cerff, 1974) and ascorbate (Schopfer, 1967) accumulation towards actinomycin D might suggest otherwise in these cases. Evidence of phytochrome-mediated activation of precursors of phenylalanine ammonia-lyase has been advanced (Attridge *et al.*, 1974) but this point is controversial (Acton & Shopfer, 1975).

The second criticism is that a growing list of responses appear unlikely to be explained by changes in gene expression. These are in general the rapid action/rapid expression responses. These are more rapid than would be expected of gene regulation and are often insensitive to inhibitors of transcription and translation (Satter & Galston, 1973). Furthermore, lipoxygenase, the only enzyme responding rapidly enough to be placed in this category, is now postulated to be controlled by phytochrome in a separate organ (Oelze-Karrow & Mohr, 1974).

The membrane hypothesis postulates that phytochrome interacts directly with cellular membranes i.e. the primary action of *Pfr* is to modify membrane properties. Indirect support for this notion derives from the fact that the majority of the most rapid responses are either surface phenomena or can be rationalized in terms of membrane changes (see 14.4.2.1). Direct evidence of an association of phytochrome with an identifiable cellular membrane is yet to be advanced however (see 14.3.4) although Evans and Smith (1976b) have recently reported the presence of phytochrome in fractions enriched for etioplast envelope membranes.

A postulated membrane locale of phytochrome action permits the formulation of a unitary hypothesis to account for all phytochrome-mediated phenomena. Both rapid and long term effects can be rationalized to have emanated from the same primary event. A single, fundamental alteration in membrane properties, perhaps simultaneously in several cellular membranes, provides an immediate mechanism for the generation and amplification of a multiplicity of secondary effects (Changeux, 1969). Such effects might include changes in ion flux, activation of membrane bound enzymes, altered compartmentalization and so on. Any or all of these might lead directly or indirectly to altered gene expression. Moreover, a phytochrome-membrane association suggests a possible basis for the dual function of the pigment in induction-reversion and HIR responses. The opportunity would exist for the molecule to function either as a reversible, stereospecific membrane effector, *via* photo-induced changes in its protein conformation; and/or as a photocoupler (Quail, 1975a) or light driven permease (Smith, 1970) *via* pigment cycling.

The speculative nature of such suggestions, however, is evidence of our ignorance in this area. Indeed the basic concept of only a *single*, primary phytochrome reaction has been challenged (Mohr, 1974). It has been suggested instead that a multiplicity of reaction partners and corresponding primary reactions must be postulated to account for both graded and cooperative responses in the same system (Mohr *et al.*, 1971). This would circumvent the second criticism of the gene regulation hypothesis and obviate any potential conflict between the two major hypotheses. The molecular mechanism of phytochrome action is clearly an open question.

FURTHER READING

Briggs W.R. & Rice H.V. (1972) Phytochrome: Chemical and physical properties and mechanism of action. *Ann. Rev. Plant Physiol.* **23**, 293–34.

Galston A.W. (1974) Plant photobiology in the last half-century. *Plant Physiol.* **54**, 427–36.

Hillman W.S. (1967) The physiology of phytochrome. *Ann. Rev. Plant Physiol.* **18**, 301–24.

Mitrakos K. & Shropshire W. Jr. (1972) *Phytochrome*. Academic Press, London and New York.

MOHR H. (1972) *Lectures on Photomorphogenesis*. Springer-Verlag, Berlin.
MOHR H. (1974) Advances in phytochrome research. In *Photochem. Photobiol.* (in press).
SATTER R.L. & GALSTON A.W. (1975) The physiological functions of phytochrome (in press).
SIEGELMAN H.W. & BUTLER W.L. (1965) Properties of phytochrome. *Ann. Rev. Plant Physiol.* **16**, 383–93.
SMITH H. (1970) Phytochrome and photomorphogenesis in plants. *Nature* **227**, 665–8.
SMITH H. (1975) *Phytochrome and Photomorphogenesis*. McGraw-Hill, U.K. Ltd., Maidenhead.

SECTION THREE
THE MANIPULATION OF PLANT CELLS

INTRODUCTION

The ultimate objectives of any scientific investigation are twofold—to increase understanding, and to create opportunities for the application of the new knowledge, ostensibly in the service of man. The two objectives may not often be consciously connected in the minds of the investigators, but history and experience have demonstrated that new information always stimulates a drive towards practical application. This is already beginning to be true for plant cell biology through the directed manipulation of plant cells grown in culture.

As is described in chapter 15, it has been possible to culture organs, tissues and cells of higher plants under aseptic conditions for many years. These techniques, developed largely through assiduous application of the trial-and-error approach, are now becoming highly sophisticated and have already yielded information of great value to our understanding of cellular processes, particularly the cell cycle. In recent years, it has become possible to remove the walls from plant cells by enzymatic procedures, yielding naked protoplasts, some of which remain viable and may be propagated further (chapter 16).

The exploitation of cell and protoplast culture techniques for purely scientific purposes is still in its infancy, but already the methods have been used commercially for the propagation of valuable stock plants, and for the propagation of virus-free plants from virus-infected stocks. What makes cultured cells and protoplasts so exciting from the practical viewpoint, however, is the potential they appear to offer for artificial genetic manipulation. It is too early, yet, to say whether protoplast fusion or gene transfers (chapter 17) will yield new and revolutionary varieties of crop plants—the prospect does exist however, and a consideration of these matters seems a fitting note on which to conclude our coverage of the molecular biology of plant cells.

CHAPTER 15
PLANT CELL CULTURE

15.1 INTRODUCTION

Almost every chapter of this book illustrates that research in plant cell physiology and biochemistry is, at each moment in time, limited by the resolving power of current experimental techniques and by the availability of appropriate experimental plant material. Here we are concerned with the development of new experimental materials (cultured organs, tissues and cells), with culture systems being evolved for their exploitation and with recent experimental work which illustrates that these culture systems are adding a new dimension to studies in plant cell physiology.

The complexity of higher plants has inevitably led workers, concerned with investigating particular aspects of plant physiology, to the use of systems which are of reduced complexity. Hence the early and continuing use by plant physiologists of isolated plant organs (e.g. seedling roots, leaves), complex tissue systems (e.g. discs cut from leaves and storage organs) organ segments (e.g. segments of coleoptiles and hypocotyls) and isolated cell organelles. However, isolated organs and organ fragments are still systems of very considerable complexity and they are, from the beginning, systems of declining viability and favourable sites for colonization by microorganisms.

The concept that the aseptic *culture* of isolated organs, tissues and cells would 'give some interesting insight into the properties and potentialities which the cell as an elementary organism possesses' and 'would provide information about the inter-relationships and complementary influences to which cells within the multicellular whole organism are exposed' was formulated as early as 1902 by Haberlandt. Then, after a lapse of more than 30 years, White (1934) described successful root cultures initiated from the root tips of tomato seedlings, and White (1939) and Gautheret (1939) demonstrated that the parenchymatous wound callus which frequently forms at the exposed surfaces of organ segments could be removed and grown indefinitely as a relatively undifferentiated tissue (callus) culture. From this period there has been rapid progress in organ and tissue culture. Root cultures have been developed from many species (Butcher & Street, 1964). Cultures of stem apices (meristem culture) and of leaf, flower and fruit primordia have been successfully established (Street, 1969). These organ cultures as experimental systems differ from isolated organs in two important respects; they are sterile (free from microorganism contamination) and are handled aseptically, and they are sytems where unimpaired viability is evidenced by growth involving both cell division and cell expansion. Such cultures have contributed to our knowledge of the specific nutritional and hormonal requirements essential for the growth and development of the separate organs of the whole plant and of their specific physiology and biosynthetic activities (Street, 1969). Callus cultures have also now been established from a very wide range of species and have been used in studies on the initiation of root and shoot primordia (Street, 1975a), on cytodifferentiation (Wetmore & Rier, 1963, Gautheret, 1966; Torrey, 1971), on the induction of division in quiescent tissue cells (Yeomann & Aitchison, 1973), on the nature of plant tumour cells (Butcher, 1973) and on the synthesis of a diversity of secondary plant products (Yeomann & Aitchison, 1973).

In 1953, Muir reported that if fragments of callus cultures of *Tagetes erecta* or *Nicotiana tabacum* were transferred to liquid medium and agitated on a reciprocal shaker, then the callus fragments broke up to give a suspension of single cells and small aggregates of cells and that this suspension contained actively dividing cells and hence could be propagated by serial subculture (Muir, Hilderbrandt & Riker, 1954). Such liquid cell suspension cultures have now been obtained from calluses of a number of species and this chapter will outline the development of more sophisticated techniques for their culture and assess their value for studies on the control of growth, metabolism and differentiation in higher plant cells.

15.2 CHANGES IN GROWTH AND METABOLISM OF PLANT CELLS IN BATCH CULTURE-CYTODIFFERENTIATION

Plant cells in batch culture, i.e. cultures in a fixed volume of culture medium, increase in biomass by cell division and cell growth until a factor in the culture environment becomes limiting and sends them into a stationary phase. When such stationary phase cells are subcultured they pass in succession through a lag phase, a short-lived period of exponential growth, a period of declining relative growth rate and then again enter stationary phase (Fig. 15.1A). Traditionally such cultures are initiated by an inoculum establishing a relatively high initial cell density and only accomplish a very limited number of divisions before entering stationary phase. For example cell cultures of sycamore (*Acer pseudoplatanus*) initiated at *ca.* 2×10^5 cells ml^{-1} will reach a final cell density of *ca.* 3×10^6 cells ml^{-1} corresponding to 4 successive doublings of the initial population.

The degree of cellular aggregation in these cell cultures depends upon the species of cell, or cell line within a species, and the culture conditions, but always shows a basically similar pattern of change during the growth cycle of a batch culture. The culture at stationary phase contains the highest proportion of free cells and mean cell volume is at its maximum value. When subcultured to new medium, the cells first embark upon a massive synthesis of new cytoplasm and associated organelles and then begin to divide. For a short time cell division proceeds at a specific growth rate (μ) which is constant and maximal (μ_{max}) for that culture environment. During this phase mean cell volume declines sharply and the proportion of cells in aggregates rises. Then the specific growth rate begins to decline (slowly at first and later at an ever increasing rate) mean cell volume increases and, associated with this cell expansion, the aggregates break up and release free cells. Associated with these growth and structural changes are changes in physiological activity. Measurements, on a per cell basis, of respiration, of the levels of individual cell constituents and of the activities of individual enzymes show that peaks of activity occur. These may be quite sharp and are not coincidental (Fig. 15.1A-D). Different metabolic patterns emerge and decline during the progress of batch culture. Thus RNA synthesis is initiated prior to cell division, proceeds for a time at a greater rate than cell number increase and then ceases whilst cell division is still proceeding (Short, Brown & Street, 1969; Nash & Davies, 1972). Free nucleotides (mainly UDP-glucose and ATP in sycamore cells) are synthesized rapidly during lag phase, presumably an essential preparation for subsequent synthesis of cell-wall polysaccharides and as an energy source for the endergonic processes of cell division, but their net synthesis ceases very shortly after the onset of division (Brown & Short, 1969). Similar transient high activity in carbohydrate oxidation by the pentose phosphate pathway during lag phase has been interpreted as providing the necessary NADPH for the massive biosynthesis achieved during the lag phase of the growth cycle (Fowler, 1971). By contrast, the very sharp peak in ethylene production in sycamore cell cultures occurs late in the cell division phase when the cells are beginning to increase in mean cell volume and may be responsible for initiating aggregate breakdown (Mackenzie & Street, 1970). An essentially similar pattern of ethylene production has been reported for cell cultures of *Rosa* spp., *Glycine max*, *Triticum monococcum*, *Melilotus alba*, *Haplopappus gracilis* and *Ruta graveolens* (La Rue & Gamborg, 1971). Large changes in the activity of phenylalanine ammonia-lyase (PAL) and in *p*-coumarate: CoA ligase occur prior to stationary phase in cultures of *Glycine max* (Hahlbrock, Kühlen & Lindl, 1971). A similar peak of PAL activity has been reported in cell cultures of *Rosa* sp. (Davies, 1972). These changes coincide with maximum production of total phenols by the cultures. Other secondary products are produced by cell cultures, the time of maximum synthesis being restricted to a phase in the growth cycle and often being markedly influenced in its intensity by the plant growth hormone composition of the culture medium (e.g. hemicelluloses and lignin by sycamore; see Carcellar, Davey, Fowler & Street, 1971; visnagin (a physiologically active furanochromone) by *Ammi visnaga*; see Kaul & Staba, 1967; caffeine by tea cell cultures; see Ogutuga & Northcote, 1970; various alkaloids by cell cultures of solanaceous plants; see Tabata, Yamanito & Hiroaka, 1971). Thus during the progress of batch culture the cells pass through a series of contrasted physiological states which encompass cells becoming meristematic, cells expressing high meristematic activity, cells undergoing expansion and becoming either metabolically quiescent or in which certain restricted metabolic pathways are emphasized.

Are these large changes in cellular structure and metabolic activity observed in batch-propagated cell cultures examples of cellular differentiation? This important question cannot at present be satisfactorily answered. Certainly cultured

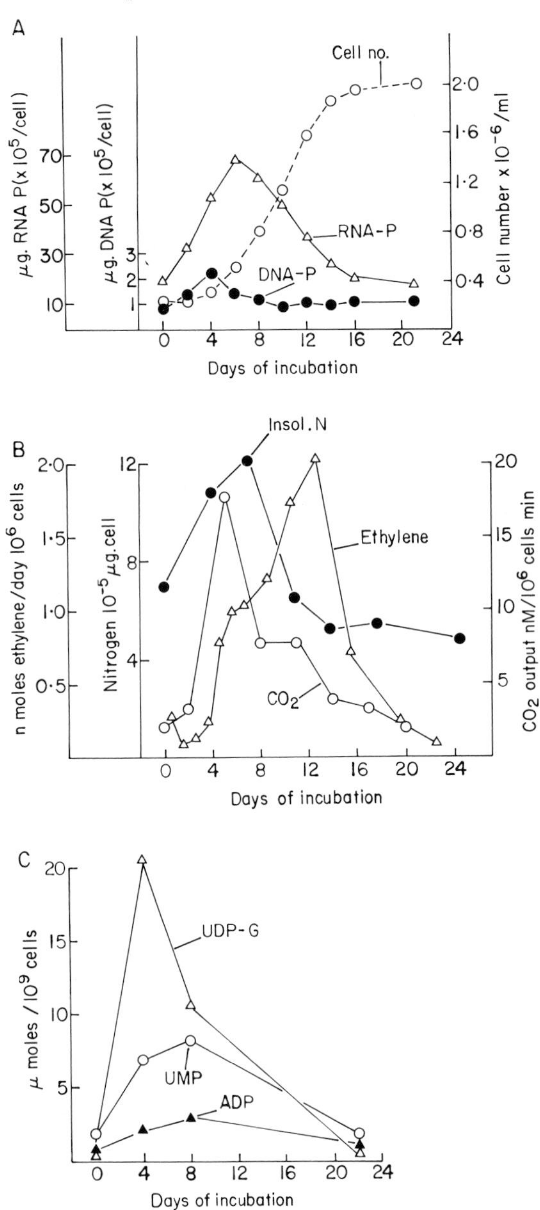

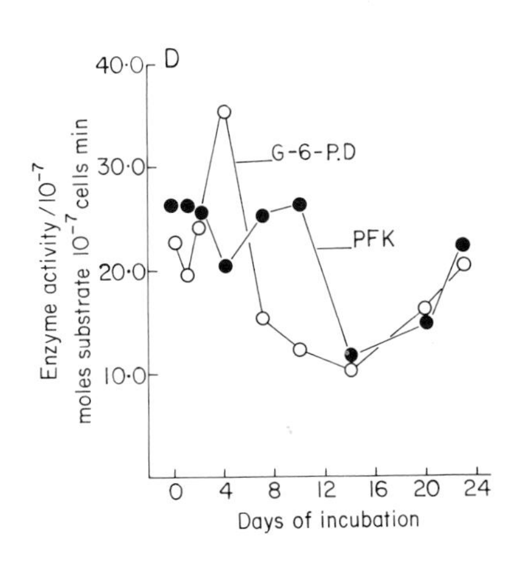

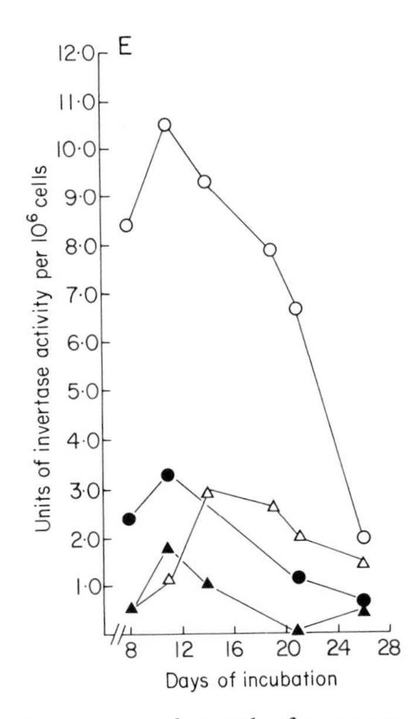

Fig. 15.1. Changes in metabolic activity during the progress of growth of sycamore (*Acer pseudoplatanus* L.) cells in batch culture. (a) Growth curve and data for total DNA and RNA per cell (from Short, Brown & Street, 1969.) (b) Data for total cellular nitrogen and respiration (from Givan & Collin, 1969) and for ethylene production (from MacKenzie & Street, 1970.) (c) Data for levels of nucleotides (ADP=adenosine diphosphate, UMP=uridine monophosphate, UDP-G=uridine diphosphate glucose) (from Brown & Short, 1969.) (d) Data for activities of glucose-6-phosphate dehydrogenase (G-6-PD) and phosphofuctokinase (PFK) (from Fowler, 1971). (e) Data for cell wall invertase assayed at pH 4·5 (○) and for soluble invertase assayed at pH 4·5 (△) and pH 7·0 (▲) (from Copping & Street, 1972.)

cells do not correspond closely, either structurally or physiologically, with particular tissue cells of the plant body. To reproduce in culture the complete pattern of differentiation of selected specialized tissue cells is thus an objective not yet realized. Nevertheless the study of the origin in culture of particular physiological states and their cytological basis *may* advance our understanding of the molecular basis of cytodifferentiation in plants.

A recent book on cytodifferentiation in animal cells advances the widely accepted concept that the changes involved are consequent upon the activation of different sets of genes in different cell types and that this activation is expressed in terms of the synthesis of enzymes and other cellular proteins (Truman, 1974). Further, the differentiation process is regarded as a relatively permanent and irreversible change. In support of this it is possible to quote studies such as those of Cahn and Cahn (1966) on the culture of retina cells; such cells continue to produce the characteristic pigment granules during prolonged culture under conditions conducive to active cell division whereas *in vivo* such cells are non-dividing. If under certain secondary conditions of culture pigment granules were lost, nevertheless when the cells were returned to the primary conditions of culture the pigmentation returned. To distinguish more minor and readily reversible changes in physiology from cytodifferentiation the former are described as modulations and are considered to reflect the operation of allosteric and other 'fine' processes of metabolic regulation. However, Truman (1974) concedes that 'it is not at all easy to draw a distinction between those enzymes which are fundamental to differentiation and those which represent very short-term modulations in the activity of cells' and in considering cytodifferentiation in liver cells states that 'differentiation and modulation do not represent distinct processes but are merely the extreme ends of a spectrum of changes that can occur'. Although from a number of plant species, cell cultures can be readily initiated from different organs (roots, stems, leaf petiole or lamina, cotyledons etc.) or from different living tissues within an organ (parenchymatous cells of pith, cortex, or mesophyll, cambial and other meristematic cells, immature vascular cells etc.) there is no very convincing evidence that they retain, in culture, characteristics of their *in vivo* origin although, as will be mentioned later in this chapter, they may not have undergone, during culture induction, the required degree of dedifferentiation necessary to express their totipotency (i.e. the capacity to generate a new plant in the way normally achieved from the fertilized egg). Such observations suggest that cytodifferentiation in higher plants, provided it has not proceeded to the point where cell death is inevitable, is a more readily reversible process than in the cells of higher animals. Hence the readily reversible physiological states observed in plant cell cultures may be basically identical with the states involved in normal cytodifferentiation. Further, it raises the possibility that cytodifferentiation does not depend on the transcriptional activity of different sets of genes for each kind of tissue cell but that the 'specialized' physiology of such cells may represent the influence of cytoplasmic factors (plant hormones?) on the stability and transport of RNA species and other aspects of the translational steps in gene expression. As discussed below, it may be possible to examine such hypotheses experimentally with plant cell cultures.

15.3 STEADY STATES OF GROWTH AND METABOLISM OF PLANT CELLS IN CONTINUOUS CULTURE

Reference was made above to the short period of exponential growth observed in batch cultures. However, even during this phase of the growth cycle, the cells do not achieve a steady state (a state of balanced growth); cell division is uncoupled from increase in cell dry weight and protein content so that the cells are changing in size and composition despite the constancy of the double time of the culture (Fig. 15.2). Such observations raised the question of whether balanced growth could be achieved in plant cell cultures if a constant culture environment could be established by developing open continuous culture systems (systems in which inflow of fresh medium is balanced by outflow of an equal volume of culture). Wilson, King & Street (1971) have developed such a system providing conditions of aeration and agitation appropriate to plant cell cultures, capable of long-term aseptic operation and functional either as a chemostat or as a turbidostat.

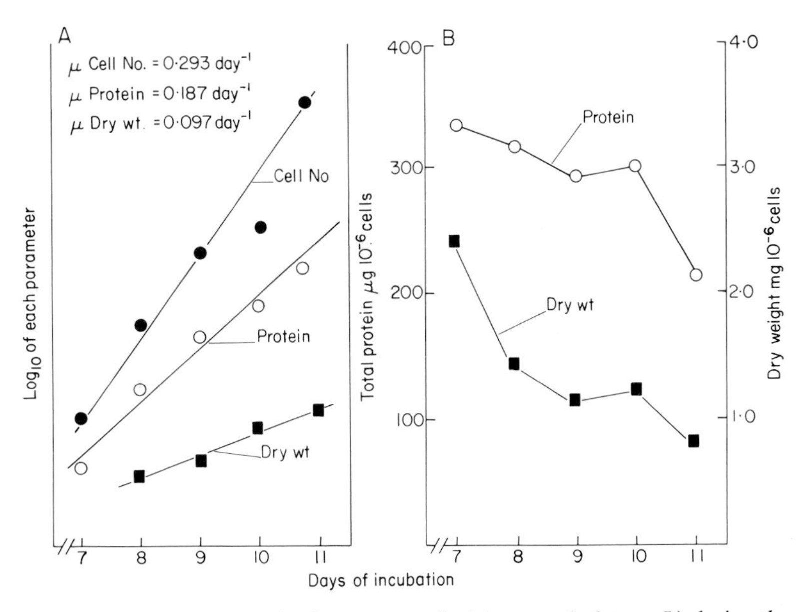

Fig. 15.2. Unbalanced growth of sycamore cells (*Acer pseudoplatanus* L) during the transient exponential growth phase achieved in batch culture. (a) Semi-logarithmic plots showing rate of change in cell number, total protein and cell dry weight per unit volume of culture. The slopes of the lines of best fit (calculated by linear regression analysis, $P < 0{\cdot}01$) were used to determine the specific growth rates (μ) for each parameter.

$$\log_e x = \mu t + \log_e x_0$$

where x_0 = initial value of parameter, x = value after time t(days). When $\log_{10} x$ plotted against t, slope $= \mu/2{\cdot}303$. (b) Changes in total protein and cell dry weight per 10^6 cells with time calculated from data in A. (From King & Street, 1973.)

15.3.1 CHEMOSTAT CULTURES

Subsequent work (King & Street, 1973; King Mansfield & Street, 1973) with this system operated as a chemostat (where equilibrium is established at a fixed rate of imput of a growth-limiting nutrient) has shown that long-term steady states of growth can be achieved with plant cell cultures and that such cultures conform to the chemostat theory developed from work with microorganisms (Monod, 1950; Novick & Sziland, 1950). In a chemostat, the relationship between cell density, x (cells per unit volume of culture), dilution rate, D (volume of new medium added per unit time expressed as a fraction of the total culture volume), specific growth rate, μ (increase in biomass per unit biomass per unit time) and time (t) is given by the equation

$$dx/dt = \mu x - Dx$$

When equilibrium is reached and a steady state established $dx/dt = 0$, $\mu = D$ and x has a value characteristic of the dilution rate. Further, the nutritive environment remains constant, each nutrient achieving an equilibrium concentration which is related to its imput concentration and the rate of its consumption by the culture. As defined above, the equilibrium achieved in a chemostat culture results from one particular nutrient (depending upon the composition of the culture medium) becoming the limiting nutrient and determining the specific growth rate (μ) of the cells.

The nature of these steady states can be illustrated from work involving sycamore cell cultures growing in a synthetic medium (Stuart & Street, 1969) in which the supply of nitrogen is the limiting factor. Data for one such steady state ($D = 0{\cdot}194$ day^{-1}) is presented in Fig. 15.3. This shows that the cells are in a balanced state of growth and metabolism (as illustrated by the values for cell number, packed cell volume, cell dry weight, protein, DNA and RNA, and oxygen demand) and that the nutrient medium within the culture vessel is constant in composition (as illustrated by constancy of culture pH and the levels of glucose, phosphate and nitrate). Such steady state cells also display constant levels of metabolites (e.g. amino acids; Street, Gould & King, 1975) and constant levels of activity of individual enzymes (e.g. enzymes concerned with carbohydrate respiration; Fowler & Clifton, 1974; and with nitrogen assimilation; Young, 1973).

Such chemostat cultures can be operated from very low growth rates (i.e. low dilution rates) to growth rates approaching the maximum growth rate (μ_{max}) for the culture medium chosen, provided dilution rate is such that the cells can still achieve a matching growth rate. Of course if dilution rate is further increased, cell density does not stabilize and the culture suffers wash-out. Cells in balanced growth but highly contrasted in growth rate (and hence in cytology and metabolism) can therefore be obtained by chemostat culture. The range of change in certain cell parameters in sycamore cell cultures at different dilution rates (and hence specific growth rates) over the range $D = 0{\cdot}06$–$0{\cdot}236$ day^{-1} (corresponding to double times over the range 280–70 hr) is illustrated in Fig. 15.4. What this means is that it is possible to stabilize at will, by fixing dilution rate at an appropriate level, the individual physiological states which have only a transient existence in batch culture.

Work with chemostat cultures has shown that the same cell population can be taken through a series of steady states and then if the dilution rate is returned to that of an earlier steady state the cells again achieve not only the new predictable growth rate but also the physiological activities earlier recorded as

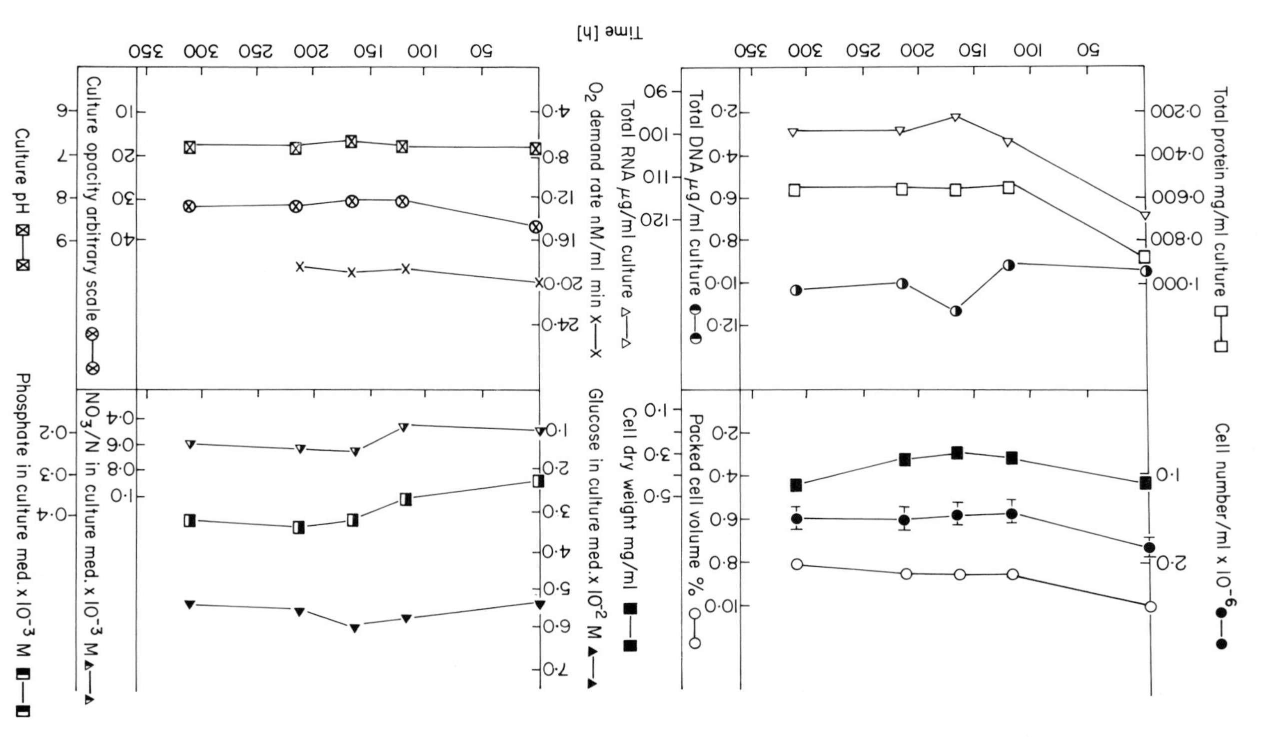

Fig. 15.3. A steady state established in a 4-litre chemostat culture of *A. pseudoplatanus* cells. The culture was diluted for 400 hrs. at a rate of 0·194 culture volumes per day. Samples were withdrawn at intervals for biomass measurements, determinations of nutrient levels in the culture medium and for respiration rate measurements. Culture opacity and pH were monitored continuously in the culture vessel. (From King & Street, 1973.)

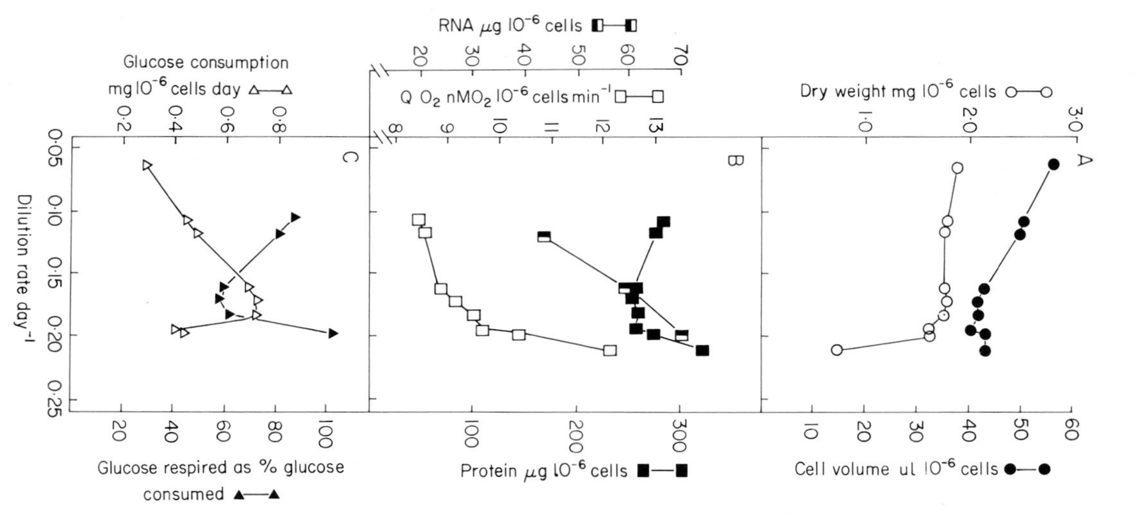

Fig. 15.4. Steady state values for parameters of cell composition and physiological activities recorded over a range of growth rates (as expressed by dilution rates) established in chemostat cultures of sycamore cells (*A. pseudoplatanus*) with nitrate/N as the limiting nutrient. The relationship between dilution rate, D (fraction of culture volume displaced per unit time), specific growth rate (μ) and doubling time (td) of the cell population in a steady state is given by

$$D = \mu = \frac{\log_e 2}{\text{td}}$$

(From King & Street, 1973.)

characteristic. Such experiments demonstate the full reversibility of the cytological and physiological changes invoked. Interest will now focus on detailed studies of the kinetics of the transition between steady states in terms of enzyme activities and metabolite levels. Preliminary studies along these lines have already shown that during these transitions there are pronounced oscillations in enzyme activity levels (characteristic for each enzyme monitored) which gradually decline in amplitude as the new steady state is established. How far this behaviour is to be explained in terms of changes in rates of enzyme synthesis and degradation has yet to be determined. Clearly study of these transitions will yield entirely new data on metabolic regulation in higher plant cells; whether it will yield the key to expose the changes underlying cytodifferentiation is less certain.

15.3.2 TURBIDOSTAT CULTURES

The constancy of culture opacity in the steady state described by Fig. 15.3 has formed the basis for a second form of continuous culture—the turbidostat system (a continuous system in which inflow of new medium occurs in response to an increase in the opacity-decrease in the light transmission of the culture). The turbidostat culture system developed for work with sycamore cell cultures is shown in Fig. 15.5 (Wilson, King & Street, 1971). Here, each time the population density exceeds a pre-selected value as determined by the optical density continuously monitored by the photocell, an electronically operated valve opens to admit a pulse of new medium. This reduces the optical density of the culture and the valve closes. These imputs of new medium are balanced by periodic harvesting of small volumes of culture in response to an electronically controlled level detector which controls the output valve. The effect is to produce a culture of constant volume and constant population density growing at a constant rate. Whereas in the chemostat growth rate must always be below μ_{max} (the culture growth being limited by the supply of a chosen nutrient) here in the turbidostat growth can safely proceed under non-limiting nutrient conditions and one can study the effect of physical factors (e.g. temperature, light regime, CO_2 tension) and growth regulators, in particular plant growth hormones, on growth rate (King, 1976), and by appropriate techniques (Gould, Baylis & Street, 1974) on the duration of the different phases of the cell cycle. This is the system where one can attempt to achieve conditions under which biomass increase expresses the maximum genetic potential for cell growth. Such a system offers an entirely new approach to studies on the molecular basis of the hormonal control of cell growth.

15.4 SYNCHRONOUS CELL CULTURES—STUDY OF THE CELL CYCLE

As indicated earlier in this chapter, plant cell cultures are routinely propagated by batch cultures initiated at a relatively high initial cell density (2×10^5 cells

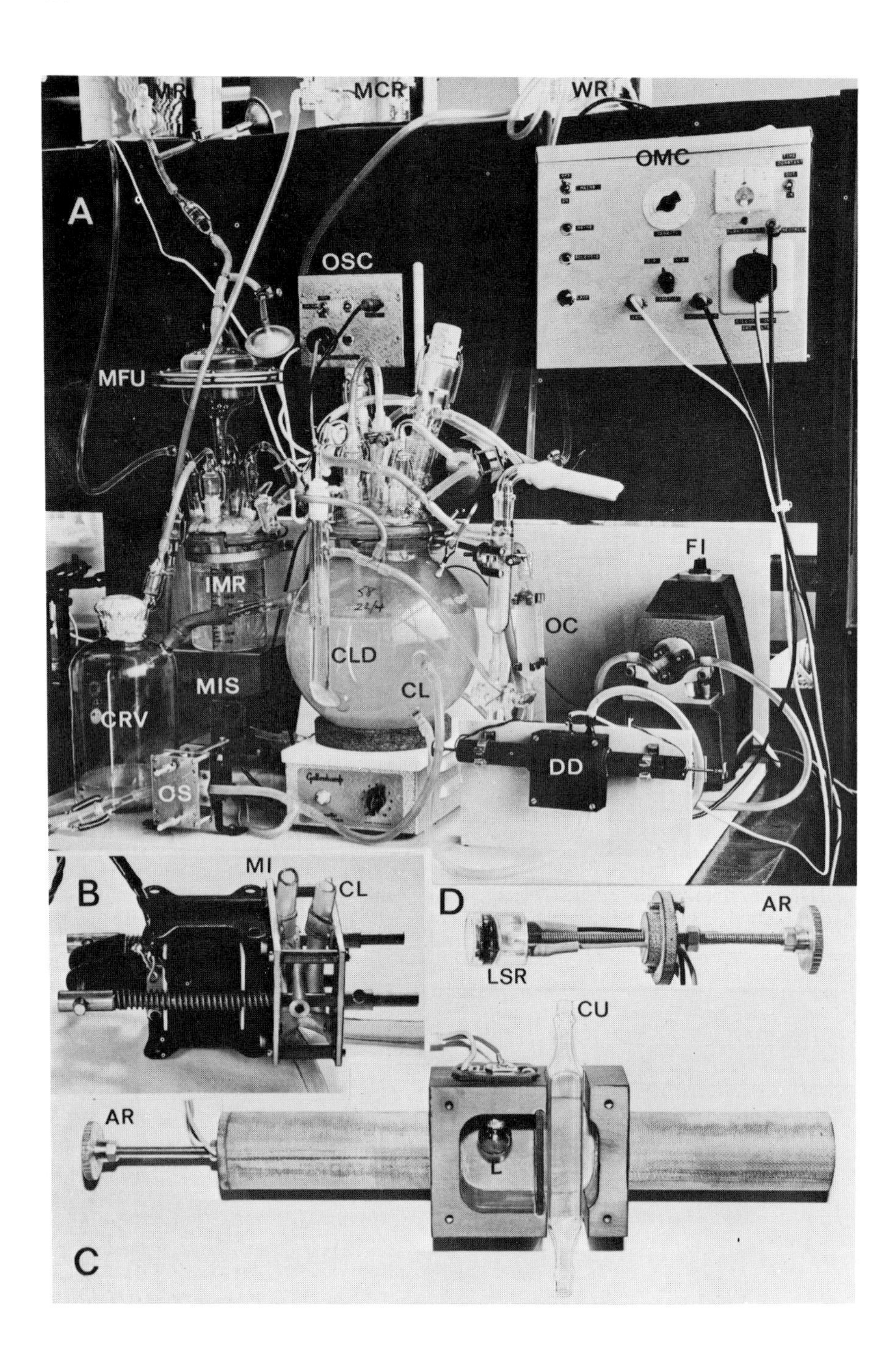
MR
MCR
WR
OMC
A
OSC
MFU
IMR
CLD
CL
MIS
CRV
OC
FI
OS
DD
B
MI
CL
D
AR
LSR
CU
AR
L
C

ml^{-1}) and subcultured when they enter stationary phase. It is, however, possible to initiate such batch cultures of sycamore cells at much lower cell densities (the minimum inocular density being *ca.* $1{\cdot}5$–$4{\cdot}0 \times 10^4$ cells ml^{-1} in the standard medium; see Stuart & Street, 1969) and to use as inoculum cells maintained in stationary phase as long as possible without suffering decline in viability. When this is done using an enriched synthetic medium (Stuart & Street, 1971) the cultures show a more extended lag phase and then embark upon a succession of highly synchronous divisions as evidenced by the data for cell counts (Fig. 15.6), mitotic index (percentage cells in a recognizable stage of mitosis) determinations and estimations of nuclear DNA content (by microdensity following Feulgen staining) (Fig. 15.7; see also Street, King & Mansfield, 1971; Gould & Street, 1975). By using for such synchronous batch cultures the 4-litre culture vessel developed for the continuous culture systems but modified by adding a stainless-steel sampling valve automatically operated by a timing device (Wilson, King &

Fig. 15.5. The turbidostat culture system of Wilson, King & Street (1971). The 4-litre culture vessel is mounted over a magnetic stirrer and is controlled in temperature by an internal water circulating coil. The culture flows through an external circulation loop (exit and entry lines of this loop labelled CL) by the action of the flow inducer (FI). This circulation loop flows through a cuvette (Fig. C. CU) in the density detector (DD). As the cells divide in the culture its opacity increases and this alters the light transmission between the lamp (L) and the light sensitive resistor (LSR) in the density detector (see Fig. C and D). When this transmission falls below a preset value the optical monitoring unit (OMC) sends an impulse to the medium imput solenoid value (MIS) and a pulse of new medium flows from the intermediate medium reservoir (IMR) fed from the main medium reservoir (MR) *via* a filter unit (MFU). The size of this imput of new medium is controlled by the observation chamber (OC) in the circulating loop since the imput solenoid valve closes only when new medium displaces culture in the cuvette within the density detector. These imputs of new medium are balanced by release of culture into the culture receiving vessel (CRV) via the outlet solenoid valve (OS) operated by an outlet solenoid control unit (OSC) responding to the electrodes in the constant level device (CLD). Thus the culture is maintained at a predetermined optical density (corresponding to a fixed cell number per unit volume) and the rate of entry of new medium (and balancing harvest of culture) is a measure of the growth rate of the culture. Samples of culture can at any time be withdrawn for cell counting and biochemical analysis *via* a sample tube located to the right of the culture vessel and after collecting the aliquot of culture this sample collector can be washed with sterile water from the reservoir WR. Periodically the excess culture collected in CRV is withdrawn and its volume measured and the outlet protected by washing with mercuric chloride solution from the reservoir MCR. The various ports in the lid of the culture vessel provide for introduction of the initial inoculum of cells, for aeration (*via* a glass sinter tube) for withdrawal of samples into the sample collector, and for flow of water through the internal temperature-controlling glass coil.

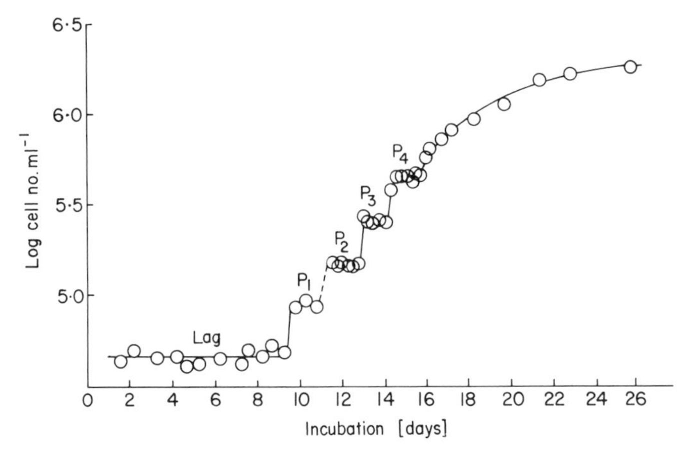

Fig. 15.6. Growth curve of a synchronous cell culture of sycamore (*A. pseudoplatanus*) showing log cell number ml^{-1} with time. The interphase plateaux separating the synchronous increases in cell number are labelled P_1-P_4. The lag phase of this culture was of 9 days duration and the culture reached stationary phase on day 24.

Street, 1971) it has been possible to show the synchrony of a number of metabolic events in the cell cycle (King *et al.*, 1974).

The period during which DNA is doubled takes place during a restricted period of the interphase between successive mitoses. Howard and Pelc (1953) termed this the S phase and distinguished the interphase period before S phase as G1 (the first gap) and the period after completion of S phase and before mitosis (M) as G2 (the second gap). Cell cleavage (cytokinesis) usually follows directly upon mitosis so that G1 can be timed from the origin of the daughter cells although, as originally defined, G1 begins as soon as nuclear division is complete. These stages in the cell cycle can be determined in exponential asynchronous cultures by determinations of doubling time (cell count data), mitotic index, labelling index (per cent nuclei and mitoses labelled following flash label with tritiated thymidine—^{3}H-Tdr), fraction of labelled mitoses (following pulse-labelling with ^{3}H-Tdr; see Quastler & Sherman, 1959; and continuous labelling with ^{3}H-Tdr; see Cleaver, 1967), microdensitomitry and autoradiography, achieved by the single slide technique of Mak (1965). These methods have been applied to cultures of sycamore cell lines (Gould, Bayliss & Street, 1974) during the phase of exponential growth in batch cultures and to steady state chemostat cultures growing at different rates (doubling times ranging from 22 to 85 hr). This has revealed that the phases S ($7 \cdot 0 \pm 0 \cdot 2$ hr), G2 ($8 \cdot 7 \pm 0 \cdot 6$ hr) and M ($2 \cdot 9 \pm 0 \cdot 3$ hr) are relatively constant whereas, according to the cell doubling time, G1 varies widely (4–60 hr). This observation that G1 varies with different cycle times, whereas S+G2 is relatively constant has also been observed in work with mammalian cells both in culture and *in vivo* (Mitchison,

1971). Certain critical events in G1 may be essential to the initiation of S phase and may be rate limiting. In the work with sycamore cells growing at reduced rates, nitrogen supply is the limiting factor and when cells enter stationary phase in batch culture, they are arrested in G1 (this arrest is important in relation to the synchrony which can be achieved by regrowth of nitrogen-starved cells; see Gould & Street, 1975).

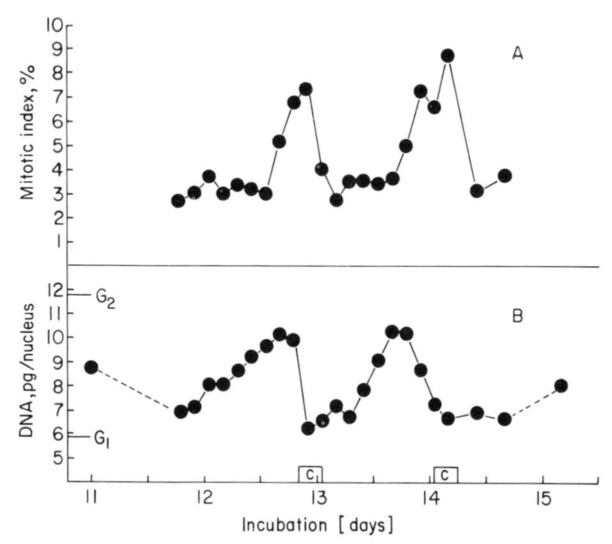

Fig. 15.7. (A) Mitotic index fluctuations for P_2 and P_3 of Fig. 15.6. (B) Average nuclear DNA contents over the same period. G1 and G2 values for nuclear DNA content of the cell line are shown for comparison. [C] represents the durations of the two successive periods of cell number increase. (Data from Gould & Street, 1975.)

The value of synchronous cultures for studies on the cell cycle is that they enable particular metabolic events to be monitored. Work along these lines with plant cells has only recently been undertaken. In a study involving synchronous sycamore cell cultures (King *et al.*, 1974) it was shown that total extractable protein and RNA rise throughout interphase but at an increased rate during S+G2. This was supported by studies of the rate of incorporation of labelled amino acids and uridine into these fractions. Respiration rate similarly rose throughout interphase but with two peaks of activity during S phase and cytokinesis. When, however, changes in the activity of extracted enzymes were studied different patterns emerged. Thymidine kinase and aspartate transcarbamoylase showed single and separated peaks of activity (Fig. 15.8), succinic dehydrogenase showed two well separated peaks of activity, and glucose-6-phosphate dehydrogenase activity rose continuously throughout interphase.

Whilst the content of total extractable histone rose parallel with total protein, study of the rates of incorporation of ^{3}H-labelled lysine and ^{14}C-labelled arginine into this fraction pointed to greatly enhanced histone turn-over associated with S-phase (Street, Gould & King, 1975). It is not known whether the changes in enzyme activity detected in these studies reflected changes in rates of enzyme synthesis. Yeoman (1974) in work on cell synchrony in explants of Jerusalem artichoke tubers has shown however, by using the deuterium labelling technique (Hu, Bock & Halvorson, 1962), that changes in the activity of glucose-6-phosphate dehydrogenase during interphase do result from changes in rate of synthesis of the enzyme.

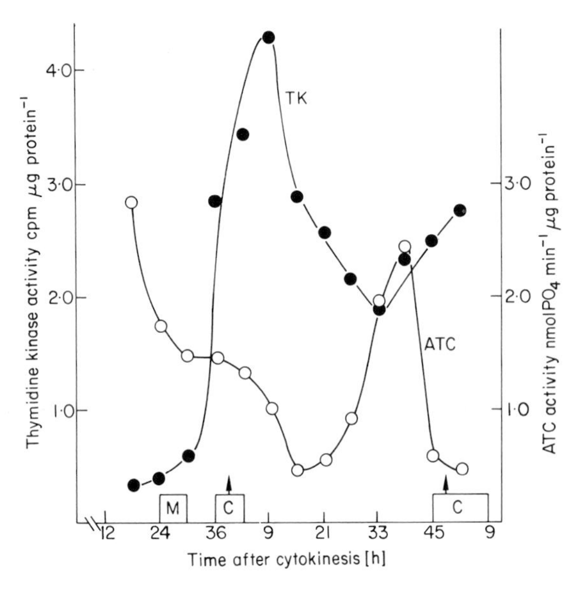

Fig. 15.8. Changes in the activities of thymidine kinase (TK) and aspartate trans-camoylase (ATC) during a cell cycle in a synchronized cell culture of sycamore (*A. pseudoplatanus*). C = duration of cytokinesis from the cell count data. M = duration of mitosis from the mitotic index data. (From King, Cox, Fowler & Street, 1974.)

If future work along these lines enables the many separate metabolic events of the cell cycle of plant cells to be chartered, we will then have a number of 'markers' (points where particular metabolic events are initiated or terminated). This will enable us to identify those events along the interphase plateau whose initiation or pace is affected by nutritional factors and plant hormones, to determine whether certain events only occur when their controlling genes (which could be 'mapped') are exposed during DNA replication and whether other processes which proceed continuously show gene dosage effects as evidenced by increases in their rate after the points in DNA replication when their controlling genes are duplicated.

15.5 MORPHOGENESIS IN CELL CULTURES—CONCEPTS OF TOTIPOTENCY AND DETERMINATION

Various forms of morphogenesis have been observed in cell suspension cultures; this morphogenesis is apparently dependent upon the development in the cultures of relative large aggregates of cells, often several hundred in number, associated in a symplast. The morphogenesis can take the form of root initiation, shoot bud initiation or the development of somatic embryos (often referred to as embryoids) (Konar, Thomas & Street, 1972). Usually only one form of morphogenesis is expressed; sometimes the situation is more complex but in these cases one form is usually dominant—carrot cell cultures can be manipulated to show predominantly root initiation or exclusively embryoid development (Kessel & Carr, 1972).

15.5.1 SOMATIC EMBRYOGENESIS

Work on embryogenesis in carrot cultures illustrates clearly the problems in cell physiology raised by the morphogenetic potential of plant cell cultures. These studies date from the report in 1958 (Steward, Mapes & Mears, 1958; Steward, 1958) of embryo-like plantlets in liquid carrot cultures and their continuing development when the cultures were plated out on agar-solidified medium. Further work quickly established the presence in such cultures of structures strikingly similar to the globular, heart-shaped and torpedo-shaped stages of normal embryology from the zygote.

Halperin and coworkers (Halperin & Wetherell, 1964; Halperin, 1967; Halperin & Jensen, 1967) and Street and coworkers (Smith & Street, 1974; McWilliam, Smith & Street, 1974) have shown that the embryos in carrot cultures arise from single cells at the surface of the cellular aggregates and are released, at various stages of development, as free-floating structures which, if they have already reached the advanced globular stage, are capable of completing their development in isolation from the parent embryogenic clump. These cultures grow actively, the embryogenic clumps proliferating and fragmenting due to enlargement and separation of cells in their interior, and do *not* form embryos in a synthetic medium containing sucrose, inorganic salts, thiamine, *meso*-inositol, kinetin and auxin (2, 4-dichlorophenoxyacetic acid). To initiate embryogenesis they are transferred to a similar medium, lacking auxin, which contains nitrate and ammonia (or urea or glutamine) and is at pH 5·0–5·4. Despite earlier claims, coconut milk is neither essential for growth nor for embryogenesis. Evidence that immature embryos have exacting nutritional requirements (isolated embryo culture—see Street, 1969) and the contrasted simplicity of the culture medium which supports prolific embryogenesis in carrot cultures suggests that the embryogenic cell aggregate fulfils a 'nurse' role to the embryogenic surface cells and that the embryos must remain attached to the aggregate to complete their early development to the point where they can survive and mature into plantlets when released.

This phenomenon of somatic embryogenesis raises a number of questions of cellular physiology. The term totipotency has been introduced to describe the embryogenic competence of the single cells from which the embryos arise. At present it is only possible to establish new plants, whether *via* embryogenesis of *via* shoot bud initiation, followed by adventitious root development, from the cell cultures of a limited (if now quite large) number of species or varieties within species. Cultures which show this morphogenetic potential support the view that the pathways of cytodifferentiation which result in living tissue cells do not involve any loss or permanent inactivation of the genome and that they are, under appropriate environmental stimuli, completely reversible. Whilst however some cell cultures remain recalcitrant, this cannot be established as a universal principle. It may be that in such cases the conditions of culture fail to provide or permit the synthesis of an essential morphogen. Halperin (1967) has, on the other hand, advanced the very interesting hypothesis that the achievement of totipotency occurs during the initiation of the carrot culture from the primary explant (storage root or seedling organ) and that embryogenesis is expressed by cell clumps derived from such 'induced' cells, the primary culture consisting of both these and 'non-induced' cells. Retention of high embryogenic capacity in the cultures will then depend upon culture conditions favouring the active proliferation of the induced cells. On this hypothesis, failure to obtain embryogenesis in culture would have as its primary cause inappropriate conditions of callus initiation; the conditions of initiation would have effectively activated cell division and growth in the explant cells but failed to achieve the necessary 'dedifferentiation' to obtain cells with the competence of the zygote (a concept already raised here in previously discussing cytodifferentiation in cell cultures).

In classical plant embryology it has been considered that the early segmentations of the zygote (at least up to the 16-celled proembryo) follow a precise and species specific sequence which has phylogenetic significance (Johansen, 1950). During these divisions the cells of the proembryo are considered to inherit different cytoplasmic potentialities from the different regions of the zygote and these differences are regarded as determining from the beginning the exact role they and their daughter cells will play in constructing the embryo and its parts. This concept is often referred to as the theory of precise mosaic organization. The early segmentations involved in somatic embryogenesis have now been followed in a limited number of species (Street, 1976). Figure 15.9 illustrates the sequence of early embryology in carrot cultures. These segmentation sequences involved in somatic embryogenesis show more uniformity one with another than is depicted in the published accounts of the zygote embryology of the species concerned. In summary, early development of somatic involves enlargement and regular segmentation in a subspherical cell walls of minimum surface. This supports the concept of D'Arcy T (1942), based upon his studies in animal embryology, that surface tensions are important in determining the early segmentations of embryology and that only at a later stage do localized growth centres emerge whose functioning gives rise to the divergences in the morphology and anatomy of embryos of different species. This regulative theory of organization regards the early segmentations as being controlled by physical factors and as not involving any 'determination' of the early formed cells. With this background, the greater diversity and specificity recorded for zygotic embryogenesis can be interpreted as resulting from the physical restrictions and polar chemical gradients imposed upon the embryo as it develops within the ovule; when these influences are removed, as in cell cultures, the embryology reverts to a more basic or 'primitive' type of segmentation. This interpretation is supported by observations on the segmentations observed in natural polyembryony and by recent reports that indeed much more variable patterns of segmentation occur in ovule embryology than has hitherto been recognized (Jensen, 1965; Brown & Morgensen, 1972).

15.5.2 POLARITY OF EMBRYOGENIC CELLS

The recognition that the embryos arising in cell culture have their origin in superficial cells of the cell aggregates raises the question of whether such cells have any unique cytological characters and whether their observation can yield information on the physical basis of cell polarity. In carrot cultures proliferating in the presence of 2, 4-D (Street & Withers, 1974) these cells are small, rich in cytoplasm and with a large diffusely-staining nucleus containing a prominent nucleolus. Small vacuoles are clustered round the nucleus and each cell contains several amyloplasts containing prominent starch grains (Fig. 15.10). Study of these cells in the electron microscope shows the presence of numerous round and oval mitochondria and Golgi bodies and the regular presence of small numbers of lipid bodies (spherosomes). As might be expected from their meristematic activity, these cells are frequently observed in mitosis (in contrast to the expanded interior cells of the aggregates) and show numerous wall microtubules and limited arrays of cytoplasmic and nuclear microfibrils. When the cultures are transferred to auxin-free medium there is a transient increase in proliferation prior to the initiation of embryogenesis and the superficial cells show a change in segmentation pattern leading to the origin of 4-celled groups (Fig. 15.10). The individual cells in these groups either initiate an embryo, or by their further division promote the growth of the aggregate or undergo expansion (and senescence?) and become involved either in the release of the developing embryo or the break-up of the proliferating embryogenic aggregate. Associated with this changed segmentation pattern the densely cytoplasmic superficial cells

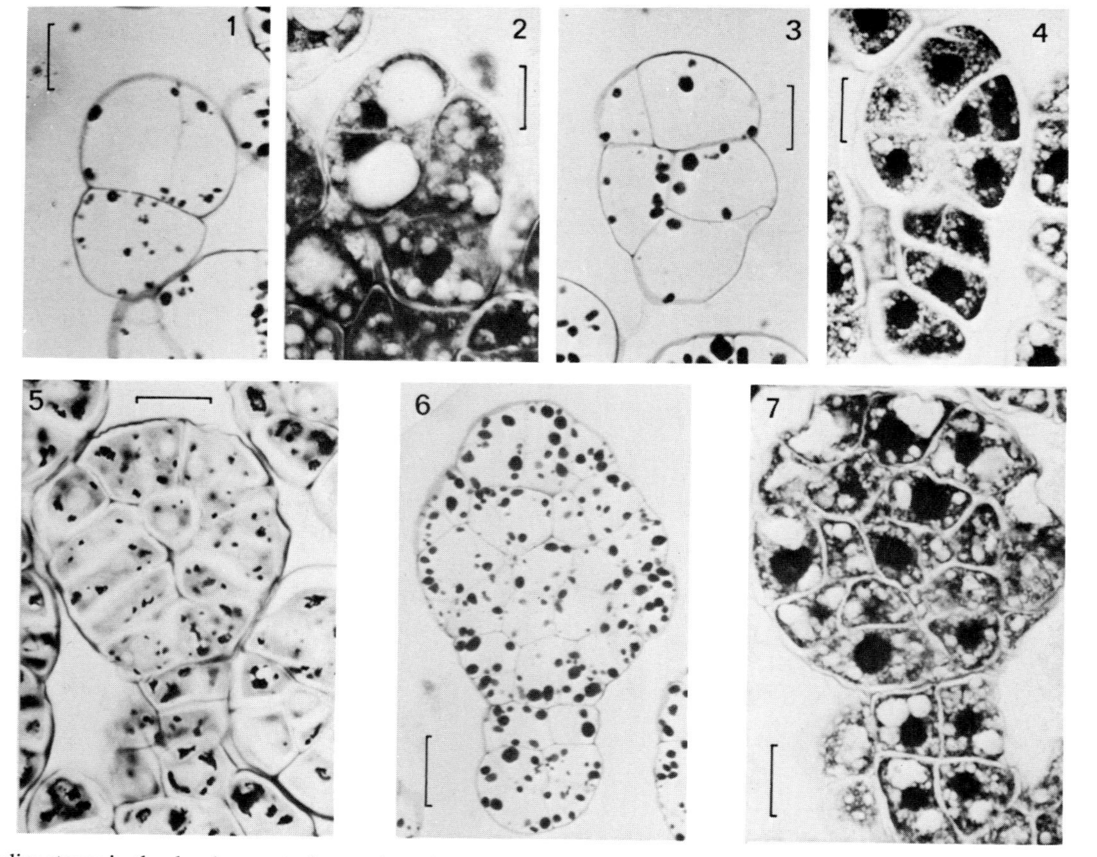

Fig. 15.9. Earlier stages in the development of somatic embryos as observed in a cell suspension culture of carrot (*Daucus carota, L*). Scale lines on 1–4 = 20 μm, on 5–7 = 25 μm. Stages 1, 3, 5, 6 stained with periodic acid—Schiff (PAS), stages 2, 4 and 7 stained with toluidine blue (TB). (From Street & Withers, 1974.)

show changes in fine structure. They have increased numbers of E.R. profiles (parallel arrays of rough E.R. profiles become particularly prominent) and of ribosomes. The Golgi bodies are more numerous and more compact. Additional large mitochondria (discs with swollen rims in outline) make their appearance and may come to occupy a considerable volume of the cytoplasm. The cells of the young proembryos show similar fine structure (Fig. 15.11). Although the first division of the embryogenic cells is by a wall parallel to the surface of the aggregate and at right angles to the longer axis of the cell (giving rise to an apical and a basal cell; the former being the first cell of the proembryo proper and the latter of the very variable suspensor), nevertheless fine structure studies do not reveal any prior asymmetry in the distribution of cytoplasm or cell organelles. Rather disappointingly, these studies have not revealed any unique features of the embryogenic cells or exposed any structural polarity. Perhaps this is not unexpected when we bear in mind the lack of any uniformity of fine structure in those angiosperm zygotes which have been studied (Jensen, 1965; Schulze & Jensen, 1969; van Went, 1970; Morgensen, 1972). Such studies only serve to emphasize that the special nature of embryogenic cells must now be approached at the level of molecular biology.

15.6 POLLEN GRAINS AS ISOLATED EMBRYOGENIC CELLS AND AS A SOURCE OF HAPLOID CELL LINES FOR MUTAGENESIS

This field of study was opened up by the pioneering studies on anther culture by Guha and Maheshwari (1964) working with *Datura*, and Bourghin & Nitsch (1967) working with *Nicotiana*. They showed that anthers, excised at an appropriate stage in pollen grain development and cultured in a simple medium gave rise to haploid embryos derived from individual pollen grains. Haploid plantlets can now be obtained by this technique from many species within the family Solanaceae. This approach extended to species in other angiosperm families has in a few cases yielded haploid callus but more frequently given a negative result. If we assume that immature pollen grains can, under appropriate stimuli, embark upon embryogenesis (sporophyte development) then the difficulty of achieving this with most excised anthers may be because conditions within the anther are too strongly promotive of the gametophyte pathway of development (i.e. that which gives functional pollen grains containing a tube nucleus and a generative cell). Recently however Nitsch (1974) in work with two Solanaceous species *Nicotiana tabacum* and *N. sylvestris* has shown that if anthers are excised at the peak of the first pollen mitosis, kept at 5°C for 48 hours and cultured for 4 days in a simple medium, then the immature pollen grains can be extracted. If this pollen grain suspension is cultured in a thin layer of liquid medium containing IAA, zeatin, glutamine and serine as supplements, some 10% of the grains embark upon embryogenesis. Although this work has so far

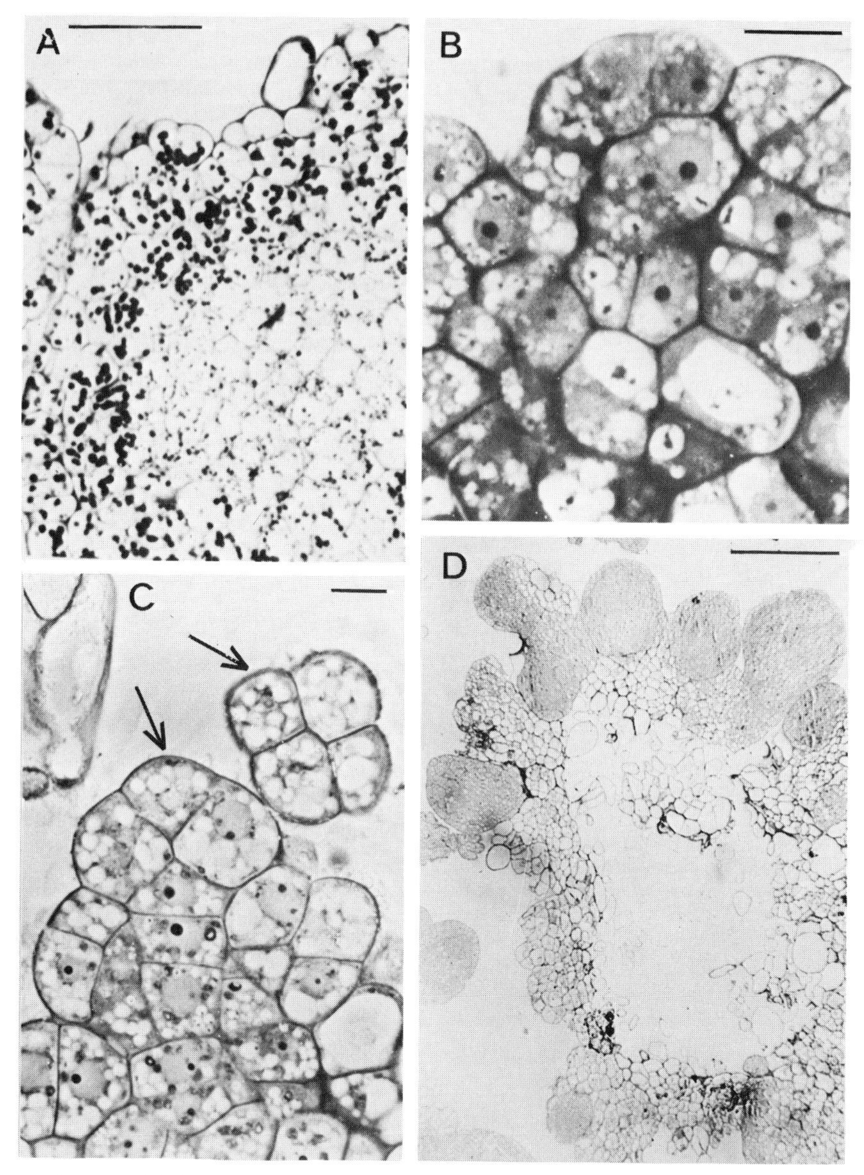

Fig. 15.10. Embryogenic cell aggregates in suspension cultures of carrot (*D. carota*). A and B from cultures in medium containing 0·1 mg l^{-1} 2,4-dichlorophenoxyacetic acid (2,4-D), C and D from cultures in media with 2,4-D omitted. (A) Section through a cell clump stained with PAS. Scale line = 200 μm. (B) Section stained with TB to show the densely cytoplasmic superficial cells with nuclei surrounded with small vacuoles and more internal cells with a large central vacuole. Scale line = 25 μm. (C) Localized superficial region of active cell division. Arrows indicate characteristic pattern of cell segmentation which appears following subculture to the auxin-free medium. Scale line = 10 μm. (D) General view of section through a cell aggregate

been restricted to Solanaceous plants, it could prove to be the key to inducing embryogenesis or haploid callus initiation in pollen grains of the many species resistant to the culture of whole anthers.

The importance of these studies is twofold. First is their potential for providing haploid plants (and by appropriate treatment homozygous diploids) and haploid cell cultures from many species and varieties. The haploid cell lines could be used as a source of mutant cell lines extending dramatically the field of the biochemical genetics of higher plants. Such mutant haploid lines could be rendered diploid by colchicine treatment and hence used to generate fertile mutants, provided their potential for morphogenesis was not impaired by the mutation. They could be preserved as cell culture lines by freezing preservation (Nag & Street, 1973, 1975a,b) as is already routinely done with mutant lines of bacteria. The use of such mutant cell lines in the continuous and synchronous culture systems previously described would permit more critical studies to be undertaken on particular metabolic sequences and their control and on the biochemical nature of the controls which operate in cell growth and division. Work along these lines is already actively proceeding (Widholm, 1974; Street, 1975b) but is currently handicapped by the need for more effective means of selecting the desired mutant cells which are present in very small numbers in the cell populations after mutagen treatment (e.g. 1 in 10^7).

The second important feature of these studies is in relation to embryogenesis. If within the pollen suspension, the pollen grains destined to embark upon embryogenesis are of characteristic size and/or density, it may be possible to isolate them by appropriate density gradient centrifugation. This would enable us to characterize embryogenic cells by the techniques of molecular biology, and to study in greater detail the segmentation patterns of early embryogenesis and the spontaneous expression of polarity in the proembryonal cell mass. Since the early stages of embryogenesis in pollen occurs within the enclosing pollen grain wall, study of the food reserves of such grains and of their biosynthetic activity may enable us to determine the special requirements of the proembryo.

This chapter has drawn attention to the availability of new experimental material and of new techniques in the field of plant cell culture. The exploitation of this approach in the molecular physiology of plants is still in its infancy. It is therefore important that the new generation of plant physiologists and biochemists should remain well informed of its present transient limitations and assured future prospects.

---

bearing numerous globular embryos at its surface. Stained with TB. Scale line = 200 μm. (A & B from Smith & Street, 1974; C & D from Street & Withers, 1974.) (See also legend to Fig. 15.9.)

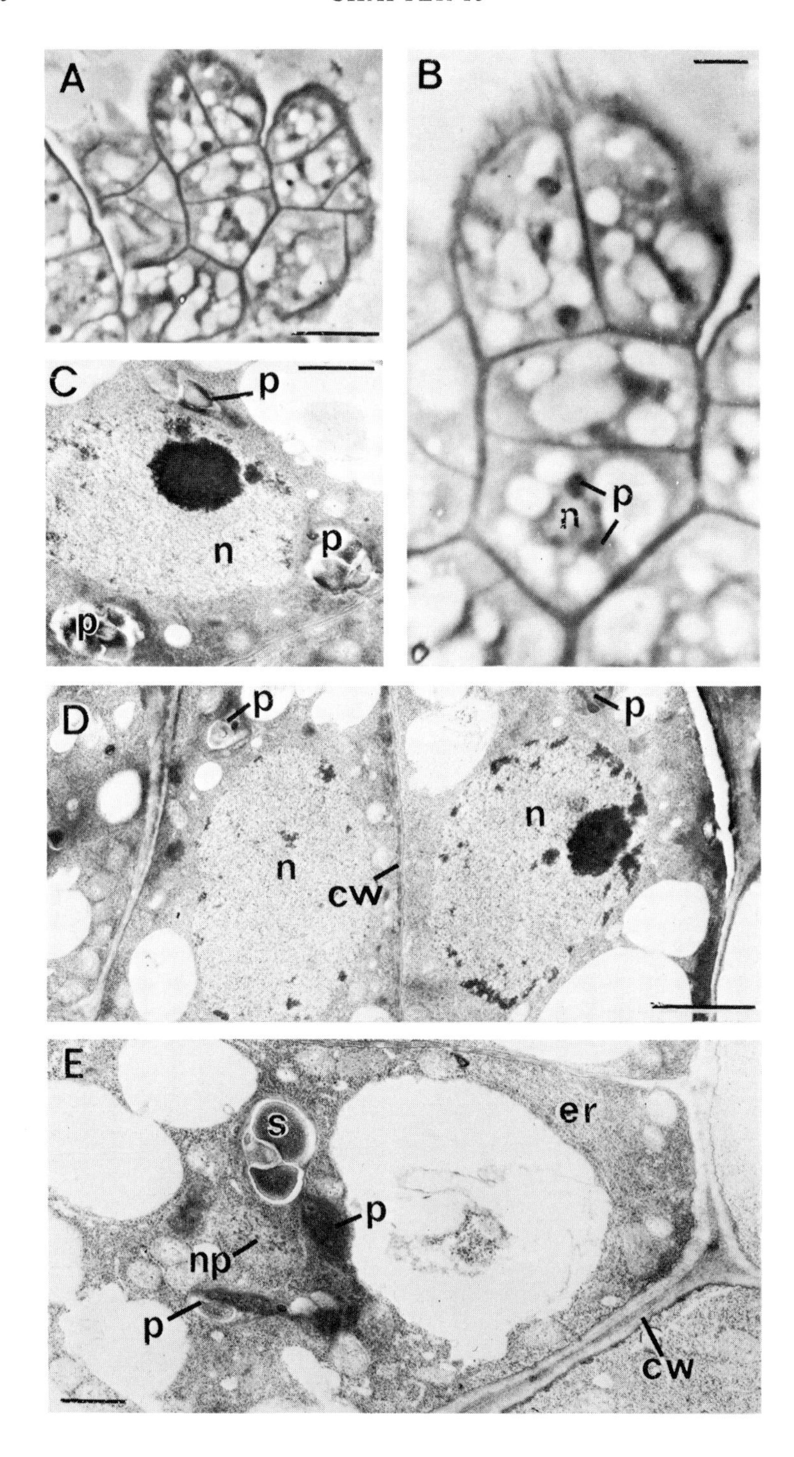
A
B
C
p
n
p
p
n
p
D
p
p
n
n
cw
E
s
er
p
np
p
cw

FURTHER READING

King P.J., Mansfield K.J. & Street H.E. (1973) Control of growth and cell division in plant cell suspension cultures. *Canad. J. Bot.* **51**, 1807–23.

Ledoux L. (Ed.) (1975) *Genetic Manipulations with Plant Material.* Plenum Press, London.

Les Cultures de Tissus de Plantes. (1971) Colloques Internationaux du C.N.R.S. No. 193, Paris.

Steward F.C. (Ed.) (1969) *Plant Physiology.* Vol. 5B. Academic Press, New York.

Street H.E. (Ed.) (1977) *Plant Tissue and Cell Culture.* Blackwell Scientific Publications, Oxford. (Second Edition.)

Street H.E. (Ed.) 1974) *Tissue Culture and Plant Science,* 1974. Academic Press, London.

Street H.E. (1976) Experimental embryogenesis—the totipotency of cultured plant cells. In *The Developmental Biology of Plants and Animals.* (eds Wareing, P.F. and Graham, C.F.). Blackwell Scientific Publications, Oxford.

Wilson S.B., King P.J. & Street, H.E. (1971) Studies on the growth in culture of plant cells. XII. A versatile system for the large scale batch and continuous cultures of plant cell suspensions. *J. exp. Bot.* **21**, 177–207.

Fig. 15.11. Light (A and B) and electron micrographs (C,D,E) of a four-celled somatic embryo of *D. carota*. (A) The embryo located within a characteristic small cell group (compare Fig. 10, C) at the surface of an embryogenic cell aggregate. Scale line=10 μm. (B) Enlarged view of embryo from A. The nucleus (n) of the basal cell is seen to be surrounded with optically dense plastids (p). Scale line=2·5 μm. (C) The nucleus of the middle cell of the embryo surrounded by amyloplasts (p). Scale line = 1 μm. (D) The two terminal (apical) cells of the embryo showing large rounded nuclei, amyloplasts and the newly-formed thin dividing wall between them. Scale line = 2 μm. E. The basal cell of the embryo separated from adjacent cells of the aggregate by a thick plasmodesmata-free wall (cw). Parallel arrays of endoplasmic reticulum (er) can be seen at the cell periphery. The nucleus (note glancing section showing nuclear pores—np) is again surrounded by amyloplasts (p). Scale line = 1 μm. (From Street & Withers, 1974.)

CHAPTER 16

THE PHYSIOLOGY OF ISOLATED PLANT PROTOPLASTS

16.1 INTRODUCTION

The isolated plant protoplast is a single cell, bounded by the plasmalemma and, when first formed, containing all the normal cell components, although being separated completely from its cell wall. From a physiological viewpoint, however, the protoplast cannot be regarded simply as a cell lacking a wall, since the mechanics of isolation, in conjunction with environmental factors, undoubtedly influence its metabolism and elicit subtle ultrastructural changes. The absence of a functional cell wall may affect the permeability of the cell membrane and lead to a general leakage of solutes from the protoplast. The protoplast is also in a transient state since most protoplasts, irrespective of their immediate cultural environment, will initiate the synthesis of a new cell wall a few hours after release, and eventually revert to a single-walled cell. In spite of these considerations, and combined with a cautious extrapolation of experimental data to the presumed situation existing in the intact cell or whole plant, protoplasts provide an important biochemical tool for the biologist.

In the absence of a cell wall, the exposed plasmalemma can be examined in great detail with respect to particle uptake, permeability, possible membrane-associated functions such as disease resistance, membrane fusion, and, during cell wall synthesis, the relationship of the plasmalemma to its cell wall.

16.2 ISOLATION AND CULTURE OF PROTOPLASTS

16.2.1 PREPARATION OF PROTOPLASTS

Protoplasts may be produced, under aseptic conditions, from a wide range of plant species either directly from the whole plant, or indirectly from *in vitro* cultured tissues. There are two basic approaches for the enzymatic isolation of protoplasts: (a) the treatment of a plant tissue with a mixture of pectinase and cellulase so as simultaneously to macerate, or separate, cells and degrade their walls (Power & Cocking, 1970); and (b) the sequential (two-step) method involving the production of isolated cells which in turn are converted into protoplasts by a cellulase treatment (Nagata & Takebe, 1970).

Since removal of the cell wall results in loss of wall pressure upon the cell, protoplasts are isolated and maintained in hypertonic plasmolytica provided by a balanced inorganic salt medium or monosaccharide sugar solution. Mannitol,

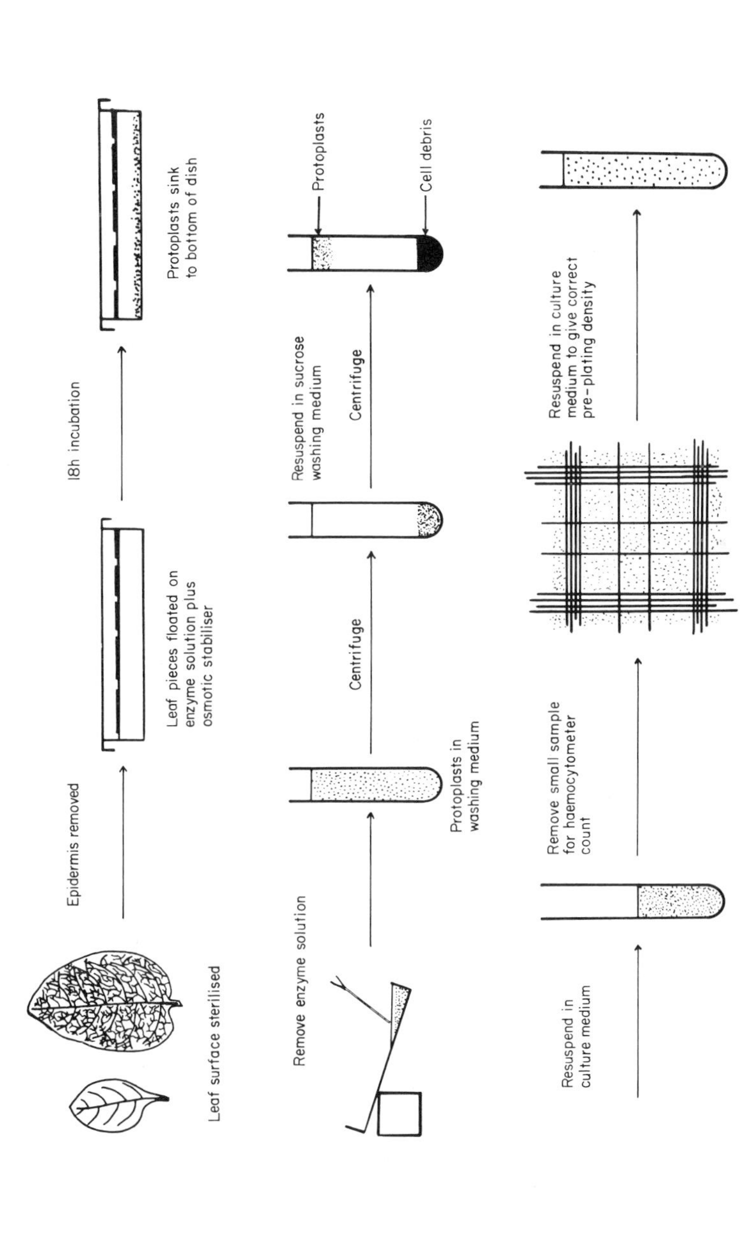

Fig. 16.1. Generally applicable scheme for the isolation, washing and counting of leaf mesophyll protoplasts. Protoplasts isolated from callus or cell suspensions are handled in the same way.

for example, is not readily transported across the plasmalemma and therefore provides a stable osmotic environment for the protoplast.

The enzymes used for the isolation of protoplasts are crude extracts of fungal origin. The pectinases are rich in polygalacturonidase activity whilst the cellulases contain hemicellulase, β-1, 4-glucanase, chitinase, lipase, nucleases and pectinase.

A generally applicable scheme for protoplast production, using the mixed enzyme procedure, is shown in Fig. 16.1. In order for the enzymes to gain access to the plant tissues, as in the case of leaf palisade and mesophyll cells, the lower epidermis must be removed by peeling or partial digestion with cutinase. Certain types of leaves, particularly of the cereals, must be sliced prior to enzyme incubation, since the epidermis cannot readily be removed. Most plant tissues and organs (roots, root nodules, coleoptile, leaf epidermis, petals, germinating pollen grains, fruit placenta, tetrads and microsporocytes) will yield protoplasts after suitable adjustment of the enzyme mixture and plasmolyticum. For example, the pollen tetrad wall consists of callose and so only an enzyme rich in β-1, 3-glucanase (snail digestive juice enzyme) will liberate protoplasts.

Protoplasts are produced from calluses and cell suspensions (Fig. 16.2b) (Wallin & Eriksson, 1973) often only during the log phase in their growth cycles, since the composition of the primary cell wall varies as secondary products, such as lignin, are deposited as the culture matures, rendering it unsusceptible to complete degradation by cellulase. In general, the production of protoplasts from an untried source will always involve a consideration of enzyme purity, pectinase to cellulase ratios, protoplast yield and viability.

Following enzyme incubation, the spherical protoplasts (Fig. 16.2a, 16.2d) must be separated from cellular debris, subprotoplasts and vacuoles. Subprotoplasts are formed during early plasmolysis when the protoplast splits into two or more subunits, some of which will be enucleate and hence non-viable. Separation can readily be achieved in a variety of ways: repeated resuspension and centrifugation of protoplasts in a washing medium; flotation of protoplasts on a hypertonic sucrose solution (Fig. 16.1); passage of the incubation mixture through a nylon sieve of suitable pore size which allows only protoplasts to pass through; or by the use of a two-phase system, such as dextran-PEG (polyethylene

Fig. 16.2. (a) Freshly isolated mesophyll protoplasts (40 μm diam.) in liquid medium (*Petunia hybrida*). (b) Protoplasts (60 μm diam.) released from cultured cells. The centrally positioned nucleus is surrounded by cytoplasmic strands (*Parthenocissus tricuspidata*). (c) Sodium nitrate induced fusion between a mesophyll (m) protoplast, containing chloroplasts, and a colourless (c) protoplast isolated from a cell suspension. Plastids are seen entering the cytoplasm of the colourless protoplast. (d) Electron micrograph of a freshly isolated mesophyll protoplast (n = nucleus; ch = chloroplast; cv = central vacuole; p = plasmalemma). (e) Uptake of *Rhizobium* (r) into vesicle within the cytoplasm of a protoplast (p = plasmalemma). (Electron micrographs provided by Dr. M.R. Davey.)

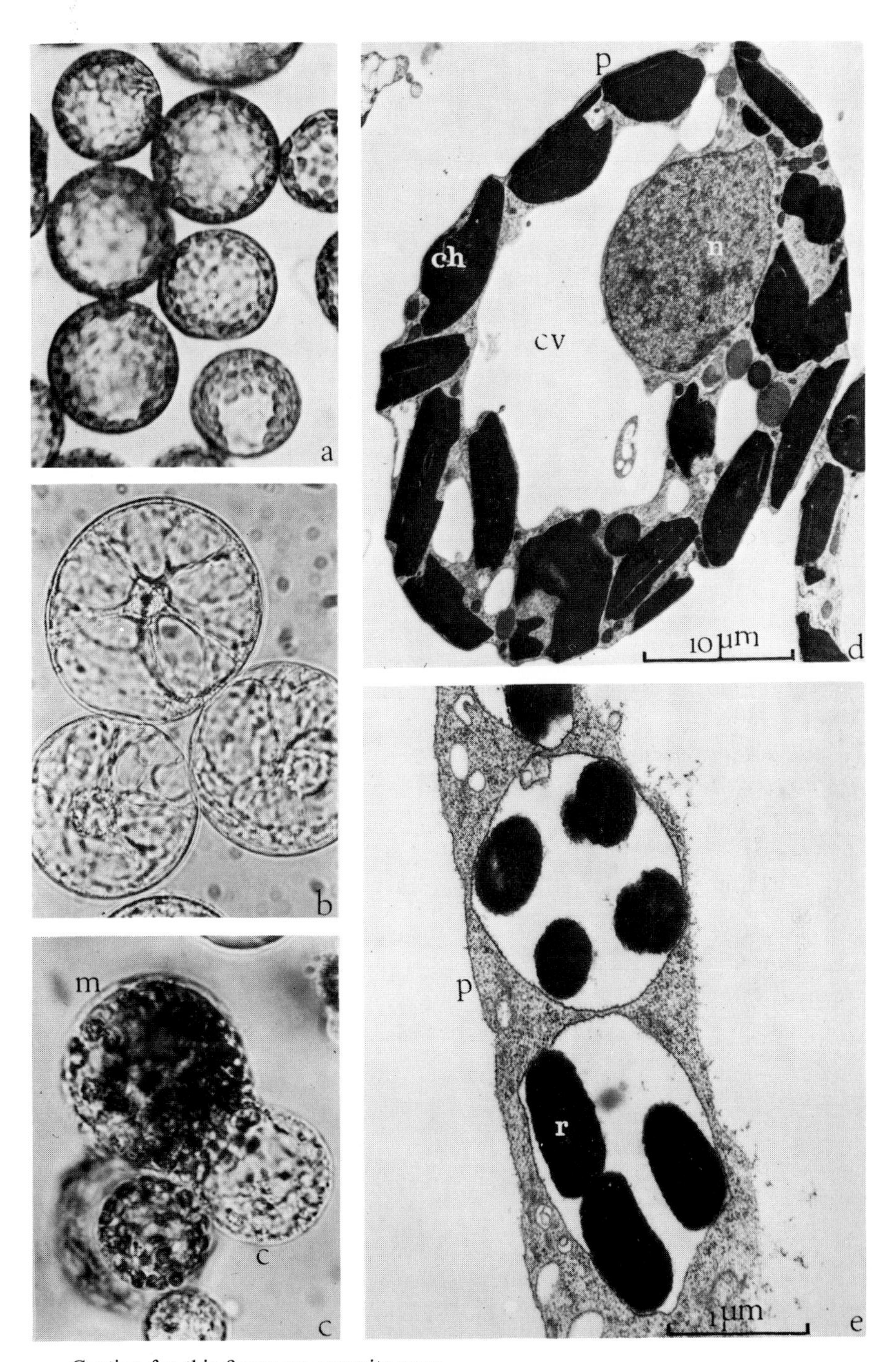

Caption for this figure on opposite page.

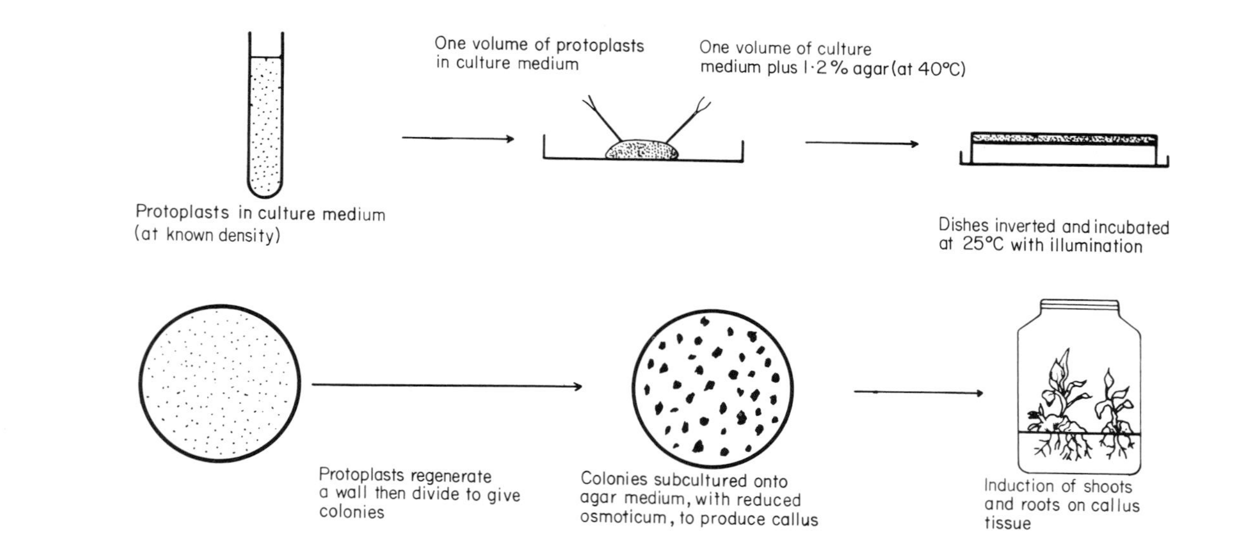

Fig. 16.3. Plating of protoplasts (whole plant or cultured cell origin) in agar solidified nutrient medium. Whole plant regeneration from protoplasts takes approximately five months.

glycol), which is based upon a density difference between protoplasts and cells (Kanai & Edwards, 1973).

Freeze etched protoplasts, collected after washing and examined in the electron microscope or treated with fluorescent brighteners which specifically bind to cellulose, reveal the complete absence of cellulose fibres on the plasmalemma surface.

16.2.2 CELL WALL AND WHOLE PLANT REGENERATION

Protoplasts are cultured in either liquid or agar solidified nutrient media (Fig. 16.3). The culture media are very similar in composition to those required for the *in vitro* culture of cells, but with the addition of an osmotic stabilizer to prevent bursting. During culture, up to 90% of protoplasts isolated from highly differentiated cells undergo a rapid and immediate process of dedifferentiation whilst at the same time initiating the synthesis of a new cell wall.

Early stages of wall synthesis are preceded by extensive infoldings of the plasmalemma together with an accumulation of pectin-like substances in vesicles found in the peripheral layer of cytoplasm. These early stages of wall synthesis, detected after 18 hours, are unaffected by the presence, in the culture medium, of protein synthesis inhibitors, suggesting that synthesis of new RNA or protein is not required for wall initiation and that residual protein and endogenous hormone levels are sufficient. Structurally, the first formed envelope is amorphous and consists of pectins, but after a few days a second inner layer of cellulose fibrils is progressively laid down on the protoplast surface, eventually producing a near normal cellulose matrix after four or five days (Burgess & Fleming, 1974).

Nuclear division and cytokinesis is concomitant with cell wall formation. Occasionally cytokinesis does not occur, perhaps due to the presence of an incomplete cell wall, and gives rise to binucleate cells which may not be capable of further division.

Most protoplasts, like suspension cultured cells, have an optimum plating density as regards their division potential. At this density (usually $5 \times 10^4 - 1 \times 10^5$ protoplasts/ml^{-1}), the plating efficiency, defined as a percentage of the original plated protoplasts that produce cell colonies after 28 days culture, can reach 70%. As division proceeds, the plasmolyticum level in the culture medium is reduced stepwise by the transfer of agar blocks, containing the dividing protoplasts, to the surface of a similar medium of lower osmotic pressure. Individual cell colonies are pricked out, 4–6 weeks after plating, and subcultured onto a regeneration medium. Callus masses thus produced may undergo differentiation, via organogenesis, eventually producing plantlets. Regenerating protoplasts of certain species, such as carrot, will form embryoids in liquid culture.

Whole plant regeneration from protoplasts can be achieved for an expanding list of species of distinct taxonomic groups including the Solanaceae (tobacco, *Petunia*) (Frearson *et al.*, 1973), Brassicaeeae (rape) (Kartha *et al.*, 1974),

Papilionaceae (pea, cowpea) and Scrophulariaceae (*Antirrhinum*). Whole plant regeneration is not restricted to monocotyledons or dicotyledons, haploid or diploid plants, or to protoplasts obtained directly from whole plant tissues. The majority of plants regenerated from protoplasts are normal and exhibit a high degree of fertility, but a small and variable (less than 0·1%) number of plants show morphological abnormalities owing to aneuploidy and polyploidy.

An example of this can be seen following the regeneration of plants from mesophyll protoplasts of diploid *Petunia*, homozygous for blue flower colour. Some regenerated plants have a tetraploid chromosome number, which is also accompanied by a flower colour change from blue to red. Since well established protoplast cultures have high plating efficiencies, and hence efficient cloning potentials, abnormal plants can arise from the expression of cell aberrations present in the tissue prior to protoplast isolation.

The nutritional and hormonal requirements of cultured protoplasts are constantly varying depending upon the stage in the regeneration process. The photosynthetic capacity and respiration rate of the protoplast is suppressed by the plasmolysing conditions. Changes in metabolic requirements during the early stages of growth could be attributed to enzyme uptake during isolation and the rigours of cell wall synthesis. These subtle changes are highlighted in the crown gall (tumour) (Scowcroft *et al.*, 1973), where protoplasts require an exogenous growth regulator supply for the initiation of division, unlike the tumour cells which are autotrophic for auxin. Following cell wall regeneration and division, the requirement for exogenous growth regulator supply is lost, suggesting that growth regulator autonomy is, at least in part, dependent upon the presence of a functional cell wall.

16.3 PROTOPLASTS AND AUXIN RESPONSES

Although protoplasts require both auxins and cytokinins for the initiation and maintenance of division following cell wall formation, the freshly isolated protoplast has facilitated studies of the direct and rapid physiological action of auxins upon the plasmalemma. Auxins such as IAA are known to influence cell elongation in the plant. Leaf or coleoptile protoplasts, incubated in the presence of physiological levels of IAA, respond by an increased permeability of the plasmalemma, probably to water. Protoplasts maintained in an isotonic plasmolyticum containing auxin, undergo a rapid expansion as the internal volume increases owing to extensive vacuolation, which often results in bursting (Hall & Cocking, 1974). The presence of anti-auxins, which preferentially bind to the site of action of the auxin, suppresses the bursting response indicating that these effects are the result of auxin action. As may be expected, protoplasts isolated from tumour cells which have high endogenous levels of auxins, require significantly higher levels of exogenously supplied auxin to elicit the bursting response. Vacuoles produced by disruption of protoplasts show no response to auxins, indicating clear differences in structure and function of the tonoplast. It is difficult to relate these responses to the auxin control of growth in the plant, but the induction of membrane convolutions, seen prior to bursting, can influence the thickness of the newly synthesized wall and hence cell shape and direction of growth.

Herbicides, such as 2,4-D, are used as growth regulators at low concentrations in culture media. There appears therefore to be a balance, for some species, between controlled growth and a herbicidal effect. Protoplasts, when incubated in the presence of herbicides such as paraquat, do not respond by expansion, but collapse as a result of disruption of the plasmalemma. It may be therefore, that control of the selective action of some herbicides is found in the plasmalemma itself.

16.4 UPTAKE PROPERTIES OF PROTOPLASTS

16.4.1 MACROMOLECULES

During isolation, and for several hours after release, protoplasts exhibit extensive pinocytic activity. This process, which may be stimulated by the plasmolysing conditions, enables macromolecules, such as ferritin and polystyrene latex particles, to enter the protoplast. Infoldings of the plasmalemma result in the formation of vesicles which contain small quantities of the surrounding medium together with any particles that may be suspended in the medium and which are less than 1 μm in size. Particles entering protoplasts in this way are often accumulated in the central vacuole. These basic studies have led on to an examination of uptake in relation to cell and ultimately whole plant modification, by the incorporation into protoplasts of biologically active particles. Certain microorganisms can be transformed by the incorporation into the genome of exogenously supplied DNA, which in turn can be expressed, thus modifying the cell. Comparable attempts at transforming eukaryotic cells, involving plant protoplasts, are in progress and it is clear that DNA will enter protoplasts but no evidence for the persistent expression of this input DNA currently exists at the protoplast/cell level.

16.4.2 CHLOROPLASTS, NUCLEI AND BACTERIA

A further extension of the uptake phenomenon has been successfully applied to whole bacteria and plant organelles such as nuclei and chloroplasts (Bonnett & Eriksson, 1974). The precise mechanism of entry of these relatively large particles may be one of membrane fusion, in the case of nuclei and chloroplasts, rather than pinocytosis. Conditions favouring organelle uptake are very similar to those that bring about protoplast fusion. Nitrogen fixing bacteria (*Rhizobium*) can enter pea leaf protoplasts during the isolation procedure (Davey & Cocking, 1972) (Fig. 16.2e) but evidence for nitrogen fixing ability of the bacterium/plant association is still awaited. Likewise transplanted nuclei may survive degradation, and following fusion with the nucleus of the recipient cell, can result in the formation of hybrid or modified cell lines. Organelles entering protoplasts as a result of uptake into vesicles may not be able to function in a strictly symbiotic manner since they are not in direct contact with the cytoplasm. However, these novel approaches to cell modification may have major implications for the plant breeder and could possibly lead to the production of more efficient economically important plant species.

16.4.3 VIRUS PARTICLES

The availability of protoplasts in large quantities has overcome many biological problems facing the investigator of host-virus relationships. In intact plants the process of infection is not synchronous and its establishment is restricted to cells with broken or damaged cell walls (Zaitlin & Beachy, 1974). Protoplasts become infected following a short incubation with intact virus or its RNA and in the presence of a polycation, such as poly-l-ornithine, which not only damages the plasmalemma but neutralizes the negative charge common to many virus particles. This allows adsorption of particles, to the plasmalemma surface, which can then enter the protoplast through damaged regions of the membrane or *via* pinocytic vesicles. The process of infection varies considerably depending upon the virus, but in general, newly synthesized material, possibly RNA, is detected after 2–3 days in vesicles closely associated with the endoplasmic reticulum and nucleus. Newly formed virus particles are also found within the nuclear membrane. However, it is impossible to eliminate influences of the plasmolyticum, enzyme carryover and the absence of a cell wall upon the infection process, which may therefore be dissimilar to that in whole plants. Some protoplasts resist infection with virus, and it may be possible to regenerate plants, as described previously, from such protoplasts, which are not susceptible to virus infection.

16.5 SOMATIC HYBRIDIZATION OF PLANTS

Crop improvement advances with the production of new hybrids, which, in turn, are selected for the expression of desired characteristics. However, in order that plant species can preserve their genotypes there exists major incompatibility barriers. The fusion of plant protoplasts of different species offers one approach that could by-pass the sexual cycle and, provided that nuclear fusion occurs in the resultant heterokaryon, new somatic hybrids may arise following regeneration.

16.5.1 INDUCED FUSION OF PROTOPLASTS

Several methods now exist for the fusion of protoplasts and all embrace the same basic concepts. It is first necessary to bring the protoplasts together so that the two plasmalemma surfaces are in contact, whilst at the same time inducing destabilized regions on the membrane. Where two such regions of adjacent protoplasts coincide, localized membrane fusion occurs, together with limited cytoplasm mixing, to produce cytoplasmic bridges. Further mixing of the cytoplasms (Fig. 16.2c) eventually results in the complete coalescence of the protoplasts which are then returned to stabilizing conditions after removal of the fusion inducing agent.

A slightly hypotonic aqueous sodium nitrate solution reduces the negative charge on the outer membrane thus allowing adhesion of protoplasts. The presence of the Na^+ ion induces localized realignment of the lipo-protein in the membrane which in turn allows fusion to occur when two such regions come into contact. Slightly deplasmolysing conditions facilitate cytoplasmic coalescence (Power & Cocking, 1971).

PEG solutions induce tight adhesion of protoplasts (Kao *et al.*, 1974) owing to electrotsatic forces. The presence of cations, such as Ca^{2+}, in the PEG solution, enhances adhesion since the calcium ion may form a molecular bridge between the negatively polarized PEG molecule and the protein of the membrane. Water is also withdrawn from the protoplast by PEG, thus reducing the turgidity of the system and allowing closer packing of protoplasts. This dramatic disruption, and possible reversal, of the charge properties of the membrane after removal of PEG, leads to membrane fusion.

Incubation of protoplasts in solutions buffered at high pH (Keller & Melchers, 1973) and containing calcium ions also induces fusion. This may resemble the action of PEG in reversing membrane charge properties in conjunction with the establishment of molecular bridges prior to membrane fusion.

Protoplasts isolated from pollen mother cells will fuse in the absence of an inducer. In such cases the membrane structure may be quite different to the plasmalemma of somatic cells since the products of pollen mother cell development, germinating pollen grains, are ultimately involved in the fertilization process and hence membrane fusion.

16.5.2 CULTURE OF FUSION PRODUCTS

Following fusion treatment, protoplasts are returned to the culture medium and cell wall formation proceeds. It is during the first mitotic division that nuclear fusion can occur as a result of common spindle formation, giving rise to the hybrid cell. Abortion of the nuclei, unidirectional chromosome elimination, or the inability to produce a nuclear hybrid may be expected between species that are not closely related genetically. Following regeneration into whole plants, the resulting cytoplasmic hybrids ('cybrids') may be of immense value to the plant breeder since many plastid characteristics and resistance to certain diseases are controlled by plasmagenes.

The frequency of true hybrid cell formation is low (1 in 10^5). This fact has led to the development of generally applicable cultural procedures aimed at preferentially selecting out hybrid cells. These techniques, currently receiving most attention by protoplast workers, are based upon an ability of the nuclei to complement in the hybrid cell. This involves the use of parental species that exhibit mutually exclusive biochemical features such as differential drug sensitivity (Cocking *et al.*, 1974), light sensitivity or auxotrophic requirements.

Following selection, somatic hybrid plants have been produced between two *Nicotiana* species. In this particular case, the selection theory was based upon prior knowledge of the cultural requirements of the sexually produced hybrid (Carlson *et al.*, 1972). Protoplasts of the parents, *N. glauca*, *N. langsdorffii*, did not grow in a culture medium that supported the growth of leaf protoplasts of the sexual hybrid. Following fusion with sodium nitrate, the protoplasts were plated in this medium and a few colonies were recovered which later developed into plants. Electrophoretic and karyotypic analysis confirmed that these plants were amphidiploid somatic hybrids.

In spite of this unambiguous proof that somatic hybridization is possible, it will only be when somatic hybridization is demonstrated for plant species which are truly sexually incompatible, that its potential for crop improvement will be fully realized.

FURTHER READING

BAJAJ Y.P.S. (1974) Potentials of protoplast culture work in agriculture. *Euphytica* **23**, 633–49.

COCKING E.C. (1972) Plant cell protoplasts—isolation and development. *Ann. Rev. Plant Physiol.* **23**, 29–50.

COCKING E.C. (1974) The isolation of plant protoplasts. *Methods in Enzymology* (eds S. Fleischer & L. Packer), Vol. XXXI part A pp. 578–83.

HEYN R.F., RÖRSCH A. & SCHILPEROORT R.A. (1974) Prospects in genetic engineering of plants. *Quart. Revs. Biophys.* **7**, 35–73.

KRUSE P.F. & PATTERSON M.K. (Eds.) (1973) *Tissue Culture—Methods and Applications*. Academic Press, London.

NICKELL L.G. & HEINZ D.J. (1974) Potential of cell and tissue culture techniques as aids in economic plant improvement. *Genes, Enzymes and Populations* (eds. A.M. Srb), pp. 109–28. Plenum Publishing Corp., New York.

SMITH H.H. (1974) Model systems for somatic cell plant genetics. *Bioscience* **24** (5), 269–76.

STREET H.E. (Ed.) (1977) *Plant Tissue and Cell Culture*, second edition. Botanical Monographs Vol. 11. Blackwell Scientific Publications, Oxford.

TEMPÉ J. (Ed.) (1973) Protoplastes et fusion de cellules somatiques végétales. *Colloques Int. C.N.R.S.*, No. 212. Paris: Editions I.N.R.A.

CHAPTER 17

GENETIC VARIATION IN CULTURED PLANT CELLS

17.1 INTRODUCTION

Genetic variability can be described as the quantity of alleles which exist among the total genetic loci of a population. In many crop plants genetic variability is rapidly diminishing (Day, 1972) as a result of increasingly specialized breeding, intensive agriculture, and wide dissemination of the most desirable cultivars. Serious consequences are uniform susceptibility to new pathogens, inability to adapt to new environments as well as the inadvertent spread of 'silent' deleterious genes. The southern corn blight, which destroyed one billion dollars worth of the U.S. corn crop in 1970, occurred in large part because 70% of the cultivated corn contained the same cytoplasmic determinant for male sterility. A pleiotropic effect of this gene is blight susceptibility.

While efforts are underway to preserve existing variability (e.g. in germ plasm collections) other approaches such as mutation breeding seek to increase variability. To this end plant cell culture is potentially well-suited. The plant cell geneticist can easily select certain mutants from among 10^4–10^5 cells per ml of suspension culture. Each cell is a potential plant which, if field tested by conventional methods, would require huge investments of time, labour and acreage. The individual cell has limited metabolic pools and can be cultured on a completely defined and relatively simple medium (see chapter 15). The plant cell geneticist thus can establish a wide range of conditions to select for specific biochemical mutants which are perhaps not so easily selected in the field. The ability to select defined genetic alterations will provide a means for the dissection of biochemical and developmental pathways and for the analysis of mechanisms of genetic expression in higher plants. This knowledge will lead to efficient selection *in vitro* of mutations which result in improved nitrate utilization, higher heavy metal tolerance, over production of a specific amino acid or seed protein, and other valuable traits (Chaleff & Carlson, 1974). Theoretically, the use of tissue cultures from anther-derived haploid plants (see chapter 15) would allow the direct selection of recessive mutations. Finally, whole plant regeneration from callus is becoming possible in an ever-increasing number of species. This is essential if the microbial geneticist's approach in plant cell culture is to complement conventional plant breeding.

The following section discusses some mutants already obtained in culture and some advances in plant cell genetics that are important for producing additional agriculturally useful variation.

17.2 INDUCED VARIATION AND SELECTION

The spectrum of mutants recovered among cultured plant cells has so far been determined primarily by the growth characteristics and requirements of these cells *in vitro*. The strict prerequisite of a high minimum density for proliferation of cells (Street, 1973) or protoplasts (Nagata & Takebe, 1971) *in vitro* makes easier the selection for mutants resistant to antibiotics and antimetabolites than the selection for auxotrophs. The isolation of auxotrophs requires an experimental system composed of haploid cells in which the wild type cells are at a disadvantage compared to the mutants. This approach was used by Carlson (1970) to isolate six auxotrophic mutants of *Nicotiana tabacum*. The selective system was adapted from a procedure used successfully with mammalian cells (Puck & Kao, 1967). In a minimal medium containing 5-bromodeoxyuridine (BUdR), actively dividing prototrophic cells will incorporate more of this light-sensitive nucleoside analogue into DNA than will arrested auxotrophs. The auxotrophs will survive subsequent illumination at a much higher frequency than the prototrophs and will form clones on a supplemented medium. The mutant tobacco calluses which were isolated grew slowly on a minimal medium. Plants were differentiated from four of the mutant clones and these also grew slowly without supplementation. In genetic analyses of these four regenerated plants, three plants transmitted the mutant phenotype as a single recessive Mendelian factor and one displayed a more complex pattern of inheritance. The recovery of only leaky mutants may be due to a lack of functional diploidization of the *N. tabacum* genome. Haploid cells obtained from this amphiploid species may contain two copies of many metabolically essential genes and thus two mutational events may be required to delete completely certain functions (Carlson, 1970).

Cell lines which exhibit increased resistance to a wide variety of antimetabolites have been isolated and studied. Haploid cultures of petunia (Binding *et al.*, 1970; Binding, 1972) and tobacco (Maliga *et al.*, 1973) resistant to streptomycin have been selected. In genetic crosses involving a diploid plant obtained from one mutant tobacco clone, drug resistance segregated as a maternally inherited character (Maliga *et al.*, 1973; Sz.-Breznovits *et al.*, 1974).

Plant cell lines containing dominant or semidominant genetic markers may prove useful in selecting hybrid cells formed by protoplast fusion. This realization has motivated the isolation of cell lines resistant to 8-azaguanine and BUdR since such mutants are employed in selecting hybrids in mammalian cell systems. Resistance to these analogue precursors of nucleic acids in cultured mammalian cells may result from deficiencies in hypoxanthine:guanosine phosphoribosyl transferase (HGPRT) and thymidine kinase activities, respectively, or in the uptake of these analogues (Kit *et al.*, 1963; Littlefield, 1964a; Breslow & Goldsby, 1969). In the presence of aminopterin, endogenous biosynthesis of thymidine and hypoxanthine is inhibited and cultured mammalian cells are dependent upon an exogenous supply of these substances (Fig. 17.1). As each

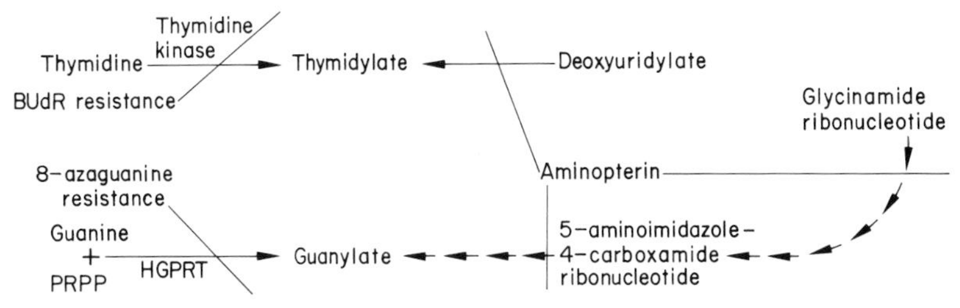

Fig. 17.1. Simplified scheme showing pathways for synthesis of guanylate and thymidylate. Endogenous biosynthetic steps blocked by aminopterin-inhibition of tetrahydrofolate synthesis are shown as well as the genetic blocks which confer resistance to BUdR and 8-azaguanine.

mutant cell line is unable to utilize one of these compounds, only complementing hybrids formed by cell fusion survive in a medium containing hypoxanthine, aminopterin, and thymidine (the HAT system) (Littlefield, 1964b). Although a low concentration of aminopterin inhibits growth of sycamore cells (Bright & Northcote, 1974) it should not necessarily be assumed that this effect is caused by the same mechanism in both animal and plant systems. The isolation of cell lines of tobacco (Lescure, 1973) and sycamore (Bright & Northcote, 1975) resistant to 8-azaguanine has been reported. The resistance phenotype is not completely stable in the sycamore cell line in which HGPRT activity is half that of the wild type level. The latter result is not unexpected in a diploid system and emphasizes the necessity of using haploid cell lines in order to recover mutants with an anticipated recessive phenotype.

Resistance to BUdR has been selected in haploid tobacco (Maliga *et al.*, 1973) and in presumably diploid sycamore (Bright & Northcote, 1974) and soybean (Ohyama, 1974) cells. The resistant sycamore and soybean cultures did not differ significantly from the wild type in uptake of thymidine or in the level of thymidine kinase activity (Bright & Northcote, 1974; Ohyama, 1974) and the basis for the resistance phenotype remains unknown. Marton and Maliga (1974) have observed transmission of BUdR resistance to progeny of plants derived from one mutant tobacco callus and unimpaired uptake and incorporation of thymidine in four resistant cell lines.

Heimer and Filner (1970) have isolated a cell line of tobacco which is insensitive to inhibition by threonine when grown in a medium containing nitrate as sole nitrogen source. By studying nitrate accumulation in this mutant under conditions in which nitrate reductase is chemically inactivated and in which nitrate uptake by wild type cells is repressed, it was concluded that regulation of nitrate uptake is altered in the resistant cell line. Such mutants should prove extremely useful in the investigation of nitrogen metabolism.

The potential agronomic importance of tissue culture techniques is illustrated by experiments in which screening of cultured somatic cells resulted in the recovery of mutants of *N. tabacum* resistant to wildfire disease (Carlson,

1973a). This disease of tobacco is caused by a bacterial pathogen, *Pseudomonas tabaci*, which produces a toxin structurally related to methionine (Stewart, 1971; Fig. 17.2). Populations of mutagenized haploid cells were plated in a medium containing an inhibitory concentration of methionine sulphoximine, an analogue of the wildfire toxin (Braun, 1955) and resistant clones were selected. Three homozygous diploid plants which were regenerated from these methionine sulphoximine-resistant calluses are less susceptible than the parent plant to the pathogenic effects of bacterial infection. One mutant plant contains wild type intracellular levels of free methionine and resistance appears to be genetically complex. In the other two resistant plants the endogenous concentration of free methionine is five times higher than in the wild type while the free pool sizes of other amino acids remain unchanged (Carlson, 1973a). The levels of free methionine in the heterozygous plants are intermediate between those in the wild type and in the homozygous mutant (Chaleff & Carlson, 1975a).

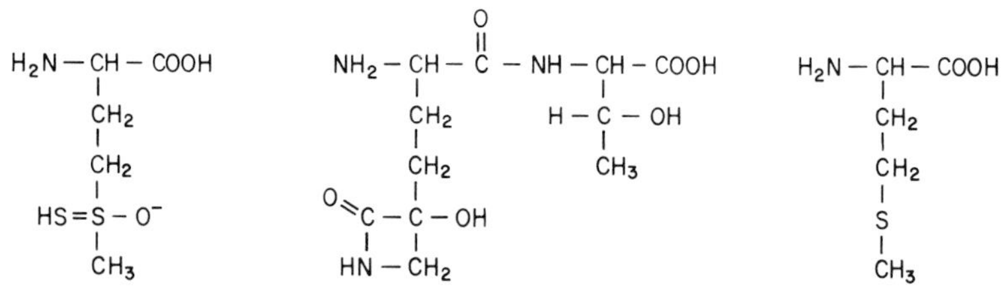

```
                                             O
                                             ‖
H2N — CH — COOH            NH2 — CH — C — NH — CH — COOH       H2N — CH — COOH
      |                           |              |                    |
      CH2                         CH2      H — C — OH                 CH2
      |                           |              |                    |
      CH2                         CH2            CH3                  CH2
      |                      O    |                                   |
 HS = S — O⁻                  ⪢C — C — OH                             S
      |                        |    |                                 |
      CH3                     HN — CH2                                CH3

 Methionine                 Wildfire toxin                        Methionine
 sulphoximine               (tabtoxin)
```

Fig. 17.2. Structures of methionine sulphoximine, wildfire toxin (Stewart, 1971), and methionine.

Intermediate methionine levels are consistent with the observed transmission of the resistance phenotype of these two plants as a single semidominant locus (Carlson, 1973a). These experiments suggest that selection for toxin resistance *in vitro* may provide a generalized procedure for generating disease resistant varieties. It is also evident that regulation of amino acid metabolism may be altered in cells selected for resistance to an amino acid analogue and that this mutational event may be expressed stably in mature plants derived from the mutant callus tissue. Additional support for this conclusion is provided by other research.

Widholm (1972a,b) has isolated presumably diploid cell lines of *N. tabacum* and *Daucus carota* which are capable of growth in the presence of a normally inhibitory concentration of 5-methyltryptophan. Crude extracts of the resistant cell lines (5-mt^r) contain a species of anthranilate synthetase which is less sensitive to feedback inhibition by tryptophan and 5-methyltryptophan than is the wild type enzyme. Endogenous levels of free tryptophan in 5-mt^r cell lines of tobacco and carrot are 15 times and 27 times higher, respectively, than the wild type levels. Plants could not be regenerated from the tobacco and carrot cells which synthesized a feedback-insensitive form of anthranilate synthetase. However, plants were obtained from analogue-resistant carrot cultures in which anthranilate synthetase activity was indistinguishable from the wild type and uptake of 5-methyltryptophan and tryptophan was reduced. Although no genetic studies with these plants have been reported, cell lines re-initiated from the plants maintained the same degree of resistance to 5-methyltryptophan as the original selected lines. Cell cultures of tobacco and carrot resistant to *p*-fluorophenylalanine were isolated also. Resistance to this analogue apparently resulted from diminished rates of transport and incorporation (Widholm, 1974).

Two attempts have been made to employ analogue resistance as a selective screen in obtaining callus cultures of cereals which overproduce amino acids essential to human nutrition. Chaleff and Carlson (1975b) recovered three cell lines of rice which were stably resistant to the lysine analogue S-2-aminoethylcysteine. Analyses of both free pool and total amino acid compositions showed that the resistant cultures contained elevated levels of several amino acids, including lysine, methionine, leucine and isoleucine. Cheng (1975) has selected cultures of *Hordeum vulgare* which are resistant to the same analogue and in which lysine levels were increased, as was the rate of incorporation of exogenously supplied lysine. Unfortunately, attempts to differentiate plantlets from the variant calluses were unsuccessful in both cases. Although much valuable information is to be obtained from research with variant cell lines, such studies are severely limited by the inability to examine genetic expression and transmission in the mature plant. Future investigations should attempt to utilize experimental systems in which plants may be differentiated from the altered cell lines.

17.3 INTRODUCED VARIATION—GENE TRANSFER

17.3.1 INTRODUCTION

The systems so far described have dealt with mutationally-induced variation of the pre-existing plant genome. Considering such factors as genome size, multiple gene copies (e.g. non-allelic isozymes), and the tendency of plants to become aneuploid and polyploid it should be apparent that the potential for induced variation in plant cells is enormous. As culturing and selective techniques improve and a better understanding of plant biochemistry is attained (largely *via* tissue culture) the spectrum of plant cell mutants recovered should widen.

In addition, manipulation of plant cells offers unique opportunities to introduce foreign genetic information not only from other plants but also from bacteria and bacterial viruses and other organisms. In nature, non-sexual gene transfer between plants by way of DNA transformation or viral-mediated transduction has not been demonstrated. Neither has gene transfer between bacteria and plants been demonstrated in nature although legume-*Rhizobium* N_2 fixing symbioses and bacterial-induced crown gall tumours may disprove this contention when more is learned about these systems. However, during the last 5 years several experimental demonstrations of gene transfer in cultured plant cells have been at least partially successful, thus holding the promise that introduced as well as induced variation may become a tool in plant breeding.

Doy, Rolfe and Gresshoff (1973a) used the term *transgenosis* to describe the asexual introduction of foreign genes into plant cells. They emphasize that such genes could be introduced using isolated nucleic acid or an intact virus, organelle, cell or protoplast. While they include whole plants and both animal and plant cells as potential recipients the treatment here is restricted to the transgenosis of cultured plant cells and protoplasts. The use of protoplasts as simultaneous donors and recipients (protoplast fusion) has been discussed in the previous chapter.

17.3.2 ISOLATED DNA-MEDIATED TRANSGENOSIS

Theoretically, isolated plant DNA is a logical choice as donor in transgenosis of plant cells. It has a better probability than bacterial or viral DNA of exhibiting homology with specific stretches of the host chromosome thus maximizing pairing and integration. The transcriptional and translational machinery of the host is probably better suited for plant DNA than for bacterial or phage DNA. However, the following transgenosis experiments point out the many technical difficulties in using isolated DNA in general and plant DNA in particular. Both plant and bacterial DNA are highly susceptible to plant cell nucleases. A plant DNA preparation in addition is much more dilute for a given gene than a bacterial preparation. Further, transgenosis by isolated plant DNA is currently hampered by a lack of suitable, biochemically characterized selective markers. Early experiments demonstrated a poor efficiency of DNA uptake by cultured cells and protoplasts. Bendich and Filner (1971) found that only 0·5% of isotopically labelled tobacco DNA or heterologous *Pseudomonas aeruginosa* bacterial DNA was taken up into the nuclear fraction of cultured tobacco cells. The rest was rapidly degraded in spite of extensive washing and pronase treatment to reduce extra-cellular DNase activity. No evidence was obtained for the integration of this DNA into the recipient genome.

Ohyama *et al.* (1973) suggested that protoplasts, which lack cell walls, may be better recipients of isolated DNA. In their experiments approximately 1% of exogenous *E. coli* DNA was found in an acid precipitable fraction of protoplasts of soybean, carrot and *Ammi visnaga*. By treating tobacco cells with cationic DEAE dextran, Heyn and Schilperoort (1973) increased 'uptake' of acid precipitable *Agrobacterium tumefasciens* DNA to 10–15% of input. Possibly DEAE Dextran and anionic DNA form a DNase resistant complex. However, only 0·145% of this DNA penetrated the cell; the rest was adsorbed to the cell wall.

Studies involving the selection of protoplasts and cultured cells which express a donor DNA marker have so far been few and only marginally successful. Holl *et al.* (1974) recently described two experimental systems. One involves the acquisition of mannitol utilization by soybean protoplasts from the bacterium *Azotobacter vinelandii*. No clear-cut DNA-mediated transgenote has yet been reported. A serious difficulty in this experimental system is the impermeability of soybean protoplasts to mannitol; thus, transgenotes would have had to acquire at least two bacterial genes—a mannitol permease as well as the gene responsible for the mannitol-fructose conversion. Such double transgenotes would probably be exceedingly rare; moreover it is difficult to envisage a bacterial permease functioning in a plant cell membrane.

In the second system described by Holl and collaborators, seeds of a mutant pea strain unable to form root nodules with *Rhizobium* bacteria were treated with DNA extracted from a strain which forms functional nodules. They observed that 1–2% of those plants grown from treated seeds were corrected for nitrogen-fixation. An analysis of this result should consider the mutagenic effects of DNA and the observation that all corrected seeds simultaneously recovered two lost functions which are encoded by independently segregating genes *viz.* nodulation and N_2-fixation. A serious restriction of this system is the limited number of seeds which can be treated. Alternatively, the authors have proposed treating cultured mutant shoot tips with DNA from the symbiosis-competent line. Transgenotes can be selected by whole plant regeneration from shoot tips on an N-free medium inoculated with *Rhizobium*. However, the cultured shoot tip consists of a heterogeneous population of partially differentiated diploid cells and it is not known if only certain types or minimal numbers of these cells must be transgenosized before the regenerated plant will fix nitrogen. As yet no transgenotes have been reported.

Ferrari, working in Widholm's laboratory (1975), has used carrot cell lines resistant ($5mt^r$) and sensitive ($5mt^s$) to 5-methyl tryptophan to study DNA transgenosis. The transfer of genetic information was monitored by observing the frequency of $5mt^r$ clones arising among a population of $5mt^s$ cells which had been treated with DNA extracted from $5mt^r$ cells. Although an apparently mutagenic effect of both $5mt^r$ and $5mt^s$ DNA contributed to altering the phenotype of the recipient cell, intact $5mt^r$ DNA was 2 to 8 times more effective than $5mt^s$ DNA in conferring resistance to 5-methyl tryptophan. However, the observation that these differences occurred in only one quarter of the experiments illustrates the difficulties of this system. An understanding of the means by which input $5mt^r$ DNA effects 5-methyl tryptophan resistance awaits a biochemical comparison between donor and acquired gene products.

Obviously refinements are needed in future DNA transgenosis systems. These would include the transfer of single gene, biochemically characterized, markers to recipient cells incapable of mutational escape from selective conditions, the means of neutralizing recipient DNAases or otherwise improving recipient cell competence, and enrichment of donor (especially plant) DNA for specific genes (e.g. by fractionation, use of organelle DNA, molecular cloning or reverse transcriptase products of purified plant messengers). Further, if isolated protoplasts are best suited for DNA uptake, callus and whole plant regeneration must be achieved from protoplasts of the many species where this is not yet possible.

17.3.3 VIRAL-MEDIATED TRANSGENOSIS

In viral-mediated transgenosis foreign genetic information is transferred to and expressed by plant cells following exposure to intact viral particles. The experiments described here deal exclusively with bacteriophage infection although the productive infection of plant protoplasts with plant viruses is, in a strict sense, also viral mediated transgenosis. In a series of experiments Doy, Gresshoff and Rolfe (1973b) took advantage of the inability of tomato callus to survive with either lactose or galactose as the sole carbon source. In one experiment, high titre preparations of the specialized transducing bacteriophages $\phi 80$ and λ were applied to tomato callus. The former ($\phi 80$ *plac*$^+$) carried the bacterial gene coding for β-galactosidase, the enzyme which catalyses hydrolysis of lactose to glucose and galactose. The latter (λ *pgal*$^+$) carried the bacterial structural genes encoding the enzymes which catalyse the conversion of galactose to utilizable glucose 1–P. Eighty per cent of the calluses treated simultaneously with both phages or with $\phi 80$ *plac*$^+$ alone were able to survive and grow slowly on lactose medium as compared with 5–10% of untreated controls.

In another experiment galactose conversion was shown to be dependent on intact bacterial *gal* genes since bacteriophage with no *gal* genes ($\phi 80$) or with a mutated gene for galactose utilization (λ *pgal*$^-$) did not elicit growth with galactose as sole carbon source in the inoculated calluses.

More convincing molecular proof of bacterial gene expression by the treated plant cells was the immunological identification of the β-galactosidase activity in the lactose utilizing callus. Antiserum specific for the *E. coli* enzyme afforded the same degree of protection against thermal inactivation of the β-galactosidase activity in extracts of both *E. coli* and the lactose utilizing callus (Table 17.1). Doy, Gresshoff and Rolfe (1973a,b) have emphasized that the definition of transgenosis implies neither a specific mechanism of viral transduction of bacterial genes nor the integration and stable transmission of the acquired genes.

Carlson (1973b) demonstrated the appearance of two early phage enzymes upon 'infection' of barley protoplasts with the lytic bacteriophage T3. The phage exposed protoplasts synthesized S-adenosyl methionine cleaving enzyme (SAMase) and RNA polymerase, two T3-encoded enzymes not normally produced by the plant. The latter enzyme was definitely assigned a bacteriophage origin since exposure to T3 containing an amber (nonsense) mutation in the RNA polymerase structural gene resulted in significant protoplast synthesis of SAMase activity only. It is interesting that the phage-borne bacterial *gal* mutation introduced into tomato callus (Doy *et al.*, 1973b) was also amber, providing

Table 17.1. β-Galactosidase activity from crude extracts of haploid tomato callus and its protection by antiserum specific for *E. coli* β-galactosidase (Doy *et al.*, 1973b.)

Callus type	Carbon source	Growth	β-galactosidase specific activity	% activity remaining after 20 min. at 60°C in presence of rabbit anti-serum specific for *E. coli* β-galactosidase
non treated	sucrose	rapid	158	5%
non treated	lactose	none	264	5%
non treated	lactose	slow (rare)	144	5%
treated with ϕ 80 *plac* and λ *p gal*	lactose	slow	6850	50–60%
non treated plus *E. coli* extract containing β-galactosidase	sucrose	fast	†	50–60%

†Adjusted to levels comparable to transgenosized cells.

circumstantial evidence that plants as well as bacteria may recognize the amber codon as a stop signal. Doy *et al.* (1973b) demonstrated that both tomato and *Arabidopsis thaliana* calluses die quite rapidly upon exposure to ϕ80 phage carrying a bacterial amber suppressor. It is, however, difficult to distinguish whether the bacterial suppressor tRNA interferes with termination of translation or simply competes with plant tRNA at sense codons.

It has been demonstrated (Johnson *et al.*, 1973; Grierson *et al.*, 1975) that sycamore cells in suspension can acquire the ability to utilize lactose from λ phage carrying the *lac* genes. Ordinary λ does not confer lactose utilizing ability. However, biochemical evidence for the presence of bacterial enzyme in cells growing on lactose has not yet been obtained. The lac^+ sycamore cell lines as well as the lactose-utilizing tomato callus lost their new function in succeeding generations (Doy, Gresshoff & Rolfe, 1973b; Grierson *et al.*, 1975). Polacco *et al.* (1974) demonstrated the expression of λ borne *lac* genes in tobacco diploid callus both by growth and biochemical criteria. However, these transgenosized lines uniformly lost all *lac* functions after several months. These results demonstrate the transient nature of bacterial and viral gene expression in plant cells.

We may tentatively summarize the advantages and drawbacks of viral-mediated transgenosis. The genome size of a transducing phage is small ($\lambda = 31 \times 10^6$ daltons) so that the desired bacterial gene is enriched about 80 fold compared with its concentration in the *E. coli* genome*. Therefore, its chances of uptake and expression are improved compared with isolated *E. coli* DNA transgenosis. Further, the phage coat may serve to protect the foreign DNA from nuclease attack and even provide some transport function. On the negative side the bacterial genes introduced probably lack homology with the plant genome thereby minimizing the probability of chromosomal integration and stable transmission. Of course it cannot be assumed that the transcription and translation of the bacterial genes by the plant host are efficient. Presumably, many phage specific genes are expressed in the recipient plant cell and this may be related to the generally deleterious effect of phage treatment on callus growth. The systems employed were far from ideal in that T3 genes conferred no selective advantage on barley protoplasts while calluses of tomato (Doy *et al.*, 1973b; Table 17.1) and tobacco (Polacco *et al.*, 1974; Fig. 17.3) and sycamore cell suspensions (Grierson *et al.*, 1975) have been shown to have plant-specific β-galactosidase activity. As a result the selective pressures in the λ *plac* system were probably not great; Doy *et al.* (1973b) reported that 5–10% of control calluses grew on lactose. The presence of a homologous plant enzyme also complicates the identification of the introduced gene.

*The *E. Coli* genome is $2{\cdot}5 \times 10^9$ daltons so that the β-galactosidase gene is

$$\frac{2{\cdot}5}{31} \times \frac{10^9}{10^6}$$

or approximately 80 fold more concentrated in the λ*plac* genome. A single gene is diluted even further in the still larger plant genome.

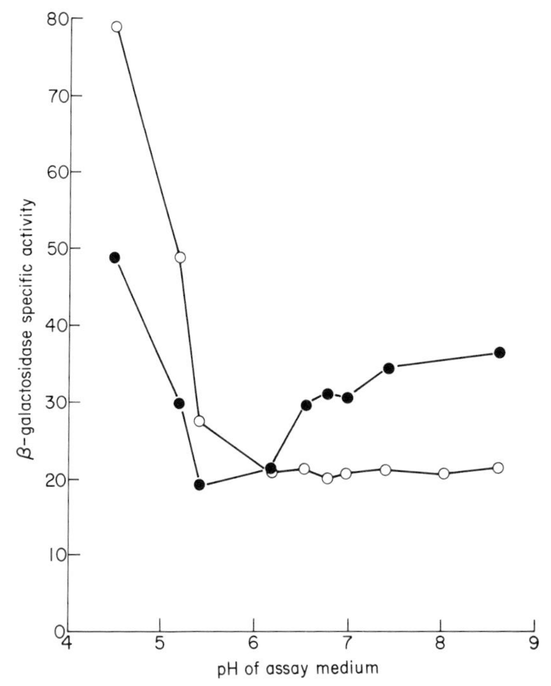

Fig. 17.3. The presence of plant specific β-galactosidase activity in extracts of tobacco callus grown on sucrose (○-○-○) and in extracts of callus exposed to lactose (●-●-●) for two weeks as sole carbon source. (From Polacco *et al.*, 1974.)

Further experiments should utilize biochemically characterized markers which confer a more rigorous selective advantage. Whole plant regeneration should be attempted under selective conditions and modes of gametic transmission of the desired trait studied. Ideally, to improve homology (and therefore integration), plant rather than bacterial genes should be introduced. In viral-mediated transgenosis the problem is to incorporate specific plant DNA stretches in a phage coat. A potential system would be the incorporation of chloroplast DNA in λ phage by techniques recently described (Thomas *et al.*, 1974); such phage could possibly be used to transgenosize albino callus to green.

By involved biochemical and genetic manipulation Thomas *et al.* prepared λ DNA (λ*gt*) with a 27% deletion (Fig. 17.4). It has lost the ability to lysogenize *E. coli* but can still replicate, produce coat protein and lyse its host. However, because the λ*gt* DNA is smaller than normal it cannot associate with the coat protein in cell lysates to form an infective particle. The λ*gt* DNA, therefore, cannot form plaques in transfection experiments, unless DNA from apparently *any* source, is biochemically inserted into the genome. Such DNA is replicated along with λ DNA. Cloning such DNA-rescued phage is equivalent, then, to cloning a small stretch of the foreign DNA incorporated in the λ genome. The legend to Fig. 17.4 provides more experimental detail.

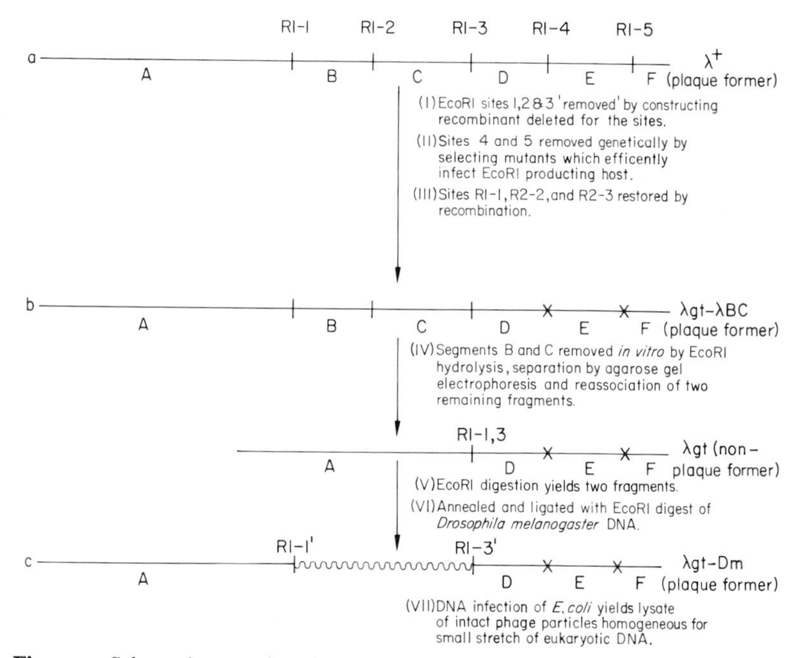

Fig. 17.4. Schematic procedure for incorporating small stretches of DNA from any source (e.g. plants) in phage λ (Thomas *et al.* 1974.) The procedure takes advantage of the properties of the *restriction endonuclease* enzyme *EcoRI*. This enzyme recognizes very specific nucleotide sequences (the six in λ are labelled R1-1, R1-2, etc.) and cleaves them to yield cohesive ends (short single stranded homologous regions) which can be annealed and rejoined *in vitro*. *EcoR1* is produced by an *E. coli* R factor; such an *E. coli* is a *restrictive* host for λ with *unmodified* RI sites. This is the basis for the selection of mutation in sites R1-4 and R1-5 in step (ii) above.

17.3.4 CONCLUSION

This chapter has reviewed recent advances in the development of techniques for modifying the genetic information of cultured cells of higher plants. We have not considered nuclear and organelle uptake by protoplasts or protoplast fusion as means of introducing genetic variation (Chaleff & Carlson, 1974). The selective systems would be similar to at least some of those already described and the techniques involved are discussed in chapter 16. It is hoped that all these experimental approaches will prove useful in the elucidation of the biological processes of higher plants. These procedures also may contribute to the development of improved crop varieties. However, successful realization of these aims will require a fuller development of the systems and methods which have been described. The differentiation and chromosome composition of cultured cells must be better controlled and the minimum cell density required for growth must be reduced. In addition, these *in vitro* techniques must be extended to a greater number of crop species.

We should not close this chapter without mention of the potential biohazards of gene transfer experiments. The experimenter should consider potential dangers to agriculture, the ecosystem and human beings. We shall draw one possible scenario: a researcher succeeds in recovering intact infective bacteriophage from viral transgenosized callus. Subsequently the virus is found able to infect whole plants and cause severe disease symptoms and necrosis—the latter due possibly to a bacterial toxin gene or a plant cellulase gene, etc. The authors believe, however, that with proper precautions such experiments have the future potential of yielding new and valuable agricultural varieties.

FURTHER READING

CARLSON P.S. (1970) Induction and isolation of auxotrophic mutants in somatic cell cultures of *Nicotiana tabacum*. *Science* **168**, 487–89.

CARLSON P.S. (1973) Methionine sulphoximine-resistant mutants of tobacco. *Science* **180**, 1366–68.

CHALEFF R.S. & CARLSON P.S. (1974) Somatic cell genetics of higher plants. *Ann. Rev. Gen.* **8**, 267–78.

DOY C.H., GRESSHOFF P.M. & ROLFE B.G. (1973a) In: *The Biochemistry of Gene Expression in Higher Organisms* (eds J. Pollak & J.W. Lee) pp. 21–37. Aust. and New Zealand Book Artarmon.

DOY C.H., GRESSHOFF P.M. & ROLFE B.G. (1973b) Biological and molecular evidence for the transgenosis of genes from bacteria to plant cells. *Proc. Nat. Acad. Sci. U.S.A.* **70**, 723–26.

JOHNSON C.B. & GRIERSON D (1974) The uptake and expression of DNA by plants. In: *Current Advances in Plant Science* (ed. H. Smith), vol. 2, No. 9, 1–12.

HEYN R.F., RÖRSCH A. & SCHILPEROORT R.A. (1974) Prospects in genetic engineering of plants. *Quarterly Review of Biophysics* 7, 35–73.

REFERENCES

ABELES F.B. (1973) *Ethylene in Plant Biology.* 302 pp. Academic Press, New York & London.

ABELES F.B. & HOLM R.E. (1966) Enhancement of RNA synthesis, protein synthesis and abscission by ethylene. *Plant Physiol.* **41**, 1337–42.

ABELES F.B., BOSSHART R.P., FORRENCE L.E. & HABIG W.H. (1970) Preparation and purification of glucanase and chitinase from bean leaves. *Plant Physiol.* **47**, 129–34.

ACTON J.G. & SCHOPFER P. (1974) Phytochrome-induced synthesis of ribonuclease *de novo* in lupin hypocotyl sections. *Biochem. J.* **142**, 449–55.

ACTON J.G. & SCHOPFER P. (1975) Control over activation or synthesis of phenylalanine ammonia-lyase by phytochrome in mustard (*Sinapis alba* L.)? A contribution to eliminate some misconceptions. *Biochim. biophys. Acta* **404**, 231–42.

ADAMS J.M. & CORY S. (1970) Untranslated nucleotide sequence at the 5′-end of R17 bacteriophage RNA. *Nature, Lond.* **227**, 570–4.

ADAMSON S.D., YAU P.M.P., HERBERT E. & ZUCKER W.V. (1972) Involvement of hemin, a stimulatory fraction from ribosomes and a protein synthesis inhibitor in the regulation of hemoglobin synthesis. *J. molec. Biol.* **63**, 247–64.

ADELMAN M.R., BORISY G.G., SHELANSKI M.L., WEISENBERG R.C. & TAYLOR E.W. (1968) Cytoplasmic filaments and tubules. *Fedn. Proc. Fedn. Am. Socs. exp. Biol.* **27**, 1186–93.

AJTKHOZHIN M.A., BEKLEMISHEV A.B. & NAZAROVA L.M. (1972) Dissociation and density characteristics of ribosomes of plant cells. *FEBS Letters* **21**, 42–4.

AJTKHOZHIN M.A., AKHANOV A.U. & DOSCHANOV Kh.I. (1973) Informosomes of germinating wheat embryos. *FEBS Letters* **31**, 104–6.

AJTKHOZHIN M.A. & AKHANOV A.U. (1974) Release of mRNP-particles of the informosome type from polyribosomes of higher plant embryos. *FEBS Letters* **41**, 275–9.

ALBERSHEIM P. (1974) Structure and growth of walls of cells in culture. *Abstr. 3rd Intl. Congress Plant and Tissue Culture*, University of Leicester, U.K.

ALBERSHEIM P. & VALENT B.S. (1974) Host-pathogen interactions VII. Plant pathogens secrete proteins which inhibit enzymes of the host capable of attacking the pathogen. *Plant Physiol.* **53**, 684–7.

ALLEN C.F. & GOOD P. (1971) Acyl lipids in photosynthetic systems. In *Methods in Enzymology*, **23**, pp. 523–47. (ed. A. San Pietro) Academic Press, New York.

ALLEN C.F., GOOD P., DAVIS H.F., CHISUM P. & FOWLER S.D. (1965) Methodology for separation of plant lipids and application to spinach leaf and chloroplast lamellae. *J. Am. Oil. Chem. Soc.* **42**, 610–14.

ALLENDE J.E. (1970) Protein biosynthesis in plant systems. *Techniques in Protein Biosynthesis*, vol. II pp. 55–100. (eds. P.N. CAMPBELL & J.R. SARGEANT). Academic Press, London & New York.

ALLENDE J.E. & BRAVO M. (1966) Amino acid incorporation and aminoacyl transfèr in a wheat embryo system. *J. Biol Chem.* **241**, 5813–8.

ALLENDE J.E., TARRAGO A., MONASTERIO O., LITVAK S., GATICA M., OJEDA J.M. & MATAMALA M. (1973) The binding of aminoacyl-tRNA to wheat ribosomes. In *Biochemical Society Symposium No.* **38**, *Nitrogen Metabolism in Plants*, pp. 77–96. (eds. T.H. GOODWIN & R.M.S. SMELLIE). Biochemical Society, London.

ANDERSON A.J. & ALBERSHEIM P. (1971) Proteins from plant cell walls which inhibit polygalacturonases secreted by plant pathogens. *Proc. natn. Acad. Sci. U.S.A.*, **68**, 1815–19.

ANDERSON G.R. JENNER E.L. & MUMFORD F.E. (1970) Optical rotatory dispersion and circular dichroism spectra of phytochrome. *Biochim. biophys. Acta* **221**, 69–73.

ANDERSON J.M., GOODCHILD D.J. & BOARDMAN N.K. (1973) Composition of the photosystems and chloroplast structure in extreme shade plants. *Biochim. biophys. Acta* **325**, 573–85.

ANDERSON J.M. & LEVINE R.P. (1974) Membrane polypeptides of some higher plant chloroplasts. *Biochim. biophys. Acta* **333**, 378–87.

ANDERSON M.A. & CHERRY J.H. (1969) Differences in leucyl-transfer RNAs and synthetase in soybean seedlings. *Proc. natn. Acad. Sci. U.S.A.* **62**, 202–9.

ANDERSON M.B. & CHERRY J.H. (1969) Differences in leucyl-transfer RNA's and synthetase in soybean seedlings. *Proc. natn. Acad. Sci. U.S.A.* **62**, 202–9.

ANDERSON W.F., BOSCH L., GROS F., GRUNBERG-MANAGO M., OCHOA S., RICH A. & STAEHELIN Th. (1974) Initiation of protein synthesis in prokaryotic and eukaryotic systems. *FEBS Letters* **48**, 1–6.

ANDREWS T.J., LORIMER G.H. & TOLBERT N.E. (1973) Ribulose diphosphate oxygenase. I. Synthesis of phosphoglycolate by Fraction-l protein of leaves. *Biochemistry* **12**, 11–18.

ARNON D.I., ALLEN M.B. & WHATLEY F.R. (1954) Photosynthesis by isolated chloroplasts. *Nature, Lond.* **174**, 394–6.

ARNON D.I., WHATLEY F.R. & ALLEN M.B. (1958) Assimilatory power in photosynthesis. *Science* **127**, 1026–34.

ARRON G. & BRADBEER J.W. (1975) The use of two-dimensional polyacrylamide gel electrophoresis in the study of the biosynthesis of ribulosebisphosphate carboxylase. *Proceedings of the 3rd International Congress on Photosynthesis Research.* (ed. M. Avron). pp. 2081–8 Elsevier, Amsterdam.

ASSELINEAU J. (1966) *The Bacterial Lipids.* Hermann, Paris.

ATKINSON A.W. Jr., JOHN P.C.L. & GUNNING B.E.S. (1974) The growth and division of the single mitochondrion and other organelles during the cell cycle of *Chlorella*, studied by quantitative stereology and three dimensional reconstruction. *Protoplasma* **81**, 77–109.

ATTRIDGE T.H. (1974) Phytochrome-mediated synthesis of ascorbic acid oxidase in mustard cotyledons *Biochim. biophys. Acta* **362**, 258–65.

ATTRIDGE T.H., JOHNSON C.B. & SMITH H. (1974) Density-labelling evidence for the phytochrome-mediated activation of phenylalanine ammonia—lyase in mustard cotyledons. *Biochim. biophys. Acta* **343**, 440–51.

AVADHANI N.G., KUAN M., VAN DER LIGN P. & PUTMAN R.J. (1973) Polyadenylic acid sequences in mitochondrial RNA. *Biochem. Biophys. Res. Commun.* **51**, 1090–96.

AVRON M. & BIALE J.B. (1957) Metabolic processes in cytoplasmic particles of the avocado fruit. III. Operation of the tricarboxylic acid cycle. *Plant Physiol.* **32**, 100–5.

BADDELEY M.S. & HANSON J.B. (1967) Uncoupling of energy-linked functions of corn mitochondria by linoleic acid and monomethyldecenylsuccinic acid. *Plant Physiol.* **42**, 1702–10.

BAGLIONI C., BLEIBERG I. & ZAUDERER M. (1971) Assembly of membrane-bound polysomes. *Nature (New Biol.)* **232**, 8–12.

BAGSHAW V., BROWN R. & YEOMAN M.M. (1969) Changes in the mitochondrial complex accompanying callus growth. *Ann. Botany* **33**, 35–44.

BAHR J. & BONNER W.D.Jr. (1973a) Cyanide-insensitive respiration. I. The steady states of skunk cabbage spadix and bean hypocotyl mitochondria. *J. Biol. Chem.* **248**, 3441–5.

BAHR J. & BONNER W.D.Jr. (1973b) Cyanide-insensitive respiration. II. Control of the alternate pathway. *J. Biol. Chem.* **248**, 3446–50.

BAILEY C.J., COBB A. & BOULTER D. (1970) A cotyledon slice system for the electron autoradiographic study of the synthesis and intracellular transport of seed storage protein of *Vicia faba*. *Planta* **95**, 103–18.

BAILEY D.S. & NORTHCOTE D.H. (1976) Phospholipid composition of the plasma membrane of the green alga, *Hydrodictyon africanum*. *Biochem. J.* **156**, 295–300.

BAJER A. (1968) Chromosome movement and fine structure of the mitotic spindle. *Aspects of Cell Motility. Symp. Soc. exp. Biol.* **22**, 285–310.

BAJER A.S. & MOLE-BAJER J. (1972) Spindle dynamics and chromosome movements. *Int. Rev. Cytol.* **Supplement 3.**

BAKER D.A., HALL J.L. (1973) Pinocytosis, ATP-ase and ion uptake by plant cells. *New Phytol.* **72**, 1281–89.

BAKER J.E., ELFVIN L.G., BIALE J.B. & HONDA S.I. (1968) Studies on ultrastructure and purification of isolated plant mitochondria. *Plant Physiol.* **43**, 2001–22.

BALDWIN J.P., BOSELEY P.G., BRADBURY E. M. & IBEL K. (1975) The subunit structure of the eukaryotic chromosome. *Nature, Lond.* **253**, 245–9.

BALTSCHEFFSKY H. BALTSCHEFFSKY M. (1974) Electron transport phosphorylation. *Ann. Rev. Biochem.* **43**, 871–97.

BALTUS F. & BRACHET J. (1963) Presence of deoxyribonucleic acid in the chloroplasts of *Acetabularia mediterranea*. *Biochim. biophys. Acta* **76**, 490–2.

BANGHAM A.D., STANDISH M.M. & WEISSMAN G. (1965). The action of steroids and streptolysin S on the permeability of phospholipid structures to cations. *J. molec. Biol.* **13**, 253–9.

BARBARESCHI D., LONGO G.P., SERVETTAZ O., ZULIAN T. & LONGO C.P. (1974) Citrate synthetase in mitochondria and glyoxysomes of maize scutellum. *Plant Physiol.* **53**, 802–7.

BARBER D.A. (1974) The absorption of ions by microorganisms and excised roots. *New Phytol.* **73**, 91–6.

BASTIA D., CHIANG K., SWIFT H. & SIERSMA P. (1971) Heterogeneity, complexity, and repitition of the chloroplast DNA of *Chlamydomonas* reinhardi. *Proc. natn. Acad. Sci. U.S.A.* **68**, 1157–61.

BAUER W.D., TALMADGE K.W., KEEGSTRA K. & ALBERSHEIM P. (1973) The structure of plant cell walls II. The hemicellulose of the walls of suspension-cultured sycamore cells. *Plant Physiol.* **51**, 174–87.

BAUR E. (1909) Das Wesen und die Erblichkeitsverhältnisse der 'Varietates albomarginatae hort' von *Pelargonium zonale*. *Z. Vererbungslehre* **1**, 330–51.

BEARDEN A.J. & MALKIN R. (1974) Primary photochemical reactions in chloroplast photosynthesis. *Quart. Rev. Biophys.* **7**, 131–77.

BEERMAN W. (1972) Chromosomes and genes. *Results and problems in cell differentiation*, 4. (ed. W. Beerman) Springer-Verlag, Berlin.

BEEVERS H. (1956) Utilization of glycerol in the tissues of the castor bean seedling. *Plant Physiol.* **31**, 440–5.

BEEVERS H. (1975) Organelles from castor bean seedlings: biochemical roles in gluconeogenesis and phospholipid biosynthesis. In *Recent Advances in Chemistry and Biochemistry of Plant Lipids*, pp. 287–299 (eds. T. Galliard & E.I. Mercer). Academic Press, New York.

BEEVERS L., LOVEYS B.R., PEARSON J.A. & WAREING P.F. (1970) Phytochrome and hormonal control of expansion and greening of etiolated wheat leaves. *Planta* **90**, 286–94.

BEEVERS L. & POULSON R. (1972) Protein synthesis in cotyledons *Pisum sativum* L. *Plant. Physiol.* **49**, 476–81.

BELL P.R. (1970) Are plastids autonomous? *Control of organelle development* pp. 109–27. (ed. P.L. Miller) *Symp. Soc. exp. Biol.* **24**. Cambridge University Press.

BENDALL D.S. & BONNER W.D.Jr. (1971) Cyanide-insensitive respiration in plant mitochondria. *Plant Physiol.* **47**, 236–45.

BENDALL D.S., DAVENPORT H.E. & HILL R. (1971) Cytochrome components in chloroplasts of the higher plants. In *Methods in Enzymology* **23** pp. 327–44. (ed. A. San Pietro). Academic Press, New York.

BENDALL D.S. & HILL R. (1956) Cytochrome components in the spadix of *Arum maculatum*. *New Phytol.* **55**, 206–12.

BENDICH A.J. & ANDERSON R.S. (1974) Novel properties of satellite DNA from musk melon. *Proc. natn. Acad. Sci. U.S.A.* **71**, 1511–15.

BENDICH A.J. & FILNER P. (1971) Uptake of exogenous DNA by pea seedlings and tobacco cells. *Mutat. Res.* **13**, 199–214.

BENNETT J. & ELLIS R.J. (1973) Solubilization of the membrane-bound deoxyribonucleic acid-dependent RNA polymerase of pea chloroplasts. *Biochem. Soc. Trans.* **1**, 892–4.

BENSON A.A. (1963) The plant sulpholipid. *Adv. in Lipid Res.* **1**, 387–94.

BERGERON J.J.M., EHRENREICH J.H., SIEKEVITZ P. & PALADE G.E. (1973) Golgi fractions prepared from rat liver homogenates. *J. Cell Biol.* **59**, 73–88.

BERIDZE T. (1972) DNA nuclear satellites of the genus *Phaseolus*. *Biochim. biophys. Acta* **262**, 393–6.

BERNHARDT D. & DARNELL J.E. (1969) t-RNA synthesis in Hela cells: a precursor to t-RNA and the effects of methionine starvation on t-RNA synthesis. *J. Molec. Biol.* **42**, 43.

BERRY D.R. & SMITH H. (1971) Red-light stimulation of prolamellar body recrystallization and thylakoid formation in barley etioplasts. *J. Cell Sci.* **8**, 185–200.

BEWLEY J.D., NEGBI M. & BLACK M. (1968) Immediate phytochrome action in lettuce seeds and its interaction with gibberellins and other germination promoters. *Planta* **78**, 351–7.

BEX J.H.M. (1972) Effects of abscisic acid on the soluble RNA polymerase activity in maize coleoptiles. *Planta*. **103**, 11–17.

BEYER R.E., PETERS G.A. & IKUMA H. (1968) Oxido-reduction states and natural homologue or ubiquinone (Coenzyme Q) in submitochondrial particles from etiolated mung bean (*Paseolus aureus*). *Plant Physiol.* **43**, 1395–400.

BICK M.D. & STREHLER B.L. (1971) Leucyl-transfer RNA synthetase changes during soybean cotyledon senescence. *Proc. natn. Acad. Sci. U.S.A.* **68**, 224–8.

BICK M.D., LIEBKE H., CHERRY J.H. & STREHLER B.L. (1970) Changes in leucyl- and tyrosyl-tRNA of soybean cotyledons during plant growth. *Biochim. biophys. Acta* **204**, 175–82.

BINDING H. (1972) Selektion in kalluskulturen mit haploiden zellen. *Z. Pflanzenzuecht.* **67**, 33–8.

BINDING H., BINDING K. & STRAUB J. (1970) Selektion in Gewebekulturen mit haploiden Zellen. *Naturwissenschaften* **57**, 138–9.

BIRNSTIEL M., SPEIRS J., PURDOM I., JONES K. & LOENING U.E. (1968) Properties and composition of the isolated ribosomal DNA satellite of *Xenopus laevis*. *Nature, Lond.* **219**, 454–63.

BISHOP J.O., MORTON J.G., ROSBASH M. & RICHARDSON M. (1974) Three abundance classes in Hela cell messenger RNA. *Nature Lond.* **250**, 199–203.

BJÖRKMAN O., BOARDMAN N.K., ANDERSON J.M., THORNE S.W., GOODCHILD D.J. & PYLIOTIS N.A. (1972) Effect of light intensity during growth of *Atriplex patula* on the capacity of photosynthetic reactions, chloroplast components and structure. *Carnegie Institution of Washington Year Book.* **71**, 115–35.

BJÖRKMAN O., GAUHL E. & NOBS M.A. (1970) Comparative studies of *Atriplex* species with and without β-carboxylation photosynthesis and their first-generation hybrid. *Carnegie Institution of Washington Year Book*, **68**, 620–33.

BLACK M. & VLITOS A.J. (1972) Possible interrelationships of phytochrome and plant hormones. *Phytochrome* pp. 517–49. (ed. by K. Mitrakos & W. Shropshire Jr.), Academic Press, New York.

BLACKMAN F.F. (1905) Optima and limiting factors. *Ann. Bot.* **19**, 281.

BLAIR G.E. & ELLIS R.J. (1973) Protein synthesis in chloroplasts I. Light-driven synthesis of the large subunit of Fraction I protein by isolated pea chloroplasts. *Biochim. biophys. Acta* **319**, 223–34.

BLAKE J.D. & RICHARDS G.N. (1971) Hemicellulose aggregation. An examination of some methods for frationation of plant hemicelluloses. *Carbohyd. Res.* **17**, 253–68.

BLAMIRE J., FLECHTNER V.R. & SAGER R. (1974) Regulation of nuclear DNA replication by the chloroplast in Chlamydomonas. *Proc. natn. Acad. Sci. U.S.A.* **71**, 2867–71.

BOARDMAN N.K. (1966) Protochlorophyll. In *The Chlorophylls* pp. 437–79. (eds. L.P. Vernon & G.R. Seeley). Academic Press, New York.

BOARDMAN N.K. (1968) The photochemical systems of photosynthesis. *Adv. Enzymol.* **30**, 1–79.

BOARDMAN N.K. (1970) Physical separation of the photosynthetic photochemical systems. *Ann. Rev. Plant Physiol* **21**, 115–40.

BOARDMAN N.K. (1971) Subchloroplast fragments: digitonin method. In *Methods in Enzymology* **23**, pp. 268–76 (ed. A. San Pietro). Academic Press, New York.

BOARDMAN N.K. (1975) Trace elements in photosynthesis. In *Trace Elements in Soil/Plant/Animal Systems* (eds. D.J.D. Nicholas & A.R. Egan). Academic Press, London.

BOARDMAN N.K., BJÖRKMAN O., ANDERSON J.M., GOODCHILD D.J. & THORNE S.W. (1974) Photosynthetic adaptation of higher plants to light intensity: relationship between chloroplast structure, composition of the photosystems and photosynthetic rates. In *Proc. 3rd Internat. Congr. Photosynthesis* pp. 1809–27 (ed. M. Avron). Elsevier, Amsterdam.

BOGIN E. & ERICKSON L.C. (1965) Activity of mitochondrial preparations obtained from Faris sweet lemon fruit. *Plant Physiol.* **40**, 566–9.

BOGORAD L., METS L.J., MULLINIX K.P., SMITH H.J. & STRAIN G.C. (1973) In *Nitrogen Metabolism in Plants* (eds. T.W. Goodwin & R.M. Smellie). *Biochem. Soc. Symp.* **38**,17–42. Biochem. Soc., London.

BOISARD J., MARMÉ D. & BRIGGS W.R. (1974) *In vivo* properties of membrane-bound phytochrome. *Plant Physiol.* **54**, 272–6.

BONNER J., DAHMUS M. E., FAMBOROUGH D., HUANG R-C.C., MARASHIGE K. & TUAN D.Y.H. (1968) The biology of isolated chromatin. *Science* **159**, 47–56.

BONNER W.D.Jr. & BENDALL D.S. (1968) Reversed electron transport in mitochondria from the spadix of *Arum maculatum*. *Biochem. J.* **109**, 47P.

BONNETT H.T. & ERIKSSON T. (1974) Transfer of algal chloroplasts into protoplasts of higher plants. *Planta, Berl.* **120**, 71–9.

BOOTHBY D. & WRIGHT S.T.C. (1962) Effects of kinetin and other plant growth regulators on starch degradation. *Nature*. **196**, 389–90.

BORISY G.G., OLMSTED J.B., MARCUM J.M. & ALLEN C. (1974) Microtubule assembly *in vitro*. *Fedn. Proc. Fedn. Am. Socs. exp. Biol.* **33**, 167–74.

BORISY G.G. & TAYLOR E.W. (1967) The mechanism of action of colchicine. Binding of colchicine-^{3}H to cellular protein. *J. Cell Biol.* **34**, 525–33.

BORST P. (1971) Size, structure and informational content of mitochondrial DNA. *Autonomy and Biogenesis of Mitochondria and Chloroplasts* pp. 260–6 (eds. N.K. Boardman, A.W. Linnane & R.M. Smellie). North Holland, Amsterdam.

BORST P. (1972) Mitochondrial nucleic acids. *Ann. Rev. Biochem.* **41**, 333–76.

BORST P., LOOS J.A., CHRIST E.J. & SLATER E.C. (1962) Uncoupling activity of long chain fatty acids. *Biochim. biophys. Acta* **62**, 509–18.

BORTHWICK H.A. (1972) History of phytochrome. *Phytochrome* pp. 3–44. (eds. K. Mitrakos & W. Shropshire Jr.). Academic Press, New York.

BORTHWICK H.A., HENDRICKS S.B., SCHNEIDER M.J., TAYLORSON R.B. & TOOLE V.K. (1969) The high-energy light action controlling plant responses and development. *Proc. natn. Acad. Sci. U.S.A.* **64**, 479–86.

BOTTOMLEY W., SPENCER D., WHEELER A.M. & WHITFIELD P.R. (1971) The effect of a range of RNA polymerase inhibitors on RNA synthesis in higher plant chloroplasts and nuclei. *Arch. Biochem. Biophys.* **143**, 269–75.

BOUDET A. (1971) Mise en evidence chez les vegetaux superieurs d'une association enzymatique bifunctionelle intervenant dans la biosynthèse de l'acide shikimique. *FEBS Letters* **14**, 257–61.

BOULTER D. (1970) Protein synthesis in plants. *Ann. Rev. Plant Physiol.* **21**, 91–114.

BOULTER D., ELLIS R.J. & YARWOOD A. (1972) Biochemistry of protein synthesis in plants. *Biol. Rev.* **47**, 113–75.

BOULTER D., EVANS I.M. & DERBYSHIRE E. (1973) Proteins of some legumes with reference to environmental factors and nutritional value. *Qual. Plant. Pl. Fds. hum. Nutr.* **XXIII**, 239–50.

BOULTER D. & LAYCOCK M.V. (1966) Molecular weight and intracellular localization of malate dehydrogenase of *Vicia faba*. *Biochem. J.* **101**, 8P.

BOURGHIN J.P. & NITSCH J.P. (1967) Obtention de *Nicotiana* haploides à partir d'etamines cultivés *in vitro*. *Ann. Physiol. vég.* **9**, 377–82.

BOWDEN L. & LORD J.M. (1976) The cellular origin of glyoxysomal proteins in germinating castor-bean endosperm. *Biochem. J.* **154**, 501–6.

BOWLES D.J. & KAUSS H. (1976) Characterization, enzymatic and lectin properties of isolated membranes from *Phaseolus aureus*. *Biochim. biophys. Acta*, **443**, 360–71.

BOWLES D.J., LEHLE L. & KAUSS H. (1977) Glucosylation of sterols and polyprenolphosphate in the Golgi apparatus of *Phaseolus aureus*. *Biochim. biophys. Acta*, in press.

BOWLES D.J. & NORTHCOTE D.H. (1972) The sites of synthesis and transport of extracellular polysaccharide in the root tissue of maize. *Biochem. J.* **130**, 1133–45.

BOWLES D.J. & NORTHCOTE D.H. (1974) The amounts and rates of export of polysaccharides found within the membrane system of maize plant root cells. *Biochem. J.* **142**, 139–44.

BOWLES D.J. & NORTHCOTE D.H. (1976) The size and distribution of polysaccharides during their synthesis within the membrane system of maize root cells. *Planta, Berl.* **128**, 101–6.

BOYER P.D. (1967) ^{18}O and related exchanges in enzymic formation and utilization of nucleoside triphosphates. *Curr. Topics Bioenergetics* **2**, 99–149.

BOYER P.D., CROSS R.L. & MOMSEN W. (1973) A new concept for energy coupling in oxidative phosphorylation based on a molecular explanation of the oxygen exchange reaction. *Proc. natn. Acad. Sci. U.S.A.* **70**, 2837–9.

BRADBEER C. & STUMPF P.K. (1960) Fat metabolism in higher plants. XIII. Phosphatidic acid synthesis and diglyceride phosphokinase activity in mitochondria from peanut cotyledons. *J. Lipid Res.* **1**, 214–20.

BRADBEER J.W. (1971) Plastid development in primary leaves of *Phaseolus vulgaris*. The effects of short blue, red, far-red, and white light treatments on dark-grown plants. *J. exp. Bot.* **22**, 382–90.

BRADBEER J.W. (1976) Chloroplast development in greening leaves. *Proceedings of the Fiftieth Anniversary Meeting of the Society for Experimental Biology*. (ed. N. Sunderland). pp. 131–43. Pergamon, Oxford.

BRADBEER J.W., GYLDENHOLM A.O., IRELAND H.M.M., SMITH J.W., REST J. & EDGE H.J.W. (1974a) Plastid development in primary leaves of *Phaseolus vulgaris*. VIII. The effects of the transfer of dark-grown plants to continuous illumination. *New Phytol.* **73**, 271–9.

BRADBEER J.W., GYLDENHOLM A.O., SMITH J.W., REST J. & EDGE H.J.W. (1974b)

Plastid development in primary leaves of *Phaseolus vulgaris*. IX. The effects of short light treatments on plastid development. *New Phytol.* **73**, 281–90.

BRADBEER J.W., GYLDENHOLM A.O., WALLIS M.E. & WHATLEY F.R. (1969) Studies on the biochemistry of chloroplast development. *Progress in Photosynthesis Research.* **1**, pp. 272–279. (ed. H. Metzner). H. Laupp Jr., Tübingen.

BRADBEER J.W., IRELAND H.M.M., SMITH J. W., REST J. & EDGE H.J.W. (1974c) Plastid development in primary leaves of *Phaseolus vulgaris*. VII. Development during growth in continuous darkness. *New Phytol.* **73**, 263–70.

BRADBEER J.W. & MONTES G. (1975) The photocontrol of chloroplast development—Ultrastructural aspects and photosynthetic activity. *Light and Plant Development*. pp. 213–227. (ed. H. Smith). Butterworths, London.

BRADBEER J.W. & RACKER E. (1961) Glycollate formation from fructase-6-P by cell-free preparations. *Fed. Proc.* **20**, 88.

BRADBEER C. & STUMPF P.K. (1959) Fat metabolism in plants XI. The conversion of fat into carbohydrate in peanut and sunflower seedlings. *J. Biol. Chem.* **234**, 498–501.

BRADBURY E.M., INGLIS R.J. & MATTHEWS H.R. (1974) Control of cell division by very lysine rich histone (Fi) phosphorylation. *Nature* **247**, 257–61.

BRADY T. & CLUTTER M.E. (1972) Cytolocalisation of ribosomal cistrons in plant polytene chromosomes. *J. Cell Biol.* **53**, 827–32.

BRANDLE E. & ZETSCHE K. (1973) Localisation of the α-amanitin sensitive RNA polymerase in nuclei of Acetabularia. *Planta.* **111**, 209–17.

BRANTON D. (1968) Structure of the photosynthetic apparatus. In *Photophysiology* **III** pp. 197–224 (ed. A.C. Giese). Academic Press, New York.

BRANTON D. & DEAMER D.W. (1972) *Membrane Structure*. Springer-Verlag, Berlin.

BRAUN A.C. (1955) A study on the mode of action of the wildfire toxin. *Phytopathology* **45**, 659–64.

BRAWERMAN G. (1974) Eukaryotic messenger RNA. *Ann. Rev. Biochem.* **43**, 621–42.

BREIDENBACH R.W. & BEEVERS H. (1967) Association of the glyoxylate cycle enzymes in a novel subcellular particle from castor bean endosperm. *Biochem. biophys. Res. Comm.* **27**, 462–9.

BREIDENBACH R.W., KAHN A. & BEEVERS H. (1968) Characterization of glyoxysomes from castor bean endosperm. *Plant Physiol.* **43**, 705–13.

BRESLOW R.E. & GOLDSBY R.A. (1969) Isolation and characterization of thymidine transport mutants of Chinese hamster cells. *Expt. Cell Res.* **55**, 339–46.

SZ. BREZNOVITS A., MARTON L. & MALIGA P. (1974) Streptomycin-resistant mutants of *Nicotiana tabacum* and *N. sylvestris*. Abstracts—*Third International Congress of Plant Tissue and Cell Culture* **69**, Leicester University, U.K.

BRIGGS W.R. & SIEGELMAN H.W. (1965) Distribution of phytochrome in etiolated seedlings. *Plant Physiol.* **40**, 934–41.

BRIGGS W.R. & RICE H.V. (1972) Phytochrome: chemical & physical properties & mechanism of action. *Ann. Rev. Plant Physiol.* **23**, 293–334.

BRIGHT S.W.J. & NORTHCOTE D.H. (1974) Protoplast regeneration from normal and bromodeoxyuridine-resistant sycamore callus. *J. Cell Science* **16**, 445–63.

BRIGHT S.W.J. & NORTHCOTE D.H. (1975) A deficiency of hypoxanthine phosphoribosyltransferase in a sycamore callus resistant to Azaguanine. *Planta*, **123**, 79–89

BRINKLEY B.R. & CARTWRIGHT J. (1971) Ultrastructural analysis of mitotic spindle elongation in mammalian cells in vitro. Direct microtubule counts. *J. Cell Biol.* **50**, 416–31.

BRITTEN R.J. & DAVIDSON E.H. (1969) Gene regulation for higher cells: a theory. *Science* **165**, 349–57.

BRITTEN R.J., GRAHAM D.E. & NEUFELD B.R. (1974) Analysis of repeating DNA sequences by reassociation. *Methods in Enzymology* **29**. pp. 363–418 (eds. S.P. Colowick & N.O. Kaplan).

BROWN D.L. & BOUCK G.B. (1973) Microtubule biogenesis and cell shape in *Ochromonas*. The role of nucleating sites in shape development. *J. Cell Biol.* **56**, 360–78.

BROWN E.G. & SHORT K.C. (1969) The changing nucleotide pattern of sycamore cells during culture in suspension. *Phytochemistry*. **8**, 1365–72.

BROWN R.C. & MOGENSEN H.L. (1972) Late ovule and early embryo development in *Quercus gambelii*. *Am. J. Bot.* **59**, 311–16.

BROWN R.M.Jr. (1969) Observations on the relationship of the Golgi apparatus to wall formation in the marine Chrysophycean alga, *Pleurochrysis scherffelii* Pringsheim. *J. Cell Biol.* **41**, 109–23.

Bruin W.J., Nelson E.B. & Tolbert N.E. (1970) Glycolate pathway in green algae. *Plant Physiol.* **46**, 386–91.

Brunton C.J. & Palmer J.M. (1973) Pathways for the oxidation of malate and reduced pyridine nucleotide by wheat mitochondria. *Eur. J. Biochem.* **39**, 283–91.

Bryan J. (1974) Biochemical properties of microtubules. *Fedn. Proc. Fedn. Am. Socs. exp. Biol.* **33**, 152–7.

Bücher Th. and Russmann W. (1964) Equilibrium and Nonequilibrium in the Glycolysis system. *Angew. Chem. Int. Ed.* **3**, 426–39.

Burg S.P. (1968) Ethylene, plant senescence and abscision. *Plant Physiol.* **43**, 1503–11.

Burg S.P., Apelbaum A., Eisinger W. & Kang B.G. (1971) Physiology and mode of action of ethylene. *Hort. Science* **6**, 359–64.

Burgess J. & Fleming E.N. (1974) Ultrastructural observations of cell wall regeneration around isolated tobacco protoplasts *J. Cell Sci.* **14**, 439–49.

Burns R.G. (1973) ^{3}H-colchicine binding. Failure to detect any binding to soluble proteins from various lower organisms. *Expl. Cell Res.* **81**, 285–92.

Butcher D.N. (1973) The origins, characteristics and culture of plant tumour cells. In *Plant Tissue and Cell Culture*, pp. 356–391 (ed. H.E. Street). Blackwell Sci. Publ., Oxford.

Butcher D.N. & Street H.E. (1964) Excised root culture. *Biol Rev.* **30**, 513–86.

Butler K.W., Smith I.C.P. & Schneider H. (1970) Sterol structure and ordering effects in spin-labelled phospholipid multibilayer structures. *Biochim. biophys. Acta* **219**, 514–17.

Butler W.L. (1961) Some photochemical properties of phytochrome. *Progress in Photobiology*, pp. 569–81. (eds. B.C. Christensen & B. Buchmann). Elsevier, Amsterdam.

Butler W.L. (1972) Photochemical properties of phytochrome *in vitro. Phytochrome* pp. 185–92. (eds. K. Mitrakos & W. Shropshire Jr.). Academic Press, New York.

Butler W.L., Hendricks S.B. & Siegelman H.W. (1964) Action spectra of phytochrome *in vitro. Photochem. Photobiol.* **3**, 521–8.

Butler W.L., Norris K.H., Siegelman H.W. & Hendricks S.B. (1959) Detection, assay & preliminary purification of the pigment controlling photoresponsive development of plants. *Proc. natn. Acad. Sci. U.S.A.* **45**, 1703–8.

Cahn R.D. & Cahn M.B. (1966) Heritability of cellular differentiation: clonal growth and expression of differentiation in retinal pigment cells *in vitro. Proc. natn. Acad. Sci. U.S.A.* **55**, 106–13.

Calvin M. & Bassham J.A. (1962) *The Photosynthesis of Carbon Compounds.* Benjamin, New York.

Cammack R. & Palmer J.M. (1973) EPR studies of iron-sulfur proteins of plant mitochondria. *Ann. N.Y. Acad. Sci.* **222**, 816–23.

Cammarano P., Pons S., Romeo A., Galdieri M. & Gualerzi C. (1972a) Characterization of unfolded and compact ribosomal subunits from plants and their relationship to those of lower and higher animals: evidence for physicochemical heterogeneity among eucaryotic ribosomes. *Biochim. biophys. Acta* **281**, 571–96.

Cammarano P., Romeo A., Gentile M., Felsani A. & Gualerzi C. (1972b) Size heterogeneity of the large ribosomal subunits and conservation of the small subunits in eucaryote evolution. *Biochim. biophys. Acta* **281**, 597–624.

Cammarano P., Felsani A., Gentile M., Gualerzi C., Romeo A. & Wolf G. (1972c) Formation of active hybrid 80-S particles from subunits of pea seedlings and mammalian liver ribosomes. *Biochim. biophys. Acta* **281**, 625–42.

Cande W.Z., Snyder J., Smith D., Summers K. & McIntosh J.R. (1974) A functional mitotic spindle prepared from mammalian cells in culture. *Proc. natn. Acad. Sci. U.S.A.* **71**, 1559–63.

Canvin D.T. & Beevers H. (1961) Sucrose synthesis from acetate in the germinating castor bean: kinetics and pathway. *J. Biol. Chem.* **236**, 988–95.

Capaldi R.A. (1974) A dynamic model of cell membranes. *Scientific American* **230**, 26–34.

Capecchi M.R. & Klein H.A. (1969) Characterization of three proteins involved in polypeptide chain termination. *Cold Spring Harbor Symp. Quant. Biol.* **34**, 469–77.

Carceller M., Davey M.R., Fowler M.W. & Street H.E. (1971). The influence of sucrose, 2,4-D, and kinetin on the growth, fine structure and lignin content of cultured sycamore cells. *Protoplasma*, **73**, 367–85.

Carlson P.S. (1970) Induction and isolation of auxotrophic mutants in somatic cell cultures of *Nicotiana tabacum. Science* **168**, 487–9.

CARLSON P.S. (1973a) Methionine sulfoximine-resistant mutants of tobacco. *Science* **180**, 1366–8.

CARLSON P.S. (1973b) The use of protoplasts for genetic research. *Proc. natn. Acad. Sci. U.S.A.* **70**, 598–602.

CARLSON P.S., SMITH H.H. & DEARING R.D. (1972) Parasexual interspecific plant hybridisation. *Proc. natn. Acad. Sci. U.S.A.* **69**, 2292–4.

CARPENTER W. & BEEVERS H. (1959) Distribution and properties of isocitratase in plants. *Plant Physiol.* **34**, 403–9.

CASELY-SMITH J.R. (1969) Endocytosis, the different energy requirements for the uptake of particles by small and large vesicles into peritoneal macrophages. *J. Micros.* **90**, 15–30.

CASELY-SMITH J.R. & CHIN J.C. (1971) The passage of cytoplasmic vesicles across endothelial and mesothelial cells. *J. Micros.* **93**, 167–89.

CASKEY C.T., BEAUDET A. & NIRENBERG M. (1968) RNA codons and protein synthesis. 15. Dissimilar responses of mammalian and bacterial transfer RNA fractions to messenger RNA codons. *J. molec. Biol.* **37**, 99–118.

CASKEY T., SCOLNICK E., TOMPKINS R., GOLDSTEIN J. & MILMAN G. (1969) Peptide chain termination, codon, protein factor, and ribosomal requirements. *Cold Spring Harbor Symp. Quant. Biol.* **34**, 479–91.

CASPAR D.L.D. & KIRSCHNER D.A. (1971) Nerve myelin structure at 10 Å resolution. *Nature (New Biol.)* **231**, 46.

CERFF R. (1974) Inhibitor-dependent, reciprocal changes in the activities of glyceraldehyde 3-phosphate dehydrogenases in *Sinapis alba* cotyledons. *Z. Pflanzenphysiol.* **73**, 109–18.

CHALEFF R.S. & CARLSON P.S. (1974) Somatic cell genetics of higher plants. *Ann. Rev. Gen.* **8**, 267–78.

CHALEFF R.S. & CARLSON P.S. (1975a) In vitro selection for mutants of higher plants. (unpublished results).

CHALEFF R.S. & CARLSON P.S. (1975b) Higher plant cells as experimental organisms. *Modification of the Information Content of Plant Cells* (eds. R. Markham, D.R. Davies, D.A. Hopwood, R. Horne). pp. 197–214 North Holland Publ. Co. Amsterdam.

CHAN P. & WILDMAN S.G. (1972) Chloroplast DNA codes for the primary structure of the large subunit of Fraction I protein. *Biochim. biophys. Acta* **277**, 677–80.

CHANCE B. (1961) The interaction of energy and electron transfer reactions in mitochondria. II. General properties of adenosine-triphosphate linked oxidation of cytochrome and reduction of pyridine nucleotide. *J. Biol. Chem.* **236**, 1544–54.

CHANCE B., BONNER W.D.Jr. & STOREY B.T. (1968) Electron transport in respiration. *Ann. Rev. Plant Physiol.* **19**, 295–320.

CHANCE B. & HACKETT D.P. (1959) The electron transfer system of skunk cabbage mitochondria. *Plant Physiol.* **34**, 33–49.

CHANCE B. & HOLLUNGER G. (1963) Inhibition of electron and energy transfer in mitochondria. IV. Inhibition of energy linked diphosphopyridine nucleotide reduction by uncoupling agents. *J. Biol. Chem.* **238**, 445–8.

CHANCE B. & WILLIAMS G.R. (1955) A method for the localization of sites for oxidative phosphorylation. *Nature* **176**, 250–4.

CHANCE B., WILSON D.F., DUTTON P.L. & ERECIŃSKA M. (1970) Energy coupling mechanisms in mitochondria: Kinetic, spectroscopic and thermodynamic properties of an energy-transducing form of cytochrome *b*. *Proc. natn. Acad. Sci. U.S.A.* **66**, 1175–82.

CHANGEUX J.P. (1969) Remarks on the symmetry & cooperative properties of biological membranes. *Symmetry & Function of Biological Systems at the Macromolecular Level. Nobel Symp.* **II** pp. 235–56. (eds. A. Engstrom & B. Strandberg). Wiley-Interscience, New York.

CHAPPELL J.B. & HAARHOFF K.N. (1967) The penetration of the mitochondrial membrane by anions and cations. *Biochemistry of Mitochondria*, pp. 75–91 (eds. E.C. Slater, L. Kanuiga & L. Wojtczak). Academic Press, London.

CHEN C.-H. & LEHNINGER A.L. (1973) Ca^{2+} transport activity in mitochondria from some plant tissues. *Arch. biochem. Biophys.* **157**, 183–96.

CHEN C.-M. & HALL R.H. (1969) Biosynthesis of N^6–(A^2–Isopentenyl) Adenosine in the Transfer Ribonucleic Acid of Cultured Tobacco Pith Tissue. *Phytochemistry* **8**, 1687–95.

CHEN D. & OSBORNE D.J. (1970) Hormones in the translational control of early germination in wheat embryos. *Nature*. **226**, 1157–60.

CHENG K.H. & COLMAN B. (1974) Measurements of photorespiration in some microscopic algae. *Planta* **115**, 207–12.

CHENIAE G.M. (1970) Photosystem-2 and O_2 evolution. *Ann. Rev. Plant Physiol.* **21**, 467–98.

CHERRY J.H. (1973) *Molecular Biology of Plants A Text-Manual.* Columbia University Press, N.Y.

CHERRY J.H. (1967) Nucleic acid biosynthesis in seed germination: Influences of auxin and growth-regulating substances. *Ann. N.Y. Acad. Sci.* **144**, 154–68.

CHERRY J.H. (1974) Regulation of RNA polymerase activity by a plasma membrane factor. *Roy. Soc. NZ Bull.* **12**, 751–7.

CHERRY J.H. & ANDERSON M.A. (1971) Cytokinin-induced changes in transfer RNA species. In *Plant Growth Substances* **1970**. pp. 181–89 (ed. D.J. Carr). Springer-Verlag, Berlin.

CHRISPEELS M.J. & HANSON J.B. (1962) The increase in ribonucleic acid content of cytoplasmic particles of soybean hypocotyl induced by 2,4-dichlorophenoxyacetic acid. *Weeds.* **10**, 123–5.

CHRISPEELS M.J. & VARNER J.E. (1967) Gibberellic acid-enhanced synthesis and release of α-amylase and ribonuclease by isolated barley aleurone layers. *Plant Physiol.* **42**, 398–406.

CHUN E.H.L., VAUGHAN M.H. & RICH A. (1963) The isolation and characterisation of DNA associated with chloroplast preparations. *J. molec. Biol.* **7**, 130–41.

CIFERRI O. (1972) Ribosome specificity of protein synthesis *in vitro. Symp. Biol. Hung.* **13**, 263–6.

CLARKSON D.T. (1974) *Ion Transport and Cell Structure in Plants.* McGraw-Hill, London.

CLEAVER J.E. (1967) *Thymidine Metabolism and Cell Kinetics.* North Holland Publ. Co., Amsterdam.

CLELAND R. (1971) Cell wall extension *Ann. Rev. Plant Physiol.* **22**, 197–222.

CLELAND R. (1973) Auxin-induced hydrogen ion excretion from *Avena* coleoptiles. *Proc. natn. Acad. Sci. U.S.A.* **70**, 3092–3.

CLUM H.H. (1967) Formation of amylase in disks of bean hypocotyl. *Plant Physiol.* **42**, 568–72.

COCKING E.C., POWER J.B., EVANS P.K., SAFWAT F., FREARSON E.M., HAYWARD C., BERRY S.F. & GEORGE D. (1974) Naturally occurring differential drug sensitivities of cultured plant protoplasts. *Plant Sci. Letters* **3**, 341–50.

COCKRELL R.S., HARRIS E.J. & PRESSMAN B. C. (1967) Synthesis of ATP driven by a potassium gradient in mitochondria. *Nature* **215**, 1487–8.

COLEMAN R.A. & PRATT L.H. (1974) Phytochrome: Immunocytochemical assay of synthesis and destruction. *Planta* **119**, 221–31.

COLLANDER R. & BARLUND H. (1933) Permeabilitatsstudien an *Chara ceratophylla. Acta Botan. Fennica* **11**, 1–14.

COLMAN B., MILLER A.G. & GRODZINSKI B. (1974) A study of the control of glycolate excretion in *Chlorella. Plant Physiol.* **53**, 395–97.

CONE R.A. (1972) Rotational diffusion of rhodopsin in the visual receptor membrane. *Nature (New Biol.)* **236**, 39–43.

COOKE R.J. & SAUNDERS P.F. (1975) Phytochrome mediated changes in extractable gibberellin activity in a cell-free system from etiolated wheat leaves. *Planta* **123**, 299–302.

COOPER T.G. & BEEVERS H. (1969a) Mitochondria and glyoxysomes from castor bean endosperm. Enzyme constituents and catalytic capacity. *J. Biol. Chem.* **244**, 3507–14.

COOPER T.G. & BEEVERS H. (1969b) β-Oxidation in glyoxysomes from castor bean endosperm. *J. Biol. Chem.* **244**, 3514–20.

COOPER W.C. & HENRY W.H. (1968) Effect of growth regulators on the coloring and abscission of citrus fruit. *Isr. J. Agr. Res.* **18**, 171–4.

COPPING L.G. & STREET H.E. (1972) Properties of the invertases of cultured sycamore cells and changes in their activity during culture growth. *Physiol. Plant.* **26**, 346–54.

CORNFORTH J.W., MILBORROW B.V., RYBACK G., ROTHWELL K. & WAIN R.L. (1966) Identification of the yellow lupin growth inhibitor as (+)-abscisin II ((+)-dormin). *Nature.* **211**, 742–3.

CORRENS C. (1909) Vererbungsversuche mit blass (gelb) grünen und bunt blattrigen Sippen bei *Mirabilis jalapa, Urtica pilulifera,* und *Lunaria annua. Z. Vererbungslehre* **1**, 291–329.

COSSINS E.A. & SINHA S.K. (1966) The interconversion of glycine and serine by plant tissue extracts. *Biochem. J.* **101**, 542–9.

COVEY S.N. & GRIERSON D. (1975) Subcellular distribution and properties of poly (A)-containing RNA from cultured plant cells. *Eur. J. Biochem.* **63**, 599–606

COX B.J. & TURNOCK G. (1973) Synthesis and processing of ribosomal RNA in cultured plant cells. *Eur. J. Biochem.* **37**, 367–76.

COX G.F. & DAVIES D.D. (1967) Nicotinamide adenine dinucleotide specific isocitrate dehydrogenase from pea mitochondria. *Biochem. J.* **105**, 729–34.

CRICK F.H.C. (1966) Codon-anticodon pairing: the wobble hypothesis. *J. molec. Biol.* **19**, 548–55.

CRICK F.H.C. (1971) General model for the chromosomes of higher organisms. *Nature* **234**, 25–7.

CRICK F.H.C. & KLUG A. (1975) Kinky helix. *Nature* **255**, 530–3.

CROTEAU R. & KOLATTUKUDY P.E. (1974) Biosynthesis of hydroxyfatty acid polymers. Enzymic synthesis of cutin from monomer acids by cell-free preparations from the epidermis of *Vicia faba* leaves. *Biochemistry.* **13**, 3193–202.

CROTTY W.J. & LEDBETTER M.C. (1973) Membrane continuities involving chloroplasts and other organelles in plant cells. *Science* **182**, 839–41.

DALGARNO K. & BIRT L.M. (1963) Free fatty acids in carrot tissue preparations and their effect on isolated carrot tissue. *Biochem. J.* **87**, 586–96.

DANIELLI J.F. & DAVSON H. (1935) A contribution to the theory of permeability of thin films. *J. Cellular. Comp. Physiol.* **5**, 495–508.

DATKO A.H. & MACLACHLAN G.A. (1968) Indoleacetic acid and the synthesis of glucanases and pectic enzymes. *Plant Physiol.* **43**, 735–42.

DATTA A. & SEN S.P. (1965) The mechanism of action of plant growth substances: growth substance stimulation of amino acid incorporation into nuclear proteins. *Biochim. biophys. Acta* **107**, 352–7.

DAVEY M.R. & COCKING E.C. (1972) Uptake of bacteria by isolated higher plant protoplasts. *Nature* **239**, 455–6.

DAVIES E., LARKINS B.A. & KNIGHT R.H. (1972) Polyribosomes from peas. *Plant Physiol.* **50**, 581–4.

DAVIES E. & MACLACHLAN G.A. (1969) Generation of cellulase activity during protein synthesis by pea microsomes *in vitro*. *Arch. Biochem. Biophys.* **128**, 595–7.

DAVIS M. (1972) Polyphenol synthesis in suspension cultures of Paul's Scarlet Rose. *Planta.* **104**, 50–65.

DAY D.A. & WISKICH J.T. (1974) The oxidation of malate and exogenous reduced nicotinamide adenine dinucleotide by isolated plant mitochondria. *Plant Physiol.* **53**, 104–9.

DAY D.A. & WISKICH J.T. (1975) Isolation and properties of the outer membrane of plant mitochondria. *Arch. Biochem. Biophys.* **171**, 117–23.

DAY P.R. (1972) Genetic variability of crops. *Ann. Rev. Phytopath.* **11**, 293–312.

DELANGE R.J., FAMBROUGH D.M., SMITH E. C. & BONNER J. (1969) Calf and pea histone IV. III. Complete amino acid sequence of pea seedling histone IV; comparison with the homologous calf thymus histone. *J. Biol. Chem.* **244**, 5669–79.

DEVAULT D. (1971) Energy transduction in electron transport. *Biochim. biophys. Acta* **226**, 193–9.

DICKERSON R.E., TAKANO T., EISENBERG D., KALLAI O.B., SAMSON L., COOPER A. & MARGOLIASH E. (1971) Ferricytochrome *c*. I. General features of the horse and bonito proteins at 2.8 A resolution. *J. Biol. Chem.* **246**, 1511–35.

DISCHE Z. (1941) Interdependence of various enzymes of the glucolytic system and the automatic regulation of their activity within the cell. *Trav. Membres. Soc. Chim. Biol.* **23**, 1140–8.

DOBBERSTEIN B., VOLKMANN D. & KLÄMBT D. (1974) The attachment of polyribosomes to membranes of the hypocotyl of *Phaseolus vulgaris*. *Biochim. biophys. Acta* **374**, 187–96.

DODGE J.D. (1973) *The Fine Structure of Algal Cells*. Academic Press, London.

DONALDSON R.P. (1976) Membrane lipid metabolism in germinating castor bean endosperm. *Plant Physiol.* **57**, 510–15.

DOUCE R. (1971) Incorporation de l'acide phosphatidique dans le cytidine diphosphate diglycéride des mitochondires isolées des inflorescences de choufleur. *C.R. Acad. Sci.* **Sér. D 272**, 3146–9.

DOUCE R., CHRISTENSEN E.L. & BONNER W. D.Jr. (1972a) Preparation of intact plant mitochondria. *Biochim. biophys. Acta* **275**, 148–60.

DOUCE R., MANNELLA C.A. & BONNER W. D.Jr. (1972b) Site of the biosynthesis of CDP-diglyceride in plant mitochondria. *Biochem. biophys. Res. Comm.* **49**, 1504–9.

DOUCE R., MANNELLA C.A. & BONNER W.D.Jr. (1973) The external NADH dehyrogenases of intact plant mitochon dria. *Biochim. biophys. Acta* **292**, 105–16.

DOY C.H., GRESSHOFF P.M. & ROLFE B.G. (1973a) In *The Biochemistry of Gene Expressing in Higher Organisms* pp. 21–37 (eds. J. Pollak & J.W. Lee). Aust. and New Zealand Book Co., Artarmon.

DOY C.H., GRESSHOFF P.M. & ROLFE B.G. (1973b) Biological and molecular evidence for the transgenosis of genes from bacteria to plant cells. *Proc. natn. Acad. Sci. U.S.A.* **70**, 723–26.

Drumm H., Brünning K. & Mohr H. (1972) Phytochrome-mediated induction of ascorbate oxidase in different organs of a dicotyledonous seedling (*Sinapis alba* L.) *Planta.* **106**, 259–7.

Drumm H. & Mohr H. (1974) The dose response curve in phytochrome-mediated anthocyanin synthesis in the mustard seedling. *Photochem. Photobiol.* **20**, 151–7.

Drury R.E., McCollum J.P. & Garrison S.A. (1968) Properties of succinate oxidation in tomato fruit mitochondria. *Plant Physiol.* **43**, 248–54.

Dublin S.B., Benedek G.B., Bancroft F.C. & Friefelder D. (1970) Molecular weights of coliphages and coliphage DNA. *J. molec. Biol.* **54**, 547–56.

Durán A., Bowers B. & Cabib E. (1975) Chitin synthetase zymogen is attached to the yeast plasma membrane. *Proc. natn. Acad. Sci. U.S.A.* **72**, 3952–5.

Dutton P.L. & Storey B.T. (1971) The respiratory chain of plant mitochondria. IX. Oxidation-reduction potentials of the cytochromes of mung bean mitochondria. *Plant Physiol.* **47**, 282–8.

Eagles C.F. & Wareing P.F. (1963) Experimental induction of dormancy in *Betula* pubescens. *Nature* **199**, 874–5.

Earnshaw M.J., Madden D.M. & Hanson J.B. (1973) Calcium accumulation by corn mitochondria. *J. exp. Bot.* **24**, 828–40.

Edgerton L.J. (1971) Apple abscission. *Hort. Science* **6**, 378–82.

Edidin M. (1974) Two dimensional diffusion in membranes. In *Transport at the Cellular Level* (eds. M.A. Sleigh & D.H. Jennings). Symposium 28, Society for Experimental Biology. Cambridge University Press.

Edidin M. & Fambrough O. (1973) Fluidity of the surface of cultured muscle fibers. Rapid lateral diffusion of marked surface antigens. *J. Cell Biol.* **57**, 27–53.

Efron D. & Marcus A. (1973) Efficient synthesis of rabbit globin in a cell-free system from wheat embryo. *FEBS Letters* **33**, 23–7.

Eisenman G., Ciani S.M. & Szabo G. (1968) Some theoretically expected and experimentally observed properties of lipid bilayer membranes containing neutral molecular carriers of ions. *Fed. Proc.* **27**, 289–1304.

El-Antably H.M.M., Wareing P.F. & Hillman J. (1967) Some physiological responses to D,L-abscisin (Dormin). *Planta* **73**, 74–90.

Ellis R.J. (1969) Chloroplast ribosomes: stereospecificity of inhibition by chloramphenicol. *Science, N.Y.* **163**, 477–8.

Ellis R.J. (1973) Fraction I protein. *Commentaries in Plant Science.* **4**, 29–38.

Ellis R.J. (1974) The biogenesis of chloroplasts: protein synthesis by isolated chloroplasts. *Biochem. Soc. Trans.* **2**, 179–82.

Ellis R.J. (1975) Inhibition of chloroplast protein synthesis by lincomycin and 2-(4-methyl-2,6-dinitro-anilino)-*N*-methyl propionamide. *Phytochem.* **14**, 89–93.

Ellis R.J. (1975a) The synthesis of chloroplast membranes in *Pisum sativum.* In *Membrane Biogenesis: Mitochondrial, Chloroplasts and Bacteria* (ed. A. Tzagoloff). Plenum Publishing Co., New York.

Ellis R.J., Blair G.E. & Hartley M.R. (1973) The nature and function of chloroplast protein synthesis. *Biochem. Soc. Symp.* **38**, 137–62.

Ellis R.J. & Hartley M.R. (1971) Sites of synthesis of chloroplast proteins. *Nature, Lond.* **233**, 193–6.

Ellis R.J. & Hartley M.R. (1974) Nucleic acids of chloroplasts. In *Biochemistry of Nucleic Acids* (ed. K. Burton). MTP International Review of Science, Biochemistry Series One, **6**, 323–48. Butterworths, London, & University Park Press, Baltimore.

Ellis R.J. & Macdonald I.R. (1970) Specificity of cycloheximide in higher plant systems. *Plant Physiol.* **46**, 227–32.

Elzam O.E. & Hodges T.K. (1968) Characterization of energy-dependent Ca^{2+} transport in maize mitochondria. *Plant Physiol.* **43**, 1108–14.

Emerson R. & Arnold W. (1932) The photochemical reaction of photosynthesis *J. Gen. Physiol.* **16**, 191–205.

Engelsma G. (1974) On the mechanism of the changes in phenylalamine ammonia lyase activity induced by ultra-violet and blue light in gherkin hypocotyls. *Plant Physiol.* **54**, 702–5.

Epstein E. (1972) *Mineral Nutrition of Plants: Principles and Perspectives.* Wiley, New York.

Erecisnka M. & Storey B.T. (1970) The respiratory chain of plant mitochondria. VII. Kinetics of flavoprotein oxidation in skunk cabbage mitochondria. *Plant Physiol.* **46**, 618–24.

ETZOLD H. (1965) Der Polarotropismus und Phototropismus der Chloronemen von *Dryopteris filix mas* (L.) Schott. *Planta* **64**, 254–80.

EVANS A. (1976) Phytochrome in etioplasts and its *in vitro* control of gibberellin efflux. In *Light and Plant Development* pp.129–41 (ed. H. Smith). Butterworths, London.

EVANS A. & SMITH H. (1976a) Localisation of phytochrome in etioplasts and its regulation *in vitro* of gibberellin levels. *Proc. natn. Acad. Sci. U.S.A.* **73,** 138–42.

EVANS A. & SMITH H. (1976b) Spectrophotometric evidence for the presence of phytochrome in the envelope membranes of barley etioplasts. *Nature*, **259,** 323–5.

EVANS G.M. (1968) Nuclear changes in flax. *Heredity* **23**, 25–38.

EVANS M.L. (1973) Rapid stimulation of plant cell elongation by hormonal and non-hormonal factors. *Bioscience*. **23**, 711–18.

EVANS M.L., RAY P.M. & REINHOLD L. (1971) Induction of coleoptile elongation by carbon dioxide. *Plant Physiol.* **47**, 335–41.

EVINS W.H. & VARNER J.E. (1971) Hormone-controlled synthesis of endoplasmic reticulum in barley aleurone cells. *Proc. natn. Acad. Sci. U.S.A.* **68**, 1631–3.

FEIERABEND J. (1969) Der Einfluss von Cytokininen auf die Bildung von Photosyntheseenzymen in Roggenkeimlingen. *Planta* **84**, 11–29.

FAHNESTOCK S., WEISSBACH H. & RICH A. (1972) Formation of a ternary complex of phenyllactyl-tRNA with transfer factor Tu and GTP. *Biochim. biophys. Acta* **269**, 62–6.

FAIZ-UR-RAHMAN, TREWAVAS A.J. & DAVIES D.D. (1974) The pasteur effect in carrot root tissue. *Planta* **118**, 195–210.

FEIERABEND J. & BEEVERS H. (1972) Developmental studies on microbodies in wheat leaves. II. Ontogeny of particulate enzyme associations. *Plant Physiol.* **49**, 33–9.

FELLENBER G. (1969) Veränderungen des Nucleoproteids von Erbsenepikotylen durch synthetische Aurine bei der Induktion der Wurzelneubildung. *Planta*. **84**, 195–8.

FERNANDEZ-MORAN H. (1962) Cell membrane ultrastructure. Low temperature electron microscopy and x-ray diffraction studies of lipoprotein components in lamellar systems. *Circulation* **26**, 1039–65.

FERRARI T. & WIDHOLM J.M. (1975) Manu-personal communication.

FILNER P., WRAY J.L. & VARNER J.E. (1969) Enzyme induction in higher plants. *Science* **165**, 358–67.

FINERAN B.A. (1973a) Organisation of the Golgi apparatus in frozen-etched root tips. *Cytobiologie* **8**, 175–93.

FINERAN B.A. (1973b) Association between endoplasmic reticulum and vacuoles in frozen-etched root tips. *J. Ultrastruct. Res.* **43**, 75–87.

FIRTEL R.A. & LODISH H.F. (1973) A small nuclear precursor of messenger RNA in the cellular slime mould *Dictyostelium discoideum*. *J. molec. Biol.* **79**, 295–314.

FLAVELL R.B., BENNETT M.D., SMITH J.B. & SMITH D.B. (1974) Genome size and the proportion of repeated nucleotide sequence DNA in plants. *Biochem. Genet.* **12**, 257–69.

FLECHTNER U.V. & SAGER R. (1973) Ethidium bromide induced selective and reversible loss of chloroplast DNA. *Nature* (*New Biol.*) **241**, 277–9.

FORER A. & BENHKE O. (1972) An actin-like component in spermatocytes of a crane fly (*Nephrotoma suturalis* Loew.). The spindle. *Chromosoma*, **39**, 145–73.

FORRESTER M.L., KROTKOV G. & NELSON C.D. (1966) Effect of oxygen on photosynthesis, photorespiration and respiration in detached leaves I. Soybean. *Plant Physiol.* **41**, 422–27.

FOWLER M.W. (1971) Studies on the growth in culture of plant cells. XIV. Carbohydrate oxidation during the growth of *Acer pseudoplatanus* L. cells in suspension culture. *J. exp. Bot.* **22**, 715–24.

FOWLER M.W. & CLIFTON A. (1974) Activities of enzymes of carbohydrate metabolism in cells of *Acer pseudoplatanus* L. maintained in continuous (chemostat) culture. *Eur. J. Biochem.* **45**, 445–50.

FOX J.E. & CHEN C.M. (1967) Characterization of labelled ribonucleic acid from tissue grown on ^{14}C-containing cytokinins. *J. Biol. Chem.* **242**, 4490–4.

FRANKE W. (1967) Mechanisms of foliar penetration of solutions. *Ann. Rev. Plant Physiol.* **18**, 281–300.

FRANKE W.W. (1974) Structure, biochemistry and functions of the nuclear envelope. *Int. Rev. Cytol. Suppl.* **4**, 71–236.

FRANKE W.W., MORRÉ D.J., DEUMLING B., CHEETHAM R.D., KARTENBECK J., JARASCH E.D. & ZENTGRAF H.W. (1971) Synthesis and turnover of membrane proteins in rat liver; an examination of the membrane flow hypothesis. *Z. Naturforsch.* **26b**, 1031–9.

FRANKLAND B. (1972) Biosynthesis and dark transformations of phytochrome. *Phytochrome* pp. 195–225. (eds. K. Mitrakos & W. Shropshire Jr.). Academic Press, New York.

FRANTZ C., ROLAND J.-C., WILLIAMSON F.A. & MORRÉ J.D. (1973) Différenciation *in vitro* des membranes des dictyosomes. *C.R. Acad. Sci. Paris* **277D**, 1471–4.

FREARSON E.M., POWER J.B. & COCKING E.C. (1973) The isolation, culture and regeneration of *Petunia* leaf protoplasts. *Dev. Biol.* **33**, 130–7.

FREDERICK S.E., GRUBER P.J. & TOLBERT N.E. (1973) The occurrence of glycolate dehydrogenase and glycolate oxidase in green plants: an evolutionary survey. *Plant Physiol.* **52**, 318–23.

FREDERICK S.E. & NEWCOMB E.H. (1969) Cytochemical localization of catalase in leaf microbodies (peroxisomes). *J. Cell Biol.* **43**, 343–53.

FREDERICQ H. (1964) Conditions determining effects of far-red and red irradiations on flowering response of *Pharbitis nil. Plant Physiol.* **39**, 812–16.

FREUDENBERG K. (1968) Constitution and biosynthesis of lignin. *Molecular Biology, Biochemistry and Biophysics* (eds. A. Kleinzeller et al.) **2**, 45–122. Springer, Berlin.

FRY W.E. & MUNCH D.C. (1975) Hydrogen cyanide detoxification by *Gloeocercospora sorghi. Physiol. Plant Pathol.*, **7**, 23–33.

FUGE H. (1974) Ultrastructure and function of the spindle apparatus microtubules and chromosomes during nuclear division. *Protoplasma* **82**, 289–320.

GALSTON A.W. (1968) Microspectrophotometric evidence for phytochrome in plant nuclei. *Proc. natn. Acad. Sci. U.S.A.* **61**, 454–60.

GALSTON A.W. (1974) Plant photobiology in the last half-century. *Plant Physiol.* **54**, 427–36.

GARDINER M. & CHRISPEELS M.J. (1975) Involvement of the Golgi apparatus in the synthesis and secretion of hydroxyproline-rich cell wall glycoproteins. *Plant Physiol.* **55**, 536–41.

GARDNER G., THOMPSON W.F. & BRIGGS W.R. (1974) Differential reactivity of the red- and far-red-absorbing forms of phytochrome to (^{14}C)N-ethyl maleimide. *Planta* **117**, 367–72.

GAREL J.P., FOURNIER A. & DAILLIE J. (1973) Functional and modulated adaptation of tRNA to fibroin biosynthesis in the silk gland of *Bombyx mori* L. In *Regulation of Transcription and Translation in Eukaryotes* pp. 323–33 (eds. E.K.F. Bautz, P. Karlson & H. Kersten). Colloquim der Gesellschaft für Biologische Chemie. Springer-Verlag, Berlin.

GASIOR E. & MOLDAVE K. (1965) Resolution of aminoacyl-transferring enzymes from rat liver by molecular sieve chromatography. *J. Biol. Chem.* **240**, 3346–52.

GAUTHERET R.J. (1939) Sur la possibilité de réaliser la culture indéfinie des tissus de tubercules de carotte. *C.r. Hebd. Séanc. Acad. Sci. Paris.* **208**, 118–21.

GAUTHERET R.J. (1966) Factors affecting differentiation of plant tissues grown *in vitro*. In *Cell Differentiation and Morphogenesis*, pp. 55–71 (ed. W. Bierman). North Holland Publ. Co., Amsterdam.

GAWADI N. (1971) Actin in the mitotic spindle. *Nature, Lond.* **234**, 410.

GEBAUER H.U., SEITZ U. & SEITZ U. (1975) Transport and processing of ribosomal RNA in plant cells after treatment with cycloheximide. *Z. Naturforsch.* **30c**, 213–18.

GEORGIEV G.P. (1969) On the structural organisation of operon and the regulation of RNA synthesis in animal cells. *J. Theoret. Biol.* **25**, 473–90.

GERHARDT B. & BEEVERS H. (1970) Developmental studies on glyoxysomes in *Ricinus* endosperm. *J. Cell Biol.* **44**, 94–103.

GHOSH K., GRISHKO A. & GHOSH H.P. (1971) Initiation of protein synthesis in eukaryotes. *Biochem. biophys. Res. Commun.* **42**, 462–8.

GIBBONS I.R. (1965) Chemical dissection of cilia. *Arch. Biol. Liège* **76**, 317–52.

GIBOR A. & IZAWA M. (1963) The DNA content of the chloroplasts of Acetabularia. *Proc. natn. Acad. Sci. U.S.A.* **50**, 1164–9.

GILBERT J.M. & ANDERSON W.F. (1970) Cell-free hemoglobin synthesis. II. Characteristics of the transfer ribonucleic acid-dependent assay system. *J. biol. Chem.* **245**, 2342–9.

GIVAN C.V. & COLLIN H.A. (1967) Studies on the growth in culture of plant cells. II Changes in respiration rate and nitrogen content associated with the growth of *Acer pseudoplatanus* cells in suspension culture. *J. exp. Bot.* **18**, 321–31.

GLAZER R.I. & SARTORELLI A.C. (1972) The differential sensitivity of free and membrane-bound polyribosomes to inhibitors of protein synthesis. *Biochem. biophys. Res. Commun.* **46**, 1418–24.

GLYNN I.M. (1967) Involvement of a membrane potential in the synthesis of ATP by mitochondria. *Nature* **216**, 1318–19.

GOLDUP A., OHKI S. & DANIELLI J.F. (1970) Black lipid films. *Recent Progress in Surface Science* **3**, 193–260.

GOLIŃSKA B. & LEGOCKI A.B. (1973) Purification and some properties of elongation Factor I from wheat germ. *Biochim. biophys. Acta* **324**, 156–70.

GORDON S.A. (1961) The intracellular distribution of phytochrome in corn seedlings. *Proceedings of the 3rd International Congress on Photobiology*. pp. 441–3. (eds. B. Chr. Christensen & B. Buchmann).

GOULD A. & STREET H.E. (1975) Kinetic aspects of synchrony in suspension cultures of *Acer pseudoplatanus* L. *J. Cell. Sci.* **17**, 337–48.

GOULD A.R., BAYLISS M.W. & STREET H.E. (1974) Studies on the growth in culture of plant cells. XVII. Analysis of the cell cycle of asynchronously dividing *Acer pseudoplatanus* L. cells in suspension culture. *J. exp. Bot.* **25**, 468–78.

GRAVES L.B., TRELEASE R.N., GRILL A. & BECKER W.M. (1972) Localization of glyoxylate cycle enzymes in glyoxysomes in *Euglena. J. Protozool.* **19**, 527–32.

GRAY J.C. & KEKWICK R.G.O. (1973) Synthesis of the small subunit of ribulose 1,5-diphosphate carboxylase on cytoplasmic ribosomes from greening bean leaves. *FEBS Letters* **38**, 67–9.

GRAY J.C. & KEKWICK R.G.O. (1974) The synthesis of the small subunit of ribulose 1,5-diphosphate carboxylase in *Phaseolus vulgaris*. *Eur. J. Biochem.* **44**, 491–500.

DE GREEF J., BUTLER W.L. & ROTH T.F. (1971) Greening of etiolated bean leaves in far-red light. *Plant Physiol.* **47**, 457–64.

DE GREEF J.A. & CAUBERGS R. (1973) Studies on greening of etiolated seedlings II. Leaf greening by phytochrome action in the embryonic axis. *Physiol. Pl.* **28**, 71–6.

GREEN P.B. (1964) Cinematic observations on the growth and division of chloroplasts in *Nitella*. *Am. J. Bot.* **51**, 334–42.

GREGORY F.G. & SEN P.K. (1937) The relation of respiration rate to the carbohydrate and nitrogen metabolism of the barley leaf as determined by nitrogen and potassium deficiency. *Ann. Bot.* **NS 1**, 521–61.

GREGORY P. & BRADBEER J.W. (1973) Plastid development in primary leaves of *Phaseolus vulgaris*. The cytochrome content of plastids during greening. *Planta.* **109**, 317–26.

GREGORY P. & BRADBEER J.W. (1975) Plastid development in primary leaves of *Phaseolus vulgaris*. The development of plastid adenosine triphosphatase during greening. *Biochem. J.* **148**, 433–8.

GREVILLE G.D. (1969) A scrutiny of Mitchell's chemiosmotic hypothesis of respiratory chain and photosynthetic phosphorylation. *Curr. Topics Bioenergetics* **3**, 1–78.

GRIERSON D. (1975) The hybridisation of ^3H-poly(uridylic) acid to DNA from *Phaseolus aureus. Roxb. Planta.* **127**, 87–91.

GRIERSON D. & COVEY S.N. unpublished results.

GRIERSON D., COVEY S.N. & GILES A.B. (1975) The effect of light on RNA metabolism in developing leaves. *Light and Plant Development* 22nd Nottingham Easter School in Agricultural Science pp. 229–53 (ed. H. Smith). Butterworth & Co., London.

GRIERSON D. & LOENING U.E. (1974) Characterisation of ribonucleic acid components from leaves of *Phaseolus aureus*. *Eur. J. Biochem.* **44**, 509–15.

GRIERSON D., MCKEE R.A., ATTRIDGE T.H. & SMITH H. (1975) Studies on the uptake and expression of foreign genetic material in higher plants. In *Modification of the Information Content of Plant Cells* pp. 91–99 (eds. R. Markham, D.R. Davies, D.A. Hopwood & R. Horne). North Holland Publ. Co., Amsterdam.

GRIERSON D., ROGERS M.E., SARTIRANA M.-L. & LOENING U.E. (1970) The synthesis of ribosomal RNA in different organisms: structure and evolution of the rRNA precursor. *Cold Spring Harbor Symp. Quant. Biol.* **35**, 589–97.

GRIMWOOD B.G. & MCDANIEL R. (1970) Variant malate dehydrogenase isozymes in mitochondrial populations. *Biochim. biophys. Acta* **220**, 410–15.

GRONER Y., POLLACK Y., BERISSI H. & REVEL M. (1972) Cistron specific translation control protein in *Escherichia coli*. *Nature (New Biol.)* **239**, 16–18.

GROOT G.S.P., FLAVELL R.A., VAN OMME G.J.B. & GRIVELL L.A. (1974) Yeast mitochondrial RNA does not contain poly(A). *Nature* **252**, 167–8.

GROSS M. & RABINOVITZ M. (1973) Partial purification of a translational repressor mediating hemin control of globin synthesis and implication of results on the site of inhibition. *Biochim. biophys. Res. Commun.* **50**, 832–8.

GROVE S.N., BRACKER C.E. & MORRÉ D.J. (1968) Cytomembrane differentiation in the endoplasmic reticulum–Golgi apparatus –vesicle complex. *Science* **161**, 171–3.

GRUBER P.J., BECKER W.M. & NEWCOMBE E.H. (1973) The development of microbodies and peroxisomal enzymes in greening bean leaves. *J. Cell Biol.* **56**, 500–18.

GRUNBERG-MANAGO M., GODEFROY-COLBURN Th., WOLFE A.D., DESSEN P., PANTALONI D., SPRINGER M., GRAFFE M., DONDON J. & KAY A. (1973) Initiation of protein synthesis in prokaryotes. In *Regulation of Transcription and Translation in Eukaryotes* pp. 213–49. (eds. E.K.F. Bautz, P. Karlson & H. Kersten), 24. Colloquium der Gesellschaft für Biologische Chemie. Springer-Verlag, Berlin.

GUALERZI C., JANDA H.G., PASSOW H. & STÖFFLER G. (1974) Studies on the protein moiety of plant ribosomes. *J. biol. Chem.* **249**, 3347–55.

GUHA S. & MAHESHWARI S.C. (1964) *In vitro* production of embryos from anthers of *Datura*. *Nature* **204**, 497.

GUILFOYLE T.J. & HANSON J.B. (1974) Greater length of ribonucleic acid synthesised by chromatin-bound polymerase from auxin-treated soybean hypocotyls. *Plant Physiol.* **53**, 110–13.

GUILFOYLE T.J., LIN C.Y., CHEN Y.M., NAGAO R.T. & KEY J.L. (1975) Enhancement of soybean RNA polymerase I by auxin. *Proc. natn. Acad. Sci. U.S.A.* **72**, 69–72.

GUMILEVSKAYA N.A., KUVAEVA E.B., CHUMIKINA L.V. & KRETOVICH V.L. (1971) Ribosomes of dry pea seeds. *Biochemistry, Acad. Nauk SSSR* **36**, 277–88.

GUNNING B.E.S. & STEER M.W. (1975) *Plant Cell Biology, an ultrastructural approach*. pp. 312. Edward Arnold, London.

GURDON J.B., LANE C.D., WOODLAND H.R. & MARBAIX G. (1971) Use of frog eggs and oocytes for the study of messenger RNA and its translation in living cells. *Nature, Lond.* **233**, 177–82.

VON GUTTENBERG H. & BEYTHIAN A. (1951) Über den Einfluss von Wirkstoffen auf die Wasserpermeabilität des Protoplasmas. *Planta.* **40**, 36–69.

HABER J.E., PELOQUIN J.G., HALVORSON H.O. & BORISY G.G. (1972) Colcemid inhibition of cell growth and the characterisation of a colcemid binding activity in *Saccharomyces cerevisiae*. *J. Cell Biol.* **58**, 355–67.

HABERLANDT G. (1890) Die Kleberschicht des Gras-Endosperms als Diastase ausscheidendes Drüsengewebe. *Ber. Dent. Botan. Ges.* **8**, 40–8.

HABERLANDT G. (1902) Kultur versuche mit isolierten Pflanzellen. *Sber. Akad. Wiss. Wien.* **111**, 69–92.

HACKETT D.P. & HASS D.W. (1958) Oxidative phosphorylation and functional cytochromes in skunk cabbage mitochondria. *Plant Physiol.* **33**, 27–32.

HACKETT D.P., HAAS D., GRIFFITHS S.K. & NIEDERPRUEM D.J. (1960) Studies on the development of cyanide resistant respiration in potato tuber slices. *Plant Physiol.* **35**, 8–19.

HAGER A., MENZEL H. & KRAUSS A. (1971) Versuche und Hypothese zur Primärwirkung des Auxins beim Streckungswachstum. *Planta* **100**, 47–75.

HAHLBROCK K., KÜHLEN E. & LINDL T. (1971) Änderungen von Enzymaktivitäten während des Wachstums von Zellsuspensionkulturen von *Glycine max:* Phenylalanin Ammonia Lyase and p-Cumarat: CoA ligase. *Planta* **99**, 311–18.

HALL D.O. (1972) Nomenclature of isolated chloroplasts. *Nature (New Biol.)* **235**, 125–6.

HALL D.O., CAMMACK R. & RAO K.K. (1972) A role for ferredoxins in the origin of life and biological evolution. In *Proc. Second Internat. Congr. Photosynthesis* pp. 1707–19 (eds. G. Forti, M. Avron & A. Melandri). Dr. W. Junk, The Hague.

HALL D.O. & RAO K.K. (1972) In *Photosynthesis*. Arnold, London.

HALL J.L. (1971) Further properties of adenosine triphosphatase & β-glycerophosphatase from barley roots. *J. exp. Bot.* **22**, 800–8.

HALL M.D. & COCKING E.C. (1974) The response of isolated *Avena* coleoptile protoplasts to indol-3-acetic acid. *Protoplasma* **79**, 225–34.

HALL R.H. (1973) Cytokinins as a probe of developmental processes. *Ann. Rev. Plant Physiol.* **24**, 415–44.

HALLIWELL B. & BUTT V.S. (1974) Oxidative decarboxylation of glycollate and glyoxyllate by leaf peroxisomes. *Biochem. J.* **138**, 217–24.

HALPERIN W. (1967) Population density effects on embryo-genesis in carrot cell cultures. *Exp. Cell Res.* **48**, 170–3.

HALPERIN W. & JENSEN W.A. (1967) Ultrastructure changes during growth and embryogenesis on carrot cell cultures. *J. Ultrastruct. Res.* **18**, 426–43.

HALPERIN W. & WETHERELL D.F. (1964) Adventive embryony in tissue cultures of wild carrot, *Daucus carota*. *Am. J. Bot.* **51**, 274–83.

HANKE J., HARTMANN K.M. & MOHR H. (1969) Die Wirkung von 'Störlicht' auf die Blütenbildung von *Sinapis alba* L. *Planta* **86**, 235–49.

HANSON J.B. & MILLER R.J. (1967) Evidence for active phosphate transport in maize mitochondria. *Proc. natn. Acad. Sci. U.S.A.* **58**, 727–34.

HARDESTY B., ARLINGHAUS R., SHAEFFER J. & SCHWEET R. (1963) Hemoglobin and polyphenylalanine synthesis with reticulocyte ribosomes. *Cold Spring Harb. Symp. Quant. Biol.* **XXVIII**, 215–22.

HARDIN J.W. & CHERRY J.H. (1972) Solubilization and partial characterization of soybean chromatin-bound RNA polymerase. *Biochem. Biophys. Res. Comm.* **48**, 299–306.

HARDIN J.W., CHERRY J.H., MORRÉ D.J. & LEMBI C.A. (1972) Enhancement of RNA polymerase activity by a factor released by auxin from plasma membrane. *Proc. natn. Acad. Sci. U.S.A.* **69**, 3146–50.

HARDIN J.W., O'BRIEN T.J. & CHERRY J.H. (1970) Stimulation of chromatin-bound RNA polymerase activity by a soluble factor. *Biochim. biophys. Acta* **224**, 667–70.

HARRIS E.J. (1969) Mitochondrial anion uptake. *The Energy Level and Metabolic Control in Mitochondria* pp. 31–44 (eds. S. Papa, J.M. Tager, E. Quagliariello & E.C. Slater). Adriatica Editrice, Bari.

HART J.W. & SABNIS D.D. (1973) Colchicine binding protein from phloem and xylem of a higher plant. *Planta, Berl.* **109**, 147–52.

HARTLEY M.R. & ELLIS R.J. (1973) Ribonucleic acid synthesis in chloroplasts. *Biochem. J.* **134**, 249–62.

HARTLEY M.R., WHEELER A.W. & ELLIS R.J. (1975) Protein synthesis in chloroplasts. V Translation of messenger RNA for the large subunit of Fraction I protein in a heterologous cell-free system. *J. molec. Biol.* **91**, 67–77.

HARTMANN M.-A., FERNE M., GIGOT C., BRANDT R. & BENVENISTE P. (1973) Isolement, caractérisation et composition en stérols de fractions subcellulaires de feuilles etiolées de Haricot. *Physiolog. Vég.* **11**, 209–30.

HARTMANN K.M. (1966) A general hypothesis to interpret 'high energy phenomena' of photomorphogenesis on the basis of phytochrome. *Photochem. Photobiol.* **5**, 349–66.

HARTMANN K.M. (1967) Ein Wirkungsspektrum der Photomorphogenese unter Hochenergiebedingungen und seine Interpretation auf der Basis des Phytochroms (Hypokotylwachstumhemmung bei *Lactuca sativa* L.). *Z. Naturf.* **22**, 1172–5.

HARTMANN K.M. & COHNEN UNSER I. (1973) Carotenoids and flavins versus phytochrome as the controlling pigment for blue-UV-mediated-photoresponses. *Z. Pflanzenphysiol.* **69**, 109–24.

HASELKORN R. & ROTHMAN-DENES L.B. (1973) Protein synthesis. *Ann. Rev. Biochem.* **42**, 397–438.

HATCH M.D. & SLACK C.R. (1969) Studies on the mechanism of activation and inactivation of pyruvate phosphate dekinase. A possible regulatory role for the enzyme in the C_4 decarboxylic acid pathway of photosynthesis. *Biochem. J.* **112**, 549–58.

HATCH M.D. & SLACK C.R. (1970) Photosynthetic CO_2-fixation pathways. *Ann. Rev. Plant Physiol.* **21**, 141–62.

HAUPT W. (1972a) Short-term phenomena controlled by phytochrome. *Phytochrome* pp. 349–68 (eds. K. Mitrakos & W. Shropshire Jr.). Academic Press, New York.

HAUPT W. (1972b) Localization of phytochrome within the cell. *Phytochrome* pp. 553–69 (eds. K. Mitrakos & W. Shropshire Jr.). Academic Press, New York.

HEATH I.B. (1975a) Colchicine and colcemid binding components of the fungus *Saprolegnia ferax*. *Protoplasma* **85**, 177–92.

HEATH I.B. (1975b) The effect of anti-mitotic agents on the growth and ultrastructure of the fungus *Saprolegnia ferax* and their ineffectiveness in disrupting hyphal microtubules. *Protoplasma* **85**, 147–76.

HEBER U. (1974) Metabolite exchange between chloroplasts and cytoplasm. *Ann. Rev. Plant Physiol.* **25**, 393–421.

HEBER U. (1975) Energy coupling in chloroplasts. *J. Bioenergetics.* **8**, 157–72.

HECHT N.B., BLEYMAN M. & WOESE C.R. (1968) The formation of 5s ribosomal ribonucleic acid in *Bacillus subtilis* by post-transcriptional modification. *Proc. natn. Acad. Sci. U.S.A.* **59**, 1278.

HECKER L.I., EGAN J., REYNOLDS R.N., NIX C.E., SCHIFF J.A. & BARNETT W.E. (1974) The sites of transcription and translation for *Euglena* chloroplastic amino-acyl-tRNA synthetases. *Proc. natn. Acad. Sci. U.S.A.* **71**, 1910–14.

HEIMER Y.M. & FILNER P. (1970) Regulation of the nitrate assimilation pathway of cultured tobacco cells. II. Properties of a variant cell line. *Biochim. biophys. Acta* **215**, 152–65.

HELDT H.W., FLIEGE R., LEHNER K., MILOVANCEV M. & WERDAN K. (1975) Metabolite movement and CO_2 fixation in spinach chloroplasts. In *Proceedings Third International Congress on Photosynthesis* pp. 1369–79 (ed. M. Avron). Elsevier, Amsterdam.

HELDT H.W., SAUER F. & RAPLEY L. (1972) Differentiation of the permeability properties of the two membranes of the chloroplast envelope. *Proceedings of the 2nd International Congress on Photosynthesis Research* pp. 1345–55 (eds. G. Forti, M. Avron & A. Melandri). Dr. W. Junk, The Hague.

HEMBERG T. (1967) Abscisin II as an inhibitor of α-amylase. *Acta Chem. Scand.* **21**, 1665–6.

HEMBERG T. (1949) Growth-inhibiting substances in terminal buds of *Fraxinus*. *Physiol. Plant.* **2**, 37–44.

HENDRICKS S.B. & BORTHWICK H.A. (1967) The function of phytochrome in regulation of plant growth. *Proc. natn. Acad. Sci. U.S.A.* **58**, 2125–30.

HENNINGSEN K.W. (1967) An action spectrum for vesicle dispersal in bean plastids. *Biochemistry of Chloroplasts*, pp. 453–57 (ed. T.W. Goodwin). Academic Press, London.

HENNINGSEN K.W. & BOARDMAN N.K. (1973) Development of photochemical activity and the appearance of the high potential form of cytochrome *b*-559 in greening barley seedlings. *Plant Physiol.* **51**, 1117–26.

HENRICH R. & RAPOPART T.A. (1974) A linear steady-state treatment of enzymatic catalysis, critique of the crossover theorem and a general procedure to identify interaction sites with an effector. *Eur. J. Biochem.* **42**, 97–105.

HEPLER P.K. & JACKSON W.T. (1974) Microtubules and early stages of cell-plate formation in the endosperm of *Haemanthus katherinae* Baker. *J. Cell Biol.* **38**, 437–46.

HEPLER P.K. & PALEVITZ B.A. (1974) Microtubules and microfilaments. *Ann. Rev. Plant Physiol.* **25**, 309–62.

HERMANN R.G., BOHNERT H.J., KOWALLIK K.V. & SCHMITT J.H. (1975) Size, conformation and purity of chloroplast DNA of some higher plants. *Biochim. biophys. Acta* **378**, 305–17.

HERTEL R. & FLORY R. (1968) Auxin movement in corn coleoptiles. *Planta.* **82**, 123–44.

HERTEL R., THOMSON K. & RUSSO V.E.A. (1972) *In vitro* auxin binding to particulate cell fractions from corn coleoptiles. *Planta* **107**, 325–40.

HESLOP-HARRISON J. (1966) Structural features of the chloroplast. *Sci. Prog.* **54**, 519–41.

HEW C.S., KROTKOV G. & CANVIN D.T. (1969) Determination of the rate of CO_2 evolution by green leaves in light. *Plant Physiol.* **44**, 662–70.

HEYDEN H.W. & ZACHAU H.G. (1971) Characterisation of RNA in fractions of calf thymus chromatin. *Biochim. biophys. Acta* **232**, 651–60.

HEYN A.N.J. (1971) Observations on the exocytosis of secretory vesicles and their products in coleoptiles of *Avena*. *J. Ultrastructure Res.* **37**, 69–81.

HEYN R.F. & SCHILPEROORT R.A. (1973) The use of protoplasts to follow the fate of *Agrobacterium tumefasciens* DNA on incubation with tobacco cells. *Colloq. Int. Cent. Nat. Rech. Sci.* **212**, 385–95.

HIATT A.J. (1961) Preparation and some properties of soluble succinic dehydrogenase from higher plants. *Plant Physiol.* **36**, 552–7.

HIATT A.J. (1962) Condensing enzyme from higher plants. *Plant Physiol.* **37**, 85–9.

HIGGINS T.J.V., MERCER J.F.B. & GOODWIN P.B. (1973) Poly(A) sequences in plant polysomal RNA. *Nature (New Biol.)* **246**, 68–70.

HIGINBOTHAM N. & ANDERSON W.P. (1974) Electrogenic pumps in higher plant cells. *Can. J. Bot.* **52**, 1011–21.

HIGINBOTHAM N., GRAVES J.S. & DAVIS R.F. (1970) Evidence for an electrogenic ion transport pump in cells of higher plants. *J. Membrane Biol.* **2**, 210–22.

HILL R. (1939) Oxygen produced by isolated chloroplasts. *Proc. Roy. Soc. London* **B.127**, 192–210.

HILL R. & BENDALL F. (1960) Function of two cytochrome components in chloroplasts: a working hypothesis. *Nature* **186**, 136–7.

HILL R.D. & BOYER P.D. (1967) Inorganic orthophosphate activation and adenosine diphosphate as the primary phosphoryl acceptor in oxidative phosphorylation. *J. Biol. Chem.* **242**, 4320–3.

HILLMAN W.S. (1967) The physiology of phytochrome. *Ann. Rev. Plant Physiol.* **18**, 301–24.

HILLMAN W.S. (1972) On the physiological significance of *in vivo* phytochrome assays. *Phytochrome* pp. 573–84 (eds. K. Mitrakos & W. Shropshire Jr.). Academic Press, New York.

HITCHCOCK C. & NICHOLS B.W. (1971) *Plant Lipid Biochemistry*. Academic Press, London and N.Y.

HODGES T.K. & HANSON J.B. (1965) Calcium accumulation by maize mitochondria. *Plant Physiol.* **40**, 101–9.

HODGES T.K., LEONARD R.T., BRACKER C.E. & KEENAN T.W. (1972) Purification of an ion-stimulated adenosine triphosphatase from plant roots: association with plasma membranes. *Proc. natn. Acad. Sci. U.S.A.* **69**, 3307–11.

HOFFMANN H.-P. & AVERS C.J. (1973) Mitochondrion of yeast: Ultrastructural evidence for one giant, branched organelle per cell. *Science* **181**, 749–51.

HOLL F.B., GAMBORG O.L., OHYAMA K. & PELCHER L. (1974) Genetic transformation in plants. *Tissue Culture and Plant Science* pp. 301–27 (ed. H.E. Street). Academic Press, London.

HOLM R.E. & ABELES F.B. (1967) Abscission: the role of RNA synthesis. *Plant Physiol.* **42**, 1094–102.

HOLM R.E., O'BRIEN T.J., KEY J.L. & CHERRY J.H. (1970) The influence of auxin and ethylene on chromatin-directed ribonucleic acid synthesis in soybean hypocotyl. *Plant Physiol.* **45**, 41–4.

HOLMES D.S., MAYFIELD J.E., SANDER G. & BONNER J. (1972) Chromosomal RNA: its properties. *Science* **177**, 72–4.

HOLZ R. & FINKELSTEIN A. (1970) Water and non-electrolyte permeability induced in thin lipid membranes by the polyene antibiotics nystatin and amphotericin B. *J. Gen. Physiol.* **56**, 125–45.

HONGLADAROM T., HONDA S.I. & WILDMAN S.G. (1965) *Organelles in living plant cells* (16 mm sound film). Educational Films Sales and Rentals, University Extension, University of California, Berkeley.

HOWARD A. & PELC S.R. (1953) Synthesis of desoxyribonucleic acid in normal and irradiated cells and its relation to chromosome breakage. *Heredity* **6** (Suppl.), 261–73.

HU A.S., BOCK R.M. & HALVORSON H.O. (1962) Separation of labeled from unlabeled proteins by equilibrium density gradient sedimentation. *Anal. Biochem.* **4**, 489–504.

HUANG A.H.C. & BEEVERS H. (1971) Isolation of microbodies from plant tissues. *Plant Physiol.* **48**, 637–41.

HUANG L., LORCH S.K., SMITH G.G. & HUANG A. (1974) Control of membrane lipid fluidity in *Acholeplasma laidlawii*. *FEBS Letters* **43**, 1–5.

HUANG R.C. (1967) Dihydrouridylic acid containing RNA from chick embryo chromatin. *Fed. Proc. Fedn. Am. Socs. exp. Biol.* **26**, 1933, p. 603.

HUFFAKER R.C. & PETERSON L.W. (1974) Protein turnover in plants and possible means of its regulation. *Ann. Rev. Plant Physiol.* **25**, 363–92.

HULME A.C., JONES J.D. & WOOLTORTON L.S.C. (1964) Mitochondrial preparations from the fruit of the apple. I. Preparation and general activity. *Phytochemistry.* **3**, 173–88.

HULME A.C., RHODES M.J.C. & WOOLTORTON L.S.C. (1971) The relationship between ethylene and the synthesis of RNA and protein in ripening apples. *Phytochemistry.* **10**, 749–56.

HUMPHREY T.J. & DAVIES D.D. (1975) A new method for the measurement of protein turnover. *Biochem. J.* **148**, 119–27.

HUTTON D. & STUMPF P.K. (1969) Fat metabolism in higher plants XXXVII. Characterization of the β-oxidation systems from maturing and germinating castor bean seeds. *Plant Physiol.* **44**, 508–16.

IHLE J.N. & DURE L.S. (1972) The temporal separation of transcription and translation and its control in cotton embryogenesis and germination. In *Plant Growth Substances* (1970) pp. 217–221 (ed. D.J. Carr). Springer-Verlag, Berlin.

IKUMA H. (1970) Necessary conditions for isolation of tightly coupled higher plant mitochondria. *Plant Physiol.* **45**, 773–81.

IKUMA H. & BONNER W.D.Jr. (1967) Properties of higher plant mitochondria. I. Isolation and some characteristics of tightly-coupled mitochondria from dark-grown mung bean hypocotyls. *Plant Physiol.* **42**, 67–75.

ILAN I. (1973) On auxin-induced pH drop and on the improbability of its involvement in the primary mechanism of auxin-induced growth promotion. *Physiol. Plant* **28**, 146–8.

IMBER D. & TAL M. (1970) Phenotypic reversion of flacca, a wilty mutant of tomato, by abscisic acid. *Science* **169**, 592–3.

INGLE J., KEY J.L. & HOLM R.E. (1965) Demonstration and characterisation of a DNA-like RNA in excised plant tissue. *J. molec. Biol.* **11**, 730–46.

INGLE J., PEARSON G.G. & SINCLAIR J. (1973) Species distribution and properties of nuclear satellite DNA in higher plants. *Nature* (*New Biol.*) **242**, 193–7.

INGLE J., POSSINGHAM J.V., WELLS R., LEAVER C.J. & LOENING U.E. (1970) The properties of chloroplast ribosomal-RNA. *Soc. Exp. Biol. Symp.* **24**, 303–25.

INGLE J. & SINCLAIR J. (1972) Ribosomal RNA genes and plant development. *Nature* **235**, 30–2.

INGLE J. & TIMMIS J.N. (1975) A role for differential replication of DNA in development. In *Modification of the Information Content of Plant Cells*, second John Innes Symposium pp. 37–52 (eds. R. Markham, D.R. Davies, D.A. Hopwood, R.W. Horne). North-Holland Pub. Co., Amsterdam.

INGLE J., TIMMIS J.N. & SINCLAIR J. (1975) The relationship between satellite deoxyribonucleic acid, ribosomal ribonucleic acid gene redundancy, and genome size in plants. *Plant Physiol.* **55**, 496–501.

INOUÉ S. & SATO H. (1967) Cell motility by labile association of molecules. The nature of mitotic spindle fibers and their role in chromosome movement. *J. gen. Physiol.* (*Suppl.*) **50**, 259–92.

IRVING A.A. (1910) The beginning of photosynthesis and the development of chlorophyll. *Ann. Bot. O.S.* **24**, 805–18.

JACKSON M. & INGLE J. (1973) The nature of polydisperse ribonucleic acid in plants. *Biochem. J.* **131**, 523–33.

JACOB F. & MONOD J. (1961) Genetic regulatory mechanisms in the synthesis of proteins. *J. molec. Biol.* **3**, 318–56.

JACOBSON A., FIRTEL R.A. & LODISH H.F. (1974) Synthesis of messenger and ribosomal RNA precursors in isolated nuclei of the cellular slime mould *Dictyostelium discoideum. J. molec. Biol.* **82**, 213–30.

JACOBSON R.N. & BONNER J. (1971) Studies of the chromosomal RNA and of the chromosomal RNA-binding protein of higher organisms. *Arch. Biochem. Biophys.* **146**, 557–63.

JAFFE M.J. (1968) Phytochrome-mediated bioelectric potentials in mung bean seedlings. *Science* **162**, 1016–17.

JAFFE M.J. (1970) Evidence for the regulation of phytochrome-mediated processes in mung bean roots by the neurohumor, acetylcholine. *Plant Physiol.* **46**, 768–77.

JAGENDORF A.T. & URIBE E. (1966) ATP formation caused by acid-base transition of spinach chloroplasts. *Proc. natn. Acad. Sci. U.S.A.* **55**, 170–7.

JAMES D.W., RABIN B.R. & WILLIAMS D.J. (1969) Role of steroid hormones in the interaction of polysomes with endoplasmic reticulum. *Nature* (*Lond.*) **224**, 371–2.

JAMIESON J.D. (1975) Membranes and secretion. In *Cell Membranes Biochemistry, Cell Biology and Pathology*, pp. 143–152 (eds. G. Weissmann & R. Claiborne). HP Publishing Co., New York.

JENSEN E.V., SUSUKI T., KAWASHIMA T., STUMPF W.E., JUNGBLUT P.W. & DESOMBRE E.R. (1968) A two-step mechanism for the interaction of estradiol with rat uterus. *Proc. natn. Acad. Sci. U.S.A.* **59**, 632–8.

JENSEN W.A. (1965) The ultrastructure and composition of the egg and central cell of *Gossypium. Am. J. Bot.* **52**, 781–97.

JEREZ C., SANDOVAL A., ALLENDE J.E., HENES C. & OFENGAND J. (1969) Specificity of the interaction of aminoacyl ribonucleic acid with a protein-guanosine triphosphate complex from wheat embryo. *Biochemistry* **8**, 3006–14.

JIMENEZ E., BALDWIN R.L., TOLBERT N.E. & WOOD W.A. (1962) Distribution of C^{14} in sucrose from glycolate-C^{14} and serine-3-C^{14} metabolism. *Arch. Biochem. Biophys.* **98**, 172–5.

JOHANSEN D.L. (1950) *Plant Embryology-Embryogeny of the Spermaphyta.* Chronica Botanica Co., Waltham, Mass.

JOHNSON C.B., GRIERSON D. & SMITH H. (1973) Expression of λplac5 DNA in cultured cells of a higher plant. *Nature* (*New Biol.*) **244**, 105–7.

JOHNSON H.M. & WILSON R.H. (1972) Sr^{2+} uptake by bean (*Phaseolus vulgaris*) mitochondria. *Biochim. biophys. Acta* **267**, 398–408.

JOHNSON K.D. & KENDE H. (1971) Hormonal control of lecithin synthesis in barley aleurone cells: regulation of the CDP-choline pathway by gibberellin. *Proc. natn. Acad. Sci. U.S.A.* **68**, 2674–7.

JONES J.D., HULME A.C. & WOOLTORTON L.S.C. (1965) Use of polyvinyl pyrrolidone in the isolation of enzymes from apple fruit. *Phytochemistry.* **4**, 659–76.

JONES R.L. (1973) Gibberellins: their physiological role. *Ann. Rev. Plant Physiol.* **24**, 571–98.

JOY K.W. & ELLIS R.J. (1975) Protein synthesis in chloroplasts. IV Polypeptides of the chloroplast envelope. *Biochim. biophys. Acta* **378**, 143–51.

JUNIPER B.E. & PASK G. (1973) Directional secretion by the Golgi bodies in maize root cells. *Planta* **109**, 225–31.

KACSER H. & BURNS J.A. (1973) The control of flux. In *Symp. Soc. expt. Biol.* **27**, pp. 65–104. Cambridge University Press.

KAGAWA T. MCGREGOR D.I. & BEEVERS H. (1973) Development of enzymes in the cotyledons of watermelon seedlings. *Plant Physiol.* **51**, 66–71.

KANABUS J. & CHERRY J.H. (1971) Isolation of an organ-specific leucyl-tRNA synthetase from soybean seedling. *Proc. natn. Acad. Sci. U.S.A.* **68**, 873–6.

KANAI R. & EDWARDS G.E. (1973) Purification of enzymatically isolated mesophyll protoplasts from C_3, C_4 and Crassulacean acid metabolism plants using an aqueous dextran-polyethylene glycol two-phase system. *Plant Physiol.* **52**, 484–90.

KAO K.N., CONSTABEL F., MICHAYLUK M.R. & GAMBORG O.L. (1974) Plant protoplast fusion and growth of intergeneric hybrid cells. *Planta* **120**, 215–27.

KARTHA K.K., MICHAYLUK M.R., KAO K.N., GAMBORG O.L. & CONSTABEL F. (1974) Callus formation and plant regeneration from mesophyll protoplasts of rape plants (*Brassica napus* L. cv. Zephyr). *Plant Sci. Letters* **3**, 265–71.

KASEMIR H., BERGFELD R. & MOHR H. (1975) Phytochrome control of prolamellar body reorganization and plastid size in mustard cotyledons. *Photochem. Photobiol.* **21**, 111–20.

KAUL B. & STABA E.J. (1967) *Amoni visnaga* L. Lam. tissue cultures: multiliter suspension growth and examination for furanochromes. *Planta Medica.* **15**, 145–56.

KAWASHIMA N. & WILDMAN S.G. (1972) Studies on fraction I protein IV mode of inheritance of primary structure in relation to whether chloroplast or nuclear DNA contains the code for a chloroplast protein. *Biochim. biophys. Acta* **262**, 42–9.

KEENAN T.W., LEONARD R.T. & HODGES T. K. (1973) Lipid composition of plasma membranes from *Avena sativa* roots. *Cytobios* **7**, 103–12.

KELLER S.J., BIEDENBACH S.A. & MEYER R.R. (1973) Partial purification of a chloroplast DNA polymerase from *Euglena gracilis*. *Biochem. biophys. Res. Comm.* **50**, 620–8.

KELLER W.A. & MELCHERS G. (1973) The effect of high pH and calcium on tobacco leaf protoplast fusion. *Z. Naturforsch.* **28c**, 737–41.

KEMP R.J. & MERCER E.I. (1968) Sterols and sterol esters of the intracellular organelles of maize shoots. *Biochem. J.* **110**, 119–25.

KENDRICK R.E. (1974) Phytochrome intermediates in freeze-dried tissue. *Nature (Lond.)* **250**, 159–61.

KENDRICK R.E. & SPRUIT C.J.P. (1972) Phytochrome properties and the molecular environment. *Plant Physiol.* **52**, 327–31.

KENDRICK R.E. & SPRUIT C.J.P. (1973) Phytochrome intermediates *in vivo*—III. Kinetic analysis of intermediate reactions at low temperature. *Photochem. Photobiol.* **18**, 153–9.

KESSEL R.H. & CARR A.H. (1972) The effect of dissolved O_2 concentration on growth and differentiation of carrot (*Daucus carota*) tissue. *J. exp. Bot.* **23**, 996–1007.

KEY J.L. (1969) Hormones and nucleic acid metabolism. *Ann. Rev. Plant Physiol.* **20**, 449–74.

KEY J.L. (1964) Ribonucleic acid and protein synthesis as essential processes for cell elongation. *Plant Physiol.* **39**, 365–70.

KEY J.L. & INGLE J. (1964) The requirement for the synthesis of DNA-like RNA for growth of excised plant tissue. *Proc. natn. Acad. Sci. U.S.A.* **52**, 1382–8.

KHAN A. (1969) Cytokinin-inhibitor antagonism in the hormonal control of α-amylase synthesis and growth in barley seed. *Physiol. Plant.* **22**, 94–103.

KIDD G.H. & PRATT L.H. (1973) Phytochrome destruction. An apparent requirement of protein synthesis in the induction of the destruction mechanism. *Plant Physiol.* **52**, 309–11.

KIERMAYER O. & DOBBERSTEIN B. (1973) Membrankomplexe dictyosomaler Herkunft als "Matrizen" für die extraplasmatische Synthese und Orientierung von Mikrofibrillen. *Protoplasma* **77**, 437–51.

KING P.J. (1976) Studies on the growth and culture of plant cells. XX. Utilization of 2,4-dichlorophenoxyacetic acid by steady-state cultures of *Acer pseudoplatanus*. *J. exp. Bot.* **27**, 1053–72.

KING P.J., COX B.J., FOWLER M.W. & STREET H.E. (1974) Metabolic events in synchronized cell cultures of *Acer pseudoplatanus*, L. *Planta.* **117**, 109–22.

KING P.J., MANSFIELD K.J. & STREET H.E. (1973) Control of growth and cell division in plant cell suspension cultures. *Canad. J. Bot.* **51**, 1807–23.

KING P.J. & STREET H.E. (1973) Growth patterns in cell cultures. In *Plant Tissue and Cell Culture*, pp. 269–337 (ed. H.E. Street). Blackwell Sci. Publ., Oxford.

KIRK B.K. & HANSON J.B. (1973) The stoichiometry of respiration driven potassium transport in corn mitochondria. *Plant Physiol.* **51**, 357–62.

Kirk J.T.O. (1963) The deoxyribonucleic acid of broad bean chloroplasts. *Biochim. biophys. Acta* **76**, 417–24.

Kirk J.T.O. (1967) Determination of the base composition of deoxyribonucleic acid by measurement of the adenine/guanine ratio. *Biochem. J.* **105**, 673–7.

Kirk J.T.O. (1971) Will the real chloroplast DNA please stand up? In *Autonomy and Biogenesis of Mitochondria and Chloroplasts.* (eds. N.K. Boardman, A.W. Linnane & R.M. Smillie). North-Holland, Amsterdam.

Kirk J.T.O. (1972) The genetic control of plastid formation: recent advances and strategies for the future. *Sub-Cell. Biochem.* **1**, 333–61.

Kirk J.T.O. & Tilney-Bassett R.A.E. (1967) *The Plastids.* W.H. Freeman & Co., London.

Kisaki T. & Tolbert N.E. (1969) Glycolate and glyoxylate metabolism by isolated peroxisomes and chloroplasts. *Plant Physiol.* **44**, 242–50.

Kit S., Dubbs D.R., Piekanski L.J. & Hsu T.S. (1963) Deletion of thymidine kinase activity from L cells resistant to bromodeoxyurdine. *Expt. Cell Research* **31**, 297–312.

Kitasato H. (1968) The influence of H^+ on the membrane potential and ion fluxes of *Nitella J. Gen. Physiol.* **52**, 60–87.

Klein W.H., Nolan C., Lazar J.M. & Clark J.M. Jr. (1972) Translation of satellite tobacco necrosis virus ribonucleic acid. I. Characterization of *in vitro* procaryotic and eucaryotic translation products. *Biochem.* **11**, 2009–14.

Kleinschmidt M.G. & McMahon V.A. (1970) Effect of growth temperature on the lipid composition of *Cyanidium caldarium.* II Glycolipid and phospholipid components. *Plant Physiol.* **46**, 290–3.

Knowland J.S. (1973) *Proceedings of the 13th International Congress of Genetics,* Berkeley, U.S.A.

Koehler D. & Varner J.E. (1973) Hormonal control of orthophosphate incorporation into phospholipids of barley aleurone layers. *Plant Physiol.* **52**, 208–14.

Kolodner K.K. & Tewari K.K. (1972) Molecular size and conformation of chloroplast deoxyribonucleic acid from pea leaves *J. biol. Chem.* **247**, 6355–64.

Komor E. & Tanner W. (1974) The hexose-proton co-transport system of *Chlorella.* pH-dependent change in K_m values and translocation constants of the uptake system. *J. gen. Physiol.* **64**, 568–81.

Konar R.N., Thomas E. & Street H.E. (1972) The diversity of morphogenesis in suspension cultures of *Atropa belladonna,* L. *Ann. Bot. N.S.* **36**, 249–58.

Kornberg H.L. & Krebs H.A. (1957) Synthesis of cell constituents from C_2-units by a modified tricarboxylic acid cycle. *Nature* **179**, 988–91.

Kornberg R.D. (1974) Chromatin structure: a repeating unit of histones and DNA. *Science* **184**, 868–71.

Koshland D.E. (1970) The molecular basis for enzyme regulation. In *The Enzymes* **1**, pp. 341–96 (ed. P.D. Boyer). Academic Press, London.

Krauspe R. & Parthier B. (1973) Chloroplast- and cytoplasm-specific aminoacyl-tRNA synthetases of *Euglena gracilis:* separation, characterization and site of synthesis. In *Biochemical Society Symposium No. 38, Nitrogen Metabolism in Plants,* pp. 111–35 (eds. T.H. Goodwin & R.M.S. Smellie). London.

Krebs H.A. (1959) In *Regulation of Cell Metabolism,* CIBA Foundation Symposium (eds. G.E.W. Wolstenholme and C.M. O'Connor). Churchill-Livingstone, London.

Krebs H.A. & Kornberg H.L. (1957) A survey of the energy transformations in living matter. *Ergeb. Physiol. biol. Chem. u. exptl. Pharmakol.* **49**, 212–98.

Krisko I., Gordon J. & Lipmann F. (1969) Studies on the interchangeability of one of the mammalian and bacterial supernatant factors in protein biosynthesis. *J. biol. Chem.* **244**, 6117–23.

Kroes H.H. (1970) A study of phytochrome, its isolation, structure and photochemical transformations. *Meded. LandbHoogesch. Wageningen* **70–18**, 1–113.

Kroman E.F. & McLick J. (1970) A dynamic stereo-chemical reaction mechanism for the ATP synthesis reaction of mitochondrial oxidative phosphorylation. *Proc. natn. Acad. Sci., U.S.A.* **67**, 1130–6.

De Kruijff B. & Demel R.A. (1974) Polyene antibiotic-sterol interactions in membranes of *Acholeplasma laidlawii* cells and lecithin liposomes. III. Molecular structure of the polyene antibiotic—cholesterol complexes. *Biochim. biophys. Acta* **339**, 57–70.

De Kruyff B., Van Dijck P.W.M., Demel R.A., Schuijff A., Brants A. & Van Deenen L.L.M. (1974) Non-random distribution of cholesterol in phosphatidyl choline bilayers. *Biochim. biophys. Acta* **356**, 1–7.

KU H.S. & LEOPOLD A.C. (1970) Mitochondrial responses to ethylene and other hydrocarbons. *Plant Physiol.* **46**, 842–4.

KUIPER P.J.C. (1968) Lipids in Grape Roots in Relation to Chloride Transport. *Plant Physiol.* **43**, 1367–71.

KUNG S.D. & WILLIAMS J.P. (1969) Chloroplast DNA from broad bean. *Biochim. biophys. Acta* **195**, 434–45.

KURIYAMA Y. & LUCK D.J.L. (1973) Ribosomal RNA synthesis in mitochondria of *Neurospora crassa*. *J. molec. Biol.* **73**, 425–37.

KURLAND C.G. (1972) Structure and function of the bacterial ribosomes. *Ann. Rev. Biochem.* **41**, 377–408.

LABAVITCH J.M. & RAY R.M. (1974) Turnover of cell wall polysaccharides in elongating pea stem segments. *Plant Physiol.* **53**, 669–73.

LAETSCH W.M. (1974) The C_4 syndrome: a structural analysis. *Ann. Rev. Plant Physiol.* **25**, 27–52.

LAMBOWITZ A.M. & BONNER W.D.Jr. (1974) The b-cytochromes of plant mitochondria. A spectrophotometric and potentiometric study. *J. Biol. Chem.* **249**, 2428–40.

LAMBOWITZ A.M., BONNER W.D.Jr. & WIKSTRÖM M.K.F. (1974) On the lack of ATP-induced midpoint potential shift for cytochrome b-566 in plant mitochondria. *Proc. natn. Acad. Sci. U.S.A.* **71**, 1183–7.

LAMPORT D.T.A. (1965) The protein content of primary cell walls. *Adv. Botan. Res.* **2**, 151–218.

LAMPORT D.T.A., KATONA L. & ROERIG S. (1973) Galactosylserine in extensin. *Biochem. J.* **133**, P9, 125.

LANCE C. & BONNER W.D.Jr. (1968) The respiratory chain components of higher plant mitochondria. *Plant Physiol.* **43**, 756–66.

LANGE H., SHROPSHIRE W. & MOHR H. (1971) An analysis of phytochrome-mediated anthocyanin synthesis. *Plant Physiol.* **47**, 649–55.

LARSSON C., COLLIN C. & ALBERTSSON P.-A. (1973) The fine structure of chloroplast stroma crystals. *J. Ultrastruct. Res.* **45**, 50–8.

LA RUE T. & GAMBORG O. (1971) Ethylene production by plant cell cultures. *Plant Physiol.* **48**, 394–8.

LATIES G.G. (1969) Dual mechanisms of salt uptake in relation to compartmentation and long distance transport. *Ann. Rev. Plant Physiol.* **20**, 89–116.

LEAVER C.J. (1975) The biogenesis of plant mitochondria. *The Chemistry and Biochemistry of Plant Proteins*, pp. 137–65. (ed. J. Harborne). Academic Press, London.

LEAVER C.J. & HARMEY M.A. (1973) Plant mitochondrial nucleic acids. *Biochem. Soc. Symp.* **38**, 175–93.

LEAVER C. & DYER J.A. (1974) Short Communications. Caution in the interpretation of plant ribosome studies. *Biochem. J.* **144**, 165–7.

LEAVER C.J. & KEY J.L. (1970) Ribosomal RNA synthesis in plants. *J. molec. Biol.* **49**, 671–80.

LEDBETTER M.C. & PORTER K.R. (1963) A 'microtubule' in plant cell fine structure. *J. Cell Biol.* **19**, 239–50.

LEDBETTER M.C. & PORTER K.R. (1970) *Introduction to the Fine Structure of Plant Cells*, p. 44. Springer-Verlag, Berlin.

LEE D.C. & WILSON R.H. (1972) Swelling in bean shoot mitochondria induced by a series of potassium salts of organic anions. *Physiol. Plant.* **27**, 195–201.

LEGOCKI A.B. (1973) Function of elongation factors in peptide synthesis. In *Biochemical Society Symposium No. 38, Nitrogen Metabolism in Plants* pp. 57–76 (eds. T.H. Goodwin & R.M.S. Smellie). London.

LEGOCKI A.B. & MARCUS A. (1970) Polypeptide synthesis in extracts of wheat germ. *J. biol. Chem.* **245**, 2814–18.

LEGON S., DARNBROUGH C.H., HUNT T. & JACKSON R.J. (1973) Initiation of eukaryotic protein synthesis: native 40S subunits bind initiator transfer ribonucleic acid before associating with messenger ribonucleic acid. *Biochem. Soc. Trans.* **1(3)**, 553–7.

LEHNINGER A.L. (1965) *Bioenergetics*. Benjamin, Inc. New York.

LEIS J.P. & KELLER E.B. (1970) Protein chain initiation by methionyl-tRNA. *Biochem. biophys. Res. Commun.* **40**, 416–21.

LEIS J.P. & KELLER E.B. (1971) N-formylmethionyl-tRNA of wheat chloroplasts. *Biochemistry* **10**, 889–94.

LEMBI C.A., MORRÉ D.J., THOMSON K. & HERTEL R. (1971) N-1-napthylphtalmic acid binding activity of plasma membrane rich fraction from maize coleoptiles. *Planta.* **99**, 37–45.

LENTON J.R., PERRY V.M. & SAUNDERS P.F. (1971) The identification and quantitative analysis of abscisic acid in plant extracts by gas–liquid chromatography. *Planta.* **96**, 271–80.

Leonard R.T. & Van Der Woude W.J. (1976) Isolation of plasma membranes from corn roots by sucrose density gradient centrifugation. *Plant Physiol.* **57**, 105–14.

Lescure A.M. (1973) Selection of markers of resistance to base analogues in somatic cell cultures of *Nicotiana tabacum. Pl. Sci. Lett.* **1**, 375–83.

Levine R.P. (1969) The analysis of photosynthesis using mutant strains of algae and higher plants. *Ann. Rev. Plant Physiol.* **20**, 523–40.

Levine R.P. & Goodenough U.W. (1970) The genetics of photosynthesis and of the chloroplast in *Chlamydomonas reinhardi. Ann. Rev. Genet.* **4**, 397–408.

Lhoste J.-M. (1972) Some physical aspects of the phytochrome photo-transformation. *Phytochrome* pp. 47–74 (eds. K. Mitrakos & W. Shropshire Jr.). Academic Press, New York.

Lichtenthaler H.K. & Park R.B. (1963) Chemical composition of chloroplast lamellae from spinach. *Nature* **198**, 1070–2.

Lieberman M. & Biale J.B. (1956) Cofactor requirement for oxidation of α-keto acids. *Plant Physiol.* **31**, 425–9.

Lieberman M. & Biale J.B. (1955) Effectiveness of EDTA in the activation of oxidations mediated by mitochondria from brocoli buds. *Plant Physiol.* **30**, 549–52.

Lin C.Y., Key J.L. & Bracker C.E. (1966) Association of D-RNA with polyribosomes in the soybean root. *Plant Physiol.* **41**, 976–82.

Lindsay J.G. & Wilson D.F. (1972) Apparent adenosine triphosphate induced ligand change in cytochrome a_3 of pigeon heart mitochondria. *Biochemistry* **11**, 4613–21.

Ling G.N. (1973) What component of the living cell is responsible for its semipermeable properties? Polarized water or lipids. *Biophys. J.* **13**, 807–15.

Linschitz H. & Kasche V. (1967) Kinetics of phytochrome conversion: Multiple pathways in the $P_r \rightarrow P_{fr}$ reaction as studied by double-flash technique. *Proc. natn. Acad. Sci. U.S.A.* **58**, 1059–64.

Linschitz H., Kasche V., Butler W.L. & Siegelman H.W. (1966) The kinetics of phytochrome conversion. *J. biol. Chem.* **241**, 3395–403.

Lipe W.N. & Crane J.C. (1966) Dormancy regulation in peach seeds. *Science.* **153**, 541–2.

Littauer U.Z. & Inouye H. (1973) Regulation of tRNA *A. Rev. Biochem.* **42**, 439–70.

Littlefield J.W. (1964a) Three degrees of quanylic acid–inosinic acid pyrophosphorylase deficiency in mouse fibroblasts. *Nature* **203**, 1142–4.

Littlefield J.W. (1964b) Selection of hybrids from matings of fibroblasts *in vitro* and their presumed recombinants. *Science* **145**, 709–10.

Lockard R.E. & Lingrel J.B. (1969) The synthesis of mouse hemoglobin β-chains in a rabbit reticulocyte cell-free system programmed with mouse reticulocyte 9s RNA. *Biochem. biophys. Res. Commun.* **37**, 204–12.

Locy R. (1974) *The purification and some properties of cytoplasmic and chloroplast tyrosyl transfer RNA synthetases and evidence for a ribonuclease specific chloroplast transfer RNA.* Ph.D. Thesis. Purdue University, December.

Lodish H.F. (1974) Model for the regulation of mRNA translation applied to haemoglobin synthesis. *Nature* (*Lond.*) **251**, 385–8.

Loening U.E. (1968) Molecular weights of ribosomal RNA in relation to evolution. *J. molec. Biol.* **38**, 355–65.

Loening U.E. (1969) The determination of the molecular weight of ribonucleic acid by polyacrylamide gel electrophoresis. The effects of changes in conformation. *Biochem. J.* **113**, 131–8.

Loening U.E. (1970) The mechanism of synthesis of ribosomal RNA. *Symp. Soc. Gen. Microbiol.* **20**, 77–106.

Loercher L. (1966) Phytochrome changes correlated to mesocotyl inhibition in etiolated *Avena* seedlings. *Plant Physiol.* **41**, 932–6.

Lord J.M. & Merrett M.J. (1970) The pathway of glycolate utilization in *Chlorella pyrenoidosa. Biochem. J.* **117**, 929–37.

Lord J.M., Kagawa T., Moore T.S. & Beevers H. (1973) Endoplasmic reticulum as the site of lecithin formation in castor bean endosperm. *J. Cell Biol.* **57**, 659–67.

Lorimer G.H. & Miller R.J. (1969) The osmotic behaviour of corn mitochondria. *Plant Physiol.* **44**, 839–44.

Lowendorf H., Slayman C.L. & Slayman C.W. (1974) Phosphate transport in *Neurospora.* Kinetic characterization of a constitutive, low affinity transport system. *Biochem. biophys. Acta* **373**, 369–82.

Lucy J.A. (1970) The fusion of biological membranes. *Nature*, **227**, 814–17.

Lundquist R.E., Lazar J.M., Klein W.H.

& CLARK M.J.Jr. (1972) Translation of satellite tobacco necrosis virus ribonucleic acid. II. Initiation of *in vitro* translation in prokaryotic and eukaryotic systems. *Biochem.* **11**(11), 2014–19.

LYONS J.M. & PRATT H.K. (1964) An effect of ethylene on swelling of isolated mitochondria. *Arch. Biochem. Biophys.* **104**, 318–24.

LYTTLETON J.W. (1962) Isolation of ribosomes from spinach chloroplasts. *Expl. Cell Res.* **26**, 312–17.

MAALE G., STEIN G. & MANS R. (1975) Effects of cordycepin and cordycepintriphosphate on polyadenylic and ribonucleic acid–synthesising enzymes from eukaryotes. *Nature* **255**, 80–2.

McCARTY R.E., DOUCE R. & BENSON A.A. (1973) Acyl lipids of highly purified plant mitochondria. *Biochim. biophys. Acta* **316**, 266–70.

McINTOSH J.R., HEPLER P.R. & VANWIE D. G. (1969) Model for mitosis. *Nature, Lond.* **224**, 659–63.

McMURRAY W.C. & DAWSON R.M.C. (1969) Phospholipid exchange reactions within the liver cell. *Biochem. J.* **112**, 91–108.

McMURROUGH I. & BARTNICKI-GARCIA S. (1973) Inhibition and activation of chitin synthesis by *Mucor rouxii* cell extracts. *Arch. Biochem. Biophys.* **158**, 812–16.

McNAB J.M., VILLEMEZ C.L. & ALBERSHEIM P. (1968) Biosynthesis of galactan by a particulate preparation from *Phaseolus aureus* seedlings. *Biochem. J.* **106**, 355–60.

McNIEL M., ALBERSHEIM P., TAIZ L. & JONES R. (1974) On the structure of the walls of barley aleurone cells. *Plant Physiol.* **53S**, 15.

McWILLIAM A.A., SMITH S.M. & STREET H. E. (1974) The origin and development of embryoids in suspension cultures of carrot (*Daucus carota*) *Ann. Bot.* N.S. **38**, 243–50.

MACGILLIVRAY A.J. & RICKOD D. (1974) The role of chromosomal proteins as gene regulators. *Biochemistry of Cell Differentiation* pp. 301–361 (ed. J. Paul). Butterworths, London.

MACKENDER R.D. & LEECH R.M. (1974) The galactolipid, phospholipid and fatty acid composition of the chloroplast envelope membranes of *Vicia faba* L. *Plant Physiol.* **53**, 496–502.

MACKENZIE J.M., COLEMAN A.R. & PRATT L.H. (1974) A specific reversible intracellular localisation of phytochrome as P_{fr}. *Plant Physiol.* **53S**, 2.

MACKENZIE I.A. & STREET H.E. (1970) Studies on the growth in culture of plant cells. VIII. The production of ethylene by suspension cultures of *Acer pseudoplatanus*, L. *J. exp. Bot.* **21**, 824–34.

MACRAE A.R. (1971a) Effect of pH on the oxidation of malate by isolated cauliflower bud mitochondria. *Phytochemistry.* **10**, 1453–8.

MACRAE A.R. (1971b) Malic enzyme activity of plant mitochondria. *Phytochemistry.* **10**, 2343–7.

MACRAE A.R. & MOORHOUSE R. (1970) The oxidation of malate by mitochondria isolated from cauliflower buds. *Eur. J. Biochem.* **16**, 96–102.

MACROBBIE E.A.C. (1969) Ion fluxes to the vacuole of *Nitella transluscens*. *J. Exp. Bot.* **20**, 236–56.

MACROBBIE E.A.C. (1971) Fluxes and Compartmentation in Plant Cells. *Ann. Rev. Plant Physiol.* **22**, 75–96.

MAHLBERG P.C., TURNER F.R., WALKINSHAW C. & VENKETSWARAN S. (1974) Ultrastructural studies of plasma membrane related secondary vacuoles in cultured cells. *Amer. J. Bot.* **61**, 730–8.

MAK S. (1965) Mammalian cell cycle analysis using microspectrophotometry combined with autoradiography. *Exp. Cell Res.* **39**, 286–9.

MALIGA P., MARTON L. & SZ. BREZNOVITS A. (1973) 5-bromodeoxyuridine-resistant cell lines from haploid tobacco. *Pl. Sci. Lett.* **1**, 119–21.

MALIGA P., SZ. BREZNOVITS A. & MARTON L. (1973) Streptomycin-resistant plants from callus culture of haploid tobacco. *Nature* (*New Biol.*) **244**, 29–30.

MALKIN R., APARICIO P.J. & ARNON D.I. (1974) The isolation and characterization of a new iron-sulphur protein from photosynthetic membranes. *Proc. natn. Acad. Sci. U.S.A.* **71**, 2362–6.

MANABE K. & FURUYA M. (1974) Phytochrome-dependent reduction of nicotinamide nucleotides in the mitochondrial fraction isolated from etiolated pea epicotyls. *Plant Physiol.* **53**, 343–7.

MANN J.D., STEINHART C.E. & MUDD S.H. (1963) Alkaloids and plant metabolism. IV. The tyramine methylpherase of barley roots. *J. Biol. Chem.* **238**, 381–5.

MANNELLA C.A. & BONNER W.D.Jr. (1975) Biochemical characteristics of the outer membranes of plant mitochondria. *Biochim. biophys. Acta* **413**, 213–25.

MANNING J.E., WOLSTENHOLME D.R., RYAN R.S., HUNTER J.A. & RICHARDS O.C. (1971) Circular chloroplast DNA from *Euglena gracilis. Proc. natn. Acad. Sci. U.S.A.* **68**, 1169–73.

MANNING J.E., WOLSTENHOLME D.R. & RICHARDS O.C. (1972) Circular DNA molecules associated with chloroplasts of *Spinacia oleracea. J. Cell Biol.* **53**, 594–601.

MANS R.J. (1967) Protein synthesis in higher plants. *Ann. Rev. Plant Physiol.* **18**, 127–46.

MANTON I. (1957) Observations with the electron microscope on the cell structure of the antheridium and spermatozoid of *Sphagnum. J. exp. Bot.* **8**, 382–400.

MANTON I. (1966) Observations on scale production in *Prymnesium parvum. J. Cell Science* **1**, 375–80.

MARCUS A., WEEKS D.P., LEIS J.P. & KELLER E.B. (1970a) Protein chain initiation by methionyl-tRNA in wheat embryo. *Proc. natn. Acad. Sci. U.S.A.* **67**, 1681–7.

MARCUS A., BEWLEY J.D. & WEEKS D.P. (1970b) Aurintricarboxylic acid and initiation factors of wheat embryo. *Science* **167**, 1735–6.

MARCUS A., WEEKS D.P. & SEAL S.N. (1973) Protein chain initiation in wheat embryo. In *Biochemical Society Symposium No. 38, Nitrogen Metabolism in Plants* pp. 97–109 (eds. T.H. Goodwin & R.M.S. Smellie). Biochemical Society, London.

MARCUS A., EFRON D. & WEEKS D.P. (1974) The wheat embryo cell-free system. *Methods in Enzymology, Nucl. Acids and Protein Synth.*, vol. XXX, part F, pp. 749–54 (eds. K. Moldave and L. Grossman). Academic Press, London.

MAREI N. & ROMANI R. (1971) Ethylene stimulated synthesis of ribosomes, ribonucleic acid, and protein in developing fig fruits. *Plant Physiol.* **48**, 806–8.

MARGULIS L. (1970) *Origin of Eukaryotic Cells.* 349 pp. Yale University Press, New Haven.

MARGULIS L. (1973) Colchicine sensitive microtubules. *Int. Rev. Cytol.* **34**, 333–61.

MARMÉ D., BOISARD J. & BRIGGS W.R. (1973) Binding properties *in vitro* of phytochrome to a membrane fraction. *Proc. natn. Acad. Sci. U.S.A.* **70**, 3861–5.

MARMÉ D., MACKENZIE J.M., BOISARD J. & BRIGGS W.R. (1974) The isolation and partial characterization of membrane vesicles containing phytochrome. *Plant Physiol.* **54**, 263–71.

MARMÉ D. & SCHÄFER E. (1972) On the localisation and orientation of phytochrome molecules in corn coleoptiles. *Z. Pflanzenphysiol.* **67**, 192–4.

MARMÉ D., SCHÄFER E., TRILLMICH F. & HERTEL R. (1971) Evidence for membrane bound phytochrome in maize coleoptiles. In *Book of Abstracts Europ. Ann. Symp. on Plant Photomorphogenesis, Athens-Eretria (Greece)*, p. 36.

MARRÈ E., LADO P., CALDOGNO R.F. & COLOMBO R. (1973) Correlation between cell enlargement in pea internode segments and decrease in the pit of the medium of incubation. *Pl. Sci. Lett.* **1**, 185–92.

MARSHALL M.O. & KATES M. (1973) Biosynthesis of phosphatidyl ethanolamine and phosphatidyl choline in spinach leaves. *FEBS Letters* **31**, 199–202.

MARTON L. & MALIGA P. (1974) Characterization of 5-bromodeoxyuridine resistant mutants of *Nicotiana tabacum* and *N. sylvestris. Abstracts Third International Congress of Plant Tissue and Cell Culture*, No. 70. University of Leicester.

MASUDA Y. (1968) Role of cell-wall degrading enzymes in cell-wall loosening in oat coleoptiles. *Planta.* **83**, 171–84.

MASUDA Y., OI S. & SATOMURA Y. (1970) Further studies on the role of cell-wall degrading enzymes in cell wall loosening in oat coleoptiles. *Pl. Cell Physiol.* **11**, 631–8.

MASUDA Y. & YAMAMOTO R. (1970 Effect of auxin on β-1,3-glucanase activity in Avena coleoptile. *Development, Growth and Differentiation.* **2**, 287–96.

MATTHEWS M.B. (1973) Mammalian messenger RNA. *Essays in Biochemistry* **9**, 59–102.

MATTHEWS M.B. (1973) Mammalian messenger RNA. *Essays in Biochemistry* **9**, 59–102.

MATTHEWS M.B. & KORNER A. (1970) Mammalian cell-free protein synthesis directed by viral ribonucleic acid. *Europ. J. Biochem.* **17**, 328–38.

MATTHYSSE A.G. & PHILLIPS C. (1969) A protein intermediary in the interaction of a hormone with the genome. *Proc. natn. Acad. Sci. U.S.A.* **63**, 877–903.

MAYO M.A. & COCKING E.C. (1969) Pinocytic uptake of Polystyrene latex particles by isolated tomato fruit protoplasts. *Protoplasma* **68**, 223–30.

MAZIA D., PETZELT CH., WILLIAMS R.O. & MEZA I. (1972) A Ca-activated ATPase in the mitotic apparatus of the sea urchin egg,

(isolated by a new method). *Expl. Cell Res.* **70**, 325–32.
MAZLIAK P. (1973) Lipid metabolism in plants. *Ann. Rev. Plant Physiol.* **24**, 287–310.
MAZLIAK P., OURSEL A., BEN ABDELKADER A. & GROSBOIS M. (1972) Biosynthesis of fatty acids in isolated plant mitochondria. *Eur. J. Biochem.* **28**, 399–411.
MENKE W. (1962) Structure and chemistry of plastids. *Ann. Rev. Plant. Physiol* **13**, 27–34.
METS L. & BOGORAD L. (1972) Altered chloroplast ribosomal proteins associated with erythromycin-resistant mutants in two genetic systems of *Chlamydomonas reinhardi. Proc. natn. Acad. Sci. U.S.A.* **69**, 3779–83.
MEYER A. (1883) Veber Krystalloide der Trophoplasten und über die Chromoplasten der Angiospermen. *Bot. Ztg.* **41**, 489–98.
MIFLIN B.J., MARKER A.F.H. & WHITTINGHAM C.P. (1966) The metabolism of glycine and glycolate by pea leaves in relation to photosynthesis. *Biochim. biophys. Acta* **120**, 266–73.
MILBORROW B.V. (1974) The chemistry and physiology of abscisic acid. *Ann. Rev. Plant Physiol.* **25**, 259–307.
MILCAREK C., PRICE R. & PENMAN S. (1974) The metabolism of a poly(A)minus mRNA fraction in Hela cells. *Cell.* **3**, 1–10.
MILLER A., KARLSSON U. & BOARDMAN N.K. (1966) Electron microscopy of ribosomes isolated from tobacco leaves. *J. molec. Biol.* **17**, 487–9.
MILLER C.O. (1969) Control of deoxyisoflavone synthesis in soybean tissue. *Planta* **87**, 26–35.
MILLER R.J., DUMFORD W.S. & KOEPPE D.E. (1970a) Effects of gramicidin on corn mitochondria. *Plant Physiol.* **46**, 471–4.
MILLER R.J., DUMFORD S.W., KOEPPE D.E. & HANSON J.B. (1970b) Divalent cation stimulation of substrate oxidation by corn mitochondria. *Plant Physiol.* **45**, 649–53.
MILLER R.J. & KOEPPE D.E. (1971) The effect of calcium and inhibitors on corn mitochondrial respiration. *Plant Physiol.* **47**, 832–5.
MILLERD A. (1953) Respiratory oxidation of pyruvate by plant mitochondria. *Arch. biochem. Biophys.* **42**, 149–63.
MILMAN G., GOLDSTEIN J., SCHOLNICK E. & CASKEY T. (1969) Peptide chain termination. III. Stimulation of *in vitro* termination. *Proc. natn. Acad. Sci. U.S.A.* **63**, 183–90.
MITCHELL P. (1961) Coupling of phosphorylation to electron and hydrogen transfer by a chemiosmotic type of mechanism. *Nature* **191**, 144–8.
MITCHELL P. (1966) Chemiosmotic coupling in oxidative and photosynthetic phosphorylation. *Biol. Rev.* **41**, 445–502.
MITCHISON J.M. (1971) *The Biology of the Cell Cycle.* Cambridge University Press.
MITRAKOS K. & SHROPSHIRE W.Jr. (1972) *Phytochrome.* Academic Press, London.
MOGENSEN H.L. (1972) Fine structure and composition of the egg apparatus before and after fertilisation in *Quercus gambelii:* the functional ovule. *Am. J. Bot.* **59**, 931–41.
MOHR H. (1966) Differential gene activation as a mode of action of phytochrome 730. *Photochem. Photobiol.* **5**, 469–83.
MOHR H. (1969) Photomorphogenesis. In *Physiology of Plant Growth and Development* pp. 509–16 (ed. M.B. Wilkins). McGraw-Hill, London.
MOHR H. (1972) *Lectures on Photomorphogenesis.* Springer-Verlag, Berlin.
MOHR H. (1974) Advances in phytochrome research. *Photochem. Photobiol.* **20**, 539–42.
MOHR H., BIENGER I. & LANGE H. (1971) Primary reaction of phytochrome. *Nature* **230**, 56–8.
MOLLENHAUER H.H. & MORRÉ D.J. (1966) Golgi apparatus and plant secretion. *Ann. Rev. Plant Physiol.* **17**, 27–46.
MOLLENHAUER H.H., MORRÉ D.J. & TOTTEN C. (1973) Intercisternal substances of the Golgi apparatus. *Protoplasma* **78**, 443–59.
MONDAL H., GANGULY A., DAS A., MANDAL R.K. & BISWAS B.B. (1972) Ribonucleic acid polymerase from eukaryotic cells. Effects of factors and rifampicin on the activity of RNA polymerase from chromatin of coconut nuclei. *Eur. J. Biochem.* **28**, 143–50.
MONDAL H., MANDAL R.K. & BISWAS B.B. (1972) RNA polymerase from eukaryotic cells. Isolation and purification of enzymes and factors from chromatin of coconut nuclei. *Eur. J. Biochem.* **25**, 463–70.
MONOD J. (1950) La technique de culture continuée. Théorie et application. *Ann. Inst. Pasteur, Paris* **79**, 390–410.
MONOD J., WYMAN J. & CHANGEUX J.-P. (1965) On the nature of allosteric transitions: a plausible model. *J. molec. Biol.* **12**, 88–118.
MONTES G. & BRADBEER J.W. (1975a) The biogenesis of the photosynthetic membranes and the development of photosynthesis in greening maize leaves. *Proceedings*

of the 3rd International Congress on Photosynthesis Research. pp. 1867–76 (ed. M. Avron). Elsevier. in press.

MONTES G. & BRADBEER J.W. (1975b) The association of chloroplasts and mitochondria in higher plants. *Pl. Sci. Lett.* **6,** 35–41.

MOON H.-M., REDFIELD B. & WEISSBACH H. (1972) Interaction of eukaryote elongation Factor EFI with quanosine nucleotides and aminoacyl-tRNA. *Proc. natn. Acad. Sci. U.S.A.* **69**, 1249–52.

MOOR H., MÜHLETHALER K., WALDNER H. & FREY-WYSSLING A. (1961) A new freezing-ultramicrotome. *J. biophys. Biochem. Cytol.* **10**, 1–13.

MOREAU F., DUPONT J. & LANCE C. (1974) Phospholipid and fatty acid composition of outer and inner membranes of plant mitochondria. *Biochim. biophys. Acta* **345**, 294–304.

MORRÉ D.J., JONES D.D. & MOLLENHAUER H.H. (1967) Golgi apparatus mediated polysaccharide secretion by outer root cap cells of *Zea mays*. I. Kinetics and secretory pathway. *Planta* **74**, 286–301.

MORRÉ D.J., MOLLENHAUER H.H. & BRACKER C.E. (1971) Origin and continuity of Golgi apparatus. In *Origin and Continuity of Cell Organelles*, pp. 82–126 (eds. J. Reinert & H. Ursprung). Springer-Verlag, Berlin.

MORRÉ D.J. & MOLLENHAUER H.H. (1974) The endomembrane concept: a functional integration of endoplasmic reticulum and Golgi apparatus. In *Dynamic Aspects of Plant Ultrastructure*, pp. 84–137 (ed. A.W. Robards). McGraw-Hill, London.

MORRÉ D.J. & VAN DER WOUDE W.J. (1974) Origin and growth of cell surface components. In *Macromolecules Regulating Growth and Development*, pp. 81–111 (eds. E.D. Hay, T.J. King & J. Papaconstantinou). Academic Press, New York.

MORRÉ D.J. (1975) Membrane biogenesis. *Ann. Rev. Plant Physiol.* **26**, 441–81.

MORRISON I.M. (1974) Structural investigations on the lignin-carbohydrate complexes of *Lolium perenne. Biochem. J.* **139**, 197–204.

MUIR W.H., HILDEBRANDT A.C. & RIKER A. J. (1954) Plant tissue cultures produced from single isolated plant cells. *Science* **119**, 877–8.

MURPHY D.B. (1975) The mechanism of microtubule-dependent movement of pigment granules in teleost chromatophores. *Ann. N.Y. Acad. Sci.* **253**, 692–701.

NACHBAUR J. & VIGNAIS P.M. (1968) Localization of phospholipase A_2 in outer membrane of mitochondria *Biochem. Biophys. Res. Commun.* **33**, 315–20.

NAG K.K. & STREET H.E. (1973) Carrot embryogenesis from frozen cultured cells. *Nature* **245**, 270–2.

NAG K.K. & STREET H.E. (1975a) Freeze preservation of cultured plant cells. I. The pretreatment phase. *Physiol. Plant.* **34**, 254–260.

NAG K.K. & STREET H.E. (1975b) Freeze preservation of cultured plant cells. II. The freezing and thawing phases. *Physiol. Plant.* **34**, 261–65.

NAGATA T. & TAKEBE I. (1970) Cell wall regeneration and cell division in isolated tobacco mesophyll protoplasts. *Planta* **92**, 301–8.

NAGATA T. & TAKEBE I. (1971) Plating of isolated tobacco mesophyll protoplasts on agar medium. *Planta* **99**, 12–20.

NAKAZATO H., VENKATESAN S. & EDMONDS S. (1975) Polyadenylic acid sequences in *E. coli* messenger RNA. *Nature* **256**, 144–6.

NASH D. & DAVIES M. (1972) Some aspects of growth and metabolism of Paul's Scarlet Rose cell suspensions. *J. exp. Bot.* **23**, 79–91.

NASHIMOTO H., HELD W., KALTSCHMIDT E. & NOMURA M. (1971) Structure and function of bacterial ribosomes. XII. Accumulation of 21S particles by some cold-sensitive mutants of *Escherichia coli. J. molec. Biol.* **62**, 121–38.

NELSON E.B. & TOLBERT N.E. (1970) Glycolate dehydrogenase in green algae. *Arch. Biochem. Biophys.* **141**, 102–10.

NELSON R.C. (1939) Studies of Production of Ethylene in the Ripening Process in Apple and Banana. *Food Res.* **4**, 173–190.

NES W.R. (1974) Role of sterols in membranes. *Lipids* **9**, 596–612.

NEWCOMB E.H. (1969) Plant microtubules. *Ann. Rev. Plant Physiol.* **20**, 253–88.

NEWMAN I.A. & BRIGGS W.R. (1972) Phytochrome-mediated electric potential changes in oat seedlings. *Plant Physiol.* **50**, 687–93.

NICHOLS R. (1968) The response of carnations (*Dianthus Caryophyllus*) to ethylene. *J. Hort. Sci.* **43**, 335–69.

NIERHAUS K.H. & DOHME F. (1974) Total reconstitution of functionally active 50S ribosomal subunits from *Escherichia coli. Proc. natn. Acad. Sci. U.S.A.* **71**, 4713–7.

NITSCH C. (1974) La Culture de pollen isolé sur milieu synthétique. *C.r. hebd. Séanc. Acad. Sci. Paris* **278**, 1031–4.

NOLAN C. & MARGOLIASH E. (1968) Comparative aspects of primary structures of proteins. *Ann. Rev. Biochem.* **37**, 727–90.

NOLL H., NOLL M., HAPKE B. & VAN DIEIJEN G. (1973) The mechanism of subunit interaction as a key to the understanding of ribosome function. In *Regulation of Transcription and Translation in Eukaryotes* pp. 257–311 (eds. E.K.F. Bautz, P. Karlson & H. Kersten), 24. Colloquium der Gesellschaft für Biologische Chemie. Springer-Verlag, Berlin.

NOLL M. (1974) Subunit structure of chromatin. *Nature* **251**, 249–51.

NONOMURA Y., BLOBEL G. & SABATINI D. (1971) Structure of liver ribosomes studied by negative staining. *J. molec. Biol.* **60**, 303–23.

NORD F.F. & FRANKLE K.W. (1928) On the mechanisms of enzyme action. II. Further evidence confirming the observations that ethylene increases the permeability of cells and acts as a protector. *J. Biol. Chem.* **79**, 27–35.

NORTHCOTE D.H. & PICKETT-HEAPS J.D. (1966) A function of the Golgi apparatus in polysaccharide synthesis and transport in the root-cap cells of wheat. *Biochem. J.* **98**, 159–67.

NORTHCOTE D.H. (1968) The organisation of the endoplasmic reticulum, the Golgi bodies and microtubules during cell division and subsequent growth. In *Plant Cell Organelles*, pp. 179–197 (ed. J.B. Pridham). Academic Press, London.

NOVICK A. & SZILAND L. (1950) Description of the chemostat. *Science* **112**, 715–16.

O'BRIEN T.J., JARVIS B.C., CHERRY J.H. & HANSON J.B. (1968) Enhancement by 2,4-dichlorophenoxyacetic acid of chromatin RNA polymerase in soybean hypocotyl tissue. *Biochim. biophys. Acta* **109**, 35–43.

O'BRIEN T.P. (1972) The cytology of cell wall formation in some eukaryotic cells. *Bot. Rev.* **38**, 87–113.

ODZUCK W. & KAUSS H. (1972) Biosynthesis of pure araban and xylan. *Phytochemistry* 11, 2489–94.

OELZE-KAROW H. & MOHR H. (1973) Quantitative correlation between spectrophotometric phytochrome assay and physiological response. *Photochem. Photobiol.* **18**, 319–30.

OELZE-KAROW H. & MOHR H. (1974) Interorgan correlation in a phytochrome mediated response in the mustard seedling. *Photochem. Photobiol.* **20**, 127–31.

OESTREICHER G., HOGUE P. & SINGER T.P. (1973) Regulation of succinate dehydrogenase in higher plants. II. Activation by substrates, reduced Coenzyme Q, nucleotides, and anions. *Plant Physiol.* **52**, 622–6.

OGUTUGA D.B.A. & NORTHCOTE D.H. (1970) Caffeine formation in tea callus tissue. *J. exp. Bot.* **21**, 258–73.

OHKUMA K., LYON J.L., ADDICOTT F.T. & SMITH O.E. (1963) Abscisin II, an abscission-accelerating substance from young cotton fruit. *Science* **142**, 1592–3.

OHNISHI T. (1973) Mechanism of electron transport and energy conservation in the site I region of the respiratory chain. *Biochim. biophys. Acta* **301**, 105–28.

OHAYAMA K. (1974) Properties of 5-bromodeoxyuridine-resistant lines of higher plant cells in liquid culture. *Exp. Cell Research* **89**, 31–8.

OHAYAMA K., GAMBORG O.L., SHYLUK J.P. & MILLER R.A. (1973) Studies on transformation: Uptake of exogenous DNA by plant protoplasts. *Colloq. Int. Cent. Nat. Rech. Sci.* **212**, 423–48.

OJALA D. & ATTARDI G. (1974) Expression of the mitochondrial genome in Hela cells. XIX. Occurrence in mitochondria of polyadenylic acid sequences, 'free' and covalently linked to mitochondrial DNA-coded RNA. *J. molec. Biol.* **82**, 151–74.

OLINS D.E. & OLINS A.L. (1974) Spheroid chromatin units (ν bodies). *Science* **183**, 330–2.

OLSON A.O. & SPENCER M. (1968) Studies on the mechanism of action of ethylene. II. *Can. J. Biochem.* **46**, 277–82.

ONGUN A. & MUDD J.B. (1968) Biosynthesis of galactolipids in plants. *J. Biol. Chem.* **243**, 1558–66.

ONGUN A., THOMSON W.W. & MUDD J.B. (1968) Lipid composition of chloroplasts isolated by aqueous and non-aqueous techniques. *J. Lipid Res.* **9**, 409–15.

OSCHMAN J.L., WALL B.J. & GUPTA B.L. (1974) Cellular basis of water transport. In *Transport at the Cellular Level* (eds. M.A. Sleigh & D.H. Jennings). Symposium of the Society for Experimental Biology. No. 28. Cambridge University Press.

OSMOND C.B. (1969) β-carboxylation, photosynthesis and photorespiration in higher plants. *Biochim. biophys. Acta* **172**, 144–9.

VAN OVERBEEK J., LOEFFLER J.E. & MASON M.I.R. Dormin (Abscisin II), inhibitor of plant dNA synthesis? *science* **156**, 1497–9.

OVERMAN A.R., LORIMER G.H. & MILLER R.J. (1970) Diffusion and osmotic transfer in corn mitochondria. *Plant Physiol.* **45**, 126–32.

OWNBY J.D., ROSS C.W. & KEY J.L. (1975) Studies on the presence of adenosine cyclic 3′:5′-monophosphate in oat coleoptiles. *Plant Physiol.* **55**, 346–51.

OXENDER D.L. (1972) Membrane Transport. *Ann. Rev. Biochem.* **41**, 777–814.

PACKER L., MEHARD C.W., MEISSNER G., ZAHLER W.L. & FLEISCHER S. (1974) The structural role of lipids in mitochondrial and sarcoplasmic reticulum membranes. Freeze-fracture electron microscopy studies. *Biochim. biophys. Acta* **363**, 159–81.

PALMER E. & SMITH O.E. (1969) Cytokinins and tuber initiation in the potato *Solanum tuberosum* L. *Nature* **221**, 279–80.

PAPAHADJOPOULOS D. (1971) Na^+ —K^+ discrimination by 'pure' phospholipid membranes. *Biochim. biophys. Acta* **241**, 154–9.

PAPAHADJOPOULOS D., COWDEN M. & KIMELBERG H. (1973) Role of cholesterol in membranes. Effects of phospholipid-protein interactions, membrane permeability and enzymatic activity. *Biochim. biophys. Acta* **330**, 8–26.

PARISH G.R. (1974) Seasonal variation in the membrane structure of differentiating shoot cambial-zone cells demonstrated by freeze-etching. *Cytobiologie* **9**, 131–43.

PARK R.B. (1962) Advances in photosynthesis. *J. Chem. Educ.* **39**, 424–9.

PARK R.B. & PON N.G. (1961) Correlation of structure with function in *Spinacea oleracea. J. molec. Biol.* **3**, 1–10.

PARK R.B. & SANE P.V. (1971) Distribution of function and structure in chloroplast lamellae. *Ann. Rev. Plant Physiol.* **22**, 395–430.

PARSONS D.F., BONNER W.D.Jr. & VERBOON J.G. (1965) Electron microscopy of isolated plant mitochondria and plastids using both the thin section and negative staining techniques. *Can. J. Botany* **43**, 647–55.

PAUL J. (1972) General theory of chromosome structure and gene activation in eukaryotes. *Nature, Lond.* **238**, 444–6.

PAYNE P.I. & DYER T.A. (1972) Plant 5.8S RNA is a component of 80S but not 70S ribosomes. *Nature* (*New Biol.*) **235**, 145–7.

PAYNE P.I. & BOULTER D. (1974) Katabolism of plant cytoplasmic ribosomes: RNA breakdown in senescent cotyledons of germinating broad-bean seedlings. *Planta* **117**, 251–8.

PAYNE E.S., BOULTER D., BROWNRIGG A., LONSDALE D., YARWOOD A. & YARWOOD J.N. (1971a) A polyuridylic acid-directed cell-free system from 60 day-old developing seeds of *Vicia faba. Phytochemistry.* **10**, 2293–8.

PAYNE E.S., BROWNRIGG A., YARWOOD A. & BOULTER D. (1971b) Changing protein synthetic machinery during development of seeds of *Vicia faba. Phytochemistry.* **10**, 2299–303.

PEARSON J.A. & WAREING P.F. (1969) Effect of abscisic acid on activity of chromatin. *Nature* **221**, 672–3.

PEGG G.F. & VESSEY J.C. (1973) Chitinase activity in *Lycopersicon esculentum* and its relationship to the *in vivo* lysis of *Verticillium albo-atrum* mycelium. *Physiol. Plant Pathol.* **3**, 207–22.

PERANI A., TIBONI O. & CIFERRI O. (1971) Absolute ribosome specificity of two sets of transfer factors isolated from *Saccharomyces fragilis. J. molec. Biol.* **55**, 107–12.

PERLMAN S., ABELSON H.T. & PENMAN S. (1973) Mitochondrial protein synthesis: RNA with the properties of eukaryotic messenger RNA. *Proc. natn. Acad. Sci. U.S.A.* **70**, 350–3.

PESTKA S. (1974) The use of inhibitors in studies of protein synthesis. *Methods in Enzymology, Nucl. Acids & Protein Synth.*, vol. XXX, part F, pp. 261–82, (eds. K. Moldave & L. Grossman). Academic Press, London.

PETERSON P.J. & FOWDEN L. (1965) Purification, properties and comparative specificities of the enzyme prolyltransfer ribonucleic acid synthetase from *Phaseolus aureus* and *Polygonatum multiflorum. Biochem. J.* **97**, 112–24.

PEVERLY J.H., MILLER R.J., MALONE C. & KOEPPE D.E. (1974) Ultrastructural evidence for calcium phosphate deposition by isolated corn shoot mitochondria. *Plant Physiol.* **54**, 408–11.

PHILIPP E.I., FRANKE W.W., KEENAN T.W., STADLER J. & JARASCH E.D. (1976) Characterization of nuclear membranes and endoplasmic reticulum isolated from plant tissue. *J. Cell Biol.* **68**, 11–29.

PHILLIPS I.D.J. & WAREING P.F. (1958) Effect of photoperiod on the level of growth inhibitors in *Acer pseudoplatanus. Naturwissenschaften.* **13**, 317.

PHILLIPS M. & WILLIAMS G.R. (1973a) Effects of 2-butylmalonate, 2-phenylsuccinate, benzylmalonate and p-iodobenzylmalonate on the oxidation of substrates by mung bean mitochondria. *Plant Physiol.* **51**, 225–8.

PHILLIPS M. & WILLIAMS G.R. (1973b) Anion transporters in plant mitochondria. *Plant Physiol.* **51**, 667–70.

PICKETT-HEAPS J.D. (1967) The effects of colchicine on the ultrastructure of dividing wheat cells, xylem differentiation and distribution of cytoplasmic microtubules. *Devel. Biol.* **15**, 206–36.

PICKETT-HEAPS J.D. & NORTHCOTE D.H. (1966) Cell division in the formation of the stomatal complex of the young leaves of wheat. *J. Cell Sci.* **1**, 121–8.

PINTO DA SILVA P. (1973) Membrane intercalated particles in human erythrocyte ghosts: sites of preferred passage of water molecules at low temperature. *Proc. natn. Acad. Sci. U.S.A.* **70**, 1339–43.

PINTO DA SILVA P. & BRANTON D. (1970) Membrane splitting in freeze-etching. Covalently bound ferritin as a membrane marker. *J. Cell Biol.* **45**, 598–605.

PLESNICAR M. & BENDALL D.S. (1971) The development of photochemical activities during greening of etiolated barley. In *Proc. Second Internat. Congr. Photosynthesis*, pp. 2367–73. (eds. G. Forti, M. Avron & A. Melandri). Dr. W. Junk, The Hague.

POLACCO J.C., PAZ E. & CARLSON P.S. (1974) Unpublished results.

POOLE R.J. (1973) The H^+ pump in red beet. In *Ion Transport in Plants* p. 129, (ed. W.P. Anderson). Academic Press, London.

POSSINGHAM J.V. & SAURER W. (1969) Changes in chloroplast number per cell during leaf development in spinach. *Planta* **86**, 186–94.

POWELL J.T. & BREW K. (1974) Glycosyltransferases in the Golgi membranes of onion stem. *Biochem. J.* **142**, 203–9.

POWER J.B. & COCKING E.C. (1970) Isolation of leaf protoplasts: macromolecule uptake and growth substance response. *J. exp. Bot.* **21**, 64–70.

POWER J.B. & COCKING E.C. (1971) Fusion of plant protoplasts. *Sci. Prog.* **59**, 181–98.

PRATT L.H. & COLEMAN R.A. (1971) Immunocytochemical localization of phytochrome. *Proc. natn. Acad. Sci. U.S.A.* **68**, 2431–5.

PRATT L.H. & COLEMAN R.A. (1974) Phytochrome distribution in etiolated grass seedlings as assayed by an indirect antibody-labelling method. *Am. J. Bot.* **61**, 195–202.

PRICHARD P.M., PICCIANO D.J., LAYCOCK D. G. & ANDERSON W.F. (1971) Translation of exogenous messenger RNA for hemoglobin on reticulocyte and liver ribosomes. *Proc. natn. Acad. Sci. U.S.A.* **68**, 2752–6.

PUCK T.T. & KAO F. (1967) Genetics of somatic mammalian cells V. Treatment with 5-bromodeoxyuridine and visible light for isolation of nutritionally deficient mutants. *Proc. Natn. Acad. Sci. U.S.A.* **58**, 1227–34.

QUASTLER H. & SHERMAN F.C. (1959) Cell population kinetics in the intestinal epithelium of the mouse. *Exp. Cell Res.* **17**, 420–38.

QUAIL P.H. (1975a) Phytochrome. *Plant Biochemistry* (eds. J. Bonner & J.E. Varner). Academic Press, New York.

QUAIL P.H. (1975b) Particle-bound phytochrome: association with a ribonucleoprotein fraction from *Cucurbita pepo* L. *Planta.* **123**, 223–34.

QUAIL P.H., MARMÉ D. & SCHÄFER E. (1973a) Particle-bound phytochrome from maize and pumpkin. *Nature* (*New Biol.*) **245**, 189–91.

QUAIL P.H., SCHÄFER E. & MARMÉ D. (1973b) Turnover of phytochrome in pumpkin cotyledons. *Plant Physiol.* **52**, 128–31.

QUAIL P.H. & SCHÄFER E. (1974) Particle-bound phytochrome: A function of light dose and steady-state level of the far-red-absorbing form. *J. Membrane Biol.* **15**, 393–404.

RABSON R., TOLBERT N.E. & KEARNEY P.C. (1962) Formation of serine and glyceric acid by the glycolate pathway. *Arch. Biochem. Biophys.* **98**, 154–63.

RACKER E. (1972) Reconstitution of a calcium pump with phospholipids and a purified Ca^{++}-adenosine triphosphatase from sarcoplasmic reticulum. *J. Biol. Chem.* **247**, 8198–200.

RACKER E., HORSTMAN L.L., KLING D. & FESSENDEN-RADEN J.M. (1969) Partial resolution of the enzymes catalyzing oxidative phosphorylation. XXI. Resolution of submitochondrial particles from bovine heart mitochondria with silicotungstate. *J. Biol. Chem.* **244**, 6668–74.

RACUSEN R.H. (1973) Membrane protein conformational changes as a mechanism for the phytochrome-induced fixed charge reversal in root cap cells of mung bean. *Plant Physiol.* **51** (Suppl), 51.

RACUSEN R. & MILLER K. (1972) Phytochrome-induced adhesion of mung bean root tips to platinum electrodes in a direct current field. *Plant Physiol.* **49**, 654–5.

RACUSEN R.H. & SATTER R.L. (1974) External and internal electrical changes in pulvinar regions of *Samanea saman*. *Plant Physiol.* **53**, S45.

RAMSHAW J.A.M., PEACOCK D., MEATYARD B.T. & BOULTER D. (1974) Phylogenetic implications of the amino acid sequence of cytochrome *c* from *Enteromorpha intestinalis*. *Phytochem.* **13**, 2783–9.

RAVEN J.A. & SMITH F.A. (1974) Significance of hydrogen ion transport in plant cells. *Can. J. Bot.* **52**, 1035–48.

RAYLE D.L. (1973) Auxin-induced hydrogen ion secretion in *Avena* coleoptiles and its implications. *Planta* **114**, 63–73.

RAYLE D.L. & CLELAND R. (1970) Enhancement of wall loosening and elongation by acid solutions. *Plant Physiol.* **46**, 250–3.

RAYLE D.L., EVANS M.L. & HERTEL R. (1970a) Action of auxin on cell elongation. *Proc. natn. Acad. Sci. U.S.A.* **65**, 184–91.

RAYLE D.L., HAUGHTON P.M. & CLELAND R. (1970b) An *in vitro* system that stimulates plant cell extension growth. *Proc. natn. Acad. Sci. U.S.A.* **67**, 1814–17.

REBHUN L.I., ROSENBAUM J., LEFEBRE P. & SMITH G. (1974) Reversible restoration of the birefringence of cold-treated, isolated mitotic apparatus of surf clam eggs with chick brain tubulin. *Nature* **249**, 113–15.

REED P.W. & LARDY H.A. (1972) A 23187: a divalent-cation ionophore. *J. Biol. Chem.* **247**, 6970–7.

REES D.A. & WIGHT A.W. (1971) Polysaccharide conformation. Part VII. Model building computations for a α-1,4 galacturonan and the kinking function of L-rhamnose residues in pectic substances. *J. Chem. Soc.* **B**, 1366–72.

REES H. & JONES R.N. (1972) The origin of the wide species variation in nuclear DNA content. *Int. Rev. Cytol.* **32**, 53–92.

REEVES S.G. & HALL D.O. (1973) The stoichiometry (ATP/2e⁻ ratio) of non-cyclic photophosphorylation in isolated spinach chloroplasts. *Biochim. biophys. Acta* **314**, 66–78.

REGEIMBAL L.O. & HARVEY R.B. (1927) The effect of ethylene on the enzymes of pineapples. *J. Amer. Chem. Soc.* **49**, 1117–18.

REGER B.J., FAIRFIELD S.A., EPLER J.L. & BARNETT W.E. (1970) Identification and origin of some chloroplast aminoacyl-tRNA synthetases and tRNAS. *Proc. natn. Acad. Sci. U.S.A.* **67**, 1207–13.

REID D.M., CLEMENTS J.B. & CARR D.J. (1968) Red light induction of gibberellin synthesis in leaves. *Nature* **217**, 580–2.

RENNER O. (1929) Artbastarde bei Pflanzen. In *Handbuch der Vererbungswissen-schaften* Vol. IIA p. 1 (eds. E. Baur & M. Hartmann). Gebr. Borntraeger, Berlin.

RICHMOND A., BLACK A. & SACHS B. (1970) A study of the hypothetical role of cytokinins in completion of tRNA. *Planta* **90**, 57–65.

RINGO D.L. (1967) Flagellar motion and fine structure of the flagellar apparatus in *Chlamydomonas*. *J. Cell Biol.* **33**, 543–71.

RIS H. & PLAUT W. (1962) Ultrastructure of DNA-containing areas in the chloroplast of *Chlamydomonas*. *J. Cell Biol.* **13**, 383–91.

ROBARDS A.W. & ROBB M.E. (1974) The entry of ions and molecules into roots: an investigation using electron-opaque tracers. *Planta* **120**, 1–12.

ROBERTS K. & NORTHCOTE D.H. (1971; Ultrastructure of the nuclear envelopes structural aspects of the interphase nucleu) of sycamore suspension culture cells. *Microscopia Acta* **71**, 102–20.

ROCHA V. & TING I.P. (1970) Preparation of cellular plant organelles from spinach leaves. *Arch. Biochem. Biophys.* **140**, 398–407.

ROGERS M.E., LOENING U.E. & FRASER R.S. S. (1970) Ribosomal RNA precursors in plants. *J. molec. Biol.* **49**, 681–92.

ROLAND J.-C. & VIAN B. (1971) Reactivité du plasmalemme végétal. Etude cytochimique. *Protoplasma* **73**, 121–37.

ROLLIN P. (1972) Phytochrome control of seed germination. *Phytochrome* pp. 229–54 (eds. K. Mitrakos & W. Shropshire Jr.), Academic Press, New York.

ROLLIT J. & MACLACHLAN G.A. (1974) Synthesis of wall glucan from sucrose by enzyme preparations from *Pisum sativum*. *Phytochem.* **13**, 367–74.

ROSE R.J., CRAN D.G. & POSSINGHAM J.V. (1974) Distribution of DNA in dividing spinach chloroplasts. *Nature* **251**, 641–2.

ROSSI C., SCARPA A. & AZZONE G.F. (1967) Ion transport in liver mitochondria. V. The effect of anions on the mechanism of K^+ uptake. *Biochemistry* **6**, 3902–10.

ROTTEM S., YASMOUV J., NE'EMAN Z. & RAZIN S. (1973) Cholesterol in Mycoplasma Membranes: Composition, Ultrastructure and Biological Properties of Membranes from *Mycoplasma Mycoides* var. Capri cells adapted to grow with low cholesterol concentrations. *Biochim. biophys. Acta* **323**, 495–508.

ROUGHAN P.C. & BATT R.D. (1969) The glycerolipid composition of leaves. *Phytochemistry* **8**, 363–9.

ROUX S.J. (1972) Chemical evidence for conformational differences between the red- and far-red-absorbing forms of oat phytochrome. *Biochemistry* **11**, 1930–6.

ROUX S.J. & YGUERABIDE J. (1973) Photoreversible conductance changes induced by phytochrome in model lipid membranes. *Proc. natn. Acad. Sci. U.S.A.* **70**, 762–4.

ROYCHOUDURY R., DATTA A. & SEN S.P. (1965) The mechanism of action of plant growth substances: The role of nuclear RNA in growth substance action. *Biochim. biophys. Acta* **107**, 346–51.

ROYCHOUDURY R. & SEN S.P. (1964) Studies on the mechanism of auxin action: Auxin regulation of nucleic acid metabolism in pea internodes and cocoanut milk nuclei. *Physiol. Plant* **17**, 352–62.

RUBINSTEIN B., DRURY K.S. & PARK R.B. (1969) Evidence for bound phytochrome in oat seedlings. *Plant Physiol.* **44**, 105–9.

RÜDIGER W. (1972) Chemistry of phytochrome chromophore. *Phytochrome*, pp. 129–41 (eds. K. Mitrakos & W. Shropshire Jr.). Academic Press, New York.

RUNECKLES V.C., SONDHEIMER E. & WALTON D.C. (1974) *The Chemistry and Biochemistry of Plant Hormones*, Vol. 7, p. 178. Recent Advances in Phytochemistry. Academic Press. N.Y.

RYSKOV A.P., MANTIEVA V.L., AVAKIAN E.R. & GEORGIEV G.P. (1971) The hybridisation properties of 5′ end sequences of giant nuclear dRNA. *FEBS Letters* **12**, 141–2.

SACHER J.A. & SALMINEN S.O. (1969) Comparative studies of effect of auxin and ethylene on permeability and synthesis of RNA and protein. *Plant Physiol.* **44**, 1371–7.

SADDLER H.D.W. (1970) The ionic relations of *Acetabularia mediterranea*. *J. exp. Bot.* **21**, 345–59.

SADLER D.M., LEFORT-TRAN M. & POUPHILE M. (1973) Structure of photosynthetic membranes of *Euglena* using X-ray diffraction. *Biochim. biophys. Acta* **298**, 620–9.

SAGER R. (1972) *Cytoplasmic Genes and Organelles*. Academic Press, London.

SAGHER D., EDELMAN M. & JAKOB K.M. (1974) Poly(A) associated RNA in plants. *Biochim. biophys. Acta* **349**, 32–8.

SANDMEIER & IVART (1972) Modification du taux des nucléotides adenyliques (ATP, ADP et AMP) par un eclairément de lumière rouge-clair (660 nm). *Photochem. Photobiol.* **16**, 51–9.

SANGER J.W. (1975) Presence of actin during chromosome movement. *Proc. natn. Acad. Sci. U.S.A.* **72**, 2451–5.

SANGER J.M. & JACKSON W.T. (1971) Fine structure study of pollen development in *Haemanthus katherinae* Baker. Microtubules and elongation of the generative cells. *J. Cell Sci.* **8**, 303–15.

SASTRY P.S. & KATES M. (1963) Lipid components of leaves. III. Isolation and characterization of mono- and di-galactosyl diglycerides and lecithin. *Biochim. biophys. Acta* **70**, 214–19.

SATIR P. (1974) The present status of the sliding microtubule model of ciliary motion. *Cilia and Flagella* pp. 131–42 (ed. M.A. Sleigh). Academic Press, London.

SATTER R. & GALSTON A.W. (1973) Leaf movements: Rosetta stone of plant behaviour? *Bioscience* **23**, 407–16.

SATTER, R.L. & GALSTON, A.W. (1976) The physiological functions of phytochrome In *Chemistry and Biochemistry of Plant Pigments* Vol. 1. pp. 681–735 (ed. T.W. Goodwin), Academic Press, London.

SAUNDERS P.F. & POULSON R.H. (1968) Biochemical studies on the possible mode of action of abscisic acid: an apparent allosteric inhibition of invertase activity. In *Biochemistry and Physiology of Plant Growth Substances* pp. 1581–91 (eds. F. Wrightman & G. Setterfield). Runge Press Ltd., Ottawa.

SCARBOROUGH G.A. (1975) Isolation and characterisation of *Neurospora crassa* plasma membranes. *J. Biol. Chem.* **250**, 1106–11.

SCHÄFER E. (1975) A new approach to explain the 'high irradiance responses' of photomorphogenesis on the basis of phytochrome. *J. math. Biol.* **2**, 41–56.

SCHÄFER E., MARCHAL B. & MARMÉ D. (1972) *In vivo* measurements of the phytochrome photostationary state in far red light. *Photochem. Photobiol.* **15**, 457–64.

SCHÄFER E. & MOHR H. (1974) Irradiance dependency of the phytochrome system in cotyledons of mustard (*Sinapis alba* L.) *J. math. Biol.* **1**, 9–15.

SCHÄFER E. & SCHMIDT W. (1974) Temperature dependence of phytochrome dark reactions. *Planta* **116**, 257–66.

SCHÄFER E., SCHMIDT W. & MOHR H. (1973) Comparative measurements of phytochrome in cotyledons and hypocotyl hooks of mustard (*Sinapis* alba L.) *Photochem. Photobiol.* **18**, 331–4.

SCHIBLER U., WYLER T. & HAGENBÜCHLE O. (1975) Changes in size and secondary structure of the ribosomal transcription unit during vertebrate evolution. *J. molec. Biol.* **94**, 503–17.

SCHIFF J.A. (1975) The control of chloroplast differentiation in *Euglena*. *Proceedings of the 3rd International Congress on Photosynthesis Research* pp. 1691–717 (ed. M. Avron). Elsevier.

SCHIMPER A.F.W. (1885) Untersuchungen über die Chlorophyllcorper und die ihnen homologen Gebilde. *Jb. wiss. Bot.* **16**, 1–247.

SCHLANGER G. & SAGER R. (1974) Localisation of five antibiotic resistances at the subunit level in chloroplast ribosomes of *Chlamydomonas*. *Proc. natn. Acad. Sci. U.S.A.* **71**, 1715–19.

SCHMIDT W., MARMÉ D., QUAIL P. & SCHÄFER E. (1973) Phytochrome: First-order phototransformation kinetics *in vivo*. *Planta* **111**, 329–36.

SCHNEIDER D.L., KAGAWA Y. & RACKER E. (1972) Chemical modification of the inner mitochondrial membrane. *J. Biol. Chem.* **247**, 4074–9.

SCHNEPF E. (1961) Quantitative Zusammenhänge zwischen der Sekretion des Fangschleimes und den Golgi–Strukturen bei *Drosophyllum lusitanicum*. *Z. Naturforsch.* **16b**, 605–10.

SCHNEPF E., RODERER G. & HERTH W. (1975) The formation of the fibrils in the Lorica of *Poteriochromonas stipitata:* tip growth, kinetics, site, orientation. *Planta* **125**, 45–62.

SCHONBAUM G.R., BONNER W.D.Jr., STOREY B.T. & BAHR J.T. (1971) Specific inhibition of the cyanide-insensitive respiratory pathway in plant mitochondria by hydroxamic acids. *Plant Physiol.* **47**, 124–8.

SCHÖNBOHM E. (1973) Kontraktile fibrillen als aktive Elemente bei der Mechanik der Chloroplastenverlagerung. *Ber. dt. bot. Ges.* **86**, 407–22.

SCHOPFER P. (1967) Weitere Untersuchungen zur phytochrominduzierten Akkumulation von Ascorbinsäure beim Senfkeimling (*Sinapis alba* L.). *Planta* **74**, 210–27.

SCHOPFER P. & MOHR H. (1972) Phytochrome-mediated induction of phenylalanine ammonia-lyase in mustard seedlings. A contribution to eliminate some misconceptions. *Plant Physiol.* **49**, 8–10.

SCHWERTNER H.A. & BIALE J.B. (1973) Lipid composition of plant mitochondria and of chloroplasts. *J. Lipid Res.* **14**, 235–42.

SCHREIER M.H. & STAEHELIN T. (1973) Functional characterization of five initiation factors for mammalian protein synthesis. In *Regulation of Transscription and Translation in Eukaryotes* pp. 335–49 (eds. E.K. F. Bautz, P. Karlson & H. Kersten), 24. Colloquium der Gesellschaft für Biologische Chemie. Springer-Verlag, Berlin.

SCHULZ SISTER R. & JENSEN W.A. (1968) *Capsella* embryogenesis: the egg, zygote and young embryo. *Am. J. Bot.* **55**, 807–19.

SCHULTZ R.D. & ASUNMAA S.K. (1970) Ordered water and the ultrastructure of the cellular plasma membrane. *Recent Prog. Surface Sci.* **3**, 291–332.

SCHWARTZ D. (1971) Genetic control of alcohol dehydrogenase—a competition model for regulation of gene action. *Genetics*. **67**, 411–25.

SCHWARZ J.H., MEYER R., EISENSTADT J.M. & BRAWERMAN G. (1967) Involvement of N-formylmethionine in initiation of protein synthesis in cell-free extracts of *Euglena gracilis*. *J. molec. Biol.* **25**, 571–4.

SCOTT N.S. & INGLE J. (1973) The genes for cytoplasmic ribosomal ribonucleic acid in higher plants. *Plant Physiol.* **51**, 677–84.

SCOTT N.S., NAIR H. & SMILLIE R.M. (1971) The effect of red irradiation on plastid ribosomal RNA synthesis in dark grown pea seedlings. *Plant. Physiol.* **47**, 385–8.

SCOWCROFT W.R., DAVEY M.R. & POWER J.B. (1973) Crown gall protoplasts—isolation, culture and ultrastructure. *Pl. Sci. Lett.* **1**, 451–6.

SEDGWICK B. (1973) The control of fatty acid biosynthesis in plants. In *Biosynthesis and its Control in Plants*, pp. 179–217 (ed. B.V. Milborrow). Academic Press, London.

SELWYN M.J. (1965) A simple test for inactivation of an enzyme during assay. *Biochem. biophys. Acta* **105**, 193–5.

SEN S., PAYNE P.I. & OSBORNE D.J. (1975) Early ribonucleic acid synthesis during the germination of rye (*secale cereale*) embryos and the relationship to early protein synthesis. *Biochem. J.* **148**, 381–7.

SHIBAOKA H. (1974) Involvement of wall microtubules in gibberellin promotion and kinetin inhibition of stem elongation. *Pl. Cell Physiol.* **15**, 255–63.

SHIH D.S. & KAESBERG P. (1973) Translation of brome mosaic viral ribonucleic acid in a cell-free system derived from wheat em, bryo. *Proc. natn. Acad. Sci. U.S.A.* **70** 1799–803.

SHINE J. & DALGARNO L. (1974) The 3′-terminal sequence of *Escherichia coli* 16S ribosomal RNA: complementarity to nonsense triplets and ribosome binding sites. *Proc. natn. Acad. Sci. U.S.A.* **71**, 1342–6.

Shore G. & MacLachan G.A. (1975) The site of cellulose synthesis. Hormone treatment alters the intracellular location of alkali-insoluble β-1,4 glucan (cellulose) synthetase activities. *J. Cell Biol.* **64**, 557–71.

Short K.C., Brown E.G. & Street H.E. (1969) Studies on the growth in culture of plant cells. VI. Nucleic acid metabolism of *Acer pseudoplatanus* L. cell suspensions. *J. exp. Bot.* **20**, 579–90.

Shropshire W.Jr. (1972) Action spectroscopy. *Phytochrome*, pp. 161–81 (eds. K. Mitrakos & W. Shropshire Jr.). Academic Press, London.

Shyamala G. & Gorski J. (1969) Estrogen receptors in the rat uterus: Studies on the interaction of cytosol and nuclear binding sites. *J. Biol. Chem.* **224**, 1097–103.

Shylk A.A. (1971) Biosynthesis of chlorophyll *b*. *Ann. Rev. Plant Physiol.* **22**, 169–81.

Siddell S.G. & Ellis R.J. (1975) Protein synthesis in chloroplasts VI Characteristics and products of *in vitro* protein synthesis in etioplasts and developing chloroplasts from pea leaves. *Biochem. J.* **146**, 675–85.

Siegelman H.W. & Butler W.L. (1965) Properties of phytochrome. *Ann. Rev. Biochem.* **43**, 805–33.

Siggel V., Renger G., Stiehl H.H. & Rumberg B. (1972) Evidence for electronic and ionic interactions between electron transport chains in chloroplasts. *Biochim. biophys. Acta* **256**, 328–35.

Silberger J. & Skoog F. (1953) Changes induced by indoleacetic acid in nucleic acid contents and growth of tobacco tissue. *Science.* **118**, 443–4.

Simon E.W. (1974) Phospholipids and Plant Membrane Permeability. *New Phytol.* **73**, 377–420.

Sinclair J., Wells R., Deumling B. & Ingle J. (1975) The complexity of satellite deoxyribonucleic acid in a higher plant. *Biochem. J.* **149**, 31–8.

Singer S.J. (1974) The molecular organisation of membranes. *Ann. Rev. Biochem.* **43**, 805–33.

Singer T.P., Oestreicher G. & Hogue P. (1973) Regulation of succinate dehydrogenase in higher plants. I. Some general characteristics of the membrane bound enzyme. *Plant Physiol.* **52**, 616–21.

Singer S.J. & Nicolson G.L. (1972) The fluid mosaic model of the structure of cell membranes. *Science* **175**, 720–31.

Skoog F. & Armstrong D.J. (1970) Cytokinins. *Ann. Rev. Plant Physiol.* **21**, 359–84.

Slater E.C. (1971) The coupling between energy-yielding and energy-utilizing reactions in mitochondria. *Quart. Rev. Biophys.* **4**, 35–71.

Slayman C.L. (1970) Movements of ions and electrogenesis in micro-organisms. *Am. Zoologist* **10**, 377–92.

Slayman C.L., Long W.S. & Lu C.Y.-H. (1973) The relationship between ATP and an electrogenic pump in the plasmamembrane of *Neurospora crassa J. Membrane Biol.* **14**, 305–8.

Slayman C.L. & Slayman C.W. (1974) Depolarization of the plasma membrane of *Neurospora* during active transport of glucose: evidence for a proton dependent cotransport system. *Proc. natn. Acad. Sci. U.S.A.* **71**, 1935–9.

Sleigh M.A. (1974) *Cilia and Flagella.* 500 pp. Academic Press, London.

Smith S.M. & Street H.E. (1974) The decline of embryogenic potential as callus and suspension cultures of carrot (*Daucus carota*), L.) are serially subcultured. *Ann. Bot.* **N.S. 38**, 223–41.

Smith D.B. & Flavell R.B. (1975) Characterisation of the wheat genome by renaturation kinetics. *Chromosoma.* **50**, 223–42.

Smith H. (1970) Phytochrome and photomorphogenesis in plants. *Nature* **227**, 665–8.

Smith I.K. & Fowden L. (1968) Studies on the specificities of the phenyl-alanyl- and tyrosyl-sRNA synthetases from plants. *Phytochemistry.* **7**, 1065–75.

Smith M.A., Criddle R.S., Peterson L. & Huffaker R.C. (1974) Synthesis and assembly of ribulose-bisphosphate carboxylase enzyme during greening of barley plants. *Arch. Biochem. Biophys.* **165**, 494–504.

Solomon A.K. & Gary-Bobo C.M. (1972) Aqueous pores in lipid bilayers and red cell membranes. *Biochim. biophys. Acta* **255**, 1019–21.

Solomon E. & Mascarenhas J.P. (1971) Auxin induced synthesis of cyclic 3′,5′-adenosine monophosphate in Avena coleoptiles. *Life Sci.* **10**, 879–85.

Solomos T., Malhotra S.S. & Spencer M. (1973) Use of dextran-40 gradients for separation of pea cotyledon mitochondria into different fractions. *Plant Physiol.* **51**, 807–9.

Spanswick R. (1972) Evidence for an electrogenic pump in *Nitella translucens*. I.

The effects of pH, potassium and sodium ions, light and temperature of the membrane potential and resistance. *Biochim. biophys. Acta* **288**, 73–89.

SPANSWICK R.M. (1974) Evidence for an electrogenic ion pump in *Nitella translucens*. II. Control of the light-stimulated component of the membrane potential. *Biochim. biophys. Acta* **332**, 387–98.

SPENCER D. (1965) Protein synthesis by isolated spinach chloroplasts. *Arch. Biochem. Biophys.* **111**, 381–90.

SPENCER D. & WHITFELD P.R. (1967) Ribonucleic acid synthesising activity of spinach chloroplasts and nuclei. *Arch. Biochem. Biophys.* **121**, 336–45.

SPRING H., TRENDELENBURG M.F., SCHEER U., FRANKE W.W. & HERTH W. (1974) Structural and biochemical studies of the primary nucleus of two green algal species, *Acetabularia mediterranea* and *Acetabularia major*. *Cytobiologie* **10**, 1–65.

SPRUIT C.J.P. (1972) Estimation of phytochrome by spectrophotometry *in vivo*: Instrumentation and interpretation. *Phytochrome*, pp. 77–104 (eds. K. Mitrakos & W. Shropshire Jr.). Academic Press, New York.

SPRUIT C.J.P. & KENDRICK R.E. (1973) Phytochrome intermediates *in vivo*. II. Characterization of intermediates by difference spectrophotometry. *Photochem. Photobiol.* **18**, 145–52.

SPRUIT C.J.P. & MANCINELLI A.L. (1969) Phytochrome in cucumber seeds. *Planta* **88**, 303–10.

STABENAU H. (1974) Localization of enzymes of glycolate metabolism in the alga *Chlorogonium elongatum*. *Plant Physiol.* **54**, 921–4.

STABENAU H. & BEEVERS H. (1974) Isolation and characterization of microbodies from the alga *Chlorogonium elongatum*. *Plant Physiol.* **53**, 866–9.

VAN STADEN J. & WAREING P.F. (1972) The effect of light on endogenous cytokin levels in seeds of *Rumex obtusifolius*. *Planta* **104**, 126–33.

STADTMAN E.R. (1970) Mechanisms of enzyme regulation in metabolism. In *The Enzymes*, V. **1**, pp. 397–459 (ed. P.D. Boyer). Academic Press, London.

STANIER R.Y. (1970) Some aspects of the biology of cells and their possible evolutionary significance. *Organization and Control in Prokaryotic and Eukaryotic Cells*, pp. 1–38. (eds. H.P. Charles & B.C.J.G. Knight). Cambridge University Press.

STAVY (RODEH) R., BEN-SHAUL Y. & GALUN E. (1973) Nuclear envelope isolation in peas. *Biochim. biophys. Acta* **323**, 167–77.

STEDMAN E. & STEDMAN E. (1950) Cell specificity of histones. *Nature* **166**, 780–1.

STEIN W.D. & DANIELLI J.F. (1956) Structure and function in red cell permeability. *Discussions Faraday Soc.* **21**, 238–51.

STEINHART C.E., MANN J.D. & MUDD J.D. (1964) Alkaloids and Plant Metabolism VII. The kinetin-produced elevation in tyramine methylpherase levels. *Plant Physiol.* **39**, 1030–8.

STEINMANN E. (1952) An electron microscope study of the lamellar structure of chloroplasts. *Exp. Cell Res.* **3**, 367–72.

STEPHENS R.E. (1970) On the apparent homology of actin and tubulin. *Science* **168**, 845–7.

STEWARD F.C. (1958) Growth and organised development of cultured cells. III. Interpretation of the growth from free cell to carrot plant. *Am. J. Bot.* **45**, 709–13.

STEWARD F.C., MAPES M.O. & MEARS K. (1958) Growth and organised development of cultured cells. II. Organisation in cultures grown from freely suspended cells. *Am. J. Bot.* **45**, 705–8.

STEWART G.R. (1969) Abscisic acid and morphogenesis in *Lemna polyrhiza* (L.) *Nature* **221**, 61–2.

STEWART P.R. & LETHAM D.S. (1973) *The Ribonucleic Acids*. Springer-Verlag, Berlin.

STOCKING C.R. (1971) Chloroplasts: nonaqueous. *Methods in Enzymology* Vol. 23A, pp. 221–8 (ed. A. San Pietro). Academic Press, London.

STOKES D.M., ANDERSON J.W. & ROWAN K.S. (1968) The isolation of mitochondria from potato tuber tissue, using sodium metabisulfite for preventing damage by phenolic compounds during extraction. *Phytochemistry*. **7**, 1509–12.

STOREY B.T. (1969) The respiratory chain of plant mitochondria. III. Oxidation rates of the cytochromes *c* and *b* in mung bean mitochondria reduced with succinate. *Plant Physiol.* **44**, 413–21.

STOREY B.T. (1970a) The respiratory chain of plant mitochondria. IV. Oxidative rates of the respiratory carriers of mung bean mitochondria in the presence of cyanide. *Plant Physiol.* **45**, 447–54.

STOREY B.T. (1970b) The respiratory chain of plant mitochondria. V. Reaction of reduced cytochromes a and a_3 in mung bean mitochondria with oxygen in the presence of cyanide. *Plant Physiol.* **45**, 455–60.

STOREY B.T. (1970c) The respiratory chain of plant mitochondria. VI. Flavoprotein components of the respiratory chain of mung bean mitochondria. *Plant Physiol.* **46**, 13–20.

STOREY B.T. (1970d) The respiratory chain of plant mitochondria. VIII. Reduction kinetics of the respiratory chain carriers of mung bean mitochondria with reduced nicotinamide adenine dinucleotide. *Plant Physiol.* **46**, 625–30.

STOREY B.T. (1971a) The respiratory chain of plant mitochondria. X. Oxidation-reduction potentials of the flavoproteins of skunk cabbage mitochondria. *Plant Physiol.* **48**, 493–7.

STOREY B.T. (1971b) The respiratory chain of plant mitochondria. XI. Electron transport from succinate to endogenous pyridine nucleotide in mung bean mitochondria. *Plant Physiol.* **48**, 694–701.

STOREY B.T. (1972) The respiratory chain of plant mitochondria. XII. Some aspects of the energy-linked reverse electron transport from the cytochromes *c* to the cytochromes *b* in mung bean mitochondria. *Plant Physiol.* **49**, 314–22.

STOREY B.T. (1973) Respiratory chain of plant mitochondria. XV. Equilibration of cytochromes *c*-549, *b*-553 and *b*-557 and ubiquinone in mung bean mitochondria. Placement of cytochrome *b*-557 and estimation of the midpoint potential of ubiquinone. *Biochim. biophys. Acta* **292**, 592–603.

STOREY B.T. (1974) The respiratory chain of plant mitochondria. XVI. Interaction of cytochrome *b*-562 with the respiratory chain of coupled and uncoupled mung bean mitochondria. Evidence for its exclusion from the main sequence of the chain. *Plant Physiol.* **53**, 840–5.

STOREY B.T. & BAHR J.T. (1969b) The respiratory chain of plant mitochondria. II. Oxidative phosphorylation in skunk cabbage mitochondria. *Plant Physiol.* **44**, 126–34.

STOREY B.T. & BAHR J.T. (1969a) The respiratory chain of plant mitochondria. I. Electron transport between succinate and oxygen in skunk cabbage mitochondria. *Plant Physiol.* **44**, 115–25.

STOREY B.T. & BAHR J.T. (1969b) The respiratory chain of plant mitochondria. II. Oxidative phosphorylation in skunk cabbage mitochondria. *Plant Physiol.* **44**, 126–34.

STOREY B.T. & BAHR J.T. (1972) The respiratory chain of plant mitochondria. XIV. Ordering of ubiquinone, flavoproteins, and cytochromes in the respiratory chain. *Plant Physiol.* **50**, 95–102.

STRASBURGER E. (1882) Ueber den Theilungsvorgang der Zellkerne und das Verhältniss der Kernteilung zur Zellteilung. *Arch. f. mikrosk. Anat. Entw. Mech.* **21**, 476–590.

STRASSER R.J. & SIRONVAL C. (1972) Induction of photosystem II activity in flashed leaves. *FEBS Letters* **28**, 56–60.

STREET H.E. (1973) Single cell clones. In *Plant Tissue and Cell Culture*, pp. 191–204. (ed. H.E. Street), Blackwell Sci. Pub., Oxford.

STREET H.E. (1969) Growth in organised and unorganised systems. Knowledge gained by culture of organs and tissue explants. *Plant Physiology*, Vol. 5B, pp. 3–224, (ed. F.C. Steward). Academic Press, New York.

STREET H.E. (1975a) The anatomy and physiology of morphogenesis-studies involving tissue and cell cultures. In *Recueil de travaux dédiés à la mémoire de Georges Morel.* (ed. R.J. Gautheret). Masson et Cie, Paris. in press.

STREET H.E. (1975b) Plant cell cultures: Present and projected applications for studies in genetics. In *Genetic Manipulations with Plant Materials* pp. 231–44 (ed. L. Ledoux). Plenum Press.

STREET H.E. (1976) Experimental embryogenesis—the totipotency of cultured plant cells. In *Developmental Biology of Plants and Animals* pp. 73–90 (eds. P.F. Wareing & C. Graham). Blackwell Sci. Pub., Oxford.

STREET H.E., GOULD A.R. & KING J. (1975) Nitrogen assimilation and protein synthesis in plant cell cultures. In *Proc. 50th Ann. Meeting S.E.B.* Pergamon Press, Oxford. in press.

STREET H.E., KING P.J. & MANSFIELD K.J. (1971) Growth control in plant cell suspension cultures. In *Les Cultures de Tissus de Plantes.* pp. 17–40. Colloq. Int. Strasbourg, No. 193. C.N.R.S. Paris.

STREET H.E. & WITHERS L.A. (1974) The anatomy of embryogenesis in culture. In *Tissue Culture and Plant Science 1974*, pp. 71–100 (ed. H.E. Street). Academic Press, London.

STRYCHARZ W.A., RANKI M. & MARCKER K.A. (1973) Translational control in eukaryotic organisms. In *Regulation of Transcription and Translation in Eukaryotes* (eds. E.K.F. Bautz, P. Karlson & H. Kersten), pp. 251–6. 24. Colloquium der Gesellschaft für Biologische Chemie. Springer-Verlag, Berlin.

STUART R. & STREET H.E. (1969) Studies on the growth in culture of plant cells. IV. The initiation of division in suspensions of stationary phase cells of *Acer pseudoplatanus*, L. *J. exp. Bot.* **20**, 556–71.

STUART R. & STREET H.E. (1971) Studies on the growth in culture of plant cells. X. Further studies on the conditioning of culture media by suspensions of *Acer pseudoplatanus*, L. *J. exp. Bot.* **22**, 96–106.

STUMPF P.K. & BARBER G.A. (1956) Fat metabolism in higher plants. VII. β-oxidation of fatty acids by peanut mitochondria. *Plant Physiol.* **31**, 304–8.

STUMPF P.K. & BARBER G.A. (1956) Fat metabolism in higher plants. VII. β-oxidation of fatty acids by peanut mitochondria. *Plant Physiol.* **31**, 304–8.

STUTZ E. (1970) The kinetic complexity of *Euglena gracilis* chloroplast DNA. *FEBS Letters* **8**, 25–8.

SUEOKA N. & KANO-SUEOKA T. (1970) Transfer RNA and cell differentiation. *Prog. nucl. Acid Res. Mol. Biol.* **10**, 23–55.

SUMIDA S. & MUDD J.B. (1968) Biosynthesis of cytidine diphosphate diglyceride by cauliflower mitochondria. *Plant Physiol.* **43**, 1162–4.

SUNSHINE G.H., WILLIAMS D.S. & RABIN B.R. (1971) Role for steroid hormones in the interaction of ribosomes with the endoplasmic membranes of rat liver. *Nature (New Biol.)* **230**, 133–6.

SYRETT P.J., MERRETT M.J. & BOCKS S.M. (1963) Enzymes of the glyoxylate cycle in *Chlorella vulgaris*. *J. exp. Bot.* **14**, 249–64.

SYRETT P.J., BOCKS S.M. & MERRETT M.J. (1964) The assimilation of acetate by *Chlorella vulgaris*. *J. exp. Bot.* **15**, 35–47.

SZABO A.S. & AVERS C.J. (1969) Some aspects of regulation of peroxisomes and mitochondria in yeast. *Ann. N.Y. Acad. Sci.* **268**, 302–12.

TABATA M., YAMAMOTO H. & HIRAOKA N. (1971) Alkaloid production in the tissue cultures of some solanaceous plants. In *Les Cultures de Tissus de Plantes*, pp. 390–402. Coll. Int. Strasbourg, 1970, No. 193. C.N.R.S. Paris.

TAIT M.J. & FRANKS F. (1971) Water in biological systems. *Nature* **230**, 91–9.

TANADA T. (1968) A rapid photoreversible response of barley root tips in the presence of 3-indoleacetic acid. *Proc. natn. Acad. Sci. U.S.A.* **59**, 376–80.

TANADA T. (1972) On the involvement of acetylcholine in phytochrome action. *Plant Physiol.* **49**, 860–1.

TANADA T. (1972) Antagonism between indoleacetic acid and abscisic acid on a rapid phytochrome-mediated process. *Nature* **236**, 460–1.

TAO K. & JAGENDORF A.T. (1973) The ratio of free to membrane-bound chloroplast ribosomes. *Biochim. biophys. Acta* **324**, 518–32.

TARRAGO A., MONASTERIO O. & ALLENDE J.E. (1970) Initiator-like properties of a methionyl-tRNA from wheat embryos. *Biochem. biophys. Res. Commun.* **41**, 765–73.

TAYLOR E.W. (1965) The mechanism of colchicine inhibition of mitosis. Kinetics of inhibition and the binding of colchicine-^{3}H. *J. Cell Biol.* **25**, 145–67.

TEWARI K.K. (1971) Genetic autonomy of extra-nuclear organelles. *Ann. Rev. Plant Physiol.* **22**, 141–68.

TEWARI K.K. & WILDMAN S.G. (1968) Function of chloroplast DNA. I. Hybridisation studies involving nuclear and chloroplast DNA with RNA from cytoplasmic ribosomes (80S) and chloroplast (70S) ribosomes. *Proc. natn. Acad. Sci. U.S.A.* **59**, 569

TEWARI K.K. & WILDMAN S.G. (1970) Information content in the chloroplast DNA. *Control of Organelle Development*, pp. 147–79 (ed. P.L. Miller). Symposium 24 of the Society for Experimental Biology. Cambridge University Press, Cambridge.

THIEN W. & SCHOPFER P. (1975) Stimulation of precursor rRNA synthesis in the cotyledons of mustard seedlings by phytochrome. *Planta* **124**, 215–17.

THOMAS J.R. & TEWARI K.K. (1974) Conservation of 70S ribosomal RNA genes in the chloroplast DNAs of higher plants. *Proc. natn. Acad. Sci. U.S.A.* **71**, 3147–51.

THOMAS M., CAMERON J.R. & DAVIS R.W. (1974) Viable molecular hybrids of bacteriophage lambda and eukaryotic DNA. *Proc. natn. Acad. Sci. U.S.A.* **71**, 4579–83.

THOMPSON W. D'ARCY (1942) *On Growth and Form*. Cambridge University Press, Cambridge.

THORNBER J.P. (1975) Chlorophyll-proteins: light harvesting and reaction centre components of plants. *Ann. Rev. Plant Physiol.* **26**, 127–58.

TILNEY L. (1971) Origin and continuity of microtubules. *Origin and Continuity of Cell Organelles*, pp. 222–260 (eds. J. Reinert & H. Ursprung). Springer-Verlag, Berlin.

TILNEY L.G. & PORTER K.R. (1965) Studies on microtubules in Heliozoa. The fine structure of *Actinosphaerium nucleofilum* (Barret) with particular reference to the axial rod structure. *Protoplasma* **60**, 317–43.

TIMELL T.E. (pers. com.)

TING I.P., SHERMAN I.W. & DUGGER W.M. Jr. (1966) Intracellular location and possible function of malic dehydrogenase isozymes from young maize root tissue. *Plant Physiol.* **41**, 1083–4.

TINKLIN I.G. & SCHWABE W.W. (1970) Lateral bud dormancy in the black currant *Ribes Nigrum* (L.). *Ann. Bot.* **34**, 691–706.

TOBIN E.M. & BRIGGS W.R. (1973) Studies on the protein conformation of phytochrome. *Photochem. Photobiol.* **18**, 487–96.

TOBIN E.M., BRIGGS W.R. & BROWN P.K. (1973) The role of hydration in the phototransformation of phytochrome. *Photochem. Photobiol.* **18**, 497–503.

TOLBERT N.E. & BURRIS R.H. (1950) Light activation of plant enzyme which oxidises glycolic acid. *J. Biol. Chem.* **186**, 791–804.

TOLBERT N.E. & COHAN M.S. (1953) Products formed from glycolic acid in plants. *J. Biol. Chem.* **204**, 649–54.

TOLBERT N.E. (1963) Glycolate pathway. *Photosynthetic Mechanisms in Green Plants*, pp. 648–62 (ed. Bessel Kok). Natl. Acad. Sci.—Natl. Res. Council, Washington, D.C.

TOLBERT N.E., OESER A., KISAKI T., HAGEMAN R.H. & YAMAZAKI R.K. (1968) Peroxisomes from spinach leaves containing enzymes related to glycolate metabolism. *J. Biol. Chem.* **243**, 5179–84.

TOLBERT N.E., OESER A., YAMAZAKI R.K., HAGEMAN R.H. & KISAKI T. (1969) A survey of plants for leaf peroxisomes. *Plant Physiol.* **44**, 135–47.

TOLBERT N.E., YAMAZAKI R.K. & OESER A. (1970) Localization and properties of hydroxypyruvate and glyoxylate reductases in spinach leaf particles. *J. Biol. Chem.* **245**, 5129–36.

TORREY J.G. (1971) Cytodifferentiation in plant tissue and cell cultures. In *Les Cultures de Tissus de Plantes*, pp. 177–86. Colloq. Int. Strasbourg, 1970, No. 193. C.N.R.S. Paris.

TRAUB P. & NOMURA M. (1968) Structure and Function of *E. coli* ribosomes, V. Reconstitution of functionally active 30S ribosomal particles from RNA and proteins. *Proc. natn. Acad. Sci. U.S.A.* **59**, 777–84.

TRAVIS R.L., KEY J.L. & ROSS C.W. (1974) Activation of 80S maize ribosomes by red light treatment of dark-grown seedlings. *Plant Physiol.* **53**, 28–31.

TREBST A. (1974) Energy conservation in photosynthetic electron transport of chloroplasts *Ann. Rev. Plant Physiol.* **25**, 423–58.

TREGUNNA E.B., KROTKOV G. & NELSON C. D. (1961) Evolution of CO_2 by tobacco leaves during the dark period following illumination with light of different intensities. *Can. J. Bot.* **39**, 1045–59.

TRELEASE R.N., BECKER W.B., GRUBER P.J. & NEWCOMB E.H. (1971) Microbodies (glyoxysomes and peroxisomes) in cucumber cotyledons. *Plant Physiol.* **48**, 461–75.

TRENDELENBURG M.F., SPRING H., SCHEER U. & FRANKE W.F. (1974) Morphology of nucleolar cistrons in a plant cell. *Acetabularia mediterranea. Proc. natn. Acad. Sci. U.S.A.* **71**, 3626–30.

TREWAVAS A.J. (1972) Determination of the rates of protein synthesis and degradation in *Lemna minor. Plant Physiol.* **49**, 40–6.

TRUMAN D.E.S. (1974) *The Biochemistry of Cytodifferentiation.* Blackwell Sci. Publ., Oxford.

TURKOVA N.S., VASILEVA L.N. & CHEREMUKHINA L.F. (1965) On the Physiology of Bending of Leaves and Stems. *Sov. Plant Physiol.* **12**, 721–6.

TWARDOWSKI T. & LEGOCKI A.B. (1973) Purification and some properties of elongation Factor 2 from wheat germ. *Biochim. biophys. Acta* **324**, 171–83.

UHRSTRÖM I. (1969) The time effect of auxin and calcium on growth and elastic molecules in hypocotyls. *Physiol. Plant.* **22**, 271–87.

UMBARGER H.E. (1956) Evidence for a Negative-Feedback Mechanism in the Biosynthesis of Isoleucine. *Science* **123**, 848.

VAIL W.J., PAPAHADJOPOULOS D. & MOSCARELLO M.A. (1974) Interaction of a hydrophobic protein with liposomes. Evidence for particles in freeze fracture as being proteins. *Biochim. biophys. Acta* **345**, 463–7.

VAN DER WOUDE W.J., LEMBI C.A., MORRÉ D.J., KINDINGER J.I. & ORDIN L. (1974) β-glucan synthetases of plasma membrane and Golgi apparatus from onion stem. *Plant Physiol.* **54**, 333–40.

Van Overbeek J., Loeffler J.E. & Mason M.I.R. (1967) Dormin (Abscisin II), inhibitor of plant DNA synthesis? *Science* **156**, 1497–9.

Van Steveninck R.F.M. (1959) Factors affecting the abscission of reproductive organs in yellow lupins (*Lupinus luteus* L.) III. Endogenous growth substances in virus-infected and healthy plants and their effect on abscission. *J. exp. Bot.* **10**, 367–76.

Van Went J.F. (1970) The ultrastructure of the egg and central cell of *Petunia*. *Act. Bot. Néerl.* **19**, 313–22.

Varner J.E. & Johri M.M. (1968) Hormonal control of enzyme synthesis. In *Biochemistry and Physiology of Plant Growth Substances*, pp. 793–814 (eds. F. Wightman & G. Setterfield). The Runge Press Ltd., Ottawa, Canada.

Verma D.P.S., Nash D.T. & Schulman H. M. (1974) Isolation and *in vitro* translation of soybean leghaemoglobin mRNA. *Nature* **251**, 74–7.

Vidaver W. & Hsiao A.I. (1972) Persistence of phytochrome-mediated germination control in lettuce seeds for 1 year following a single monochromatic light flash. *Can. J. Bot.* **50**, 687–9.

Vigil E.L. (1970) Cytochemical and developmental changes in microbodies (glyoxysomes) and related organelles of castor bean endosperm. *J. Cell Biol.* **46**, 435–54.

Villemez C.L., Lin T.-Y. & Hassis W.Z. (1965) Biosynthesis of the polygalacturonic acid chain of pectin by a particulate enzyme preparation from *Phaseolus aureus* seedlings. *Proc. natn. Acad. Sci. U.S.A.* **54**, 1626–32.

Villemez C.L. (1974) The relation of plant enzyme-catalysed β-(1,4)-glucan synthesis to cellulose biosynthesis *in vivo*. In *Plant Carbohydrate Biochemistry*, pp. 183–189 (ed. J.B. Pridham). Academic Press, London.

Villiers T.A. (1968) An autoradiographic study of the effect of the plant hormone abscisic acid on nucleic acid protein metabolism. *Planta* **82**, 342–54.

Vince D. (1972) Phytochrome and flowering. *Phytochrome*, pp. 257–91 (eds. K. Mitrakos & W. Shropshire Jr.). Academic Press, London.

Waley S.G. (1964) A note on the kinetics of multi-enzyme systems. *Biochem. J.* **91**, 514–17.

Walker D.A. (1964) Improved rates of carbon dioxide fixation by illuminated chloroplasts. *Biochem. J.* **92**, 22C–23C.

Walker D.A. (1971) Chloroplasts (and grana): Aqueous (including high carbon fixation ability). *Methods in Enzymology* (ed. A. San Pietro), **23A**, 211–20.

Walker D.A. & Beevers H. (1956) Some requirements for pyruvate oxidation by plant mitochondrial preparations. *Biochem. J.* **62**, 120–7.

Walker G.C., Leonard N.J., Armstrong D.J., Mural N. & Skoog F. (1974) The mode of incorporation of 6-benzylaminopurine into tobacco callus transfer ribonucleic acid. *Plant Physiol.* **54**, 737–43.

Walker J.R.L. & Hulme A.C. (1965) The inhibition of the phenolase from apple peel by polyvinyl pyrrolidone. *Phytochemistry.* **4**, 677–85.

Wallace W. (1974) Purification and properties of a nitrate reductase—inactivating enzyme. *Biochem. biophys. Acta* **341**, 265–76.

Wallin A. & Eriksson T. (1973) Protoplast cultures from cell suspensions of *Daucus carota*. *Physiol. Plant.* **28**, 33–9.

Walton T.J. & Kolattukudy P.E. (1972) Determination of the structure of cutin monomers by a novel depolymerization procedure and combined gas chromatography and mass spectrometry. *Biochemistry*, **11**, 1885–97.

Wang J.H. (1970) Oxidative and phosynthetic phosphorylation mechanisms. *Science* **167**, 25–30.

Wara-Aswapati O. (1973) *The development of photosynthetic activity and of some photosynthetic enzymes in greening leaves*. Ph.D. Thesis. University of London.

Wara-Aswapati O., Kemble R. & Bradbeer J.W. (1977) The activation of the phosphoribulokinase and the triosephosphate dehydrogenase ($NADP^+$) of bean chloroplasts. In preparation.

Warner F.D. (1974) The fine structure of the ciliary and flagellar axoneme. *Cilia and Flagella*, pp. 11–38 (ed. M.A. Sleigh). Academic Press, London.

Warner H.L. & Leopold A.C. (1971) Timing of growth regulator responses in peas. *Biochem. biophys. Res. Comm.* **44**, 989–94.

Watt W.D. & Fogg G.E. (1966) The kinetics of extracellular glycollate production by

Chlorella pyrenoidosa. J. exp. Bot. **17**, 117–34.

WEAIRE P.J. & KEKWICK R.G.O. (1975) The synthesis of fatty acids in avocado mesocarp and cauliflower bud tissue. *Biochem. J.* **146**, 425–37.

WEEKS D.P., VERMA D.P.S., SEAL S.N. & MARCUS A. (1972) Role of ribosomal subunits in eukaryotic protein chain initiation. *Nature* **236**, 167–8.

WEIER T.E. & BENSON A.A. (1967) The molecular organisation of the chloroplast membranes. *Amer. J. Bot.* **54**, 389–402.

WEIER T.E. & BROWN D.L. (1970) Formation of the prolamellar body in 8-day, dark-grown seedlings. *Amer. J. Bot.* **57**, 267–75.

WEINBACH E.C. & GARBUS J. (1966) The rapid restproation of respiratory control to uncoupled mitochondria. *J. Biol. Chem.* **241**, 3708–13.

WEINTRAUB R.L. & LAWSON V.R. (1972) Mechanism of phytochrome-mediated effects of light on cell growth. *Book of Abstracts, VI. International Congress on Photobiology, Bochum*, 1972, No. 161.

WEISENBERG R.C. (1972) Microtubule formation *in vitro* in solutions containing low calcium concentrations. *Science* **177**, 1104–5.

WEISENSEEL M.H. & SMEIBIDL E. (1973) Phytochrome controls the water permeability in *Mougeotia. Z. Pflanzenphysiol.* **70**, 420–31.

WELLBURN A.R. & WELLBURN F.A.M. (1971) A new method for the isolation of etioplasts with intact envelopes. *J. exp. Bot.* **22**, 972–9.

WELLBURN F.A.M. & WELLBURN A.R. (1973) Response of etioplasts *in situ* and in isolated suspension to pre-illumination with various combinations of red, far-red and blue light. *New Phytol.* **72**, 55–60.

WELLS G.N. & BEEVERS L. (1973) Protein synthesis in cotyledons of *Pisum sativum* L. II. The requirements for initiation with plant messenger RNA. *Pl. Sci. Lett.* **1**, 281–6.

WELLS R. & BIRNSTIEL M. (1969) Kinetic complexity of chloroplastal deoxyribonucleic acid and mitochondrial deoxyribonucleic acid from higher plants. *Biochem. J.* **112**, 777–86.

WELLS R. & SAGER R. (1971) Denaturation and the renaturation kinetics of chloroplast DNA from Chlamydomonas reinhardi. *J. molec. Biol.* **58**, 511–22.

WENT F.W. (1934) A test method of rhizocaline. The root-forming substance. *K. Akad. Wentenschap. Amsterdam Proc. Sect. Sci.* **37**, 445–55.

WEST S.H., HANSON J.B. & KEY J.L. (1960) Effect of 2,4-dichlorophenoxyacetic acid on the nucleic-acid and protein content of seedling tissue. *Weeds* **8**, 333–40.

WETMORE R.H. & RIER J.P. (1963) Experimental induction of vascular tissues in callus of angiosperms. *Am. J. Bot.* **50**, 418–30.

VON WETTSTEIN D. HENNINGSEN K.W., BOYNTON J.E., KANNANGARA G.C. & NIELSEN O.F. (1971) The genic control of chloroplast development in barley. *Autonomy and Biogenesis of Mitochondriaa nd Chloroplasts*, pp. 205–23 (eds. N.K. Boardman, A.W. Linnane & R.M. Smillie). North Holland, Amsterdam.

WETTSTEIN F.O., STAEHELIN T. & NOLL H. (1963) Ribosomal aggregate engaged in protein synthesis: characterisation of the ergosome. *Nature* **197**, 430–5.

WHALEY W.G., DAUWALDER M. & KEPHART J.E. (1972) Golgi apparatus: influence on cell surfaces. *Science* **175**, 596–9.

WHEELER A. & HARTLEY M.R. (pers comm).

WHITE J.M. & PIKE C.S. (1974) Rapid phytochrome-mediated changes in adenosine-5′-triphosphate content of etiolated bean buds. *Plant Physiol.* **53**, 76–9.

WHITE P.R. (1934) Potentially unlimited growth of excised tomato roots in a liquid medium. *Plant Physiol.* **9**, 585–600.

WHITE P.R. (1939) Controlled differentiation in a plant tissue culture. *Bull Torrey Bot. Club.* **66**, 505–13.

WHITFIELD P.R. & SPENCER D. (1968) Buoyant density of tobacco and spinach chloroplast DNA. *Biochim. biophys. Acta* **157**, 333–43.

WIAME J.M., STALON V., PIERARD A. & MESSENGUY F. (1973) Control of transcarbamoylation in micro-organisms. *S.E.B. Sym.* **27**, 333–63.

WIDHOLM J.M. (1972a) Cultured *Nicotiana tabacum* cells with an altered anthianilate synthetase which is less sensitive to feedback inhibition. *Biochim. biophys. Acta* **261**, 52–8.

WIDHOLM J.M. (1972b) Anthianilate synthetase from 5-methyltryptophan susceptible and resistant cultured *Daucus carota* cells. *Biochim. biophys. Acta* **279**, 48–57.

WIDHOLM J.M. (1974) Selection and characteristics of biochemical mutants of cultured plant cells. *Tissue Culture and Plant Sciences*, pp. 287–99, (ed. H.E. Street). Academic Press, London.

WIDHOLM J.M. (1974) Selection and characteristics of biochemical mutants of cultured plant cells. In *Tissue Culture and Plant Science*, pp. 287–99, (ed. H.E. Street). Academic Press, London.

WILDMAN S.G. (1967) The organization of grana-containing chloroplasts in relation to location of some enzymatic systems concerned with photosynthesis, protein synthesis, and ribonucleic acid synthesis. *Biochemistry of Chloroplasts*, pp. 295–319. (ed. T.W. Goodwin). Academic Press, London.

WILDMAN S.G., JOPE C. & ATCHISON B.A. (1974) Role of mitochondria in the origin of chloroplast starch grains. Description of the phenomenon. *Plant Physiol.* **54**, 231–7.

WILDNER G.F., ZILG H. & CRIDDLE R.S. (1972) Light effect on isolated ribulose diphosphate carboxylase activity. *Proceedings of the 2nd International Congress on Photosynthesis Research*. pp. 1825–30 (eds. G. Forti, M. Avron & A. Melandri). Dr. W. Junk, The Hague.

WILLIAMSON F.A., MORRÉ D.J. & JAFFE M.J. (1975) Association of phytochrome with rough-surfaced endoplasmic reticulem fractions from soybean hypocotyls. *Plant Physiol.* **56** ,738–43.

WILSON D.F. & DUTTON P.L. (1970a) The oxidation reduction potentials of cytochrome $a+a_2$ in intact rat liver mitochondria. *Arch. Biochem. Biophys.* **136**, 583–4.

WILSON D.F. & DUTTON P.L. (1970b) Energy dependent changes in the oxidation-reduction potential of cytochrome b. *Biochem. Biophys. Res. Comm.* **39**, 59–64.

WILSON J.M. & CRAWFORD R.M.M. (1974) The acclimatization of plants to chilling temperatures in relation to the fatty acid composition of leaf polar lipids. *New Phytol.* **73**, 805–20.

WILSON L & BRYAN J. (1974) Biochemical and pharmacological properties of microtubules. *Adv. Cell molec. Biol.* **3**, 21–72.

WILSON L. & FRIEDKIN M. (1967) The biochemical events of mitosis. The *in vivo* and *in vitro* binding of colchicine in grasshopper embryos and its possible relation to inhibition of mitosis. *Biochemistry* **6**, 3126–35.

WILSON L. & MEZA I. (1973) The mechanism of action of colchicine. Colchicine binding properties of sea urchin sperm tail outer doublet tubulin. *J. Cell Biol.* **58**, 709–19.

WILSON R.H., DEVER J., HARPER W. & FRY R. (1972) The effects of valinomycin on respiration and volume changes in plant mitochondria. *Pl. Cell Physiol.* **13**, 1103–11.

WILSON R.H. & HANSON J.B. (1969) The effect of respiratory inhibitors on NADH, succinate and malate oxidation in corn mitochondria. *Plant Physiol.* **44**, 1335–41.

WILSON R.H., HANSON J.B. & MOLLENHAUER H.H. (1969) Active swelling and acetate uptake in corn mitochondria. *Biochemistry* **8**, 1203–13.

WILSON R.H., THURSTON E.L. & MITCHELL R. (1973) Ultrastructural transformations in bean inner mitochondrial membranes. *Plant Physiol.* **51**, 26–30.

WILSON S.B. (1970a) Energy conservation associated with cyanide-insensitive respiration in plant mitochondria. *Biochim. biophys. Acta* **223**, 383–7.

WILSON S.B. (1970b) Phosphorylation associated with cyanide-insensitive respiration in plant mitochondria. *Biochem. J.* **116**, 20P–1P.

WILSON S.B., KING P.J. & STREET H.E. (1971) Studies on the growth in culture of plant cells. XII. A versatile system for the large scale batch or continuous culture of plant cell suspensions. *J. exp. Bot.* **21**, 177–207.

WILSON S.B. & MOORE A.L. (1973) Cyanide-insensitive oxidation of ascorbate plus tetramethylphenylene-diamine mixture by mung bean (*Phaseolus aureus*) mitochondria. *Biochem. Soc. Trans.* **1**, 881.

WINTERMANS J.F.G.M. (1960) Concentrations of phosphatides and glycolipids in leaves and chloroplasts. *Biochim. biophys. Acta* **44**, 49–54.

WISKICH J.T. (1966) Respiratory control by isolated apple mitochondria. *Nature* **212**, 641–2.

WISKICH J.T. (1974) Substrate transport into plant mitochondria. *Australian J. Plant Physiol.* **1**, 177–81.

WISKICH J.T. & BONNER W.D.Jr. (1963) Preparation and properties of sweet potato mitochondria. *Plant Physiol.* **38**, 594–604.

WISKICH J.T., MORTON R.K. & ROBERTSON R.M. (1960) The respiratory chain of beet root mitochondria. *Australian J. Biol. Sci.* **13**, 109–22.

WITHROW R.B., KLEIN W.H. & ELSTAD V.B. (1957) Action spectra of photomorphogenic induction and its photoinactivation. *Plant Physiol.* **32**, 453–62.

WITTMANN H.G. (1973) Structure and function of bacterial ribosomes. In *Regulation of Transcription and Translation in Eukaryotes* pp. 211–12, (eds. E.K.F. Bautz, P. Karlson & H. Kersten). 24. Colloquium der Gesellschaft für Biologische Chemie. Springer-Verlag, Berlin.

WONG K.F. & DENNIS D.T. (1973) Aspartokinase in *Lemna minor* L. Studies on the *in vivo* regulation of the enzyme. *Plant Physiol.* **51**, 327–31.

WOOD A. & PALEG L.G. (1974) Alteration of Liposomal membrane fluidity by gibberellic acid. *Aust. J. Plant Physiol.* **1**, 31–40.

WOOD A. & PALEG L.G. (1972) The influence of GA_3 on the permeability of model membrane systems. *Plant Physiol.* **50**, 103–8.

WOOD A., PALEG L.G. & SPOTSWOOD T.M. (1974) Hormone-phospholipid interaction: a possible hormonal mechanism of action in the control of membrane permeability. *Aust. J. Plant Physiol.* **1**, 167–9.

WOODIN T.S. & NISHIOKA L. (1973) Evidence for three enzymes of chorismate mintase in alfalfa. *Biochem. biophys. Acta* **309**, 211–23.

WOODING F.B.P. (1968) Radioautographic and chemical studies of incorporation into sycamore vascular tissue walls. *J. Cell. Sci.* **3**, 71–80.

WOODING F.B.P. (1969) P-protein and microtubule systems in *Nicotiana. Planta* **85**, 284–98.

WRIGGLESWORTH J.M., PACKER L. & BRANTON D. (1970) Organisation of mitochondrial structure revealed by freeze etching. *Biochim. biophys. Acta* **205**, 125–35.

WRIGHT D.J. & BOULTER D. (1974) Purification and subunit structure of legumin of *Vicia faba* L. (Broad Bean). *Biochem. J.* **141**, 413–18.

WRIGHT S.T.C. (1969) An increase in the 'Inhibitors β' content of detached wheat leaves following a period of wilting. *Planta* **86**, 10–20.

WRIGHT S.T.C. & HIRON R.W.P. (1969) (+)-Abscisic acid, the growth inhibitor induced in detached leaves by a period of wilting. *Nature* **224**, 291–8.

YANG S.F. & STUMPF P.K. (1965) Fat metabolism in higher plants. XXI. Biosynthesis of fatty acids by avocado mesocarp systems. *Biochim. biophys. Acta* **98**, 19–26.

YARWOOD A., BOULTER D. & YARWOOD J.N. (1971a) Methionyl-tRNAs and initiation of protein synthesis in *Vicia faba* (L.). *Biochem. biophys. Res. Commun.* **44**, 353–61.

YARWOOD A., PAYNE E.S., YARWOOD J.N. & BOULTER D. (1971b) Aminoacyl-tRNA binding and peptide chain elongation on 80S plant ribosomes. *Phytochemistry.* **10**, 2305–11.

YATES R.A. & PARDEE A.B. (1956) Control of pyrimidine biosynthesis in *Escherichia coli* by a feed-back mechanism. *J. Biol. Chem.* **221**, 757–70.

YEOMAN M.M. (1974) Division synchrony in cultured cells. In *Tissue Culture and Plant Science 1974* pp. 1–18. (ed. H.E. Street), Academic Press, London.

YEOMAN M.M. & AITCHISON P.A. (1973) Growth patterns in tissue (callus) cultures. In *Plant Tissue and Cell Culture*, pp. 240–268 (ed. H.E. Street). Blackwell Sci. Publ., Oxford.

YOMO H. & VARNER J.E. (1971) Hormonal control of a secretory tissue. (eds. A. Monroy & A. Moscona). *Curr. Top. Develop. Biol.* Academic Press, New York **6**, 111–44.

YOUNG M. (1973) Studies on the growth in culture of plant cells. XVI. Nitrogen assimilation during nitrogen-limited growth of *Acer pseudoplatanus*, L. cells in chemostat culture. *J. exp. Bot.* **24**, 1172–85.

YU R. (1975) Distribution of phytochrome in subcellular fractions from maize coleoptiles following glutaraldehyde prefixation. *Aust. J. Plant Physiol.* **2**, 273–80.

YUNGHANS H. & JAFFE M.J. (1972) Rapid respiratory changes due to red light or acetylcholine during the early events of phytochrome-mediated photomorphogenesis. *Plant Physiol.* **49**, 1–7.

ZAITLIN M. & BEACHY R.N. (1974) The use of protoplasts and separated cells in plant virus research *Adv. Virus Res.* **19**, 1–32.

ZALIK S. & JONES B.L. (1973) Protein biosynthesis. *Ann. Rev. Plant Physiol.* **24**, 47–68.

ZELITCH I. & OCHOA S. (1953) Oxidation and reduction of glycolic and glyoxylic acids in plants. I. Glycolic acid oxidase. *J. Biol. Chem.* **201**, 707–18.

ZELITCH I. (1968) Investigations on photorespiration with a sensitive ^{14}C-assay. *Plant Physiol.* **43**, 1829–37.

INDEX